D1570952

Methods in Rock Magnetism and Palaeomagnetism

Techniques and instrumentation

Methods in Rock Magnetism and Palaeomagnetism

Techniques and instrumentation

D. W. Collinson

Department of Geophysics and Planetary Physics
University of Newcastle upon Tyne, UK

LONDON NEW YORK
CHAPMAN AND HALL

First published 1983 by
Chapman and Hall Ltd
11 New Fetter Lane, London EC4P 4EE
Published in the USA by
Chapman and Hall
733 Third Avenue, New York NY10017

Printed in Great Britain by J. W. Arrowsmith Ltd., Bristol

ISBN 0 412 22980 3

British Library Cataloguing in Publication Data

Collinson, D. W.
Methods in rock magnetism and palaeomagnetism techniques and instrumentation.
1. Magnetism, Terrestrial
I. Title
538′.7′028 QC815.5

ISBN 0–412–22980–3

Library of Congress Cataloging in Publication Data

Collinson, D. W. (David Wright)
Methods in rock magnetism and palaeomagnetism.

Bibliography: p.
Includes index.
1. Rocks—Magnetic properties.
2. Paleomagnetism. I. Title.
QE431.6.M3C64 1982 538′.7 82–9544
ISBN 0–412–22980–3

To J.M.C., for her help and encouragement during the writing of this book, and F.J.C., who would have preferred the book to remain unwritten but showed continued if distracting interest in it.

Contents

Preface

During the last 30 years the study of the magnetic properties of rocks and minerals has substantially contributed to several fields of science. Perhaps the best known and most significant advances have resulted from the study of palaeomagnetism, which led to quantitative confirmation of continental drift and polar wandering through interpretation of the direction of remanent magnetism observed in rocks of different ages from different continents. Palaeomagnetism has also, through observations of reversals of magnetization, ancient secular variation and ancient field intensities provided data relevant to the origin of the geomagnetic field, and other investigations have contributed significantly to large-scale and local geological studies, the dating of archaeological events and artefacts and more recently to lunar and meteoritic studies. Rock and mineral magnetism has proved to be an interesting study in its own right through the complex magnetic properties and interactions observed in the iron–titanium oxide and iron sulphide minerals, as well as contributing to our understanding of remanent magnetism and magnetization processes in rocks.

Simultaneous with the development of these studies has been the development of instruments and techniques for the wide range of investigations involved.

The need for reliable measurements of the remanent magnetization of more weakly magnetized rocks has resulted in the development of sensitive magnetometers and improvements in the techniques for the removal of unwanted magnetizations. Much effort has been expended in improving methods for deriving ancient magnetic field intensities from terrestrial and extraterrestrial rocks and in devising instruments and techniques for identifying and characterizing magnetic minerals. Along with these developments have been those concerning new methods of presentation of palaeomagnetic data and of techniques for deriving the most reliable estimate of ancient magnetic field directions and pole positions.

Details of many of these instruments and techniques are in the literature but scattered rather widely, and others remain unpublished yet are deserving of wider knowledge. It is the purpose of this book to bring together for easy reference the range of instruments and techniques currently in use, and also to describe some of those used in the past and now less used but which may still be of interest and of use for particular investigations. It is hoped that the book will be useful both to those already working in the subject and to those with either a geological or physical background embarking on studies of any of the many aspects of the magnetism of rocks.

Although equipment for routine palaeomagnetic and rock-magnetic studies is now commercially available, information is given in the book to enable such equipment to be built in the laboratory and also more specialized equipment for particular studies. The emphasis is on the basic principles of the various instruments and techniques. Signal processing and the use of computers and microprocessors is not described in detail as information on these aspects of instrumentation is available from other sources.

Workers in palaeomagnetism and rock magnetism have been among the most reluctant converts to SI units, but an increasing number are now using them. In the author's opinion the use of c.g.s. is now no longer justified and SI units are used in the book. A table of conversion factors between c.g.s. and SI units is given in Appendix 1, together with some notes on the SI units used in the book.

I am grateful to many of my colleagues both in this country and abroad for discussion and information on various topics. Particular thanks are due to M. J. Gross, F. J. Lowes, L. Molyneux, J. H. Parry, W. O'Reilly, J. S. Rathore, P. V. Sharma, A. de Sa, A. Stephenson, M. Stupavsky and D. H. Tarling. I am grateful to Mrs L. Boon, Miss S. Bell, Miss A. Codling and Mrs L. Whiteford for jointly typing the manuscript and to Mrs D. Cooper for drawing the diagrams.

D. W. Collinson
Newcastle upon Tyne
England

Chapter One

Rock magnetism and magnetic minerals

1.1 Introduction

Rock magnetism is the term commonly applied to the study of the magnetic properties of rocks and minerals, how these properties depend on factors such as grain size and shape, temperature and pressure, and the origin and characteristics of the different types of remanent magnetizations which rocks and magnetic minerals can acquire. Although all minerals possess some magnetic properties, even if only paramagnetic or diamagnetic, the term 'magnetic minerals' is used here only for those minerals which are capable of carrying remanent magnetism.

Rock magnetic investigations and techniques are basically an extension to rocks and minerals of studies of the classical magnetic materials, namely iron and other transition elements and their alloys. However, metallic iron is only very rarely encountered in terrestrial rocks (although it is the dominant magnetic mineral in lunar samples and some meteorites), and it is the ferrimagnetic and antiferromagnetic iron–titanium oxide minerals which are of chief concern in and add interest and complexity to rock magnetism. In addition, composite minerals of the titanomagnetite and titanohaematite series, in which there is a gradation in or discontinuous change of magnetic properties with composition, and the presence of particles covering a wide range of grain sizes associated with superparamagnetic, single domain and multidomain behaviour are other factors which contribute a variety of interesting phenomena.

Although rock magnetism is an interesting and informative study in its own right, many investigations are directed towards identifying the mineral(s) carrying the natural remanent magnetization (NRM) of rocks and towards a better understanding of the origin of the NRM and its reliability as a palaeomagnetic indicator.

The magnetic mineral content of rocks is typically only 1–5 % by weight and it may not be easy to separate them out in a pure form for investigation, particularly if they are fine grained. An alternative approach pursued by several workers is the investigation of laboratory-prepared minerals, notably the titanomagnetite series and haematite. One result of this work has been to emphasize the dependence of magnetic properties on the presence of small amounts of impurities, non-stoichiometry and, particularly in the case of haematite, the mode of formation of the mineral.

The magnetic properties of rocks, including their NRM, can also often provide information on their formation and history and on the constituent minerals and their physical state. For example, the temperature to which the country rock is raised during intrusion of a nearby dyke or sill can be estimated from the blocking temperature of the PTRM (partial thermoremanent magnetization) acquired, and magnetite can be detected in rocks by magnetic measurements at a level (0.01–0.001%) well below that detectable by conventional methods such as X-ray diffraction or optical reflectance. These and other applications, including determination of the intensity of the ancient geomagnetic field, are described later in this book.

A branch of rock magnetism of potential practical use is the effect of pressure on magnetic properties as a possible indicator for earthquake prediction. The principle is the surface observation of changes in the magnetic anomaly pattern due to a build-up of stress in the underlying rocks (Stacey, 1964). Depending on the Königsberger ratio of the rocks (Section 10.2), changes in either initial susceptibility or remanent magnetism could be the cause of a changed anomaly pattern. In a similar way, changes in induced or permanent magnetism of crustal rocks arising from temperature and pressure changes due to movement of magma have been proposed as an indicator for predicting volcanic eruptions (Stacey, Barr and Robson, 1965).

Although the study of the magnetism of meteorites has been mainly concerned with their NRM and its origin, they are of some interest for rock magnetism because of the contribution of iron and iron–nickel (as well as magnetite) to their magnetism. An understanding of the magnetic properties of the former two materials and their variation with temperature is clearly necessary for a further understanding of meteoritic NRM and its implications.

A recent development in rock magnetism is the investigation of the magnetic properties of lunar rocks and dust returned by the *Apollo* missions of 1969–72 and the unmanned Russian *Luna* landing and sample return programme. The presence of iron (or dilute nickel–iron) in a wide range of grain sizes as the dominant magnetic mineral, uncertainty as to the origin of the iron in some rocks and of some of the rocks themselves (the breccias) and the limited amount of material available for investigation all combine to make lunar rock magnetism a fascinating and demanding study. For an overview of the field the reader is referred to the review by Fuller (1974), and for more detailed accounts to the proceedings of the annual lunar science conferences, published as supplements of *Geochimica et Cosmochimica Acta*.

Some of the instruments and techniques employed in rock magnetism were developed from those used in classical magnetic studies and some have been developed specifically for use in rock magnetism. In the former category are various types of Faraday balance and vibration magnetometer for the measurement of induced magnetization, high field susceptibility and Curie points, while the phenomena of rotational hysteresis and piezomagnetism have been mainly studied only in rocks and minerals and have required the development of new instrumentation for their successful investigation. The need for this stems partly from the greater sensitivity often required for measurements on rock samples compared with other magnetic minerals, either because of the weak magnetic properties of many rocks or the availability of only restricted amounts of separated minerals.

1.2 Magnetic minerals

In this section a brief account is given of the minerals of chief interest in the rock magnetism and palaeomagnetism of terrestrial and extraterrestrial material. It is not intended as a comprehensive description of the minerals and their properties, but rather as a guide to the occurrence and abundance of magnetic and other minerals of interest in rocks, their most important magnetic properties, and other properties relevant to magnetic studies. The references given are intended to guide the reader to the literature, where more detailed information and accounts of relevant research are available. Useful general references to the mineralogy of rock magnetism are Nicholls (1955), Nagata (1961) and O'Reilly (1976).

For those who are not familiar with them, the terms used to describe magnetic properties are defined in Section 3.1.

1.2.1 Iron and nickel–iron

(*a*) *Iron*

Native iron is rare in terrestrial rocks, but is the most important magnetic constituent of many meteorites and all lunar rocks. In these materials it occurs either in the pure form or with a varying nickel content. A terrestrial occurrence in some Newfoundland ophiolites is reported by Deutsch, Rao, Laurent and Seguin (1972), based on studies of their magnetic properties.

Iron exists in at least two allotropic forms, one of which is the body-centred cubic structure (α-iron), stable at room temperature and up to 910°C. The Curie point of iron is 778°C, above which it is paramagnetic, and between 778° and 910°C it is known as β-iron although there is no structural change from the α-form. Above 910°C the structure is a face-centred cubic lattice (γ-iron) (Adcock and Bristow, 1935). The relative density of iron is 7.85, and its electrical conductivity is $1.0 \times 10^3\,\Omega^{-1}\,\mathrm{m}^{-1}$.

Iron is the classic ferromagnetic material, in which neighbouring atomic moments are aligned parallel to one another by strong exchange forces. Its spontaneous (saturation) magnetization is 218 $A\,m^2\,kg^{-1}$. The intrinsic susceptibility is very high ($\sim 10^{-2}\,m^3\,kg^{-1}$) but the effective initial susceptibility of an assembly of non-interacting grains (as occurs in lunar rocks or stony meteorites) is governed by their demagnetizing factor and is typically $\sim 5 \times 10^{-4}\,m^3\,kg^{-1}$ (Section 2.3).

Iron is magnetically anisotropic with easy and hard directions of magnetization along the (100) (cube edge) and (111) (cube diagonal) respectively. The coercive force of large multidomain grains is very low (1–5 mT), but is much higher in single domain and small multidomain particles. Néel (1949) derived theoretical values for the superparamagnetic–single domain and single domain–multidomain boundaries in spherical grains of 160 Å (0.016 μm) and 320 Å (0.032 μm) respectively, but Butler and Banerjee (1975a) and Wasilewski (1981) suggest there is no stable single-domain size range for spherical grains, but may be for elongated grains.

The magnetism of iron and its alloys and their many industrial applications have been comprehensively documented by Bozorth (1951) and in the electrical engineering literature, in particular in the *Transactions of the Institute of Electrical and Electronic Engineers* (*Magnetics*). Research into the magnetic properties of finely divided iron has been promoted through the recent interest in the magnetism of lunar rocks and dust returned by the American *Apollo* and Russian *Luna* missions. These materials contain grains of iron (or dilute nickel–iron) up to ~1 mm in diameter which often span the superparamagnetic to multidomain range and carry remanent magnetism with a wide range of stabilities. The grain-size distribution of iron in lunar dust has been examined through magnetic studies by Stephenson (1971b), Dunlop *et al.* (1973) and Gose and Carnes (1973).

Studies of laboratory-prepared analogues of lunar iron are described by Pearce, Hoye, Strangway, Walker and Taylor (1976). The reader is also referred to papers in the proceedings of the lunar science conferences, published annually from 1970, in which results of lunar magnetic studies are reported, and to the review by Fuller (1974).

(*b*) *Nickel–iron*

Indigenous nickel–iron alloys are found only very rarely on Earth, but are an important constituent of some meteorites and often contribute the major part of their magnetic properties. Nickel is ferromagnetic with a saturation magnetization of 57 $A\,m^2\,kg^{-1}$ and a Curie point of 357°C. α-Iron can hold up to ~ 20% of nickel (which has a face-centred cubic structure) in the lattice, and this form of nickel–iron in meteorites, with typically 5–10% of nickel, is known as kamacite: nickel–iron with a higher proportion of nickel is termed taenite, which has the face-centred cubic structure of γ-iron. Plessite is an intimate mixture of kamacite and taenite crystallites which occurs in some meteorites.

The nickel–iron content of meteorites is typically ~100% in the irons, ~50% in the stony irons, 5–25% in the chondrites and ~1% or less in the achondrites. Of the irons, the octahedrites, with an average 8% Ni, contain kamacite and subsidiary taenite and show the characteristic triangular Widmannstätten pattern (intersecting plates of kamacite with taenite margins) on etching a polished surface: the hexahedrites, with typically 5% Ni contain kamacite only and show a system of lines (Neumann lines) on etching. The cobalt content of meteoritic metal is usually 5–15% of the nickel content. Among the comprehensive texts and reviews on meteorites, that of Wasson (1974) contains a summary of their magnetic properties.

The magnetic properties of nickel–iron show a marked discontinuity at ~27% (atomic) nickel content, at which composition the alloy essentially loses its ferromagnetic properties. The apparent Curie point of the alloys also varies in a complex way with nickel content, but there is a distinction to be made here between a true Curie point and the $\alpha \rightarrow \gamma$ transition temperature, above which alloys with 27% Ni are paramagnetic. The effect of increasing nickel content in iron is to steadily lower the transition temperature (910°C in pure iron) until at a nickel content of ~4–5% it is just below the iron Curie point of 778°C (Pickles and Sucksmith, 1940). Thus, although alloys in the range 5–27% Ni lose their ferromagnetic properties at lower temperatures (e.g. 705°C and 625°C at 9% and 16% Ni respectively) these are not true Curie points. Among other features, the $\alpha \rightarrow \gamma$ transition in these alloys is not thermally reversible, i.e. on cooling, the $\gamma \rightarrow \alpha$ transition occurs at a lower temperature. With a nickel content in the range ~30–50% there is a true Curie point in γ-phase taenite (Hoselitz and Sucksmith, 1943).

Because of the above effects and the wide range of grain size of the nickel iron in meteorites, considerable variation in their magnetic properties is observed. As expected, the remanent magnetism of the irons is of low stability because of the coarsely crystalline nature of the metal. Some of the chondrites show a rather stable remanence carried by kamacite, for example the Cook and Farrington meteorites (Stacey, Lovering and Parry, 1961) and the Al Rais meteorite (Watson, Larson, Herndon and Rowe, 1975): these chondrites also exhibit the thermal hysteresis associated with the $\gamma \rightarrow \alpha$ transition in kamacite. The stability of the remanence suggests the presence of very fine particles or perhaps a microstructure in the nickel–iron. Stacey *et al.* (1961) point out that if the meteorites acquired their magnetic remanence by cooling in an ambient field, then this remanence may in fact be a type of CRM (chemical remanent magnetization) rather than TRM (thermoremanent magnetization) if it was acquired at the $\gamma \rightarrow \alpha$ transition rather than by cooling through a true Curie point or blocking temperature. Such a magnetization process may also have implications for determinations of the field intensity in which the meteorite remanence was acquired.

1.2.2 The titanomagnetite series

This important series of magnetic minerals consists of solid solutions or intergrowths of different compositions of the end members, magnetite

(Fe_3O_4) and ulvöspinel (Fe_2TiO_4). Magnetite-rich to approximately equimolecular compositions are important in palaeomagnetism as the carriers of NRM in a wide range of igneous rocks, some sedimentary rocks (e.g. limestones) and varved clays and lake and sea-bottom sediments, and magnetite is the magnetic constituent of some meteorites.

(*a*) *Magnetite* (Fe_3O_4)

Magnetite is a cubic mineral with the inverse spinel structure, i.e. the cations occupy two different lattices, A and B, in the crystal, with Fe^{2+} and Fe^{3+} ions in the latter and Fe^{3+} only in the former. In the normal spinel structure, divalent cations are on one lattice and trivalent cations on the other. In magnetite there are two cations on B sites for each one on A and the atomic moments are oppositely directed. Thus the unit cell has a net magnetic moment: this is ferrimagnetism, of which magnetite is one of the best-known examples.

Magnetite is a dark black mineral with a theoretical relative density of 5.20, although natural samples lie in the range 5.16–5.22. It is optically isotropic in polished section, with a reflectivity of ~21% in air: the electrical conductivity of natural magnetite is very variable, but is commonly in the range $\sim 10^2$–$10^4\,\Omega^{-1}\,m^{-1}$ (Parkhomenko, 1967).

With the exception of iron, magnetite has the strongest magnetic properties, with a saturation magnetization variously quoted in the range 90–93 $A\,m^2\,kg^{-1}$: its Curie point is 578°C. It is magnetically anisotropic with easy and hard directions of magnetization along (111) and (100) respectively. Based on the theory of Néel (1955), the single domain–multidomain boundary for spherical grains is at about 280 Å diameter: this is close to the value reported for the superparamagnetic boundary (Stacey, 1963; Dunlop, 1973a), i.e. the diameter below which particles are rendered magnetically unstable through thermal agitation (Section 2.4). Thus the direct transition from multidomain to superparamagnetic behaviour in equidimensional magnetite particles appears to be a possibility, and stable NRM may be carried by elongated grains or pseudo-single-domain grains (Stacey, 1963; Butler and Banerjee, 1975b). The intrinsic susceptibility of magnetite is very high ($\sim 10^{-2}\,m^3\,kg^{-1}$) but the effective initial susceptibility of an assembly of non-interacting grains (i.e. as in a rock sample), as with iron, is governed by their demagnetizing factor and is typically $\sim 5.0 \times 10^{-4}\,m^3\,kg^{-1}$ (Section 2.3).

The coercive force of multidomain magnetite is low, usually in the range 2–20 mT, and the coercivity of remanence usually lies in the range 10–50 mT. Both these properties are particle-size dependent, their magnitude increasing with decreasing size (Stacey, 1963; Parry, 1965; Dunlop, 1973b). Saturation magnetization is achieved in fields of 50–150 mT with a saturation remanence of typically 1–20 $A\,m^2\,kg^{-1}$, again dependent on grain size.

Between about −145°C and −155°C magnetite undergoes a structural change from the cubic to orthorhombic form (the Verwey transition), and marked changes in mechanical and electrical properties occur (Verwey and Haayman, 1941). This transition occurs in the same temperature range in which K_1, the first magnetocrystalline anisotropy constant, passes through zero when magnetite is cooled from room temperature, and it is likely that the two phenomena are connected (O'Reilly, 1976).

(*b*) *Ulvöspinel* (Fe_2TiO_4)

Ulvöspinel (sometimes known as ulvite) is the other end member of the titanomagnetite series. Like magnetite it has the inverse spinel structure, but the Fe^{3+} cations on the A and B sites in magnetite are replaced by Fe^{2+} and Ti^{4+} respectively. Thus, since Ti^{4+} contributes no magnetic moment, and the Fe^{2+} cations on each lattice are oppositely directed, ulvöspinel is weakly ferrimagnetic with a Néel temperature (the temperature above which the atomic ordering is destroyed) of 120 K (−153°C). It is paramagnetic at room temperature, with a theoretical susceptibility of $\sim 1 \times 10^{-6}\,m^3\,kg^{-1}$, based on its ferrous iron content.

Ulvöspinel is rare in terrestrial rocks, occurring chiefly as exsolution blebs or lamellae in some magnetite ores: there is some magnetic evidence of its occurrence in lunar rocks (Runcorn, Collinson, O'Reilly, Stephenson, Battey, Manson and Readman, 1971). Its name is taken from Södra Ulvön, in northern Sweden, where a magnetite ore containing an estimated 52% of discrete ulvöspinel is found (Deer, Howie and Zussman, 1962). It has an X-ray relative density of 4.7–4.8, and can be synthesized by sintering an intimate mixture of the Fe_2O_3 and TiO_2, or produced from a melt (Hauptman and Stephenson, 1968). It is optically isotropic, with a reflectivity slightly less than that of magnetite. Ulvöspinel oxidizes to ilmenite and magnetite: the ilmenite lamellae are parallel to (111) in the magnetite, whereas those of ulvöspinel in magnetite ores are parallel to (100) (Ramdohr, 1969).

(*c*) *Titanomagnetites* ($xFe_2TiO_4 \cdot (1-x)Fe_3O_4$)

Complete solid solution of magnetite and ulvöspinel only occurs above 600°C and below this temperature a varying degree of exsolution takes place, resulting in a structure of intergrown lamellae except in a restricted range of compositions near $x = 0$ and $x = 1.0$ (Basta, 1960).

There is a steady gradation of properties with change in bulk composition. The Curie temperature decreases almost linearly with increasing x, from 578°C for $x = 0$ to −153°C for ulvöspinel ($x = 1$). At $x \simeq 0.80$, the Curie point is at room temperature. Other magnetic properties behave in a similar but less linear fashion, e.g. saturation magnetization and saturation remanence. Although measurements have not been reported, initial susceptibility may be expected to change with x in a rather complex way. With high magnetite content, susceptibility will be governed by the particle demagnetizing factor (Section 2.3), and be essentially constant. As x increases, the intrinsic susceptibility of a particle decreases and the demagnetizing effect assumes less importance, the bulk susceptibility then depending more on the intrinsic susceptibility and the proportion of magnetite in a particle. There may also be interactive effects between magnetite-rich and ulvöspinel-rich lamellae in exsolution structures.

A considerable amount of research on the magnetic properties of titanomagnetites has been carried out on synthetic material of known composition. Methods of producing single-crystal and polycrystalline specimens are described in several of the papers cited in this section. Jensen and Shive (1973) summarize their own and other investigations of cation distribution and its association with saturation magnetization, and Fletcher and O'Reilly (1974) describe investigations of crystal anisotropy and its temperature dependence. Other studies include magnetostriction (Shive and Butler, 1969), indirect investigations of the superparamagnetic size limit (O'Donovan, 1975), also discussed by Day (1977), domain structure and size (Butler and Banerjee, 1975b; Soffel, 1977) and rotational hysteresis (Manson, 1971). Thermomagnetic curves (J_i–T), Curie points and TRM of titanomagnetites are discussed by Stephenson (1969, 1972), Lewis (1968), Hauptman (1974), O'Donovan and O'Reilly (1977b), Tucker and O'Reilly (1980). This is by no means a comprehensive list, but will guide the interested reader to further references.

Most natural titanomagnetites are more or less oxidized and also contain cations other than Fe^{2+}, Fe^{3+} and Ti^{4+}. Magnesium and aluminium are the commonest additional cations present in the lattice, which influence the magnetic properties (Richards, O'Donovan, Hauptman, O'Reilly and Creer, 1973). During the 1960s, the oxidation state of natural titanomagnetites generated considerable interest when evidence emerged of a correlation between oxidation state of lavas and polarity of their NRM, reversed samples showing generally more advanced oxidation (Ade-Hall, 1964; Wilson and Watkins, 1967). Some of the apparent correlations have been disputed by other workers analysing the same data, and in investigations of other lava sequences no correlation is observed (Watkins and Haggerty, 1968; Ade-Hall and Watkins, 1970). The association between NRM polarity and oxidation state is difficult to perceive, and the observations still await a satisfactory explanation.

(*d*) *Titanomaghemites*

If finely divided titanomagnetite undergoes prolonged heating in air at low or moderate temperatures ($< \sim 300$°C), titanomaghemite is produced. This process, maghemitization, has recently become of some interest because of the occurrence of titanomaghemites in submarine basalts (Ryall and Ade-Hall, 1975; Prevot, Lecaille and Mankinen, 1981) and the possible effect of the process on observed sea-floor magnetic anomalies. The simplest example of maghemitization is the low-temperature oxidation of magnetite to maghemite, γ-Fe_2O_3. The properties of the latter are described in Section 1.2.4.

Synthetic titanomaghemites can be prepared by crushing (or ball-milling) titanomagnetite to submicron size in water or methanol for several hours before heating (Sakamoto, Ince and O'Reilly, 1968; O'Donovan and O'Reilly, 1977a).

Titanomaghemite as a carrier of unstable NRM in igneous (dolerite) rocks was reported by Akimoto and Kushiro (1960). Natural and synthetic maghemites show a wide range of magnetic properties according to composition. Determination of Curie temperatures, which rise with increasing oxidation, are complicated by the fact that the inversion temperature of titanomaghemites is ~350°C, above which haematite and other products appear depending on the original composition (Moskowitz, 1981).

Self-reversal of magnetization in natural titanomaghemites is reported by Schult (1968a, 1976) and Nishida and Sasajima (1974). Other investigations of the magnetic properties of this series and of maghemitization in general are reported by Havard and Lewis (1965), Ozima and Sakamoto (1971), Readman and O'Reilly (1972), O'Donovan and O'Reilly (1978), Manson, O'Donovan and O'Reilly (1979) and Keefer and Shive (1981).

1.2.3 The titanohaematite series

These minerals consist of different proportions of the two end members, haematite (α-Fe_2O_3) and ilmenite ($FeTiO_3$), generally represented by $xFeTiO_3 \cdot (1-x)Fe_2O_3$. Their properties vary markedly with composition, from the complex antiferromagnetism of haematite, through the ferrimagnetism of an intermediate range of compositions to the paramagnetism (at room temperature) of ilmenite. A certain range of composition exhibit the property of self-reversal of TRM, which attracted some attention to the series at the time when the origin of reversals of NRM was being discussed.

(*a*) *Ilmenite* ($FeTiO_3$)

Ilmenite is paramagnetic at temperatures above about 60 K (−213°C), below which it is antiferromagnetic. It melts at 1470°C and has an X-ray relative density of 4.80. The paramagnetic susceptibility of ilmenite is variously quoted in the range $(1.0–1.2) \times 10^{-6}\,m^3\,kg^{-1}$, and its theoretical susceptibility based on ferrous oxide content, is $0.99 \times 10^{-6}\,m^3\,kg^{-1}$. Naturally occurring ilmenite often contains traces of magnetite or haematite, detectable by the existence of NRM after removal from an applied field.

Ilmenite is a blackish mineral, appearing greyish-white in polished section: its reflectivity (~18% in air) is somewhat less than that of magnetite and haematite and it is optically anisotropic. Ilmenite occurs commonly in terrestrial igneous rocks, and also in lunar samples. It is highly resistant to weathering, and therefore persists in some sediments. It also occurs in association with its alteration products, magnetite, haematite, and the oxide, rutile (TiO_2).

(*b*) *Haematite* (α-Fe_2O_3)

Haematite is of major importance as the carrier of NRM in a wide range of sedimentary rocks. It occurs in two forms, as black, polycrystalline particles (specularite) and in a finely divided state as a coating on other matrix particles and in interstices in the rock. The colour of the latter form (pigment) varies from yellow-brown through orange to red and purple and gives the wide range of 'red' sandstones their distinctive colours. Although haematite occurs as an alteration product in igneous and metamorphic rocks, it is only very rarely responsible for any significant magnetic properties in those rocks because of the usual presence of the much more strongly magnetic titanomagnetites. Haematite also carries the NRM in many types of archaeological material and in some lake sediments. Martite is a form of haematite formed by alteration of, and pseudomorphous after, magnetite. The magnetic properties of haematite are generally very variable, through impurities in the natural mineral and apparently arising from the preparation method in synthetic samples, and grain size is also important.

Haematite has the corundum structure with rhombohedral symmetry, and the massive form is extremely hard. Its relative density is about 5.3, and in polished section it has a higher reflectivity (21–28% in air) than magnetite or ilmenite, and it is optically anisotropic. The true melting point is ~1750°C, but in air it dissociates to magnetite at ~1400°C.

Although haematite is antiferromagnetic, i.e. the atomic Fe^{3+} moments on the A and B sub-lattices are equal and antiparallel, there is a slight departure from exact antiparallelism ('canting') of the moments resulting in a weak permanent magnetism (Dzyaloshinsky, 1958). The saturation magnetization is 0.2–0.4 $\mathrm{A\,m^2\,kg^{-1}}$, and the Curie point 680°C. The coercivity of remanence is variable but can be very high (0.1–0.5T) in fine-grained material. The initial mass susceptibility of natural material is in the range $(60–600) \times 10^{-8}\,\mathrm{m^3\,kg^{-1}}$, with a high field susceptibility of $25 \times 10^{-8}\,\mathrm{m^3\,kg^{-1}}$. The susceptibility is isotropic at room temperature. Haematite appears to possess single-domain structure up to grain diameter of ~2.0 mm, with the super-paramagnetic boundary at ~0.2 μm (Stacey, 1963; Strangway, McMahon, Honea and Larson, 1967).

The magnetic properties of haematite undergo a transition at about −20°C, the Morin transition (Morin, 1950). Below this temperature the spin-canted ferromagnetism disappears and the direction of the antiferromagnetic moments changes from within the basal plane of the crystal to the ternary axis. The susceptibility decreases and becomes anisotropic. The remanent magnetism decreases sharply, but exhibits 'memory' and partially recovers on rewarming through the critical temperature. However, the change in remanence is not always observed in haematite-bearing sediments, and the transition temperature is lowered if impurities such as Mn^{2+}, Ti^{4+}, Al^{3+} and Mg^{2+} are present in the lattice (Haigh, 1957; Ishikawa and Akimoto, 1958).

Another type of ferromagnetism has been observed in haematite below the Néel temperature ($\sim 725°C$), above which it is paramagnetic. This is 'defect' ferromagnetism, associated with interactions between impurity ions and/or lattice defects and the basic antiferromagnetism (Dunlop, 1970). Because of its sensitivity to structure, it can be altered by thermal effects or mechanical stress, a feature which once caused some concern regarding the stability of NRM in red sediments. However, there is now evidence that the spin-canted ferromagnetism is harder than the defect remanence and the latter is preferentially removed during thermal demagnetization.

Magnetostriction in haematite has been investigated by Urquhart and Goldman (1956) and Anderson, Birss and Scott (1964), and rotational hysteresis by Vlasov, Kovalenko and Fedoseeva (1967b), Day, O'Reilly and Banerjee (1970) and Cowan and O'Reilly (1972). Comprehensive accounts of the magnetic properties of haematite are given by Fuller (1970) and Dunlop (1971).

(*c*) *Titanohaematites* ($xFeTiO_3 \cdot (1-x)Fe_2O_3$)

These minerals are also known as haemo–ilmenites (ilmenite-rich) or ilmeno haematites (haematite-rich). Complete solid solution is possible for all x-values above $\sim 1000°C$, but at lower temperatures intermediate compositions consist of intergrowths of $FeTiO_3$ and α-Fe_2O_3, except where $x < 0.1$ and $x > 0.9$. Titanohaematites occur sparsely in most volcanic rocks but more commonly in gneisses and granitic and metamorphic rocks (Balsley and Buddington, 1958). They also occur in minor amounts in some sediments but are not normally an important contributor to their NRM.

A marked change in magnetic properties occurs at $x \simeq 0.5$. In the range $0 \leqslant x \leqslant 0.5$ the properties are similar to those of the end-member haematite, a basic antiferromagnetism with a weak superimposed ferromagnetism. From $0.5 \leqslant x \leqslant 1.0$, strong ferrimagnetism is observed (Ishikawa and Akimoto, 1958) with a saturation magnetization as high as 20–30 $A\,m^2\,kg^{-1}$ (Akimoto, 1955). Curie points are typically in the range $-100°C$ to $+200°C$. Throughout the series there is an almost linear dependence of Curie point on x-value, from $\sim 675°C$ ($x = 0$) to $-200°C$ ($x = 1.0$). The compositions which are ferrimagnetic above room temperature, and thus are of interest in palaeomagnetism, lie in the range $0.50 \leqslant x \leqslant 0.80$. The coercive force of samples with compositions in the antiferromagnetic region is much higher than is observed in the ferrimagnetic region (Nagata, 1961).

Compositions in the range $0.45 \leqslant x \leqslant 0.60$ show the phenomenon of self-reversal of magnetization (acquisition of TRM antiparallel to the applied field), first observed in a Japanese dacite pumice containing titanohaematite (Nagata, Uyeda and Akimoto, 1952; Ishikawa and Syono, 1963). The phenomenon has also been observed in natural samples in the range $0.15 \leqslant x \leqslant 0.25$ (Carmichael, 1961).

Studies of various magnetic properties of the titanohaematite series have been carried out by Uyeda (1957), Ishikawa (1962), Westcott-Lewis and Parry (1971), and Hoffman (1975).

1.2.4 Maghemite (γ-Fe_2O_3)

Maghemite has the same chemical composition as haematite, but it has a cubic spinel structure, with a lattice defect such that an Fe^{3+} position is vacant. It is a metastable mineral and converts to haematite at temperatures above about 300°C, although the inversion temperature varies among samples. Impurity ions in the lattice raise the inversion temperature (Bénard, 1939; de Boer and Selwood, 1954). It is usually a brownish colour and occurs in a fine-grained form through low-temperature oxidation of magnetite or dehydration of lepidocrocite (γ-FeOOH). It occurs naturally in minor amounts as a weathering product in red sandstones and some basalts, and also in oceanic muds to which it is therefore likely to contribute a chemical remanence. Wilson (1961a, b) reports an investigation of the magnetic properties of lateritic lavas containing maghemite. There is some evidence from experiments carried out on Mars by the two *Viking* landers that maghemite is present in significant amounts in the Martian surface (Hargraves, Collinson, Arvidson and Spitzer, 1977).

The X-ray relative density of maghemite is about 4.9, but natural samples are usually less dense. If of large enough grain size, it can be polished and has a reflectivity (isotropic) of ~20% in air. It is strongly magnetic, with a saturation magnetization, 80–85 A m^2 kg^{-1}, comparable to that of magnetite. Because of its low temperature inversion to haematite, the true Curie point of maghemite is not known. Indirect methods, involving addition of impurities to raise the inversion temperatures and extrapolating the observed Curie points to the value corresponding to zero impurity content, give about 675°C, i.e. the same as haematite (Michel and Chaudron, 1935). However, other techniques (Brown and Johnson, 1962; Banerjee, 1965; Banerjee and Bartholin, 1970) give values between 545° and 750°C. Aspects of the magnetism of maghemite and the magnetite–maghemite–haematite transition are discussed by Wilson (1961a), Bagin (1966), Banerjee (1966), Rybak (1971), and Stacey and Banerjee (1974).

1.2.5 Pyrrhotite

Of the iron sulphide minerals, troilite (FeS), pyrite (FeS_2) and pyrrhotite (FeS_{1+x}), only pyrrhotite is ferrimagnetic. The quoted values of x for ferrimagnetism vary somewhat, but appear to be in the range $\sim 0.10 < x < 0.14$ at room temperature.

Pyrrhotite is a brassy-coloured mineral with metallic lustre which tarnishes on exposure to air. Its relative density is 4.6–4.8, depending on composition. It is strongly anisotropic optically (reflectivity in air ~38–44%), with a distinctive yellowish colour, although troilite and pyrite are very similar. It occurs in basic igneous rocks, in metamorphic rocks, e.g. slates and also in massive form in ore bodies. Troilite occurs in some meteorites and in lunar rocks.

The magnetic properties of natural and synthetic pyrrhotite have been investigated by Fuller (1963), Schwarz (1968a) and Schwarz and Vaughan (1972). Useful reviews are by Ward (1970) and Schwarz (1975). Depending on composition, saturation magnetization is in the approximate range 1–20 $\mathrm{A\,m^2\,kg^{-1}}$; exact measurements are difficult to obtain because of the high magnetocrystalline anisotropy of pyrrhotite. Where measured, initial bulk susceptibility is in the range $10^{-5} - 10^{-7}\,\mathrm{m^3\,kg^{-1}}$ (Fuller, 1963). The Curie point of the $x = 0.14$ composition is 320°C. Everitt (1962a) reports the occurrence of self-reversal of TRM in a shale containing pyrrhotite.

1.2.6 Goethite (α-FeOOH)

Goethite is one of several hydrated iron oxides (iron oxyhydroxides) occurring in nature, and is the only one which can carry a remanent magnetization. It is a fairly common mineral, formed as a weathering product of iron-bearing minerals and also by direct precipitation from iron-bearing solutions. On moderate heating it converts to haematite ($\sim$350°C). Its relative density is $\sim$4.3 and it has a rather low reflectivity in air (14–17%) and is optically anisotropic.

Goethite is antiferromagnetic but spin compensation is imperfect, allowing a small net moment (Stacey and Banerjee, 1974). It has a Néel point of 100°C–120°C and can acquire a TRM on cooling through this temperature. Strangway *et al.* (1968a) report magnetic studies on natural goethite samples and the change of these properties on conversion to haematite, and Hedley (1968) has investigated the acquisition of chemical remanent magnetization during the dehydration of synthetic goethite and other hydrated iron oxides. Creer (1962a) has also studied certain magnetic properties of these minerals, and Vlasov, Kovalenko and Chikhacher (1967a) report on investigations of superparamagnetism in goethite. Heller and Channell (1979) adduce evidence that goethite carries a very stable NRM in some limestones from Germany.

Chapter Two

Initial magnetic susceptibility

2.1 Introduction

Magnetic susceptibility is defined by the relation

$$M_i = kH \quad (2.1)$$

or alternatively

$$J_i = \chi H = \chi B/\mu_0 \quad (2.2)$$

where a magnetization per unit volume M_i or per unit mass J_i is induced in a material of volume susceptibility k or mass susceptibility χ by an applied field $B = \mu_0 H$. In SI units, using $B = \mu_0 (H + M)$, k is dimensionless and the units of χ are $m^3 kg^{-1}$. In this chapter we are initially concerned with isotropic susceptibility, i.e. susceptibility that is independent of the axis along which it is measured in a material and J_i and B are parallel.

The magnetic state of the material may or may not be irreversibly changed by the application of B and susceptibility may be independent of or vary with B. The major classes of materials are paramagnetic and diamagnetic, in which the effect of the applied field is reversible and for practical purposes susceptibility is independent of the magnitude of the field; ferromagnetism is associated with irreversibility and dependence of susceptibility on the magnitude of the applied field. However, except in special cases, these latter materials behave reversibly in sufficiently low applied fields ($< \sim 1$ mT), with χ independent of B, and initial susceptibility is the property of minerals and rocks in this low-field region.

2.2 Diamagnetism and paramagnetism

Diamagnetic susceptibility is a fundamental property of all atoms and molecules whether they have a resultant magnetic moment or not. It arises from the effect of an applied field on orbiting electrons, which, with magnetic moments due to electron spin, are the source of magnetism in matter. The effect of the field is to cause the electron orbits to precess, and, following Lenz's law, the effect of the precessing orbits is equivalent to the generation of an orbital moment opposing the applied field and thus diamagnetic susceptibility is a negative quantity. Since it arises within atoms, its magnitude, of the order of $1 \times 10^{-8}\,\mathrm{m^3\,kg^{-1}}$, is essentially independent of temperature. Common minerals which exhibit diamagnetism are quartz (SiO_2), calcite ($CaCO_3$) and dolomite ($CaCO_3 \cdot MgCO_3$) (Table 2.1).

If an atom has a resultant magnetic moment it will also have a paramagnetic susceptibility, which is normally much stronger than its diamagnetism. Paramagnetism arises from the partial alignment of atomic magnetic moments along the applied field direction: the magnetic torque works against the randomizing influence of thermal motions, which cause complete randomness again when the field is removed.

The degree of alignment of the magnetic moments along the field direction is expressed by the Langevin function $\mathrm{L}(a)$ where

$$\mathrm{L}(a) = \coth a - \frac{1}{a} \tag{2.3}$$

and $a = Bp/kT$, where B is the applied field, p is the atomic magnetic moment, k is Boltzmann's constant and T the absolute temperature. (This derivation is based on classical physics, but essentially the same result is obtained from quantum mechanical considerations.) $\mathrm{L}(a)$ is the ratio of the total magnetic moment of the partially aligned atoms to the magnetic moment resulting from complete alignment, equal to Np, where N is the number of atoms. Thus $\mathrm{L}(a) = N\bar{p}/Np = \bar{p}/p$, where $\bar{p}$ is the apparent average value of the atomic moment in the field B at temperature T. If a is small, i.e. for weak fields at high temperatures, $\mathrm{L}(a) \simeq a/3$ and

$$\frac{\bar{p}}{p} = \frac{Bp}{3kT} \tag{2.4}$$

If there are N atoms per unit volume

$$N\bar{p} = \frac{NBp^2}{3kT} = M \tag{2.5}$$

or

$$\frac{M}{H} = \frac{N\mu_0 p^2}{3kT} \tag{2.6}$$

where M is the volume magnetization and M/H the volume susceptibility. Thus, under the above conditions, the latter is independent of H and proportional to T^{-1} – this is the Curie law of paramagnetic susceptibility. With sufficiently high fields and low temperatures ($B/T \gg \sim 1\,\mathrm{T\,K^{-1}}$), paramagnetic susceptibility becomes field dependent as perfect alignment of the atomic moments along B is approached. Typical values of paramagnetic susceptibility are $0.1\text{–}1.0 \times 10^{-6}\,\mathrm{m^3\,kg^{-1}}$.

Among the more important paramagnetic minerals are ilmenite ($FeTiO_3$), biotite, siderite ($FeCO_3$) and the clay minerals such as illite, chamosite,

Table 2.1 Initial mass susceptibility of materials of interest in rock and palaeomagnetism

	Initial mass susceptibility ($10^{-8}\,m^3\,kg^{-1}$)
Rocks and minerals	
Quartz	−0.62
Calcium carbonate	−0.48
Magnesium carbonate	−0.48
Pyrrhotite	~10–1000
Ilmenite	100–115
Haematite	60–600
Magnetite[a] Maghemite[a]	5.7×10^4
Illite	~15
Montmorillonite	~13
Nontronite	~65
Chamosite	~88
Biotite[b]	79
Orthopyroxenes[b]	50
Igneous rocks	$\sim 10^2$–10^4
Red sediments	~0.5–5
Limestones	~0.2–2
Chemicals	
Copper sulphate ($CuSO_4 \cdot 5H_2O$)	7.4
Ferrous ammonium sulphate ($FeSO_4(NH_4)_2SO_4\,6H_2O$)	40.5
Ferrous sulphate ($FeSO_4 \cdot 7H_2O$)	51.7
Manganese sulphate ($MnSO_4 \cdot 4H_2O$)	81.0
Nickel sulphate ($NiSO_4 \cdot 7H_2O$)	20.1
Ferrous chloride ($FeCl_2$)	145
Sodium chloride (NaCl)	−0.64
Miscellaneous	
Air[c]	0.35×10^{-6}
water	−0.90
Aluminium[d]	0.82
Copper[d]	−0.11
Platinum[d]	1.22
Mercury	−0.21
Plastic construction materials (PVC, Perspex, etc.)[d]	~ −0.5

[a] Observed value for isolated, spherical grains.
[b] Average values.
[c] Value given is volume susceptibility.
[d] Commercial materials often contain ferromagnetic impurities, and also may become contaminated when cut or turned.

montmorillonite and nontronite (Table 2.1). The major contributors of the paramagnetism of minerals are ferrous (Fe^{2+}) and ferric (Fe^{3+}) ions, and a knowledge of their magnetic moments and concentration in a mineral or rock enables their contribution to the observed susceptibility to be estimated. Assuming Bohr magneton numbers of 5.3 and 5.8 for Fe^{2+} and Fe^{3+} ions respectively (i.e. magnetic moments 5.3 and 5.8 times that of the hydrogen atom, 9.3×10^{-24} A m^2), each 1 % by weight to paramagnetic FeO and Fe_2O_3 will contribute 2.07×10^{-8} and 2.26×10^{-8} m^3 kg^{-1}, respectively, to the initial susceptibility of the mineral or rock. Table 2.2 shows that there is reasonable agreement between observed and calculated values of susceptibility for the minerals listed.

Table 2.2 Observed and calculated susceptibilities of clay and other minerals

			Susceptibility ($\times 10^{-8}$ m^3 kg^{-1})	
Mineral	FeO (%)	Fe_2O_3 (%)	Observed	Calculated
Illite	1.4	4.7	15	14
Montmorillonite	2.8	3.0	14	13
Nontronite	0.2	28.0	66	64
Chamosite	14.0	40.0	89	120
Biotites*	19.2	7.9	80	63
Orthopyroxenes†	25.0	1.0	50	52
Ilmenite	46.4	–	100–114	98
Siderite	62.0	–	–	128

* Mean values from Syono (1960).
† Mean value from Akimoto, Horai and Boku (1958). Other values measured by the author.

2.3 Ferromagnetic susceptibility

The magnitude of initial susceptibility of ferro-, ferri- and antiferromagnetic minerals covers a wide range. The most important contribution to χ is from reversible domain wall movements, domains aligned near the applied field direction temporarily growing at the expense of others: in fields of ~ 1 mT or less, χ is independent of B. In some minerals, such as maghemite (γ-Fe_2O_3) and magnetite (and in the lunar rocks, iron), the 'intrinsic' volume susceptibility k_i is very high ($> \sim 10$): this is the value that would be obtained from a measurement technique in which, for instance, the induced magnetization in a ring of the mineral was measured. However, in a 'normal' sample, or a particle in a rock, the applied field is modified by a back or demagnetizing field due to effective magnetic poles which appear at each end of the sample due to its magnetization. The magnitude of this field is given by $-NM$, where N is the demagnetizing factor of the particle (which depends only on its shape) and M is its (volume) magnetization. Consider an isolated, approximately equidimensional particle of intrinsic volume susceptibility k_i and demagnetization factor N: if the applied field is H, then the field H_i inside the particle is given by

$$\begin{aligned} H_i &= H - NM \\ &= H - Nk_i H_i \end{aligned} \tag{2.7}$$

The observed initial volume susceptibility k is the ratio of M to the applied field H

$$k = \frac{k_i H_i}{H} \tag{2.8}$$

Combining Equations 2.7 and 2.8

$$k = \frac{k_i H_i}{H_i(1 + Nk_i)} = \frac{k_i}{1 + Nk_i} \tag{2.9}$$

For $k_i < \sim 1.0$ the value of k is only slightly less than k_i, but as k_i increases Nk_i becomes $\gg 1$ and $k \rightarrow N^{-1}$. Thus, for an assembly of non-interacting particles of high intrinsic susceptibility, the observed initial susceptibility will depend only on N. For a sphere, $N = 1/3$ and $k \simeq 3.3$, and for a random assembly of ellipsoidal grains of major/minor axis ratio 1.5:1, Stacey (1963) gives $N = 0.31$ and $k \approx 3.2$. Thus, an assembly of roughly equidimensional grains of multidomain magnetite or maghemite (relative density ~ 5.2) will have a value of initial mass susceptibility of $\sim 6.0 \times 10^{-4}$ $m^3\,kg^{-1}$ per unit mass of mineral. This value increases for an assembly of aligned elongated particles, e.g. for an ellipsoidal shape of axial ratio 3:1, $N = 0.109$ and $k = 9.2\,m^3\,kg^{-1}$ parallel to the long axis. Some values of N for ellipsoids have been tabulated by Stoner (1945) and are given in Appendix 2.

The intrinsic initial susceptibility of haematite is variable, depending on its origin and grain size, but appears to be in the range $3\text{–}30 \times 10^{-3}$. Demagnetizing fields are therefore negligible and it is the intrinsic initial susceptibility that is observed and measured.

2.4 Superparamagnetism

In an assembly of single-domain grains of different volumes, those grains for which the ambient temperature is above the blocking temperature (T_B) have low relaxation times and thermal fluctuations will continually re-orient their magnetic moments (Section 11.2.1). In zero ambient fields these grains will contribute no net moment, but they will rapidly equilibrate with an applied field B, behaving like the atomic moments in paramagnetic materials. A net moment along B will appear, controlled by the Boltzmann function (pB/kT). The grain moment p is much larger than the corresponding atomic moments, leading to the possibility of high values of susceptibility: hence the name 'superparamagnetism' (SPM) for the phenomenon. Another feature of SPM is that saturation is approached in much lower fields than those associated with paramagnetism.

If the temperature of the assembly is raised, T_B of larger grains will be exceeded and they become superparamagnetic, while the induced moment contributed by those grains already in that state will decay according to the Curie law, i.e. as T^{-1}. Thus, depending on the distribution of grain volumes, a variety of χ–T curves are possible.

Theory (Stacey, 1963; Dunlop 1973a) and experiment (Dunlop, 1973a) indicate an upper limit to the diameter of superparamagnetic spherical grains of magnetite at 20°C of 0.035–0.050 μm (350–500 Å): because of shape anisotropy, elongated grains have a much smaller limiting volume. Stephenson (1971a) points out that the method of susceptibility measurement may determine whether grains of a given volume will exhibit superparamagnetism. For example, for a susceptibility measurement in a 1 kHz alternating field, he deduces an upper diameter of 160 Å for spherical magnetite particles.

Stacey and Banerjee (1974) and Stephenson (1971a) have calculated the susceptibility of a material containing a volume fraction f ($f \ll 1$) of non-interacting superparamagnetic magnetite grains. The former authors take a grain volume of 10^{-17} cm^3 ($d = \sim 270$ Å) and a relaxation time $\tau = 1$ sec, which leads to $k = 226f$, i.e. $k = 0.23$ for $f = 0.001$; this compares with $k \approx 3 \times 10^{-3}$ for the same fraction of multidomain grains. Stephenson finds $k = 63f$ for magnetite for measurements at 1 kHz and $20f$ for iron grains, the limiting superparamagnetic diameter of which is ~ 107 Å.

Superparamagnetism in minerals other than iron and magnetite has not been extensively studied. Stacey (1963) gives 1800 Å as a theoretical upper limit of the diameter of haematite grains, although there is some doubt about the value of the anisotropy constant used in the calculation, and Creer (1961) detected superparamagnetism in haematite grains (estimated diameter 20 Å) in a red marl (Section 13.6). Hedley (1968) and Banerjee (1971), from experimental evidence, put the upper limit at only 200–300 Å. Some theoretical calculations on superparamagnetism in titanomagnetites are described by O'Donovan (1975) (see also O'Reilly, 1976).

One of the best-documented occurrences of superparamagnetism in natural materials is in the fines (materials < 1 mm in grain size) returned from the surface of the Moon (Nagata, Ishikawa, Kinoshita, Kono, Syono and Fisher, 1970). The dominant magnetic mineral in the lunar rocks and fines is iron, and measurements of initial susceptibility of the fines leads to an excessively high iron content when derived on the assumption of non-interacting approximately equidimensional iron grains (Equation 2.9).

2.5 The Hopkinson effect

An increase with temperature (up to near the Curie point) of induced magnetization, and therefore susceptibility, in a weak field was first reported in iron by Hopkinson (1889). The increase is also observed in magnetic minerals and is designated the Hopkinson effect.

In an assembly of uniaxial randomly oriented single-domain grains, the initial susceptibility is proportional to J_s/B_c, where J_s is the spontaneous magnetization and B_c the remanent coercive force, i.e. the back field required to reduce the saturated IRM (isothermal remanent magnetization) to zero. Thus, if B_c decreases faster than J_s with increasing temperature, susceptibility will increase, and this is observed in many rocks and their extracted minerals (Nagata, 1961; Radhakrishnamurty and Likhite, 1970; Dunlop, 1974). The effect may or may not be observed in multidomain material, depending on the extent to which the observed susceptibility depends on the self-demagnetizing field of the grains (Section 2.3). In the extreme case of $k = 1/N$, initial susceptibility is independent of temperature until the Curie temperature is approached.

Rocks generally contain magnetic particles possessing a range of blocking temperatures, resulting in a broad Hopkinson peak. In large multidomain grains, the blocking temperature is near the Curie point and a rather narrow peak occurs (Dunlop, 1974). Observed susceptibility enhancement factors, i.e. the increase referred to the room-temperature value, are typically in the range 1.5–3.0 in rocks and minerals. Some examples are shown in Fig. 2.1. Radhakrishnamurty and Likhite (1970) have investigated the Hopkinson effect in basalts and Dunlop (1974) proposes it as the cause, through the higher temperature of rocks at depth, of magnetic anomalies which appear to arise from a higher magnetic susceptibility than is commonly found in surface rocks.

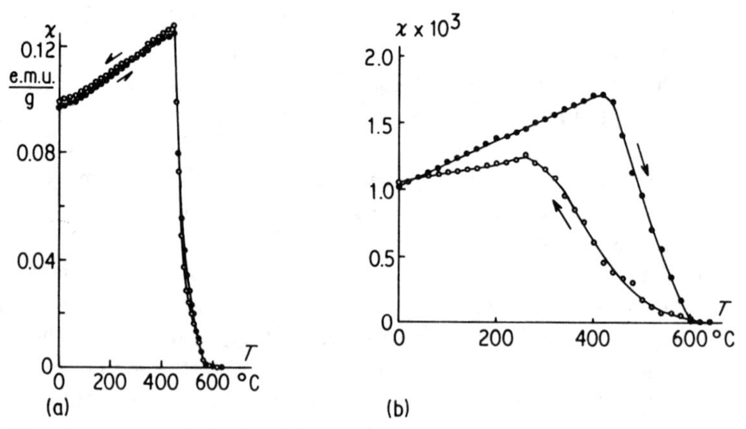

Fig. 2.1 Examples of the Hopkinson effect: (a) in a ferrimagnetic phenocryst extracted from a rhyolite, and (b) in a lava from Mt Usame, Japan. (Reproduced with permission from Nagata, 1961)

2.6 Measurement of initial susceptibility

Initial susceptibility is normally measured in an applied field $B \leqslant 1\,\mathrm{mT}$, in either direct or alternating fields.

2.6.1 Direct field measurements

The method is based on the application of Equations 2.1 and 2.2, and is most easily carried out on an astatic (or parastatic) magnetometer, preferably one with a surrounding coil system for nulling or modifying the ambient field over the sample. The deflections d_t of the magnetometer due to a sample rotated to different azimuths θ in the applied horizontal field perpendicular to the axis of the detecting magnet give a curve of the form

$$d_t = d_r \sin\theta + d_s \tag{2.10}$$

where d_r is the amplitude of the deflection due to remanence and d_s the deflection due to susceptibility. The sample is assumed to be in position (a) or (b) of Fig. 9.9. Assuming a magnetically homogeneous sample of regular shape, d_s can be obtained by adding the deflections at $\theta°$ and $(\theta + 180)°$, when the remanence term in Equation 2.10 vanishes. An alternative method is to take readings in zero and an applied ambient field at constant θ, and obtain d_s by subtraction, allowing for any change in sensitivity that the applied field produces. The former technique can be used with a component of the horizontal geomagnetic field as the applied field, but the latter requires a coil system for first mulling the geomagnetic field and then applying the inducing field. If position (c) is employed (Fig. 9.9) with an astatic system the applied field is vertical, either the vertical component of the Earth's field or from a coil. It is also possible to measure susceptibility in a vertical field with the sample in the off-centre position of Blackett's method of NRM measurement (Section 9.2.4). In using any of the above methods it is advisable to take readings with the sample in several orientations relative to the magnet system to average out the effects of inhomogeneity of both NRM and susceptibility, and of anisotropic susceptibility if present.

Samples with weak induced magnetization placed close to the magnet system in positions (a) or (b) or in the off-centre positions may also possess a significant component of magnetization induced by the field of the sensing magnet. However, since this magnetization is parallel to the magnetic axis of the magnet system, no torque on the system will result from the field arising from the induced component in homogeneous isotropic samples.

The sensitivity of the above methods depend on the magnetometer sensitivity and the magnitude of the applied field B. If the Earth's field is used, $B = 20$–$50\,\mu\mathrm{T}$. The minimum value of χ that is measurable is approximately numerically equal to (μ_0/B) times the weakest measurable specific NRM in a sample of the same size and shape.

The maximum field that can be applied with a coil system is chiefly governed by the astaticism of the magnet system (i.e. by the extent to which the field deflects the system) and the loss of sensitivity incurred. This field is normally in the range 10–100 μT. Direct field methods are unsatisfactory for rocks in which the Königsberger ratio Q_n is $>$ 10 (Section 10.2) because the ratio d_s/d_r is then small, although it may be possible to measure susceptibility along an axis in the sample nearly perpendicular to the NRM. Alternatively, crushed rock can be used, which eliminates or much reduces the NRM of the sample and also any inhomogeneity in NRM.

A method based on that of Bozorth (1951) may be used in special investigations. Similar small solenoids are placed on either side of the magnet system of an astatic magnetometer and carry currents such that the net torque exerted on the system by the solenoid fields is zero. The induced magnetization in a sample placed in one of the solenoids can then be measured either directly by the deflection of the magnet system or by the current required in an auxiliary coil to reduce the deflection to zero.

Large, irregularly shaped samples can be best measured in positions (a) or (c), or by using a fluxgate field detecting system (Section 9.2.5). The latter system can also be used for conventional samples, although the sensitivity achieved will not be comparable with that of a good astatic magnetometer.

2.6.2 Alternating field measurements

(a) Bridge circuits

Methods involving alternating applied fields have the advantage of compactness, ease of measurement and independence of the NRM of the sample. Several different approaches have been employed. The magnetic induction field B, the total number of lines of force per unit area inside magnetized matter, is related to the (volume) magnetization M by

$$B = \mu_0(H + M)$$

where H is the external magnetic field. When M is the induced magnetization in the field H, then $M = kH$, and Equation 2.8 can be written

$$B = \mu_0(H + kH) \qquad (2.11)$$

or

$$B = \mu H \qquad (2.12)$$

where μ is the permeability of the material. Combining Equations 2.11 and 2.12 we get

$$\mu = \mu_0(1 + k) \qquad (2.13)$$

where k is the volume susceptibility of the material. It is also useful to define relative permeability $\mu/\mu_0 = 1 + k = \mu_r$. Thus, k is related to μ_r, and since the latter property is closely associated with the characteristics of alternating current circuits containing inductive elements, Equation 2.13 forms a basis for measuring initial susceptibility. As an example, consider a solenoidal coil in which the winding thickness is very small. The inductance may be defined as the flux linked with the coil (the product of total lines of force through the coil and number of turns) when unit current flows in it, and

$$L = \frac{\mu_r \mu_0 N^2 A}{l} \tag{2.14}$$

where N is the number of turns on a coil of length l and cross-sectional area A immersed in a medium of relative permeability μ_r. Thus, in air of relative permeability μ_a

$$L_1 = \frac{\mu_a \mu_0 N A}{l} \tag{2.15}$$

and in a medium of relative permeability μ_s

$$L_2 = \frac{\mu_s \mu_0 N^2 A}{l} \tag{2.16}$$

The change in inductance on immersion in the medium is given by

$$\Delta L = L_2 - L_1 = \frac{\mu_0 N^2 A}{l}(\mu_s - \mu_a) \tag{2.17}$$

$$= C(k_s - k_a) \tag{2.18}$$

where C is a constant. Thus, $\Delta L \propto k_s - k_a$, where k_s and k_a are the volume susceptibilities of the medium and air respectively.

Air is paramagnetic, with a volume susceptibility at STP of 3.6×10^{-7} and the great majority of rocks have volume susceptibilities in the range 10^{-5}–10^{-1}: therefore, for all but the most weakly magnetic rocks, or unless very accurate measurements in the lower range of suceptibility are required, k_a can be neglected and $\Delta L \propto k_s$. For the above range of susceptibilities, Equation 2.18 gives fractional changes $\Delta L/L$ of $\sim 10^{-5}$–10^{-1}. In practice, of course, the coil is not 'immersed' in the rock sample, but the sample is inserted into the coil with only imperfect flux linkage, and the practical change in inductance is less. A practical application of Equation 2.18 for susceptibility measurement is the change of balance conditions when the sample is inserted into a coil forming one arm of an a.c. bridge.

A method based on the transformer bridge is described by Girdler (1961) and Fuller (1967). The basic circuit is shown in Fig. 2.2. The bridge arms consist of two identical air-cored coils and the two sides of a centre-tapped Mumetal-cored coil, i.e. a transformer of ratio 1:1. The chief advantage of this type of bridge is that stray impedances affecting the balance point are very small. Resistive balance is achieved by the adjustment of R_3 and R_4 or by

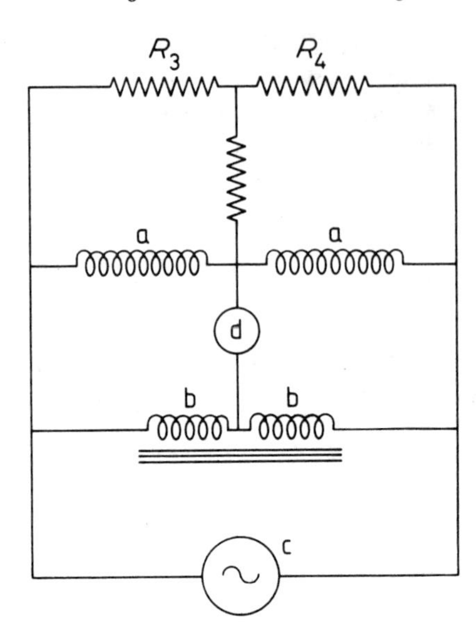

Fig. 2.2 Transformer bridge circuit (Girdler, 1961). a, air-cored coils; b, Mumetal-cored coils; c, oscillator; d, detector.

moving a closed wire loop parallel to the axis of one of the coils, and inductive balance is obtained by positioning a small ferrite slug inside one of the coils. The change in inductance of the other coil when a rock sample is placed in it unbalances the bridge and it is rebalanced by adjustment to the ferrite slug, the necessary displacement being proportional to the susceptibility of the sample. The resistive balance is unaffected. The coils are designed to provide a peak applied field over the sample of $\sim 150\,\mu$T uniform to 0.5%, and the bridge frequency is 1440 Hz. Sensitivity is about $6 \times 10^{-9}\,\text{m}^3\,\text{kg}^{-1}$ for cylindrical samples 3.5 cm in diameter and 3.5 cm high.

There are some disadvantages in the use of the ferrite slug for rebalancing (e.g. imperfect linearity and the generation of mechanical noise) and Fuller (1967) used the amplified bridge output as a measure of the sample susceptibility. Provided the change in inductance or insertion of the sample is not too great, the resulting out-of-balance current is proportional to ΔL and thus to the sample susceptibility.

At the time of writing a commercially available susceptibility meter, based on a transformer bridge design of Jelinek (1973), is made by the Institute of Applied Geophysics (UGF), Prague. The operating frequency is 970 Hz and the sensor is an air-cored coil. Cubic samples 2.0 cm on a side are measured in a peak field of 140 μT. The bridge is a null reading instrument, and the out-of-balance current is compensated by an equal and opposite current derived from a third arm to the bridge connected across the input and output. The sample susceptibility is read off on a potentiometer dial controlling the attenuator and provides five ranges of full-scale value of 10^{-7}–5×10^{-5} m^3 kg^{-1} for a standard sample of 2.0 cm cube. The minimum measurable susceptibility in a 2.0 cm cube is stated to be somewhat less than 10^{-10} m^3 kg^{-1}. The bridge is calibrated in volume SI units.

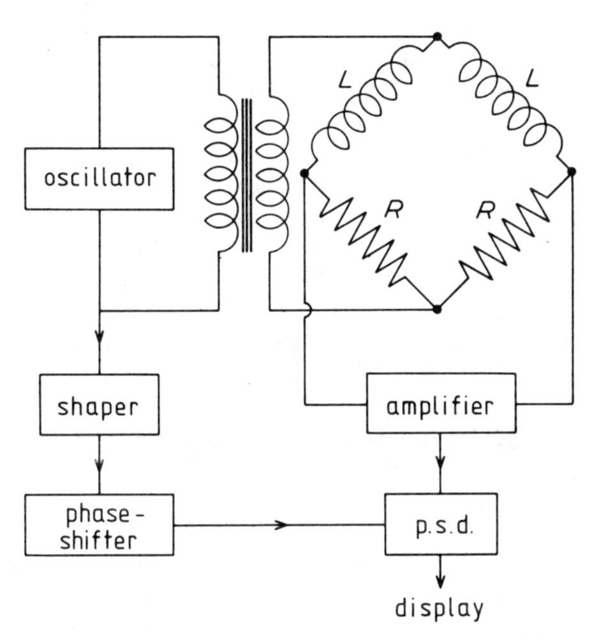

Fig. 2.3 Equal impedance bridge (Stephenson and de Sa, 1970). The sample is inserted into one of the coils, *L*.

The bridge of Stephenson and de Sa (1970) has two air-cored coils and two resistances forming an equal impedance bridge (Fig. 2.3), and is adapted for high- and low-temperature measurements. Assuming a zero impedance input of e.m.f. E and a detector of infinite impedance, the out-of-balance signal of such a bridge is $E\delta/4(1+\delta/2)$ when the impedance of one arm changes by a factor $1+\delta$. Therefore, for $\delta \approx 0.02$ the output signal is proportional to δ to within $\sim 1\%$. The output is amplified and converted to d.c. by a phase-sensitive detector.

For a pure inductance change on sample insertion (a valid assumption in practice), there is no phase change in the output waveform. Once the reference signal for the phase-sensitive detector, derived from the input oscillator, is adjusted correctly, it will remain so for subsequent measurements, assuming that the circuit characteristics remain constant.

The output signal is conveniently recorded on a micro- or milliameter, and the phase-sensitive detector provides improved signal recovery from random noise when samples of low susceptibility are measured. The original bridge of Stephenson and de Sa was designed for the measurement of small synthetic titanomagnetite samples, and had a noise level of $\sim 1 \times 10^{-11}$ m^3 total susceptibility. At the time of writing, a similar bridge for the measurement of 2.5 cm samples is marketed by Highmoor Electronics Ltd, Salford, England.

The design features of bridges for susceptibility measurements are generally the same as for conventional a.c. bridges. Where high sensitivity is required, special attention may be necessary to the screening of leads and shielding of components, and to the prevention of temperature variations. In general, maximum sensitivity, i.e. maximum out-of-balance signal for a given fractional change in coil inductance, is achieved when the arms are of approximately equal impedance. The coils are required to provide, from an input of a few volts, a field of $\sim 100\,\mu$T r.m.s. of reasonable uniformity ($\pm 2\%$) over the sample, and to have a high $Q(=\omega L/R$, where R is the coil resistance). The optimum size of the coils is in part a compromise between the provision of sufficient field homogeneity for a given sample size and the need for a reasonable sample filling factor (sample volume/coil volume) to maximize the inductance change produced by the sample. With modern techniques of signal processing and recovery it is not difficult to achieve sufficient sensitivity for the great majority of susceptibility measurements.

(*b*) *Balanced transformers*

The principle of these circuits is the change in mutual inductance between two coils when a susceptible sample is placed relative to them such that the flux linkage in one coil due to an alternating current in the other is altered. The induced voltage in the secondary coil is monitored when the sample is placed in one of the coils or in a magnetic circuit of which the coil and its core form a part.

In the method of Christie and Symons (1969), two sensing coils connected in series opposition are each surrounded coaxially by an exciting coil carrying an alternating current of frequency 1 kHz. The residual output of the sensing coils is balanced to zero by trimming resistors and capacitors. A paramagnetic sample inserted into one of the inner coils increases the flux linking the coil and a net output voltage appears across the coils. From the analysis in Section 2.6.2(a), it is easy to show that the increase in flux linked with the sensing coil on sample insertion, and therefore the increase in the net induced voltage, is proportional to $(k_s - k_a)$. The exciting coils are tuned to resonance with a small capacitor and are supplied from the oscillator of a commercial lock-in amplifier, whose phase-sensitive detector is used to process and display the out-of-balance signal. The noise level corresponds to a minimum measurable susceptibility of about 2×10^{-10} m^3 kg^{-1} in a cylindrical sample 3.2 cm diameter by 2.8 cm high.

An alternative approach is due originally to Bruckshaw and Robertson (1948). Two coaxial coils wound one outside the other and containing an equal number of area-turns are connected in series opposition. In an external, uniform alternating field, equal antiphase voltages are generated in the coils, and there is no net voltage output from the system. When a sample is placed in or near the inner coil, the change in flux linkage with that coil is greater than with the outer coil and a net induced voltage appears across the coils. This type of instrument has close affinities with the spinner magnetometer, and the exciting field can be considered as inducing a sinusoidally varying magnetic moment in the sample which then induces the alternating voltage in the coil system. The r.m.s. output voltage V is of the form

$$V = \omega(C_1 - C_2)(k_s - k_a)vB \text{ V} \tag{2.19}$$

where v is the sample volume, B is the inducing field, and ω its angular frequency. C_1 and C_2 are constants involving the size of the coils, their geometry and number of turns, and can be calculated from the expressions given in Section 9.3.1. for spinner magnetometers.

Bruckshaw and Robertson (1948) used an inner coil 1.5 cm long of internal and external radius 2.0 and 4.7 cm respectively on which the outer coil of external radius 5.9 cm was wound. The coils contained 22, 100 and 9400 turns. The output voltages induced in the applied field of 50 μT at 50 Hz (produced by a Helmholtz pair) were balanced by adjustment of turns and then moving the system a small distance along the Helmholtz coil axis into a region where the slight non-uniformity of field optimized the cancellation of the two voltages. Susceptibility was measured by a potentiometer system in which the out-of-balance signal was nulled by a fraction of the output of a third coil wound on the double coil. Using an amplifier and tuned vibration galvanometer as detector, susceptibilities of $\sim 4 \times 10^{-8}$ m^3 kg^{-1} could be measured in a 2.0 cm cubic sample, the limiting factor being amplifier noise rather than imperfect balancing of the double coil.

A similar double pick-up coil and Helmholtz pair were used by Likhite and Radhakrishnamurty (1965). The output signal was amplified and read directly, and the equipment had a noise level of $\sim 10^{-10}$ m^3 kg^{-1} for an 8 cm^3 sample, at 400 Hz. It could also be used for measurement at different frequencies (Section 2.6.5).

A simplified calculation indicates the requirements for good sensitivity and signal/noise ratio in this type of instrument, where 'noise' refers to the signal arising from imperfect balancing of the double coil.

Consider the voltages V_p, V_b induced in a conducting loop of radius r by an oscillating dipole moment P at its centre, and by a uniform alternating field $B = \mu_0 H$ acting parallel to the coil axis. Then the flux ϕ linking the loop due to P is $\mu_0 P/2r$ and

$$V_p = \frac{d\phi}{dt} = \frac{\mu_0 P\omega}{2r} \tag{2.20}$$

and

$$V_b = \omega B \pi r^2 \tag{2.21}$$

where ω is the angular frequency of P and B. Then

$$\frac{V_p}{V_b} = \frac{\mu_0 P}{2\pi B r^3} \tag{2.22}$$

or, since $P = kvH$, where v is the sample volume

$$\frac{V_p}{V_b} = \frac{kv}{2\pi r^3} \tag{2.23}$$

Achievement of a high signal/noise ratio depends on maximizing V_p/V_b, since V_b is systematic noise which cannot be removed by, for instance, phase-sensitive detection. As expected, V_p/V_b increases in general as r decreases, but if the sample approximates to a sphere which fills the loop $v = (4/3)(\pi r^3)$ and

$$\frac{V_p}{V_b} = \frac{2k}{3} \tag{2.24}$$

If $k \approx 10^{-5}$, $V_p/V_b \approx 10^{-5}$ also. This implies that the voltage induced in the inner coil in the applied field must be compensated to approximately one part in 10^6 if a volume susceptibility of $\sim 1 \times 10^{-5}$ is to be measured satisfactorily (i.e. $V_p/V_b \approx 10$). Although this is a simplified analysis, it indicates the order of compensation required for the measurement of sediments. With modern low-noise amplification, compensation is the factor most likely to limit the effective sensitivity. Using a subsidiary winding connected in a potentiometric arrangement with the double coil, Likhite and Radhakrishnamurty quote a compensation factor of 10^6 at 400 Hz. Provided that sufficient compensation can be achieved, the sensitivity of these instruments increases with frequency. Some of the design features of double-coil pick-up systems, described in connection with spinner magnetometer, are also relevant to these susceptibility instruments.

High sensitivity can be achieved in a more compact instrument by high permeability ferrite-cored transformers. Two annular cores containing identical gaps are each wound with excitation and pick-up windings, the latter being connected in series opposition (Fig. 2.4). The flux linking each pick-up coil depends inversely on the reluctance S of the gap. In air, assuming no leakage flux

$$\frac{1}{S_a} = \frac{\mu_0 \mu_a A}{l} \tag{2.25}$$

where l and A are the width and cross-sectional area of the gap respectively, and μ_a the relative permeability of air (~ 1). If a sample of relative permeability μ_s is placed in and fills the gap

$$\frac{1}{S_s} = \frac{\mu_0 \mu_s A}{l} \tag{2.26}$$

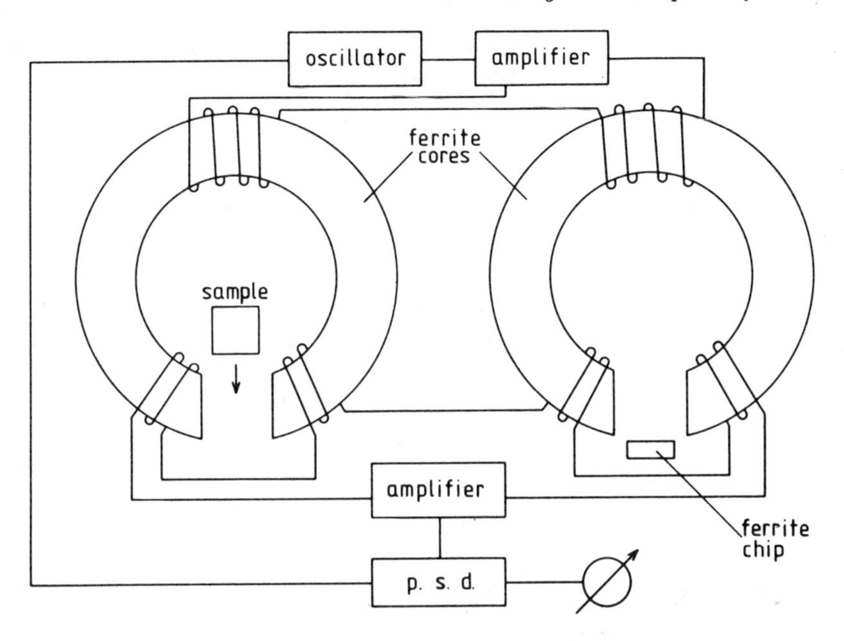

Fig. 2.4 Balanced transformer circuit with ferrite-cored coils.

and

$$\Delta(1/S) = \frac{\mu_0 A(\mu_s - \mu_a)}{l} \tag{2.27}$$

and the change in flux linkage and therefore output voltage is proportional to $(\mu_s - \mu_a) = (k_s - k_a)$, provided the reluctance of the air gap and sample is much greater than that of the core, which is normally the case.

The net output voltage of the two pick-up coils is brought to zero by adjusting a small piece of ferrite and a strip of good electrical conductor in or near the gap of the compensating transformer until inductive and resistive balance is achieved. An out-of-balance voltage then appears when the sample is inserted in the gap of the other transformer. In the author's equipment (Collinson, Molyneux and Stone, 1963), the mean diameter of the cores (of Ferroxcube) is 9.0 cm with a cross-section of 2.0 cm and 2.3 cm and gap width 2.7 cm. The excitation and pick-up windings are of 100 and 20 turns respectively and the excitation current (800 Hz) is a few milliamps, which provides a gap field of $\sim 200\,\mu$T. In the instrument of Christie and Symons (1969), the cores are somewhat larger, with a 5.0 cm gap, and $\sim$4000 and 300 turns on the excitation and pick-up windings. The number of turns on the excitation windings is governed by the current available: the number of ampere-turns is chosen such that the core operates in the high initial permeability region ($\mu_r \approx$ 100–1000). It is also desirable that the gap field is in the range 50–500 μT. Equation 3.5 can be used to estimate the gap field generated per ampere-turn in the excitation winding.

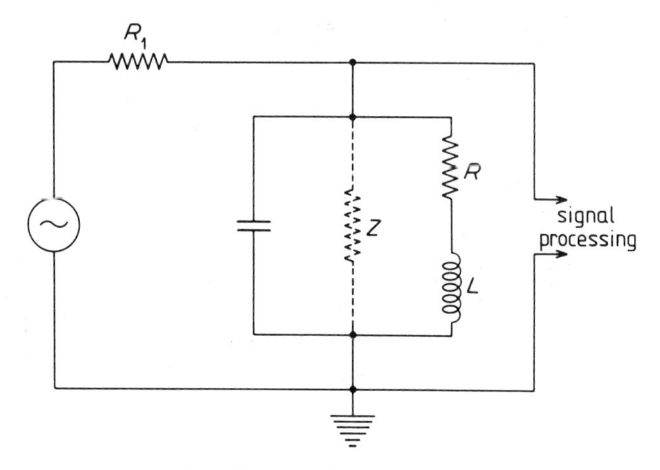

Fig. 2.5 Simplified diagram of the radio-frequency method of Cooke and de Sa (1981). L is the sample coil (a single-layer solenoid, $L°1\ \mu$H), Z is the resistance presented to the external circuit and R is the effective coil resistance.

The radio frequency method of Cooke and de Sa (1981) is capable of separating the real (χ') and imaginary (χ'') parts of initial susceptibility. The former arises from magnetic changes in the sample arising from the inducing field and the latter from eddy current and hysteresis losses, which are normally negligible in rocks. Figure 2.5 is a simplified circuit diagram. The oscillator drives the *LCR* circuit at its resonant frequency (5 MHz) and the tuned circuit then appears as a real resistance Z. If a sample is inserted into the coil, χ' alters the coil inductance and the resonant frequency, producing a phase shift between oscillator and tuned circuit signals, and χ'' effectively increases the coil resistance, resulting in a decrease in signal amplitude via the voltage divider ratio $Z{:}R_1$. Two feedback circuits are used to restore the circuit to its initial state, the feedback signals being proportional to χ' and χ''.

The instrument is designed to measure small ($\sim$1 g) powdered samples and the noise level for χ' is equivalent to a total susceptibility of $\sim 1.0 \times 10^{-11}$ m^3.

2.6.3 Ballistic methods

Ballistic methods of measuring initial susceptibility are essentially the same as those for measuring NRM, described in Section 9.5.3. A long solenoid or Helmholtz coil system provides the inducing field and the charge transfer in a pick-up coil placed coaxially inside the solenoid or coil system on insertion of the sample is a measure of the sum of the induced magnetization and the component of NRM along the coil axis. By reversing the sample or applied field the contribution of the NRM can be eliminated. Because of some lack of sensitivity and the long time constants associated with sensitive conventional ballistic galvanometers, the method is not generally favoured, although modern techniques of pulse amplification should, in principle, enable reasonable time constants and sensitivity to be achieved. Another disadvantage is one associated with all detecting techniques which respond both to remanent and induced magnetization, namely a possible low ratio of induced to remanent signal when rocks of high Königsberger ratio are measured.

An advantage of the ballistic method over the Bruckshaw–Robertson alternating field device is that the inducing field does not have to be compensated by an additional coil connected in series opposition to the pick-up coil.

The ballistic magnetometer of Daly (1967), described in Chapter 9 in connection with NRM measurement, can be used for initial susceptibility measurements. As in NRM measurements, the sample is transferred from one coil to the other with an axial field of up to 10 mT applied by an external solenoid. In a sample of mass ~ 200 g, specific susceptibility down to $\sim 10^{-11}$ m^3 kg^{-1} can be measured, in an applied field of ~ 10 mT, although the measurement of time is lengthy at this sensitivity (see also Section 2.6.5 concerning measurements of very weak susceptibility). Nagata (1967b) also describes ballistic magnetometers with which initial susceptibility can be measured.

The modern equivalent of the ballistic magnetometers is the superconducting or SQUID (superconducting quantum interference device) instrument (Section 9.4), and low-field susceptibility can be measured in these instruments by allowing the superconducting shield to trap a weak field within itself, rather than using the shield to provide the normal field-free environment for NRM determinations. The signals due to NRM and susceptibility are separated as in analogous techniques, by reversing the sample orientation relative to the detector and obtaining signals proportional to (kH + NRM) and (kH − NRM). Since the SQUID detector is cooled below the critical temperature in the presence of the magnetic field, it does not record the field but only changes in it produced by insertion of the sample into the pick-up coils. The properties of superconducting shields ensure that the trapped field is very stable.

Hanneken, Carnes and Vant-Hull (1976) describe a method of susceptibility measurement with the cryogenic magnetometer using a solenoid-generated applied field. To avoid the large signal arising from the flux due to a single coil, a 'flux-compensator' coil is used consisting of an inner coil and an outer coaxial Helmholtz pair connected in series opposition with the inner coil. By adjusting the windings ratio, the total magnetic flux due to the system can be made very small while still providing a small axial field at the sample measuring position at the centre of the inner coil.

2.6.4 Calibration

The calibration of susceptibility meters on their more sensitive ranges can conveniently be carried out with pure (AnalaR) chemicals contained in holders of standard rock sample shape and size, and linearity can be checked at the same time. A useful range of susceptibility is provided by copper sulphate ($CuSO_4 \cdot 5H_2O$), ferrous sulphate ($FeSO_4 \cdot 7H_2O$) and ferrous chloride ($FeCl_2$) with mass susceptibilities (paramagnetic) of 7.4, 51.7, and $145 \times 10^{-8}\,m^3\,kg^{-1}$, respectively (these values vary slightly (1–2%) from different authorities). Values for some other chemicals are given in Table 2.1. Day-to-day monitoring of instrument sensitivity is probably best carried out with a calibrated (isotropic) rock sample, since most chemicals are either hygroscopic or deliquescent and require efficient sealing if their susceptibility is to remain constant.

There are no synthetic materials with well-defined susceptibilities much above about $250 \times 10^{-8}\,m^3\,kg^{-1}$; where higher values are required, the author has used various concentrations of powdered magnetite in aluminium oxide. From Equation 2.9, $\chi \approx 6.0 \times 10^{-4} p$, where p is the mass fraction of magnetite ($p \ll 1$), and suitably dilute, homogenized mixtures will cover the range 10^{-6}–$10^{-4}\,m^3\,kg^{-1}$: these are satisfactory for checking linearity in the instrument, if not for highly accurate calibration, because of uncertainty in the mean particle demagnetizing factor. Christie and Symons (1969) calibrate their bridge with rock samples whose susceptibility has been determined on an astatic magnetometer: this method may be satisfactory, but the present author has found discrepancies of up to 10% in the measured susceptibility of a variety of rocks determined on a bridge instrument and an astatic magnetometer (unpublished data), and these differences have not been explained.

An absolute calibration method for inductance bridge instruments, using a small coil loaded with an impedance, has been described by Zahn (1963). The following is a simplified analysis of this method.

If a small coil carrying N turns of mean area A is placed in a uniform alternating magnetic field B of angular frequency ω, an e.m.f. E is generated given by

$$E = NA\omega B \tag{2.28}$$

If the coil is loaded with an external impedance Z, the current in the coil is

$$i = \frac{E}{Z} = \frac{NA\omega B}{Z} \tag{2.29}$$

assuming that the coil impedance is negligible. The magnetic moment P of the coil is

$$P = NAi = \frac{(NA)^2 \omega B}{Z} \tag{2.30}$$

Thus, the effective total susceptibility S of the coil is

$$S = \frac{\mu_0 P}{B} = \frac{(NA)^2 \omega}{Z} \tag{2.31}$$

To simulate positive, real susceptibility, the load must be purely inductive, $Z = L\omega$, and

$$S = \frac{(NA)^2}{L} \tag{2.32}$$

If $N = 50$, $A = 5\,\text{cm}^2$, and $L = 1H, S = 7.85 \times 10^{-10}\,\text{m}^3$, a typical value encountered in a standard (2.5 cm) sample of red sediment or weakly magnetic igneous rock. Different values of L will provide a range of equivalent susceptibilities, or the system can be used to produce an apparent 'diamagnetic' susceptibility equal and opposite to that of a sample, thus providing a null method of measurement. However, the calibrated variable inductance which would be needed for the absolute null method is not easy to construct, and also $S \propto L^{-1}$, which is undesirable from the point of view of scale linearity. Zahn shows that the variable inductance can be replaced by a fixed inductance in parallel with a variable capacity. If the Q-value of the inductance is sufficiently large ($\gg \sim 100$) the load effectively maintains the property of being purely inductive and the apparent susceptibility of the coil is given by

$$S = (NA)^2 \omega^2 \Delta C \tag{2.33}$$

ΔC is the change in capacity required to null the signal when the sample is inserted. If the initial setting of the capacity is such that the parallel LC circuit is resonant, then this value provides a 'zero' for absolute susceptibility measurement. For optimum performance and accuracy phase-sensitive detection of the output is desirable, to eliminate possible effects of harmonics and any imaginary part of the apparent coil susceptibility due to a resistive component in the circuit.

2.6.5 Frequency dependence of initial susceptibility

The variation of initial susceptibility of rocks with the frequency of the inducing field is of some interest, both through the basic physics involved and because of the association of a.f. susceptibility with magnetic viscosity and therefore also with palaeomagnetic stability (Néel, 1955; Vincenz, 1965). Vincenz calculated a decrease of susceptibility in multidomain magnetite of about 1% per decade of increase in frequency, although Bhathal and Stacey (1969) suggested a third of this value is a better estimate, if domain wall movement is the major contribution to viscosity. For the iron grains present in lunar fines, Stephenson (1971b) derived an expected decrease of about 14% per decade of frequency increase.

Any of the methods of measuring susceptibility in an alternating field can, in principle, be used to measure the variation of susceptibility with frequency, although there may be practical difficulties. For instance, in bridge circuits because of dependence of the coil inductance on frequency, and in balanced transformer circuits because of the need to preserve good voltage compensation in the double-coil system over a range of frequencies.

The latter problem was encountered by Likhite and Radhakrishnamurty (1965) and they used a small variable capacity across one of the double-coil windings to compensate for the variation with frequency of the phase difference between the two voltages. This variation arises from the self-capacity of the coils, and the authors found that good compensation was not possible above about 1500 Hz because of the approach to a resonant frequency in the circuit formed by a coil inductance and its self-capacity. The authors report measurements of a.f. susceptibility in basalt samples in the range 50–2000 Hz, using a modified double-coil system. The effects of self-capacity were much reduced by Bhathal and Stacey (1969), who used single-layer coils with separated turns. Their sensor was a balanced mutual inductance consisting of a long primary (exciting) solenoid with two balanced secondary coils inside. The change in mutual inductance when a sample is inserted in one of the secondary coils is measured in terms of the change required in a standard, variable mutual inductance to rebalance a Hartshorn bridge (Hague, 1962) in which the mutual inductances form two arms. The standard inductance exhibited some frequency dependence above 2000 Hz, but satisfactory measurements of high accuracy ($\sim \pm 0.1\%$) were possible in the range 100–1000 Hz. No frequency dependence of susceptibility was observed in the basalt samples which were tested, or in laboratory-prepared, sized magnetite powders (20–200 μm) dispersed in plaster of Paris in volume concentrations up to $\sim 10\%$. The inducing field was 18 μT.

Another method by which susceptibility may be measured at different frequencies is by the use of a marginal oscillator (de Sa, 1968). In a parallel *LCR* circuit, the resonant frequency ω_0 can be expressed as $\omega_0 = R_1/QL$, where $R_1 = Q^2 R$. The resistance and inductance of the coil are R and L respectively, and Q is the quality factor. Thus, $\omega_0 \propto L^{-1}$, and small changes in L produced by inserting a rock sample into the coil can be detected, provided a high value of Q is used. By incorporating a network having a negative resistance characteristic, Q can be increased (Fig. 4.3). Negative resistance is generated with the *LCR* circuit and non-inverting amplification of high input and low output impedance. By adjusting the gain of the amplifier the circuit is maintained in a marginally oscillating condition with high Q, and $\Delta\omega_0 \propto \Delta L \propto \chi$.

This circuit is suitable for measurements on igneous rocks and synthetic magnetite and titanomagnetite samples. By using different coils, the frequency range that can be used is approximately 10 kHz–10 MHz.

Stephenson (1971b) used a similar technique to test the prediction of a 14% decrease in susceptibility per decade of frequency in a sample of lunar fines.

The linearity of the circuit was checked by observing the change of frequency produced by different concentrations of magnetite in a constant volume sample. The lunar samples were compared with three reference samples (iron powder, haematite, and magnetite), the susceptibility of which was measured at the different frequencies. The ratios of the reference susceptibilities remained constant at the measurement frequencies, which, in view of their different physical and magnetic properties, indicated that their susceptibility was independent of frequency. The lunar and reference samples were measured at 1.5 kHz on a bridge circuit and between 0.3 and 11 MHz with the marginal oscillator and the theoretical prediction of frequency dependence of χ in the lunar fines was confirmed.

2.7 Initial susceptibility of rocks

Because initial susceptibility of ferromagnetic mineral particles depends on size, shape, internal stresses and composition, the initial susceptibility of rocks, with possible multiple magnetic mineral content, is often not of itself a particularly informative property. In particular rocks it may be indicative of the amount of magnetic mineral present, for instance where magnetite is the only important contribution to susceptibility.

In igneous rocks, magnetite and titanomagnetite contribute the major part of the observed susceptibility and dia- and paramagnetic material will not be apparent in measurements of χ. In many sediments, particularly redbeds in which there are no strongly magnetic minerals, paramagnetic constituents such as clay minerals and even diamagnetic quartz and carbonates may contribute significantly to χ, together with the commonly occurring haematite (Table 2.1).

A quantity which is sometimes of use in interpreting magnetic properties of rocks is the Königsberger ratio, Q. This was originally defined by Königsberger (1938) as the ratio of TRM intensity J_H acquired in a field $H = B/\mu_0$ in an (igneous) rock to the induced intensity in the same field at 20°C

$$Q_t = \frac{\mu_0 J_H}{\chi B} \tag{2.34}$$

However, a more practical and commonly used ratio is

$$Q_n = \frac{\mu_0 J_R}{\chi B_E} \tag{2.35}$$

i.e. relating the NRM intensity J_R to the induced magnetization in the geomagnetic field, B_E: this is usually taken as 50 μT.

It is clear that the significance of Q_n will be very dependent on rock type and mineral content and will generally be more informative in igneous rocks, in which NRM and χ are likely to be contributed by the same mineral, than in sediments. Its use as an indicator of stability of NRM is discussed in Section 10.2, and as a normalizing parameter in geomagnetic field intensity investigations in Section 14.2. Q_n is of most direct interest in the interpretation of magnetic anomalies where it is used to estimate the relative contribution of induced and remanent magnetization in the underlying rocks.

The change in initial susceptibility of a rock is sometimes useful as an indicator of chemical or physical alteration of the magnetic minerals, for example during heating. Examples of such alteration are the oxidation of magnetite to haematite and the breakdown of clay minerals when some sediments are heated, resulting in the formation of magnetite. The latter example is of practical use during thermal cleaning, when magnetite formation is a possible source of spurious NRM (Section 11.2). Assuming the formation of multidomain magnetite with $\chi \approx 6.0 \times 10^{-4}\ \mathrm{m^3\,kg^{-1}}$ (Section 2.3), one part in 10^4 by weight of magnetite will contribute an initial susceptibility of $\sim 6.0 \times 10^{-8}\ \mathrm{m^3\,kg^{-1}}$. Since χ of many red sandstones and siltstones is in the range $1\text{–}3 \times 10^{-8}\ \mathrm{m^3\,kg^{-1}}$, this is clearly a sensitive test. The formation of maghemite, for example by dehydration of lepidocrocite (γ-FeOOH), can be equally easily detected, or the inversion of maghemite to haematite, when a substantial decrease in χ occurs.

Very little, if any, work has been reported on comparisons of susceptibilities measured in the same samples by direct and alternating field methods. Cox and Doell (1962) drew attention to the possibility of different results being obtained, and the present author (unpublished data) has observed differences of up to 20% in sediments measured in a $\sim 10\ \mu\mathrm{T}$ direct field and a $\sim 200\ \mu\mathrm{T}$, 1000 Hz alternating field. These differences have not yet been satisfactorily explained. Bhathal and Stacey (1969) found agreement to within $\pm \sim 1\%$ when their basalt samples were measured on a.c. and d.c. equipment.

In the study of the NRM of lake and oceanic sediments, initial susceptibility is sometimes found to be a useful parameter for stratigraphic correlation in different cores from the same locality, and has also been tried as a normalizing factor for palaeointensity investigations in sediment cores (Section 14.2). For correlation purposes, the cores are either sampled at close intervals down their length, or their susceptibility can be measured continuously as a function of depth with equipment designed by L. Molyneux and now manufactured by Digico (Molyneux and Thompson, 1973). The sensing head is a solenoid 10 cm long with a central pick-up coil 1 cm long. One such coil system fits over the core and is traversed along its length, providing an exciting field of $\sim 100\ \mu\mathrm{T}$ at 10 kHz, while a similar system, away from the core, is used as a reference. The out-of-balance signal from the coils is computer-processed in a manner analogous to that used in the Digico magnetometer and the susceptibility at chosen intervals along the core is printed out. A typical noise level for a single traverse down a core 6.0 cm in diameter is $\sim 1.0 \times 10^{-9}\ \mathrm{m^3\,kg^{-1}}$, which can be improved, if necessary, by doing multiple traverses. The effective sampling length of the device is ~6.0 cm. It is calibrated by inserting a long tube containing a chemical of known susceptibility into the sense coil.

A variety of field susceptibility meters are available from the geophysical instrument manufacturers. They include small sample types, in-situ meters, in which the sensor is placed against the rock surface in an outcrop, and meters designed for scanning down the length of a sediment core. The noise level of these meters is usually around $1.0 \times 10^{-8}\,m^3\,kg^{-1}$.

2.8 Anisotropic susceptibility

2.8.1 Introduction

Anisotropy of magnetic susceptibility in minerals, i.e. dependence of susceptibility on the direction along which it is measured in a sample, arises from either fundamental anisotropy in the crystal structure (magnetocrystalline anisotropy) or from non-sphericity of shape of mineral particles. The former is a property of all the common magnetic minerals, but shape anisotropy due to aligned, elongated particles or to a planar or linear distribution of particles is the most common cause of anisotropy of initial susceptibility observed in rocks. The reason for this is that there is rarely any reason for alignment of crystallographic axes of equidimensional magnetic particles in a rock, but elongated or platy grains can be aligned, for instance by flow in a partially solidified igneous rock or by preferential horizontal settling of platy or elongated grains in sediments. In metamorphic rocks stress is often the aligning agent and a special form of anisotropy may occur, termed textural anisotropy, caused by the occurrence of planes or 'strings' of interacting magnetic particles. Textural anisotropy is not common and was first observed by Grabovsky and Brodskaya (1958) in a magnetite-rich foliated quartzite, and Stacey (1960a) and Fuller (1961) have investigated it in laboratory samples.

Thus, magnetic anisotropy has come to be used as a profitable tool for the detection of rock fabrics which may not otherwise be apparent. However, susceptibility anisotropy which is unrelated to rock structure can also be observed, namely statistical anisotropy. This arises when there is a statistical departure from complete randomness in the alignment of elongated grains in a rock sample in which no anisotropy related to fabric occurs. In such a case the anisotropy is randomly directed among different samples, and will appear as 'noise' in samples in which there is true structural anisotropy.

Shape anisotropy arises from the dependence of demagnetizing factor N on particle shape. In an applied field H the effective field, H_i, inside a particle (Section 2.3) is

$$H_i = H - NM \tag{2.36}$$

where M is the volume intensity of induced magnetization and N is the demagnetizing factor of the particle along the direction of the applied field.

Following the analysis of Section 2.3

$$H_i = H - Nk_i H_i$$

leading to

$$k_0 = \frac{k_i}{1 + Nk_i} \tag{2.37}$$

where k_i and k_0 are the intrinsic and observed susceptibility, respectively. The dependence of k_0 on N implies variation in k_0 according to the direction in which it is measured in a non-spherical particle. For shape anisotropy to be significant in an elongated particle $Nk_i > \sim 0.01$. This condition is easily met in iron and the titanomagnetite minerals, and elongated grains of the latter are the most common cause of the observed anisotropy of initial susceptibility in igneous rocks. In low fields, magnetocrystalline anisotropy is not observed in the cubic minerals, and therefore does not contribute to initial susceptibility anisotropy of rocks, although it does contribute in high field measurements (Section 3.3).

In haematite and pyrrhotite, $Nk_i < \sim 10^{-3}$ and anisotropy due to the magnetic effects of shape is negligible. However, haematite often occurs in the form of 'platelets' whose crystallographic and easy magnetic axes are in the plane of the particles. Thus there can be susceptibility anisotropy of magnetocrystalline origin in haematite-bearing sediments, brought about by preferential settling of particles with their planes horizontal or by stress- or flow-induced alignment in other rocks. In some metamorphic rocks, pyrrhotite particles, also of platy form, are aligned by stress, e.g. in the Welsh slates studied by Fuller (1963).

Quantitatively anisotropy is conveniently described by the ratio of maximum to minimum susceptibility (the anisotropy factor) observed in a rock sample. Typical values are 1.01–1.20 in igneous rocks and sediments, and up to 2.0 in some markedly structured metamorphic rocks such as slates.

Another aspect of magnetic anisotropy in minerals is its possible influence in deviating from the ambient field direction the direction of thermo- or chemical remanent magnetization acquired by minerals, thus undermining one of the basic assumptions of palaeomagnetic interpretation. Such a deviation is possible because in an anisotropic mineral the induced magnetization is not in general parallel to the applied field, and TRM or CRM will tend to be acquired along the 'easy' (maximum susceptibility) axis in a particle. This aspect of magnetic anisotropy has been studied by Stacey (1960b), Uyeda, Fuller, Belshe and Girdler (1963), Fuller (1963) and Irving and Park (1973). The deviation of a TRM direction from the ambient field is expected to be rather small ($< \sim 5°$) in most non-metamorphosed igneous rocks with anisotropy factors not exceeding ~ 1.2, but can be substantial if more marked anisotropy is present.

2.8.2 Measurement of anisotropic susceptibility

(a) Introduction

Before describing measurement techniques it is desirable to establish some of the physical aspects of susceptibility anisotropy. We are concerned here with initial susceptibility, and it is assumed throughout that induced magnetization is a linear function of the applied field, and that the field is small, i.e. $< \sim 5\,\mathrm{mT}$. To maintain simplicity in the equations used in the following analyses, volume susceptibility k, volume magnetization M, and the field H are used.

In a magnetically isotropic medium, induced magnetization is parallel to the applied field. In an anisotropic medium the magnetization is not in general parallel to the field. Thus, if the field H and induced magnetization M have components H_x, H_y, H_z and M_x, M_y, M_z along x, y and z, the latter can be expressed in the form

$$\begin{aligned} M_x &= k_{xx}H_x + k_{xy}H_y + k_{xz}H_z \\ M_y &= k_{yx}H_x + k_{yy}H_y + k_{yz}H_z \\ M_z &= k_{zx}H_x + k_{zy}H_y + k_{zz}H_z \end{aligned} \tag{2.38}$$

where, for instance, k_{xy} is the susceptibility appropriate to the contribution to the magnetization M_x along x resulting from H_y acting along y. Equation 2.38 can be written in the form

$$\mathbf{M} = k_{ij}\mathbf{H}_j \, (i = 1, 2, 3) \tag{2.39}$$

where k_{ij} is a second-order susceptibility tensor. It can be shown by considering the energy density in an anisotropic sample in an applied field that $k_{xy} = k_{yx}$, $k_{yz} = k_{zy}$ and $k_{zx} = k_{xz}$, and therefore the susceptibility tensor is symmetrical and anisotropic susceptibility is characterized by six constants corresponding to the remaining components of the tensor.

A more practical and instructive way of representing Equation 2.39 is by a geometrical surface, which for susceptibility is an ellipsoid. There are, in fact, two ellipsoids which are useful in discussing anisotropic susceptibility, namely the representation quadric and the magnitude ellipsoid.

In an anisotropic sample, three orthogonal axes can be defined along which the susceptibility takes maximum, minimum and intermediate values. These axes coincide with the principal axes of the ellipsoid, and the susceptibilities are called principal susceptibilities.

The representation quadric is of the form

$$k_a x_1^2 + k_b x_2^2 + k_c x_3^2 = 1 \tag{2.40}$$

where the ellipsoid is referred to its principal axes and k_a, k_b, k_c are the principal susceptibilities. An alternative form of Equation 2.40 is

$$\frac{x_1^2}{(1/\sqrt{k_a})^2} + \frac{x_2^2}{(1/\sqrt{k_b})^2} + \frac{x_3^2}{(1/\sqrt{k_c})^2} = 1 \tag{2.41}$$

The denominator in each term is the length of the semi-axes of the ellipsoid, and thus the representation quadric is 'inverted' in the sense that the axes of maximum and minimum susceptibility are the shortest and longest axes respectively of the ellipsoid.

The representation quadric has two useful purposes. The length of the radius vector in any direction from the origin is equal to the inverse square root of the susceptibility in that direction, and for a given direction of H the direction of M is parallel to the normal to the tangent plane to the ellipsoid at the point where the radius vector parallel to H meets the ellipsoid surface (Fig. 2.6(a)).

The magnitude ellipsoid is of the form

$$\frac{x_1^2}{k_a^2} + \frac{x_2^2}{k_b^2} + \frac{x_3^2}{k_c^2} = 1 \qquad (2.42)$$

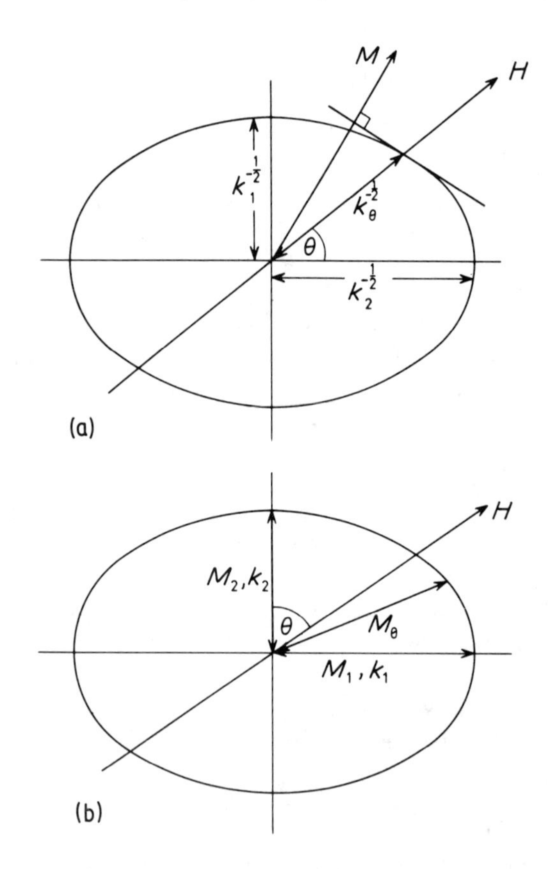

Fig. 2.6 Diagram of the representation quadric (a) and magnitude ellipsoid (b) in one of the principal planes. k_1 and k_2 are the maximum and minimum susceptibilities in the plane.

The ellipsoid has semi-axes proportional in length to k_a, k_b, k_c. The length of the radius vector in any direction from the origin is proportional to the magnitude of the induced magnetization M in that direction resulting from a field H applied along another direction (Fig. 2.6(b)). The direction of M is given by the radius-normal property of the representation quadric. It is common practice to describe the anisotropic susceptibility of a sample in terms of the principal susceptibilities and the orientation of the susceptibility ellipsoid relative to convenient axes in the sample. Three angles are required for the latter which, with the three principal susceptibilities, comprise the six quantities required to completely define the anisotropy ellipsoid.

Consider the simple case of a two-dimensional disc-shaped sample with the principal axes of susceptibility in its plane making angles of θ and $(90-\theta)$ with the applied field H, also in the plane of the disc (Fig. 2.7). The components of induced magnetization along the principal axes are

$$M_1 = k_1 H \cos\theta \tag{2.43}$$

$$M_2 = k_2 H \sin\theta \tag{2.44}$$

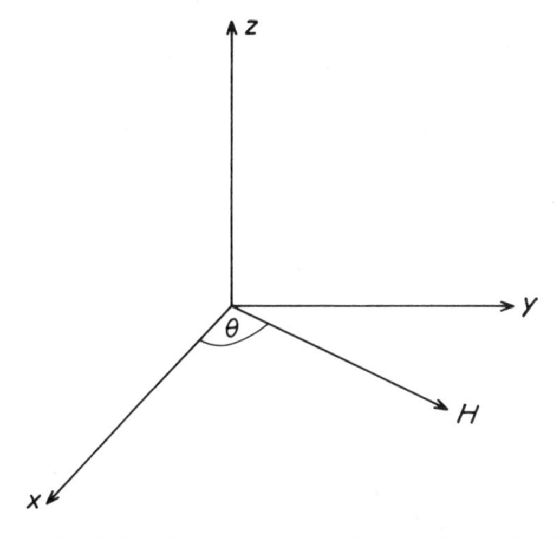

Fig. 2.7 Geometry for induced magnetization in x–y plane of anisotropic sample due to applied field H in the plane.

where k_1 and k_2 are the principal susceptibilities in the disc plane. A detector on the axis of the disc, measuring the field perpendicular to H and parallel to the disc plane due to the induced magnetization (e.g. an astatic magnetometer or fluxgate sensor) will record a field proportional to

$$\begin{aligned} & M_1 \sin\theta - M_2 \cos\theta \\ &= Hk_1 \cos\theta \sin\theta - Hk_2 \sin\theta \cos\theta \\ &= \tfrac{1}{2} H (k_1 - k_2) \sin 2\theta \end{aligned} \tag{2.45}$$

If the disc is rotated about its axis to different azimuths, a sin 2θ signal will be observed, or a sinusoidal signal of twice the rotation frequency in the case of continuous rotation. A similar result is obtained if the detector senses field parallel to H, except that in general the sin 2θ signal is superimposed on a constant signal due to H or a field related to H. Note that if the anisotropy cannot be represented by a second-order tensor, higher harmonics will also be present in the waveform.

By rotating a cylindrical sample about different mutually perpendicular axes, values of quantities such as $k_{xx}-k_{yy}$, $k_{yy}-k_{zz}$, $k_{zz}-k_{xx}$, k_{xy}, k_{yz}, k_{zx} can be measured, from which the shape of the susceptibility ellipsoid and its orientation relative to sample axes can be determined. A separate measurement of k_{xx}, k_{yy} or k_{zz}, by a conventional bridge or other technique, is then required to determine the size of the ellipsoid and thus the magnitude of the principal susceptibilities.

The calculation of susceptibility ellipsoids from data obtained from the different measurement technique is briefly described in Section 2.8.2(h).

(*b*) *Direct-reading instruments*

One of the simplest methods of anisotropy determination is the measurement of susceptibility along different axes in a rock sample, using one of the techniques for measuring isotropic susceptibility. The sensitivity of the method depends on both the mean susceptibility of the sample and the degree of anisotropy.

If we consider an applied field H in the x–y plane (Fig. 2.7), the induced magnetization M_H when H makes an angle θ with the x-axis is given by

$$M_H = M_x \cos\theta + M_y \sin\theta$$

where

$$\begin{aligned} M_x &= k_{xx}H\cos\theta + k_{xy}H\sin\theta \\ M_y &= k_{yy}H\sin\theta + k_{yx}H\cos\theta \end{aligned} \tag{2.46}$$

Since

$$k_{xy} = k_{yx}$$

$$\begin{aligned} M_H &= k_{xx}H\cos^2\theta + 2k_{xy}H\sin\theta\cos\theta + k_{yy}H\sin^2\theta \\ &= H[k_{xy}\sin 2\theta + \tfrac{1}{2}(k_{xx}+k_{yy}) + \tfrac{1}{2}(k_{xx}-k_{yy})\cos 2\theta] \end{aligned} \tag{2.47}$$

As θ goes from $0 \rightarrow 2\pi$, M (and the output signal) consists of a sinusoidal 2θ term superimposed on a steady signal due to the mean susceptibility in the $x-y$ plane. The amplitude of the sinusoidal term depends essentially on $(k_{xx}-k_{yy})$. Measurements can be made using the normal output signals from the instrument for different values of θ or alternatively, if the output due to the constant term in Equation 2.47 can be suppressed, the output corresponding to the 2θ terms only can be used. In the former case, $k_{xx}(\theta = 0)$, $k_{xy} + \frac{1}{2}(k_{xx}+k_{yy})$ $(\theta = 45°)$, k_{yy} $(\theta = 90°)$ and $k(k_{xx}+k_{yy}) - k_{xy}$ $(\theta = 135°)$ are determined, plus corresponding terms in the x–z, y–z planes. With the constant signal suppressed, $(k_{xx} - k_{yy})$ and k_{xy} are determined, plus corresponding terms from the other two planes.

The advantage of suppressing the constant signal is that increased sensitivity can be achieved provided that random noise on the signal is not already limiting the sensitivity. The minimum detectable difference between any of the susceptibility terms (e.g. $k_{xx} - k_{yy}$) depends on the error to be attached to an absolute reading of k. If an instrument can measure $k \approx 1.0 \times 10^{-5}$ with an accuracy of $\sim \pm 1\%$, then the minimum detectable susceptibility difference based on a single reading along each axis cannot be less than $\sim 1 \times 10^{-7}$ and at this level will be subject to a large error.

For operational and computational simplicity, it is convenient to choose reference axes x, y and z where x, y are in the diametral plane and z is along the axis of a cylindrical sample, i.e. the same reference axes as used for NRM measurements.

As (1967a) developed a method of determining anisotropic susceptibility of rock samples on an astatic magnetometer simultaneously with the measurement of their remanent magnetism. The rock samples are measured in position (d) (Fig. 9.9), in which the astatic system responds to the vertical component of the sample magnetization, and the inducing fields are the vertical and North components of the local geomagnetic field. Twenty-four readings are taken for a determination of anisotropy (and remanence), i.e. four 'upright' and four 'inverted' readings at 90° azimuths (sample rotation about a vertical axis), with each of the three mutually orthogonal reference axes set vertical (see Section 9.2 for details of astatic magnetometer measurements). Suitable combination of these 24 readings gives values of X, Y, Z, the remanent magnetization components, and k_{xx}, k_{yy}, k_{zz}, k_{xy}, k_{xz} and k_{yz}, the susceptibility components.

The main disadvantages of this method are the time required and low sensitivity, which is limited by the moderate sensitivity of astatic magnetometers and which is also low if the induced magnetization in the sample is weak relative to the remanent magnetization. The technique has been used by van der Voo and Klootwijk (1972) to measure the anisotropy of samples of the Flamanville granite, north-west France. Typical susceptibilities of these rocks are around $3 \times 10^{-7}\,\mathrm{m^3\,kg^{-1}}$ with anisotropy factors up to 1.3, giving signals well above the equivalent noise level of the magnetometer of $\sim 1.0 \times 10^{-9}\,\mathrm{m^3\,kg^{-1}}$.

In practice, the direct measurement of the variation of M_H with θ does not provide a very convenient or sensitive technique and the use of instruments for measuring isotropic susceptibility, described in Section 2.6, is more satisfactory. Girdler (1961) describes the use of the transformer bridge, and the balanced instruments of Stone (1967a) and Christie and Symons (1969) have also been used.

For measurements in the x–y plane in the sample, it is placed in the coil or core-gap with the applied field direction parallel to the x-reference direction in the sample, i.e. $\theta = 0$. Then with the constant signal suppressed, readings are taken at $\theta = 0°$, 45°, 90° and 135°, giving values. of $\frac{1}{2}(k_{xx} - k_{yy})$, k_{xy}, $-\frac{1}{2}(k_{xx} - k_{yy})$, $-k_{xy}$, respectively. Corresponding readings taken in the x–z and y–z planes in the sample give corresponding terms involving k_{zz}, k_{yz} and k_{xz}.

Graham (1967a) describes a balanced transformer bridge incorporating an automatic system which records the axis of maximum susceptibility and the susceptibility difference in the plane of measurement of a rock sample. The bridge operates at 3.0 kHz and the sample is rotated at 5 Hz in a 100 μT field in the detector coil of the bridge. The out-of-balance signal from the bridge, due to the rotating anisotropic sample, is modulated sinusoidally at 10 Hz and a phase-comparison circuit then defines the principal susceptibility axis and the amplitude of the signal is proportional to the susceptibility difference. The noise level of the instrument under optimum conditions is equivalent to a susceptibility difference of $\sim 4 \times 10^{-12}$ m^3 kg^{-1} in a 2.5 cm sample.

(c) *The torsion balance*

The basic principle of the torsion (or torque) method is measurement of the torque exerted on an anisotropic sample in an applied field, due to the non-parallelism of the field and the induced magnetization. If the applied field is direct, the torque arises from the interaction of both the remanent and induced magnetization with the field. In an alternating applied field the torque due to remanence changes sign over each half-cycle whereas that due to the susceptibility maintains the same sign. If the sample is mounted on a vertical suspension of torsional constant such that the suspended system cannot respond to the alternating torque, any torque observed will be due to anisotropic susceptibility.

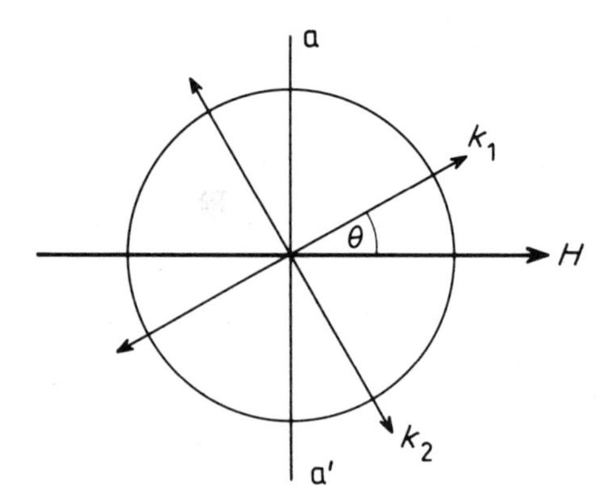

Fig. 2.8 Determination of torque about the z-axis acting on anisotropic sample in an applied field. k_1 and k_2 are the maximum and minimum susceptibilities in the x–y plane.

Consider a sample suspended on a vertical suspension subjected to a horizontal applied field H, which makes angles θ and $(90° - \theta)$ with the axes of maximum and minimum susceptibility in the horizontal plane through the sample (Fig. 2.8). The components of induced magnetization along the axes are

$$M_1 = Hk_1 \cos\theta$$
$$M_2 = Hk_2 \sin\theta \tag{2.48}$$

where k_1 and k_2 are the maximum and minimum (principal) susceptibilities in the plane. The torque on each induced component is $BP = \mu_0 HP$, where P is the induced magnetic moment perpendicular to H. Therefore, the effective torque T acting on the sample is

$$T = B(P_1 \sin\theta - P_2 \cos\theta) \tag{2.49}$$
$$= B(vHk_1 \sin\theta \cos\theta - vHk_2 \cos\theta \sin\theta)$$
$$= \tfrac{1}{2}\mu_0 H^2 v(k_1 - k_2)\sin 2\theta \tag{2.50}$$

where v is the volume of the sample. Thus, as the field direction is rotated through 360°, the deflections of the system describe a sin 2θ curve, the amplitude of which is proportional to $(k_1 - k_2)$.

In terms of the susceptibilities k_{xx}, k_{yy} along reference directions x and y in the sample, the corresponding expression is

$$T = \mu_0 H^2 v[\tfrac{1}{2}(k_{yy} - k_{xx})\sin 2\theta + k_{xy}\cos 2\theta] \tag{2.51}$$

Most torsional balances are direct-reading instruments, using a lamp and scale with a mirror on the suspended system. If l is the length of the light path and a deflection d is observed on the scale, the angular deflection is $d/2l$ radians corresponding to a torque of $d\sigma/2l$, where σ is the torsional constant of the suspension. Thus, at equilibrium, equating this torque with that given by Equation 2.50

$$k_1 - k_2 = \frac{d\sigma}{\mu_0 l H^2 v \sin 2\theta} = \frac{\mu_0 d\sigma}{l B^2 v \sin 2\theta} \tag{2.52}$$

For maximum sensitivity i.e. $(k_1 - k_2)/d$ as small as possible, σ/v is the important variable, since B is limited to ~2.0 mT (reversible susceptibility range) and l by available space and good optics. For a suspension fibre of circular cross-section of radius r and fixed length, σ varies as r^4 and its tensile strength, on which the maximum value of v depends, as r^2. Thus, if r is reduced by a factor n, σ/v decreases by a factor of approximately n^2 (assuming holder mass $\ll$ sample mass), and it is therefore advantageous to measure small samples with the value of r chosen such that the suspension can support the sample without approaching its elastic limit or giving rise to torsional hysteresis. A rule of thumb appears to be that the load on the suspension should not exceed 20% of the tensile strength. A similar calculation shows that, on the same assumptions, the time constant increases with decreasing sample size, but only slowly. For instance, if r is halved, the time constant increases by a factor of ~ 1.3. King and Rees (1962) consider design features for maximizing the signal/Brownian noise ratio, and they also allow for the mass of the sample holder. However, it is likely that other sources of noise will dominate over Brownian noise, and designing for maximum sensitivity is probably a more useful exercise.

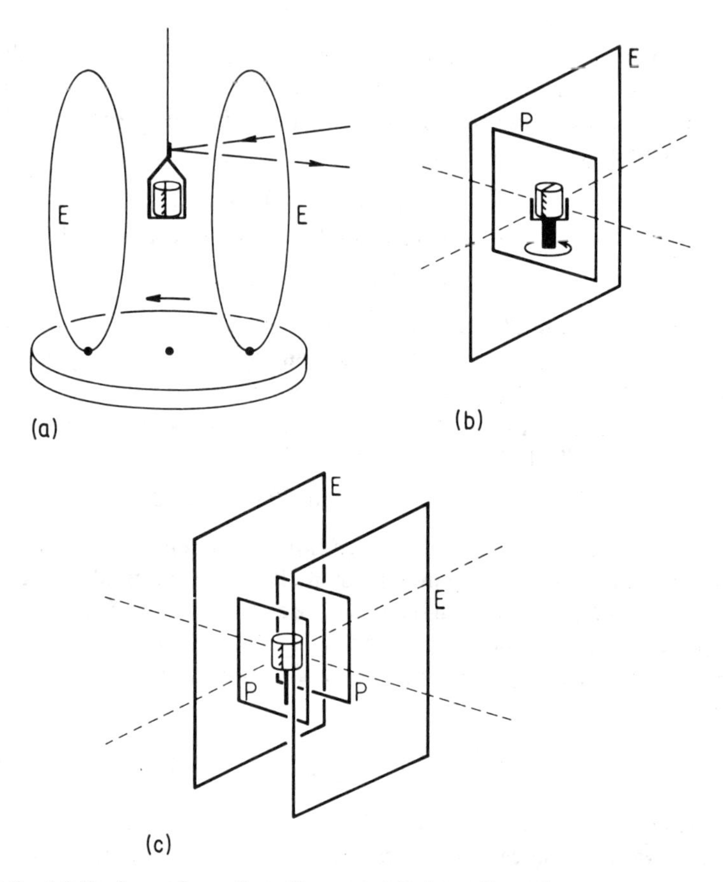

Fig. 2.9 (a) Basic configuration of a torsion balance for anisotropy measurements; (b) original coil system in the Digico anisotropy meter; (c) the modified coil system. Each diagram shows the exciting coil(s) (E) and pick-up coil(s) (P).

The basic torsion instrument (Fig. 2.9(a)) consists of a suitable holder containing the sample and suspended on a torsion fibre, the system being placed inside a draught-proof case. A Helmholtz coil pair with its axis horizontal and mounted on a rotating table surrounds the system with the sample at its centre. The size of the coils is chosen to provide a 50–60 Hz alternating field uniform to, say, $< \pm 0.5\%$ over the sample. Damping of the suspended system is provided by a horizontal disc attached to the underside of the holder and immersed in a low viscosity oil.

Although Ising (1942) and Granar (1958) used torsion balances for their pioneering studies of anisotropy in rocks, they give few details of their instruments. King and Rees (1962) describe two instruments. A sensitive one measures 1 cm samples suspended on a 12 μm phosphor-bronze fibre 12 cm long ($\sigma \approx 8 \times 10^{-10}$ N m/radian). With an applied field of 2.0 mT and a simple photoelectric amplifier to amplify small deflections, the minimum detectable (mass) susceptibility difference is $\sim 3 \times 10^{-13}$ m^3 kg^{-1}. The time constant is ~ 60 s. A less sensitive instrument for routine measurements uses applied fields of up to 8.0 mT and a stiffer fibre, and susceptibility differences of $\sim 1.5 \times 10^{-9}$ m^3 kg^{-1} can be measured. Stone (1963, 1967b) uses a larger sample (4.5 cm^3) suspended on a 10 cm phosphor-bronze strip 0.28 mm $\times$ 0.020 mm, and a field of 4.0 mT. With a 2 m light path, the sensitivity is $\sim 5 \times 10^{-10}$ m^3 kg^{-1} per mm deflection.

Both instruments use a motor drive to rotate the coils to the measuring positions, 0°–180° at 45° intervals (King and Rees) and 0–360° at 45° intervals (Stone), readings being taken in three mutually perpendicular planes in the sample. These readings enable the shape of the susceptibility ellipsoid to be determined, but for the determination of magnitudes of the principal susceptibilities k_a, k_b and k_c, it is necessary to measure the mean susceptibility or one of k_{xx}, k_{yy} and k_{zz}.

Torsion instruments usually show noise of mechanical rather than magnetic origin, and they should be mounted as far as possible in a low-vibration environment. Second and higher even harmonics of the applied field must be avoided, since they can result in a net direct field which will exert a torque on the sample due to its remanent magnetism and which may also promote the acquisition of ARM (anhysteretic remanent magnetization) (Section 6.5). It may be necessary to null the geomagnetic field over the sample for the same reasons. Further discussion of sources of errors in torsion balance measurements is given by King and Rees (1962) and King (1967).

(*d*) *The spinner magnetometer*

The torsion method is sensitive but time consuming, and the development of spinner magnetometers afforded the possibility of much more rapid measurement of anisotropy while maintaining adequate sensitivity.

Assuming no anisotropy effects due to sample shape, applying a direct field to a sample with isotropic susceptibility during measurement of NRM with a spinner magnetometer does not give rise to any additional signal. If the sample possesses anisotropic susceptibility, the general effect is an output signal of twice the rotation frequency in addition to the signal (at the rotation frequency) due to NRM. For the x–y plane in the sample, the anisotropy signal is of the form

$$P = C\,[(k_{yy} - k_{xx})^2 + 4k_{xy}{}^2]^{\frac{1}{2}} \sin 2\,(\omega t - \delta) \tag{2.53}$$

where C is a constant involving the applied field, sample mass or volume and magnetometer calibration factor, ω is the rotation frequency, and P is an apparent magnetic moment. The applied field is in the (horizontal) x–y plane in the sample and rotation is about the (vertical) z-axis. If the sample is placed in the holder in the same way as for a determination of the NRM component in the x–y plane and the magnetometer output gives the angle this component makes with the reference (x) direction, then δ in Equation 2.53 is the angle obtained during an anisotropy measurement in the x–y plane. In Equation 2.53 ω is half the rotation frequency normally used for NRM measurements and it is assumed that the NRM signal is filtered out. The quantities $(k_{yy} - k_{xx})$ and k_{xy} are obtained from (Noltimier, 1971)

$$k_{yy} - k_{xx} = \frac{P}{C} \cos 2\delta \tag{2.54}$$

and

$$k_{xy} = \frac{P}{2C} \sin 2\delta \tag{2.55}$$

where P and δ are the measured apparent induced moment and the phase angle respectively. Corresponding terms containing k_{zz}, k_{zy} and k_{zx} are obtained by spinning the sample about the x and y axes.

Noltimier (1971) also describes a measurement procedure for anisotropy measurements on spinners in which the output is in the form of 'in-phase' and 'quadrature' data, i.e. the magnetization components parallel and perpendicular to the reference direction of the sample holder. Since the in-phase and quadrature signals arise from terms such as $k_{yy} - k_{xx}$ and k_{yy} respectively, the necessary relations are rather simple

$$k_{yy} - k_{xx} = P_i/C \tag{2.56}$$

and

$$k_{xy} = -P_q/2C \tag{2.57}$$

where P_i and P_q are the apparent in-phase and quadrature moments measured when the sample rotates about the z-axis. In this type of spinner magnetometer, the outputs are usually obtained by means of two reference signals at shaft frequency and 90° out of phase, which are compared with the signal from the rock in separate phase-sensitive detectors. For anisotropy measurements at half normal frequency, the frequency of the reference signals must be doubled and the phase difference between them, as generated at the shaft, halved (to 45°) to preserve the correct phase and frequency relationships in the detector circuits. The constant C should also allow for the reduced sensitivity of the magnetometer at the lower rotation frequency.

Another method of measurement is available with the PAR spinner magnetometer, which has two Helmholtz coil pairs around the sample with their axes vertical and parallel to the pick-up coil axis respectively. During NRM measurement the coils are used to null the geomagnetic field at the sample and both can be used for applying fields for anisotropy measurements. Briefly, the procedure is as follows. The sample is spun at normal speed in zero field with its z-axis vertical, and the in-phase and quadrature NRM moments P_x and P_y are recorded. A vertical field is then applied and the horizontal moments P_x' and P_y' are recorded, which now contain components of induced magnetization arising from anisotropy. Then the susceptibility terms k_{xy} and k_{yz} are determined from

$$k_{xz} = (P_x' - P_x)/vH_z \tag{2.58}$$

and

$$k_{yz} = (P_y' - P_y)/vH_z \tag{2.59}$$

where v is the sample volume and H_z the applied field. The vertical field is then nulled and the sample, in the same orientation, is rotated at half speed with the horizontal field applied. The susceptibility difference in the x–y plane is then given by $k_{yy} - k_{xx} = P_x/vH_z$, where P_x is the recorded in-phase moment, and also $k_{xy} = P_y/vH_x$, where P_y is the quadrature moment. By changing the orientation of the sample in the holder and repeating the last step, $k_{xx} - k_{zz}$ and $k_{yy} - k_{zz}$ can be determined. The first part of this method may not be satisfactory if $P_x' - P_x$ and $P_y' - P_y$ are small, in which case k_{xz} and k_{yz} may be better determined by the above method for k_{xy}.

The Schonstedt spinner magnetometer is available with a coil system inside the shield for anisotropy measurements. A pair of rectangular coils, approximately in the Helmholtz configuration, are placed around the fluxgate detector with the coil axis vertical and perpendicular to the fluxgate axis. The inducing field is 50 μT and a six-spin measuring procedure is used to obtain the matrix elements of the susceptibility tensor from the in-phase and quadrature signals. The instrument appears to give good results if the NRM intensity of the rock and its anisotropy are low, so that measurements can be made with the sample in the closest position to the fluxgate detector. A possible problem with anisotropic samples possessing strong NRM is leakage of the remanence signal (at half the anisotropy signal frequency) into the anisotropy signals, particularly when there is marked inhomogeneity of NRM in the samples which contributes a second harmonic signal.

A novel but technologically complex anisotropy meter of the spinner type is described by Boetzkes and Gough (1975). The sample spins at ~50 Hz in the gap of a permanent magnet in a highly uniform field of ~0.12 T. Coils in the gap detect the induced magnetization in the sample parallel and perpendicular to the applied field, and the anisotropy signal magnitudes are read via a phase-sensitive detection circuit, the reference signals for the in-phase and quadrature components being obtained photoelectrically from the rotor shaft, as in spinner magnetometers for remanence measurement. The effective noise level is set by variable signals arising from the shaft and sample holder, and is equivalent to a susceptibility difference of $\sim 5 \times 10^{-13}\,m^3\,kg^{-1}$ in the plane of measurement in a standard (2.5 cm) sample. This corresponds to a susceptibility difference of 0.01% in a weakly susceptible sediment ($\chi \approx 5 \times 10^{-9}\,m^3\,kg^{-1}$). Thus, the instrument is extremely sensitive but this sensitivity is achieved partly by the use of a strong applied field which in some samples may give anomalous results through non-linearity of magnetization and applied field, and partly by the solution of several complex technological problems. Full details of these problems, alternative detection methods, and other aspects of the instrument are given by Boetzkes (1973).

(*e*) *The Digico anisotropy meter*

The Digico instrument for measuring susceptibility anisotropy employs a different technique. The sample, rotating at ~7 Hz about a vertical axis, is surrounded by two coils with their axes horizontal and mutually perpendicular (Fig. 2.9). One coil carries alternating current at a frequency of ~10 kHz, generating an applied field of ~0.5 mT r.m.s. at the sample, and the other is the pick-up coil (Rathore, 1975). If the coils are accurately orthogonal there is no coupling between them and no signal appears across the pick-up coil, and no signal appears if a sample possessing isotropic susceptibility is rotated in the coils. If the sample is anisotropic in the plane perpendicular to the rotation axis a sinusoidal signal, proportional in amplitude to the difference in the principal susceptibilities in the horizontal plane in the sample, is generated in the pick-up coil at twice the sample rotation frequency. The signal waveform derived from each revolution of the sample is stored digitally in the computer as with a Digico remanence measurement and an average waveform obtained over a chosen number of spins. The waveform is then Fourier-analysed in the computer. After measurements have been made in the z–y and z–x planes, phase and amplitude data of the sin 2θ components are used, in conjunction with a measurement of the axial susceptibility of the cylindrical sample (k_{zz}) on the susceptibility meter supplied with the equipment, to compute and print out the magnitudes of the maximum, intermediate and minimum susceptibilities and the orientation of the principal axes of the ellipsoid relative to reference directions in the sample.

The currently available system of excitation and pick-up coils consists of square Helmholtz pairs, 16 cm and 8 cm on a side for excitation and pick-up respectively (Fig. 2.9 (c)). The coils are mounted independently of the rotation system to reduce noise of mechanical origin. Using a measurement time corresponding to 2^9 spins (~75 s), the noise level is equivalent to a susceptibility difference in the plane of measurement in a standard sample of $\sim 3 \times 10^{-11}\,m^3\,kg^{-1}$. At this level of sensitivity, there may be a spurious signal from the sample holder, and there is provision for subtracting any systematic signals from this source from those obtained from a sample.

(*f*) *Other methods*

The ballistic magnetometer described by Daly (1967) (Section 9.5.3) is also equipped for measuring anisotropic susceptibility. Coils are provided for applying an axial direct field, with which the axial susceptibility, say k_{xx}, is measured. A set of coils is also provided for applying field perpendicular to the axis, enabling terms such as k_{xz} to be measured. The magnetometer can accommodate samples up to 5 cm on a side, in which a susceptibility of $\sim 4 \times 10^{-11}\ \mathrm{m^3\,kg^{-1}}$ can be measured, using a field of ~ 10 mT.

The use of the cryogenic magnetometer is described by Scriba and Heller (1978). A field of 0.1 mT is trapped in the shield, parallel to the axis of one of the two pairs of pick-up coils which detect magnetization in the horizontal plane. Since NRM and induced magnetization is measured, it is necessary to choose appropriate sample orientations to eliminate the remanence signals, i.e. rotation of the sample through 180° about a vertical axis. With the three-axis detection system, it is in principle only necessary to insert the sample once and take readings at azimuths of 0°, 90°, 180° and 270°, from which both NRM and anisotropy can be determined. However, as with routine NRM measurements, the quality of the data can be improved by taking readings with the sample at different orientations relative to the sensing coils.

(*g*) *Calibration of anisotropy meters*

Where anisotropy is measured by direct measurement of susceptibility along different axes in a sample, no further calibration is required for anisotropy determination.

Torsion balances respond to differences in susceptibility in the plane of measurement and require calibration for determining the magnitude of the difference and the direction of maximum and minimum susceptibilities in the plane. The former calibration, essentially for the meter sensitivity, can be done theoretically by equating the magnetic torque on the system (Equation 2.51) to the restoring torque exerted by the suspension, which can be determined from its dimensions and the shear modulus of the material, or experimentally by timing torsional oscillations of a mass of known moment of inertia suspended on it. The directional calibration can be checked by placing a small loop of copper wire in the sample holder with its axis horizontal and parallel to the reference direction. The induced currents in the loop generate a variable induced magnetic moment as the field coils are rotated round the sample, the loop showing an apparent maximum and zero susceptibility when the field coil axis is respectively parallel and perpendicular to the loop axis. The maximum torque on the loop should be observed when the coil and loop axes are at 45°.

Noltimier (1964, 1971) describes the use of a small wire loop with the spinner magnetometer. If the self-inductance of the loop is neglected and it spins in an applied direct field H with angular velocity ω, a time-varying magnetic moment is generated by the induced current in the coil. The component P_1 of this moment along the applied field (and pick-up coil) axis is given by

$$P_L = \frac{\mu_0 A^2 \omega H}{2R} \sin 2\omega t \tag{2.60}$$

where A is the area of the coil and R its resistance. The amplitude S_L of the output signal of the spinner is proportional to ωP_L. The amplitude S_R of the output due to an anisotropic rock sample is proportional to $\frac{1}{2}vH(k_1 - k_2)\omega$, assuming that the coupling between the loop and pick-up coil and the sample and pick-up coil are the same. k_1 and k_2 are principal susceptibilities in the measurement plane and v is the sample volume. Thus

$$\frac{S_L}{S_R} = \frac{\mu_0 A^2 \omega}{Rv(k_1 - k_2)} \tag{2.61}$$

from which $(k_1 - k_2)$ can be determined from the known quantities. The phase of the loop signal is the same as in the torsion balance, i.e. the maximum signal occurs when the loop and pick-up coil axes are at 45°. Thus, in a spinner with in-phase and quadrature output, the former should be zero and the latter a maximum (corresponding to P_L) when the loop axis is set parallel or perpendicular to the holder reference mark, and vice versa when the loop is set at 45° to the reference direction (Noltimier, 1971).

Noltimier (1964) points out that a closer electromagnetic analogue of an anisotropic sample is provided by two loops of slightly different resistance mounted at right angles to each other when, since R occurs in the denominator of Equation 2.60, the signal amplitude is proportional to the difference in the loop conductivities.

An alternative but probably less accurate method of phase and magnitude calibration for anisotropy meters, also investigated by Noltimier (1971), is the use of a thin, high-permeability plate. Such a plate is magnetically highly anisotropic owing to its shape, the observed susceptibility k' along any axis being given by

$$k' = k_i/(1 + Nk_i) \tag{2.62}$$

where k_i is the intrinsic susceptibility of the material and N the demagnetizing factor along the axis. Since k_i is very high (typically $> 10^3$), Equation 2.62 approximates to $k' = N^{-1}$. Then, for a square plate with x and y axes parallel to the sides and the z axis normal to the plate, the apparent susceptibility terms are $k_x' - k_y' = 0, k_y' - k_z' = k_x' - k_z' = N_x^{-1} - N_z^{-1}$. From the side length and thickness of the plate the values of N can be determined with reasonable accuracy from tables (or see Appendix 2), and the computed value of $k_x' - k_z'$ compared with that deduced from the magnetometer output (Equation 2.56). A modification of this method is to use a strip of magnetic recording tape embedded in a non-magnetic synthetic sample and measure the susceptibility difference along and across the tape on a susceptibility meter. This type of artificial sample is commonly used for phase calibration.

For checking the performance of the Digico meter, a small cylindrical multiturn coil in series with a resistance and rotating in the sample holder can be used. Appropriate choice of R ensures that inductive effects can be ignored ($R \gg \omega L$), when the expression relating apparent susceptibility difference and coil constants is

$$k = 2\mu_0 \omega n^2 A/Rl \tag{2.63}$$

where n, l and A are the number of turns, length and area of the coil, R is the total resistance, and ω the excitation frequency. Thus, if the meter output is proportional to the susceptibility difference in the plane of measurement, the Digico pick-up coil output should be proportional to R^{-1}. This has been demonstrated by Rathore (1975) and in principle the method can also be used for magnitude calibration, using a single-layer coil of the same dimensions as the rock samples.

It may be noted that magnitude calibrations by the methods described will not be precise if, as may well be the case, the flux linkage between the artificial sample and sensor differs from that between rock sample and sensor.

(h) *Calculation of susceptibility ellipsoids*

A detailed account of the mathematics involved in deriving the orientation and principal susceptibilities of the susceptibility ellipsoid from the readings obtained with different types of instrument is beyond the scope of this book, and for the analytical methods the reader is referred to the references given below.

The axes in the sample to which the susceptibility measurements are referred are normally the same as those used for NRM measurements, since the axes of the principal susceptibilities in the sample are not initially known. Thus, the quantities that are measured are of the following type: k_{xx}, k_{yy}, k_{zz}, $k_{xx}-k_{yy}$, $k_{yy}-k_{zz}$, $k_{zz}-k_{xx}$ and k_{xy}, k_{yz}, k_{zx}. These terms are then used to derive the principal susceptibilities k_a, k_b and k_c and the orientation relative to the x, y, z axes in the sample of the ellipsoid they define.

Referring to Equations 2.38 and 2.39, we have

$$\mathbf{M}_i = k_{ij}\mathbf{H}_j \quad (i,j = x, y, z)$$

where the components of M are given by

$$M_x = k_{xx}H_x + k_{xy} + k_{xz}H_z$$
$$M_y = k_{yx}H_x + k_{yy}H_y + k_{yz}H_z$$
$$M_z = k_{zx}H_x + k_{zy}H_y + k_{zz}H_z$$

In general **M** and **H** are not parallel, except when **H** is directed along one of the principal axes of susceptibility of the sample. In this case

$$\mathbf{M} = \lambda\mathbf{H} \tag{2.64}$$

where λ is a constant scalar quantity. Combining Equations 2.38 and 2.64, a set of equations is obtained which, in determinant form, must satisfy

$$\begin{vmatrix} (k_{xx}-\lambda) & k_{xy} & k_{xz} \\ k_{yx} & (k_{yy}-\lambda) & k_{yz} \\ k_{zx} & k_{zy} & (k_{zz}-\lambda) \end{vmatrix} = 0 \tag{2.65}$$

By expanding the determinant, a cubic equation in λ is obtained, the roots of which are k_a, k_b and k_c, the principal susceptibilities. For the usual case where differences are determined, we can put $\lambda = \lambda_1 + k_{xx}$, when Equation 2.65 transforms to

$$\begin{vmatrix} (k_{xx}-\lambda_1-k_{xx}) & k_{xy} & k_{xz} \\ k_{yx} & (k_{yy}-\lambda_1-k_{xx}) & k_{yz} \\ k_{zx} & k_{zy} & (k_{zz}-\lambda_1-k_{xx}) \end{vmatrix} = 0 \tag{2.66}$$

which can be solved for λ_1 and the principal susceptibilities determined by measuring k_{xx}. The direction cosines of the principal axes relative to the sample reference axes can then be obtained by substituting the three values of λ_1 ($= k_a$, k_b and k_c) successively into Equation 2.64.

Further details of the calculation of susceptibility ellipsoids are given by Girdler (1961) for bridge measurements, and Granar (1958), King and Rees (1962), Stone (1963) and Runcorn (1967a) for torsion balance and spinner measurements. A useful account of matrix algebra and its applications is given by Nye (1957). Methods of estimating the precision of a set of anisotropy determinations is given by Stone (1967b) and Noltimier (1972), and King (1967) discusses sources of error in torsion balance measurements.

2.8.3 General comments

The measurement of anisotropic susceptibility and the accurate determination of principal susceptibilities and the corresponding ellipsoids is not entirely straightforward, and experience has shown that in general the agreement obtained from measurements of samples on different instruments is not always good. In general the orientation of the ellipsoid, i.e. the axes of the three principal susceptibilities in the sample are more accurately defined by different instruments than are the magnitudes of the susceptibilities.

Among possible causes for disagreement are inaccurate calibration, non-uniformity in the applied field, inhomogeneity of susceptibility in samples and sample shape. There are two aspects of sample shape to consider, namely a contribution to the anisotropy signal from the non-spherical shape of the sample and spurious signals arising from the effect of sample shape on the sample–sensor flux linkage.

Apparent anisotropy due to sample shape can almost always be ignored. The observed susceptibility along a direction in a sample, in which the demagnetizing factor is N, is

$$k' = k_i/(1 + Nk_i) \tag{2.67}$$

where k_i is the intrinsic susceptibility. Defining an anisotropy factor $P = k_a'/k_b'$, where directions a and b are along the axis of and across a cylindrical sample, then

$$P = \frac{1 + k_i N_b}{1 + k_i N_a} \tag{2.68}$$

Since $N_a + 2N_b = 1$, P can be expressed in the form

$$P = \frac{1 + \frac{1}{2}k_i(1 - N_a)}{1 + k_i N_a} \tag{2.69}$$

The susceptibility k of rocks is rarely above 10^{-2}, and therefore for the worst case, with $k = 10^{-2}$ and $N_a = 0\,(a/b \to \infty)$, $P = 1.005$, i.e. 0.5 % anisotropy. For a cylindrical sample with $a/b \approx 1.0$, $N_a \approx 0.31$, $P \approx 1.0004$ %, if $k = 10^{-2}$.

It can be seen from the above that anisotropy due to sample shape can normally be ignored, but there remains the problem of spurious 2θ signals due to non-dipole components of the induced magnetization. This problem should not occur in torque magnetometers assuming that the applied field is uniform over the sample, but probably contribute to the discrepancies observed in comparisons between various instruments. The best approach is therefore to carry out tests on the particular equipment to be used to determine experimentally the optimum sample shape that reduces spurious anisotropy signals to a minimum. For this purpose it is necessary to use samples of low intrinsic anisotropy and reasonable bulk susceptibility so that the spurious signals are relatively enhanced.

Tests on the Digico instrument with the original pick-up and excitation coil geometry (Fig. 2.9(b)) have been carried out by Scriba and Heller (1978) and F. Addison (private communication). Two methods were used by the former authors. Fractional susceptibility differences (e.g. $(k_1 - k_2)/k$) were measured in the $x-y$, $y-z$, $z-x$ planes in cylindrical 2.5 cm diameter samples with varying length–diameter ratios. Minimum values of $\Delta k/k$ were observed for $l/d = 0.90$ ($\Delta k/k \approx 0.5$ %), with $\Delta k/k < \sim 1$ % for $0.85 < l/d < 0.95$. Since the observed directions of the principal susceptibility axes will also be affected the sample shape, Scriba and Heller also examined the dispersion and precision parameter of the mean principal susceptibility directions measured in three separate tonalite samples with different values of l/d, and found the optimum value of l/d again to be close to 0.90. For the same samples measured with the cryogenic magnetometer, the optimum value found was 0.86, reflecting the different pick-up coil configuration in the two instruments. In another Digico instrument with the same excitation and pick-up coil geometry, Addison found 0.86 ± 0.02 was the best length/diameter ratio in 14 samples comprising a wide range of rock types. This is close to the value obtained by Rathore (1975), derived theoretically to provide a constant demagnetizing length per revolution of the sample. With the modified coils (Fig. 2.9(c)) the shape dependence for $0.80\ l/d < 1.00$ appears to be almost negligible.

Even if shape anisotropy is not expected to be significant, it is probably good practice to choose the optimum l/d ratio to reduce the shape effect to a minimum if no shape effect is important. Sharma (1966) calculates the axial and radial demagnetizing factors for cylindrical samples and shows they are equal when $l/d \approx 0.90$. Noltimier (1971) calculates the value of l/d for which the sample magnetization most closely approaches a dipole, and finds $l/d = 0.86$.

It is clear, therefore, that for precise anisotropy measurements, it is necessary to choose a suitable dimension ratio for the samples according to which instrument is being used. Given a uniform exciting field, the torsion balance is expected to be at least affected by sample shape and to give the 'best' results, but has hitherto suffered from the lengthy measurement time required. Spinner magnetometer measurements are more rapid but may be subject to errors arising from second harmonic remanence signals, particularly when the ratio NRM/susceptibility is high, and, together with the Digico equipment, from sample–sensor shape effects. Comparisons of anisotropy measurements carried out on the samples on different instruments are reported by Kent and Lowrie (1975) and Ellwood (1978). In general, it may be said that, provided the sample dimension ratio is around 0.85–0.90, and a number of samples are measured in which the anisotropy axes are scattered relative to the sample reference axes, then the average derived orientation of the susceptibility ellipsoid will, in most cases, be close to the 'true' orientation arising from rock fabric when measurements are made on any of the currently available instruments. The precision of the measured mean axis is likely to vary between instruments, as are the mean magnitudes of the principal susceptibilities.

The susceptibility ellipsoid is triaxial, with dimensions defined by the magnitudes of the principal susceptibilities. These lie along the three orthogonal axes of symmetry of the ellipsoid and are designated the maximum, intermediate and minimum susceptibilities k_{max}, k_{int} and k_{min}. These quantities are combined in various ways to describe different features of the ellipsoid and of the magnetic fabric it represents. Parameters P_1, P_2 and P_3 are defined by

$$P_1 = k_{max}/k_{int} = \text{lineation } (L)$$
$$P_2 = k_{max}/k_{min} = \text{anisotropy factor}$$
$$P_3 = k_{int}/k_{min} = \text{foliation } (F)$$

L is a measure of the extent of linear parallel orientation of particles contributing to the susceptibility, and F of their planar distribution. The ratio P_3/P_1 is termed the eccentricity E of the ellipsoid

$$E = (k_{int})^2/k_{max}\,k_{min}$$

If $E > 1$, the ellipsoid is oblate, and if $k_{int} \approx k_{max}$ the ellipsoid is disc-shaped. If $E < 1$, the ellipsoid is prolate, and as k_{int} approaches k_{min} the ellipsoid becomes increasingly cigar-shaped. These two ranges of E correspond to the dominance of foliation and lineation respectively.

Table 2.3 Values of P_1, P_2, P_3 and E obtained from sites around the Carnmellis granite, Cornwall

Site no.	Locality	Minium axis declination°	Minium axis inclination°	P_1	P_2	P_3	E
33	SW 762 374	73.0	−24.8	1.0244	1.7227	1.6819	1.6424
32	771 363	64.5	−29.6	1.0443	1.1830	1.1334	1.0865
18	668 253	82.9	−32.9	1.0520	1.1677	1.1100	1.0552
31	757 279	168.1	−41.6	1.0253	1.6145	1.5744	1.5356
17	707 263	167.9	−50.9	1.0532	1.1677	1.1087	1.0528
23	654 276	200.7	−72.1	1.0909	1.2518	1.1485	1.0542
24	652 270	253.7	−65.3	1.0363	1.2187	1.1756	1.1343
29	624 290	244.3	−61.8	1.0632	1.7617	1.6543	1.5541
30	644 296	240.8	−27.0	1.0785	1.4854	1.3786	1.2802

The granite boss is intruded into sedimentary beds, and the orientation of the maximum and minimum susceptibility axes reflects the structural fabric at each site due to the intrusion.
Data from Rathore (1980).

Other combinations of k_{max}, k_{int} and k_{min} are sometimes used in the literature to define L and F, but the above are the most common in igneous and metamorphic studies. Rees (1966) and Hamilton and Rees (1971) discuss other combinations applicable to the structure of natural and laboratory-deposited sediments. Examples of P and E values measured in samples from around the Carnmenellis granite in south-west England are shown in Table 2.3. Some investigations of the interpretation of susceptibility anisotropy in terms of geological strain and of applications to geological structures are described by Graham (1954), the pioneer of magnetic fabric studies, Halvorsen (1973), Heller (1973) and Rathore (1979, 1980).

Other anisotropy investigations are Rees (1961, 1965) and Hamilton and Rees (1970) on sedimentary fabrics, Uyeda *et al.* (1963), Fuller (1963) and Khan (1962) on anisotropy effects in rocks and minerals, and Hrouda and Janak (1971) on anisotropy in red sediments. Bhathal (1971) has reviewed earlier work in the field and discusses underlying principles.

Chapter Three

High-field measurements

3.1 Introduction

In this chapter we consider measurements involving the application to rock samples of direct magnetic fields of ~ 2 mT and upwards. Some of the characteristic magnetic properties involved in these measurements are illustrated in Fig. 3.1: in the description that follows the magnetization is assumed to be low, and demagnetizing fields (Section 2.3) negligible.

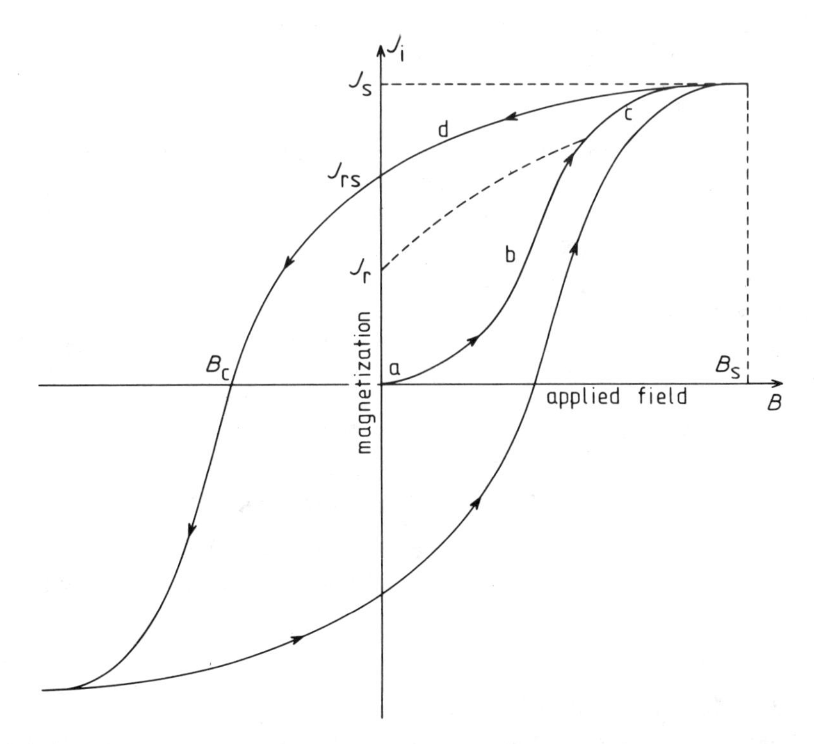

Fig. 3.1 J_i–B loop showing some important magnetic quantities. See text for explanation of symbols.

Consider a ferromagnetic sample of multidomain material initially in a demagnetized state. If an increasing field $+B$ (B is the applied field $= \mu_0 H$) is now applied, a magnetization J_i will be induced in the sample, typically according to part (a) of the curve (Fig. 3.1). For small B (up to ~ 1 mT) the growth of J_i is usually linear with B and is in the initial susceptibility range discussed in Section 2.3, where J_i is due to reversible domain-wall movement. As B increases further, irreversible domain-wall movement through potential barriers takes place, domains whose directions are near to that of B growing at the expense of others (part (b)). Finally, when wall movement is nearly complete domain rotation occurs (c) and ultimately, in a sufficiently high field B_s, saturation magnetization J_s is achieved. The slope of the curve at any point is proportional to the 'incremental susceptibility' and the ratio of J_i to B/μ_0 at the point is the high field mass susceptibility.

If B is reduced to zero when J_s is reached the magnetization follows curve (d) and when $B = 0$ the saturation isothermal remanence J_{rs} remains: if B is reduced to zero before J_s is reached J_i will follow a curve such as the dotted line in Fig. 3.1, and an isothermal remanence (IRM) J_r remains at $B = 0$.

If B now increases in the negative direction J_i falls to zero at a value of $-B$ known as the coercive force, B_c: as $-B$ increases, saturation magnetization is reached at $-B_s$. If the field is then taken from $-B_s \rightarrow 0 \rightarrow B_s$ the curve is completed in a symmetrical manner. The complete cycle is a 'hysteresis loop', and the area enclosed by it is a measure of the work done in taking unit mass of the magnetic material round the cycle.

In practice, a more useful measure of coercivity in rock magnetism is 'coercivity of remanence' B_{cr}. This is the field which, when applied antiparallel to J_r and then removed, leaves $J_r = 0$. The coercivity of maximum remanence is the back field required to reduce the saturated remanence J_{rs} to zero.

Measurements in applied fields provide information on the nature and quantity of the magnetic minerals in rocks and the presence of dia-, para-, ferro- and superparamagnetism, their coercivity spectra and domain structure. They can also be used to monitor chemical and physical changes that occur in rocks, e.g. through heating.

3.2 IRM and B_{cr} measurements

These are the simplest applied field measurements, for which only the applied field, obtained from a coil or electromagnet, and a magnetometer are required.

3.2.1 Coils

Field in the range 1–100 mT are conveniently generated in a cylindrical coil (solenoid). For a finite solenoid with a single-layer winding of n turns per metre

$$B_0 = \mu_0 ni \cos\theta_0 \qquad (3.1)$$

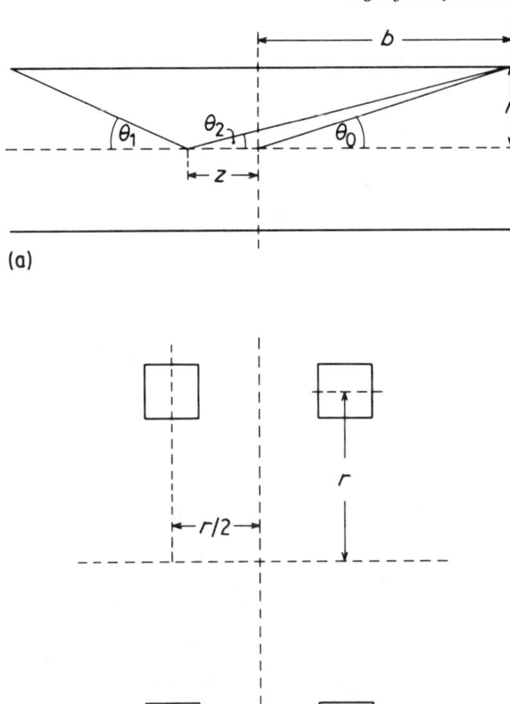

Fig. 3.2 (a) Solenoidal coil showing dimensions and angles used to calculate axial field and field gradient; (b) pseudo-Helmholtz coil system. In the system described in the text the winding cross-section is 1.5 cm square and $r = 5.0$ cm.

where B_0 is the field at the centre when a current i flows through the coil and θ_0 is as is shown in Fig. 3.2(a). If there are N layers and the depth of the windings is small relative to the mean coil diameter, we can write

$$B/i = \mu_0 n N \cos\theta \tag{3.2}$$

where θ is now derived from the mean radius of the windings. Equation 3.2 is sufficiently exact for most purposes, at least for designing a solenoid if not for calibrating it. The field variation along the axis of a single-layer solenoid can be calculated from

$$B_z = \frac{\mu_0 n i}{2}(\cos\theta_1 + \cos\theta_2) \tag{3.3}$$

where B_z is the axial field at a distance z from the centre and θ_1 and θ_2 are as shown in Fig. 3.2(a). The variation in B_z at a fractional distance z/b from the solenoid centre for different values of r/b is shown in Fig. 3.3.

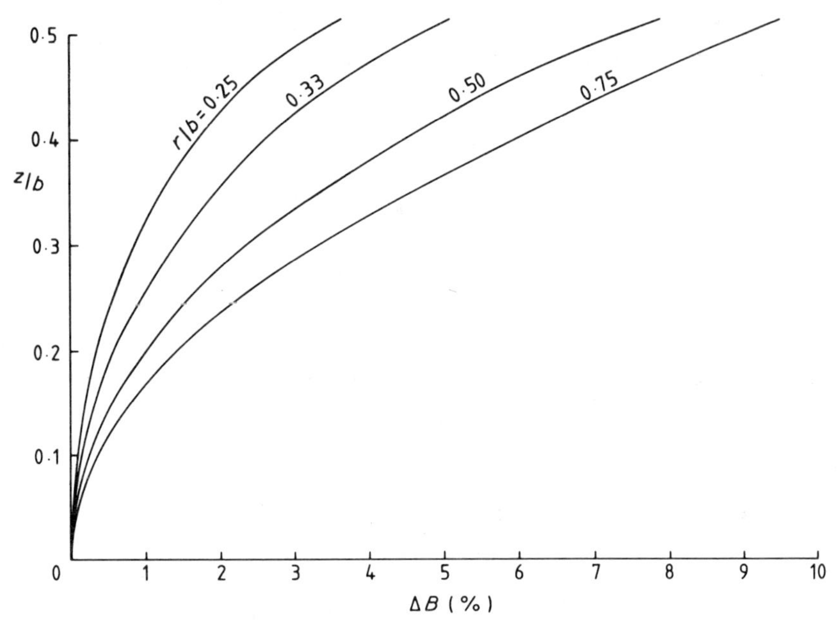

Fig. 3.3 Graphs showing axial variation of B for solenoids with different diameter/length ratios. Ordinates scale is distance from centre expressed as fraction of half-length of coil.

As an example, a solenoid of 15 cm length and 7 cm outside diameter wound with six layers of 20 SWG (0.91 mm) wire ($n \approx 10^3$) gives a central field of $\sim$6.7 mT A^{-1} which is sensibly uniform ($< \pm 1\%$) over a 2.5 cm cylindrical sample. The coil resistance is 5.6 Ω and it will take $\sim$10 A for short periods without serious overheating. An alternative design, giving easier sample access, is a 'pseudo-Helmholtz' coil pair (Fig. 3.2(b)). With the dimensions shown and the winding space filled with 24 SWG wire (0.56 mm diameter) the central field is $\sim$11 mT A^{-1} or $\sim$5.5 mT A^{-1} depending on whether the coils are connected in series or parallel. The respective resistances are 28 Ω and 7 Ω: the low resistance configuration is sometimes useful if there is a high current–low voltage power supply available. Commercial power supplies are available for coils, although they usually provide an unnecessarily stable current for many IRM investigations. A satisfactory alternative is a simple full-wave diode rectifier with capacity smoothing and series resistance, if required, working off the mains via a Variac-type voltage regulator. If there is a ripple in the rectifier output of fractional magnitude p (ripple amplitude/d.c. level), the maximum applied field with a current i in the coil will be $B = Ci\,(1 + p)$ instead of Ci (as indicated by an ammeter), where C is the coil constant: $p < 0.03$ is acceptable for many investigations. For more precise work, improved smoothing can be added to the rectifier and the dependence of ripple magnitude on the current drawn from the rectifier should be investigated. The ripple magnitude can be measured by displaying the voltage across the coil on an oscilloscope.

Solenoids and other coil systems can be calibrated with a commercial field measuring instrument of the fluxgate or Hall effect type: if not available, a simple method for coils with suitable access (e.g. the Helmholtz configuration) is to time the oscillations of a small magnet at the coil centre when a known current is flowing (see Appendix 6). Alternatively, the coil may be supplied with alternating current of known magnitude and frequency. The field may then be derived from the voltage developed across a small search coil of known area-turns placed at the coil centre (Appendix 6).

3.2.2 Electromagnets

For fields in excess of 100 mT an electromagnet is commonly used. Several types are commercially available, and their design and construction will not be discussed in detail here. The following is intended as a guide to some of their more important features.

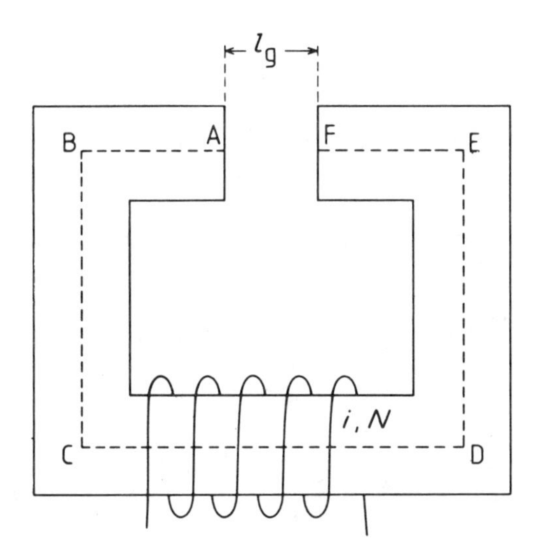

Fig. 3.4 Idealized magnetic circuit with gap width l_g and flux path-length ABCDEF.

An approximate but instructive method of calculating performance is provided by an analogy with Ohm's law. For the magnetic circuit shown in Fig. 3.4.

$$\frac{\int H \, dl}{(l_c/A\mu_c) + (l_g/A\mu_g)} = \phi \tag{3.4}$$

where $\int H \, dl$ is designated the magneto-motive force (cf. voltage in Ohm's law),

A is the cross-sectional area of the core, l_c, μ_c and l_g, μ_g are the median length and permeability of the core and air gap respectively, and ϕ is the flux in the core and gap (cf. current). The quantity $l/A\mu$ is the reluctance of any part of a magnetic circuit and is the analogue of electrical resistance. The flux density in the gap, or the magnetic field B, is ϕ/A. If there are N turns wound on the core, carrying current i, then $\int H\,dl = Ni$. Equation 3.4 does not take into account leakage of flux out of the core and at the edge of the gap: this is more marked than in the electrical case, because the ratio of the permeabilities of iron and air is $\sim 10^3$–10^4, while that of the conductivities of copper and air is $\sim 10^{20}$.

It is clear from Equation 3.4 that the field in the gap will be approximately linearly dependent on the gap width when $l_g \ll$ the core diameter. The permeability of the core should be high: a typical relative permeability μ_r for soft iron below saturation is ~ 1000, hence the dominance of the reluctance of the air gap ($\mu_r = 1$) and the rapid fall-off of the gap field as l_g increases. Equation 3.4 can be used to derive an approximate relation for calculating the field obtained from a simple electromagnet with a narrow gap and plane pole faces, and the core not near saturation

$$Ni \approx Bl_g/\mu_0 \tag{3.5}$$

The maximum field obtainable depends chiefly on the saturation volume magnetization M_s of the core material, which corresponds to an H-field $= M_s = B/\mu_0$ in the gap near the core faces: for iron this value of B is about 2.1 T. However, the maximum field in the gap can be increased in two ways, by tapering the pole faces to form truncated cones, which concentrate the field, and by making them of an iron–cobalt alloy ($\sim 30\%$ Co) which has M_s equivalent to ~ 2.4 T, about 15% higher than iron.

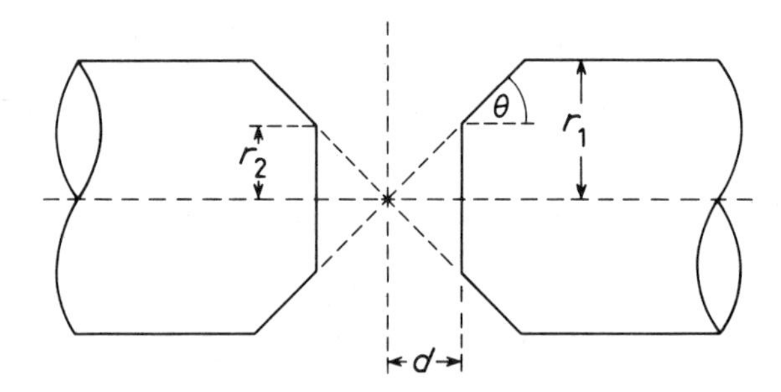

Fig. 3.5 Configuration of conical pole-pieces with coinciding apices.

The complete theory of conical pole-pieces is complex, but the case in which the apices of the cones coincide (Fig. 3.5) is mathematically simpler and indicates the range of field enhancement that is possible. The field at the gap centre and on the axis with saturated pole-pieces is

$$B = \mu_0 M_s \left(1 - \cos\theta + \sin^2\theta \cos\theta \ln\frac{r_1}{r_2}\right) \tag{3.6}$$

where θ is the semi-angle of the cone and r_1 and r_2 are the core and gap face radii respectively.

For plane pole-pieces

$$B = \mu_0 M_s (1 - \cos\theta) \tag{3.7}$$

where (as also in Equation 3.6) $\theta = \tan^{-1} r_2/d$, where $2d$ is the gap width. As $d \to 0$, $B \to \mu_0 M_s$. The last term in Equation 3.6 is the contribution to the central field of the magnetic charges on the conical surface: it can be shown that this term takes a maximum value for $\theta = 55°$. In practice $\theta = 60°$ is often chosen, because of incomplete saturation of the pole-pieces. Equation 3.6 then becomes

$$B/(\mu_0 M_s) = f = 0.50 + 0.375 \ln\frac{r_1}{r_2} \tag{3.8}$$

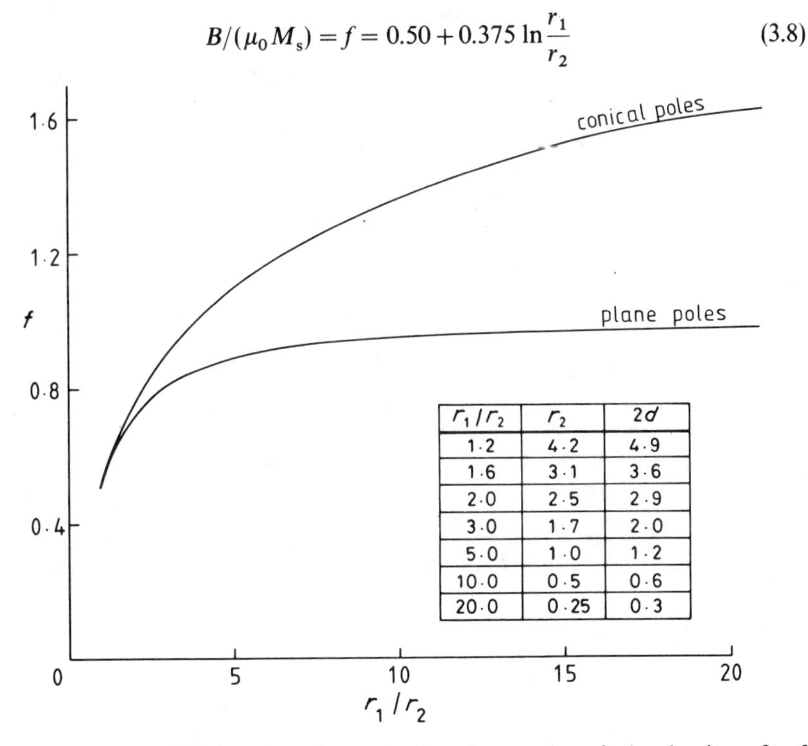

r_1/r_2	r_2	$2d$
1.2	4.2	4.9
1.6	3.1	3.6
2.0	2.5	2.9
3.0	1.7	2.0
5.0	1.0	1.2
10.0	0.5	0.6
20.0	0.25	0.3

Fig. 3.6 Values of f plotted against r_1/r_2 for plane and conical pole-pieces for $\theta = 60°$.

Figure 3.6 shows f as a function of r_1/r_2 and for the same gap width with plane poles. The table accompanying Fig. 3.6 gives the gap width ($2d$) and radius r_2

for the plotted values or r_1/r_2 for $r_1 = 5.0$ cm. Note that a property of coinciding apices is that $2d/r_2 = 2\cot\theta =$ constant.

If a larger ratio of r_2/d is required than is provided by pole pieces with coincident apices of optimum taper, the ratio and the field can be increased by reducing the gap width. It will increase approximately as $\beta = \tan^{-1}(r_2/d)$ and can be increased further by removal of as much material as possible forward of the 60° line for coinciding apices in the new gap.

Most high performance electromagnets have their windings placed symmetrically about and as close as possible to the gap. In this position they contribute an additional field, proportional to the current in the coils. For low fields the additional field is small compared with that contributed by the core, but when saturation is reached any subsequent increase in field with current is due to the windings.

Figure 3.7 shows the performance of some commercially available electromagnets.

The homogeneity of the field of an electromagnet depends on the pole-piece configuration, gap width/diameter ratio and the field strength, the latter because of the different magnetic state of the pole-pieces as B changes and the differing contributions of the windings to the field. For circular cores the field is cylindrically symmetrical about the gap axis. For plane pole-pieces the field B_z parallel to the gap axis at a distance z from the centre along the axis and at the radial distance r from the axis, near $z = 0$ is given by the series

$$B_z(z) = \mu_0 M\left[F_0 + F_2\left(\frac{z}{d}\right)^2 + F_4\left(\frac{z}{d}\right)^4 \ldots\right] \tag{3.9}$$

and

$$B_z(r) = \mu_0 M\left[F_0 - \frac{1}{2}F_2\left(\frac{r}{d}\right)^2 + \frac{3}{8}F_4\left(\frac{r}{d}\right)^4 \ldots\right] \tag{3.10}$$

where M is the volume magnetization, assumed uniform, of plane pole-pieces and $F_0 = (1-\cos\theta)$ (see Equation 3.7). Kroon (1968) has plotted the values of F_2 and F_4 as a function of $\theta = r_1/d$, where r_1 is the pole radius and $2d$ the gap width

$$F_2 = \frac{3}{2}(\cos^3\theta - \cos^5\theta) \tag{3.11}$$

$$F_4 = \frac{5}{8}(3\cos^5\theta - 10\cos^7\theta + 7\cos^9\theta) \tag{3.12}$$

Some representative values of field variation are given in Table 3.1, showing percentage field change between $z, r = 0$ and $z = d/2$, $r = r_2/2$.

Expressions for field homogeneity between tapered poles with coinciding apices are also given by Kroon (1968): some values for gaps with the same ratio of r_2/d as for r_1/d for plane poles are also shown in Table 3.1. An assumption in the development of these expressions is $(r_2/r_1)^2 \ll 1$, and the field variations shown are for $r_2/r_1 = 0.33$. It can be seen that for a given gap shape the field is

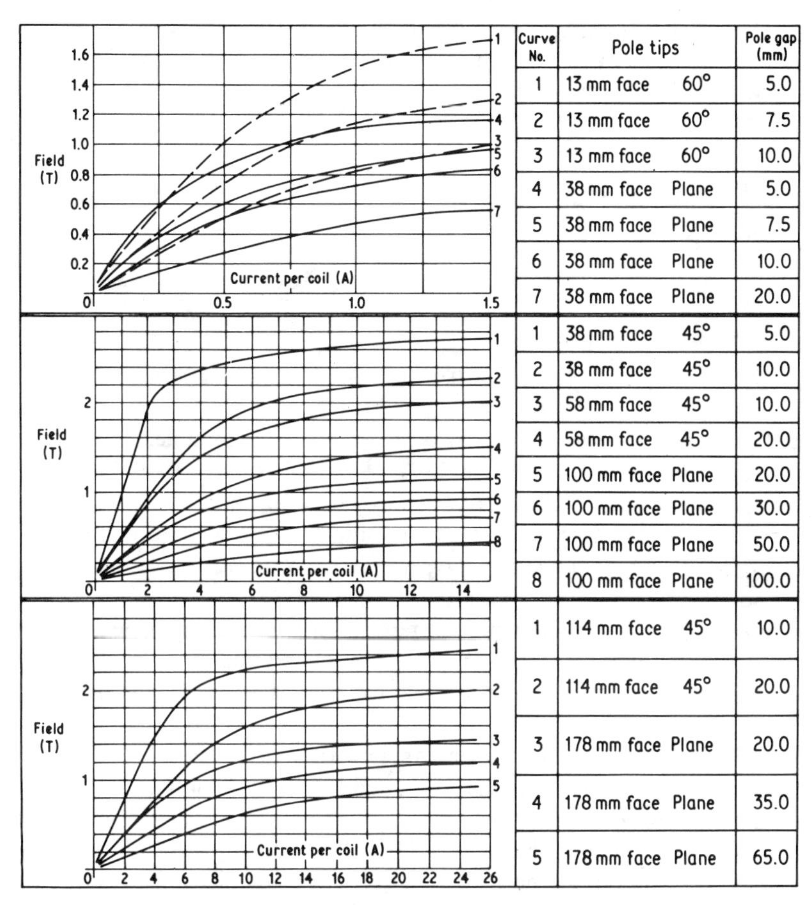

Curve No.	Pole tips		Pole gap (mm)
1	13 mm face	60°	5.0
2	13 mm face	60°	7.5
3	13 mm face	60°	10.0
4	38 mm face	Plane	5.0
5	38 mm face	Plane	7.5
6	38 mm face	Plane	10.0
7	38 mm face	Plane	20.0
1	38 mm face	45°	5.0
2	38 mm face	45°	10.0
3	58 mm face	45°	10.0
4	58 mm face	45°	20.0
5	100 mm face	Plane	20.0
6	100 mm face	Plane	30.0
7	100 mm face	Plane	50.0
8	100 mm face	Plane	100.0
1	114 mm face	45°	10.0
2	114 mm face	45°	20.0
3	178 mm face	Plane	20.0
4	178 mm face	Plane	35.0
5	178 mm face	Plane	65.0

Fig. 3.7 Typical performance of commercial electromagnets. (Reproduced by courtesy of Oxford Instruments Company Ltd)

much more homogeneous with tapered pole-pieces, and to a first approximation maximum uniformity occurs at $r_2/d \approx 2$, above which value B_z increases with r. Table 3.1 should only be used as a rough guide to field homogeneity. The theory on which the tables are based assumes uniform pole magnetization and ignores the field contributed by the magnet windings. The latter will only be important in high fields and will not in general seriously affect the field homogeneity, particularly in small gaps. Kroon (1968) gives further information on this point.

For measurements of IRM and remanent coercivity the effects of field inhomogeneity may be minimized by using small samples (~ 0.1–$0.5\ \text{cm}^3$): usually IRM > NRM, and there will be ample total moment for measurement. If a field magnetometer with a sufficiently small probe (e.g. some Hall-effect

Table 3.1 Percentage variation of the axial field at $z = d/2$ (near $r = 0$) and at $r = r_2/2$ (near $z = 0$) from the axial field at $z = r = 0$, for plane and tapered poles

		Plane poles			
r_1/d	10.0	5.0	3.0	2.0	1.0
$(\Delta B_z)_z$	+0.04	+0.3	+1.6	+4.9	+22.5
$(\Delta B_z)_r$	−2.4	−5.2	−8.2	−10.6	−11.2
		Tapered poles			
r_1/d	10.0	5.0	3.0	2.0	1.0
$(\Delta B_z)_z$	<0.01	−0.06	−2.8	<0.01	+6.3
$(\Delta B_z)_r$	<0.1	−0.6	−1.2	−0.3	+3.5

instruments or a fluxmeter with small search coil) is available, field variations $> \pm 2\%$ over any volume can be directly measured. A simple and not inappropriate method of approximately checking on non-uniformity is to measure the IRM of a small sample or magnetic particle at different points in the gap, partially demagnetizing between readings and checking on repeatability.

Several low voltage–high current power supplies are available for electromagnets, but for many applications a laboratory-built supply can be used, basically of the type already described for coils. The large inductance of the magnet assists with smoothing. It is important to connect a diode or similar device across the terminals of the magnet windings to protect the power supply against the high induced voltage developed across the windings in the event of a sudden break in the current.

A word of warning may not be out of place concerning the attractive force between the poles of an electromagnet. The force is given approximately by

$$F = \frac{B^2 A}{2\mu_0} \tag{3.13}$$

where A is the area of the pole faces and B the field in the gap. Thus, for $B = 1$ T in a gap of radius 2.5 cm, $F \approx 80$ kg wt. This points to the necessity of firmly securing adjustable pole-pieces.

For further details of electromagnet design and performance the reader is referred to accounts by Bitter (1936), the comprehensive account by Kroon (1968) and other references given at the end of this section.

3.2.3 Iron-free solenoids

The fields obtainable with laboratory electromagnets of reasonable size are adequate for the majority of investigations in rock magnetism. For higher fields (> 2.5 T) iron-free solenoids are used, as electromagnets become impracticable because of the core diameter required to enable large values of

r_1/r_2 and reasonable gap volumes to be achieved. Only a brief account is given here of the principles of high field solenoids and the reader is referred to the references given for further details. Some related information on high field coil design is also given in Section 11.1.

To obtain high magnetic fields in air-cored coils entails large power inputs. An approximate relation quoted by Olsen (1964) gives

$$B = 10\left(\frac{10^{-3}\,W}{\text{inner radius (cm)}}\right)^{\frac{1}{2}} \tag{3.14}$$

where W is the power required in kW. Thus 3.0 T in a coil of inner radius 1.0 cm requires ~100 kW, and the design of such coils is as much concerned with the removal of Joule heat and the control of mechanical forces as the efficient generation of the field. Water cooling is commonly used, and, depending on the efficiency of heat transfer, 100 kW will heat a flow of ~1 litre/s through ~25°C: too high a temperature rise is undesirable because of the increase in the coil resistance.

Another useful relation for the design of high field coils is that due to Cockroft (1928)

$$B = 10^{-6}\,G\left(\frac{W\sigma}{r_i\rho}\right)^{\frac{1}{2}} \tag{3.15}$$

where W watts is the power for a coil of inner radius r_i (metres) wound with wire of resistivity $\rho\,\Omega$ m. G is a factor which depends only on the shape of the coil and σ is the packing factor, the proportion of winding space occupied by conductor. For a cylindrical coil with cylindrical bore and uniform current density G has a maximum value of 0.18, and is not critically dependent on shape. If α is the ratio of the outer to inner radius and β the ratio of the half-length of the coil to the inner radius, the optimum shape ($0.17 < G < 0.18$) corresponds to $2.5 \ll \alpha \ll 4$ and $1.5 \ll \beta \ll 3$. G_{max} can be increased by having non-uniform current density in the coil. For instance, with the disc-shaped conductors of Bitter (1936) in which the current density varies as $1/r$, $G_{max} \approx 0.21$. It is apparent from the foregoing discussion that there is heavy expense involved in power supplies and cooling installations for high-field solenoids. Creer (1967) describes the construction and use of a 5.0 T solenoid.

Some attempts have been made to reduce the power requirement for high field solenoids by cooling in liquid nitrogen. At this temperature (77 K) the resistivity of copper is decreased by a factor of between 8 and 9, and from Equation 3.14 the theoretical power saving for a given field is given by the square root of this factor. Cooling is by evaporation of the nitrogen, which is one of the chief drawbacks of the system since large quantities are boiled off even in moderate field generation. Fritz and Johnson (1950) describe such a solenoid which generates 0.4–0.5 T in a cylindrical bore of radius 3.8 cm with a power input of ~3 kW: about 1.2 l/min of liquid nitrogen are boiled off.

The ultimate solution to the problem of Joule heat dissipation is the superconducting or cryogenic solenoid. The phenomenon of superconductivity is the sudden drop to zero of the resistance of a conductor at a very low 'critical' temperature. Thus, in principle, once started a current can exist indefinitely in a superconducting coil without power input and without Joule heating. An early difficulty in employing the effect to produce high fields was the inhibiting of the superconducting state by the magnetic fields itself, as low as 0.1 T in some cases. However, this has been overcome in certain alloys (e.g. niobium/tin), which can carry the high current densities required for strong fields without losing their superconducting properties. Commercial cryogenic magnets for the laboratory are now available, giving fields of up to 15 T in a 3 cm bore.

Some useful references to iron-free solenoids are Daniels (1950), Creer (1967) and Kroon (1968). Cryogenic solenoids are discussed by Berlincourt (1963) and Olsen (1964) and the generation of high magnetic fields in general by Kolm, Lax, Bitter and Mills (1962), Parkinson and Mulhall (1967) and Montgomery (1963).

3.2.4 Measurement of the IRM of single particles

It may sometimes be of interest to measure the remanent magnetic properties of individual magnetic particles extracted from a rock. Techniques have been described in classical magnetic studies which may be adaptable to rock magnetism. Yu and Morrish (1956) and Muirhead (1962) describe sensitive torsion magnetometers in which the particle under investigation is mounted on one end of a short horizontal arm, the other end being supported by a very fine vertical quartz fibre. The particle is at the centre of a coil pair which is used first to magnetize the particle and then, with the coils connected in series opposition, to provide a field gradient and a force on the particle which causes an angular deflection of the system proportional to the particle magnetic moment. IRM measurements can also be made on some of the small-scale magnetometers described in Section 9.6.

3.3 Measurements of J_i, high field susceptibility and hysteresis curves

Instruments for these measurements are required to measure induced magnetic moments of $\sim 10^{-9}$–10^{-3} A m^2 in an ambient inducing field of $\sim$0.01–2.0 T. Direct measurement and display of hysteresis loops is also described. The instruments described in these sections are concerned with room temperature measurements, but they are often designed for high- and low-temperature investigations as well and their use for these purposes is described in the next chapter.

3.3.1 Magnetic balances

The principle of magnetic balances is the measurement of the translational force experienced by a magnetized body in a non-uniform magnetic field. Such balances have long been used in conventional magnetic studies, originating with the early work of Faraday, Gouy and Curie on the measurement of susceptibility.

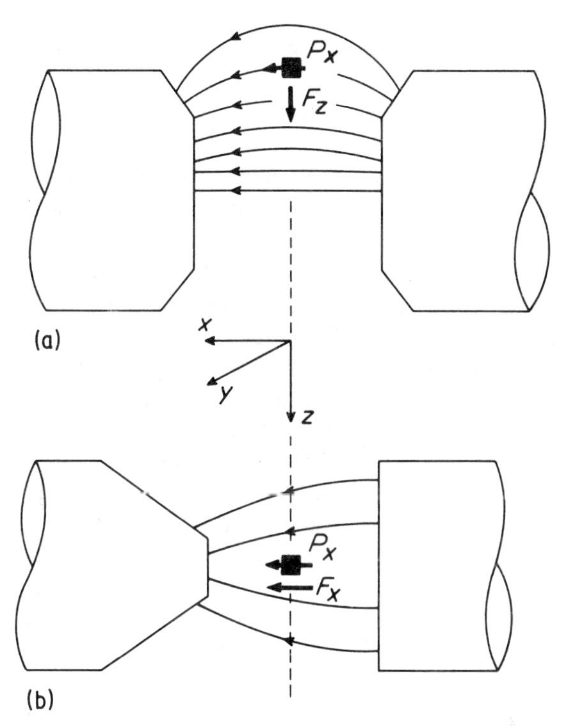

Fig. 3.8 Pole configurations for use in magnetic balances. In (a) the force and field gradient are perpendicular to the induced moment and in (b) they are parallel to the moment.

On the assumption that the moment of the sample can be regarded as dipolar, the force on it in the z-direction (Fig. 3.8(a)) in an inhomogeneous field is of the form

$$F_z = \frac{(m\chi - m_a\chi_a)}{\mu_0}\left(B_x\frac{dB_x}{dz} + B_y\frac{dB_y}{dz} + B_z\frac{dB_z}{dz}\right) + \frac{m}{\mu_0}\left(J_x\frac{dB_x}{dz} + J_y\frac{dB_y}{dz} + J_z\frac{dB_z}{dz}\right) \tag{3.16}$$

where m and χ are the mass and (isotropic) mass susceptibility of the sample, and m_a and χ_a are the mass and the susceptibility of air displaced by the sample. B_x, B_y, B_z, dB_x/dz are the applied fields and field gradients along the axes defined by the subscripts, and J_x, J_y, J_z are components of permanent magnetization of the sample. In the most general case there are equivalent expressions for F_x and F_y. In a practical balance F is measured along one axis only, and the sample position and gradient geometry is chosen such that two of each of the three terms associated with induced and permanent magnetization are small. The quantity $m_a \chi_a$ is, except in very rare cases, negligible compared with $m\chi$ (see Section 2.6.2). Equation 3.16 thus reduces to

$$F_z = \frac{m\chi B}{\mu_0} \frac{dB_x}{dz} + \frac{mJ_x}{\mu_0} \frac{dB_x}{dz} \tag{3.17}$$

for the commonly used gradient geometry shown in Fig. 3.8(a). The field, field gradient and force vector for another configuration are shown in Fig. 3.8(b). For small (< 5 mm) roughly equidimensional samples the assumption of dipolar moment is a valid one in most practical instruments.

Since in rock samples $J_i = \chi B/u_0$, even in a field of ~100 mT J_i is almost always much greater than the NRM and the second term of Equation 3.17 is usually negligible compared with the first when measuring a J_i–B curve with B increasing, and always is if, as in common practice, powdered samples are used. However, the IRM remaining in the sample as B is reduced may then cause the second term to become significant. To estimate the range of forces to be measured, assuming typical values of B and dB/dx obtainable with electromagnets, we consider the extreme cases of the measurement of the lower part of the J_i–H curve of a low-susceptibility red sandstone and the saturation magnetization J_s of an igneous rock containing 5% by weight of magnetite. In the former case we take $B = 100$ mT, $dB/dz = 2.0$ T m^{-1} and $\chi = 1 \times 10^{-8}$ m^3 kg^{-1}, and in the latter $dB/dz = 5.0$ T m^{-1} and $J_s \simeq 100$ A m^2 kg^{-1} for magnetite. For a typical sample mass of 0.2 g the forces are ~4.0×10^{-7} N and ~5.0×10^{-3} N respectively (approximately 40 μg wt and 0.5 g wt).

A magnetic balance consists essentially of two parts, a system for providing a variable field and field gradient and a force measuring device of appropriate sensitivity for the type of samples to be measured.

For determining J_i–B curves (where B is the applied field) on a wide range of rocks, a range of B up to at least 1.0 T is desirable. The field gradient depends partly on the sensitivity required and is usually in the range 1.0–20 T m^{-1}. The combination of field and gradient is most conveniently produced by an electromagnet with, if necessary, shaped pole-pieces to give an increased gradient which at any point will increase as B increases. The maximum values of B and dB/dz for a given magnet current will not, in general, occur at the same point in the field. The simplest arrangement to give a modest gradient is to use the fringing field near the edge of a gap with plane pole-pieces. Sucksmith (1929) placed the centre of the sample midway between and on a line joining the circumference of plane pole-pieces 3 cm apart and 10 cm in diameter, where a maximum value of $B(dB/dz)$ of ~10 T^2 m^{-1} was obtained.

Stronger gradients and a volume in which the product $B(dB/dz)$ is constant (sometimes a useful feature, see below) can be obtained with shaped pole-pieces, for instance the design of Garber, Henry and Hoeve (1960) and Heyding, Taylor and Hair (1961) (Fig. 3.9). Where a significant volume in which $B(dB/dz)$ is constant is not required, a simplified form of the Fig. 3.9 design may be used which will still provide a maximum value of $B(dB/dz)$ close to the optimum for a given gap width. Using the Heyding design the present author obtains 800 mT and 16.0 T m^{-1} and 1.4 T and 20.0 T m^{-1} in a 3.0 cm gap with 10 cm and 17.8 cm pole diameter electromagnets respectively. These values are controlled by the gap width, which is normally set by the diameter of the protective tube round the sample, and, in high-temperature measurements, by the furnace diameter.

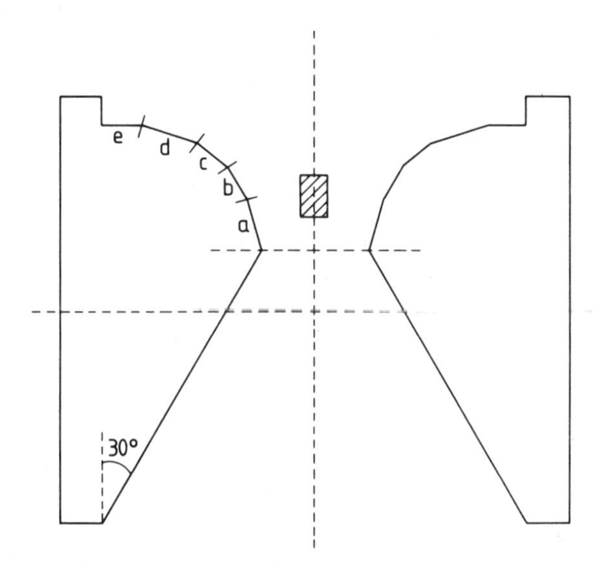

Fig. 3.9 Shaped pole-pieces for producing a region of constant $B(dB/dz)$, shown by the hatched area. The flats machined on the circular pole faces are at the following angles with the vertical: 13°(a), 28°(b), 51°(c), 74°(d) and 85°(e). (After Heyding *et al.*, 1961)

It is possible to have a range of absolute and relative values of B and dB/dz in one instrument by having pairs of differently shaped, detachable pole-pieces which dovetail on to the core. In an appropriately constructed magnet they can be lifted out and replaced without disturbing the sample. An alternative but less versatile arrangement is to calibrate the field and field gradient at different positions on the z-axis.

For some applications it may be advantageous to use an air-cored coil system to provide the field and field gradient, which can then be altered independently of each other when a range of magnitudes is required. Coils are also free of hysteresis, which sometimes causes problems in calibrating an electromagnet field against current and in setting $B = 0$, since then $i \neq 0$.

The field is produced in a solenoid of the type described in Section 11.1 and the field gradient may be generated in a 'Helmholtz' arrangement of two coils connected in series opposition and mounted coaxially with and either inside or outside the field solenoid.

Differentiating the expression for the field on the axis of a short coil of small winding cross-section, with N turns of mean radius r carrying a current of i A (Equation 5.1) gives

$$\frac{dB_z}{dz} = \frac{3\mu_0 Nizr^2}{2(r^2+z^2)^{5/2}} \tag{3.18}$$

where z is the distance along the axis from the coil centre. Maximum gradient is obtained at the centre of a Helmholtz pair, in which the coil separation equals the coil radius (Section 5.1.1), the coils being connected in series opposition. Substituting for the Helmholtz condition $z = r/2$ in Equation 3.18 and adding the gradients from each coil

$$\frac{dB_z}{dz} = \frac{48}{25\sqrt{5}}\left(\frac{\mu_0 Ni}{r^2}\right) \tag{3.19}$$

at the centre of the system. Thus, 1000 ampere-turns with $r = 5$ cm gives $dB/dz \approx 0.43\ \mathrm{T\,m^{-1}}$, indicating that only moderate gradients are achieved in such systems. However, if the above-mentioned advantages are important, and the samples are such that the forces acting can be satisfactorily measured, the system has its uses. McKeehan (1934) used a solenoid with gradient coils ($r = 5$ cm) mounted on its outer surface, obtaining ~160 mT with 800 W input, and a gradient of ~$0.035\ \mathrm{T\,m^{-1}\,A^{-1}}$. This combination provided sufficient force on small iron–cobalt crystals to measure their susceptibility. Equation 3.19 shows that $dB/dz \propto r^{-2}$, and Parry (1957) describes a composite solenoid in which the gradient coils are wound on the inner cylindrical coil former inside the field windings, thus minimizing r. The water-cooled system was designed to produce a maximum 700 mT and $3.5\ \mathrm{T\,m^{-1}}$ in a bore of diameter 1.2 cm with a total current input of 30 A.

In magnetic balances the magnitude of B and dB/dz acting on the sample depends on its position on the line along which the force acts and a null method of force measurement is necessary, i.e. one in which a force equal and opposite to the magnetic force is applied and measured, the sample remaining in a fixed position.

Methods of force measurement that are or have been used include a vertical helical quartz spring, vertical on horizontal torsional suspension, pendulum suspension and the principle of the chemical balance. With the quartz spring the sample hangs from the lower end and the upper end is supported by a device for accurately measuring small distances, such as a micrometer screw gauge. The sample displacement due to the magnetic force F is counteracted by screwing up the gauge a distance x until the sample returns to its zero position as judged by the setting of a suitable sighting mark viewed in a low-power microscope. Then $F = kx$, where k is the spring constant. In principle a torsional suspension can be used in a similar way, with the sample at the end of a horizontal arm supported by a horizontal or vertical torsion wire. A torque T opposing that exerted by the field gradient is applied by twisting the suspension through an angle θ and $T = \sigma\theta$, where σ is the torsional constant of the system. The principle of the pendulum suspension is the deflection by a horizontal force F of a mass m hanging on a suspension of length l where $F = mgx/l$ or $x = Fl/mg$ where x is the deflection of the mass and g is the acceleration due to gravity. For use in a null instrument a modification known as a horizontal pendulum is normally used, and is described below.

The above methods suffer from a restricted sensitivity range and long time constants if high sensitivity is required, and the advantages of electronic feedback and the associated wider range of sensitivities, short time constant and the capability of automatic recording are apparent. The method has been applied successfully to torsional and pendulum systems, and is available commercially in the form of a microbalance. Schwarz (1968b) describes an instrument based on the Cahn electrobalance, which has microgram sensitivity (equivalent force $\sim 10^{-8}$ N) and thus ample sensitivity, and in which the feedback current, proportional to F, is used for automatic recording. With a maximum field of 790 mT, the sensitivity range is 5×10^{-9}–10^{-3} A m^2 per centimetre of recording chart. Cox and Doell (1967) describe a similar instrument, using a range of detachable pole-pieces to vary field/field gradient ratios.

The present author and A. de Sa have constructed a horizontal torsion balance in which a light aluminium beam is supported by a horizontal phosphor-bronze wire (Fig. 3.10). One end of the beam carries the suspension and sample and the other an inverted U-shaped yoke to the ends of which are fixed small magnets with their axes vertical and their moments antiparallel. Gradient coils surround the magnets and current from the feedback system passing through them keeps the beam, and therefore the sample, in a fixed position. A capacity transducer, one plate of which is fixed to the beam and the other to the base-plate, controls the feedback current (Creer and de Sa, 1970) from which a voltage is derived which is proportional to the force on the sample. The coil and magnet arrangement could equally well be used for manual restoration of the null position. The pole-pieces follow Heyding, Taylor and Hair (1961) and 1.4 T and 20 T m^{-1} are obtained with a 17.8 cm water-cooled magnet in a 3.0 cm gap. The astatic arrangement of the magnets together with small trimming magnets elsewhere on the beam ensures that the effect of the main magnet field on the beam is very small. The sensitivity with $dB_x/dz = 20$ T m^{-1} is 5×10^{-8} A m^2 per millivolt output with a noise level of ~ 0.1 mV. The force sensitivity of the balance is 1.0×10^{-6} N/mV. A modified version of this instrument using a variable differential transformer rather than a capacity transducer is described by de Sa (1981).

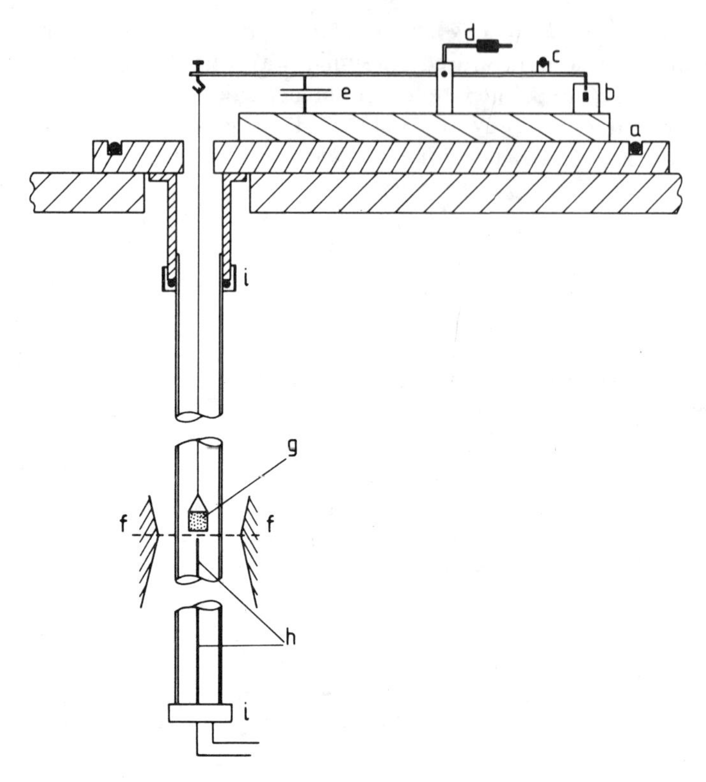

Fig. 3.10 Diagram of magnetic balance with torsional suspension of the beam, and coil and magnet restoring force. a, O-ring seal; b, gradient coils; c and d, riders and fine adjustment for levelling the beam; e, capacity transducer; f, magnet; g, sample; h, thermocouple; i, vacuum seals.

Details of two instruments based on quartz fibre torsion balances are given by Clark (1967) and Domen (1968).

There are two features of vertical suspension balances which sometimes cause problems, namely a horizontal force on the sample tending to pull it to one or other of the pole-pieces and the inability of such a balance to distinguish between magnetic and gravitational force. The latter is seen when a sample loses water or otherwise changes weight during heating, resulting in a spurious deflection. A 1 % mass loss in a 0,2 g sample corresponds to $F = -2 \times 10^{-5}$ N. This is equivalent to the magnetic force on the sample if $B = 500$ mT, $dB/dx = 10\,\text{Tm}^{-1}$ and $\chi = 2 \times 10^{-8}\,\text{m}^3\,\text{kg}^{-1}$, a typical value for a red sandstone.

The field B_x in the magnet gap has a gradient dB_x/dx which changes sign, and is zero, at the gap centre. An accurately centred sample experiences no horizontal force, but in practice this is difficult to achieve and the sample tends to pull towards one of the poles as B_x and dB_x/dx increases. The sample will be pulled out of the region in which B_x and dB_x/dz have been calibrated, and may ultimately contact the protective tube that it and the suspension hangs on. Both these effects can lead to measurement errors.

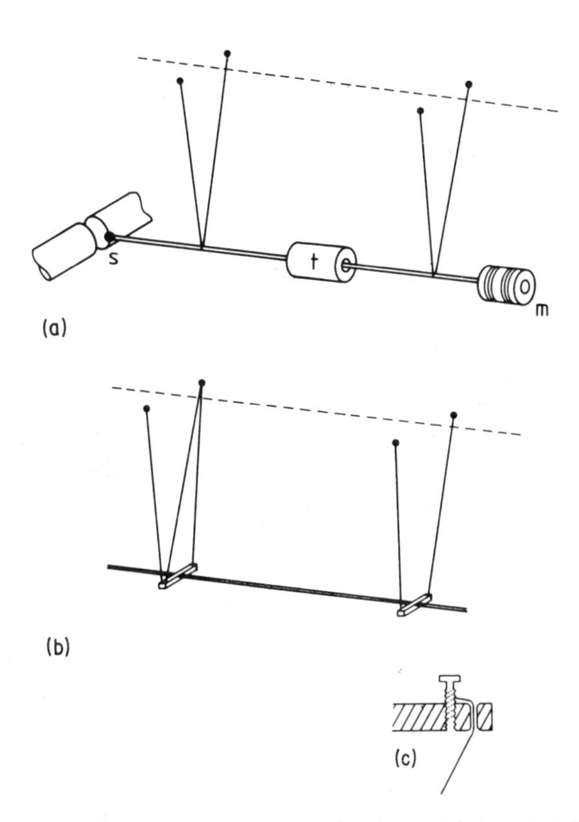

Fig. 3.11 Horizontal balance: (a) basic design, with suspended horizontal beam carrying sample (s) at one end, the coil and magnet system (m) for providing the restoring force at the other, and the transducer (t) at the centre; (b) alternative method of beam suspension; (c) method of adjusting suspension lengths.

Although the transverse force is rarely a serious factor with measurements on sediments and loss of weight can be distinguished from change in magnetic force by taking zero field readings, the effects of both are eliminated by using a horizontal pendulum suspension. This takes the form of a light horizontal metal beam (non-ferrous), with the sample at one end, supported horizontally by a five-string suspension (Fig. 3.11). This permits only longitudinal motion of the beam along the magnetic force axis (now horizontal) with a gravitational restoring force, and transverse motion is prevented. A fractional change (p) in the mass of the sample (m_s) results in a fractional change in the restoring force of $p(m_s/m_t)$, where m_t is the total mass of the suspended system: thus the effect of a mass change is normally negligible, since $m_s/m_t \approx 0.1$. Note that the magnetic force is not significantly affected by loss of water. Creer and de Sa (1970) describe a horizontal balance using a capacity transducer and coil and magnets for the feedback restoring force.

In a translation balance designed by Hummervoll (1976) most of the feedback system is provided by a commercial fluxgate magnetometer. Two astatically-arranged fluxgate probes are placed near the sense magnet on the beam and the magnetometer output, when the beam moves, is passed to the coil providing the restoring force via a d.c. amplifier.

(*a*) *Calibration*

To use Equation 3.17 for determining J_i and for plotting J_i–B curves it is necessary to know the sensitivity of the force-measuring system and also the values of B and dB/dx at the sample position in the gap for different values of the current in the electromagnet.

Force sensitivity and linearity in a vertical balance is measured by hanging known masses m on the system and recording the output V of the feedback device, or the magnitude of the response required if a manual system is used. Then the force constant is mg/V N per unit output, where g is the acceleration due to gravity. With a horizontal balance it is less easy to determine the force constant, but it can be done as follows. If the base on which the horizontal beam is supported is tipped through a small angle θ about a horizontal axis perpendicular to the beam axis, the beam will move a distance $x = l\theta$, where l is the vertical distance of the centre of gravity of the beam below the plane of the suspension points. The force required along the beam axis to restore the beam to its original position is $F = m_t g\theta = m_t gx/l$ where m_t is the total mass of the beam: thus, force/unit output signal can be determined. θ can be measured (and easily varied) if blocks of different and known thickness, placed under one end of the base, are used to tip the system. Thus, if required, the range of proportionality between output signal and force can be tested by varying θ.

The optimum sample position on the line along which the magnetic force acts can be determined by putting a sample (natural or synthetic) in the holder to provide a reasonable deflection with values of field and gradient somewhere near the middle of the range, and adjusting the sample position until maximum force is obtained. It is then desirable to arrange a system of sighting wires or some other arrangement to ensure that the sample can be repeatedly set in the same position to ± 0.05 cm or better.

The field may be calibrated against magnet current by using a Hall-effect magnetometer, provided, as is often the case, the detector dimension perpendicular to the field is comparable to that of the sample. If the centres of the detector and sample position coincide, the magnetometer will record the mean field acting on the sample with sufficient accuracy. An alternative method, more suitable for vertical balances, was used by Cox and Doell (1967). They measured the force acting on a straight wire set perpendicular to the field and carrying a current i: the force is given by

$$F_z = Bil \tag{3.20}$$

where l is the length of the wire: F_z is perpendicular to both B and i. However, there are practical difficulties with this method involving the true value of l, the avoidance of forces along the F_z axis arising from the remainder of the current loop, and leading the current in and out of the loop without affecting the force reading.

The field gradient is calibrated by measuring the force on a known magnetic moment in the sample holder. Cox and Doell (1967) used a small coil carrying a known current, with the coil axis parallel to the field: the coil constant, i.e. magnetic moment/unit current, can be determined by measuring the voltage developed across the coil in a known alternative magnetic field (Appendix 6). In an alternative method the sample holder is filled with a pure chemical of known (paramagnetic) susceptibility (Table 2.1) and Equation 3.17 applied, using the previously calibrated values of B. Although this method depends on a knowledge of B it has the advantage of greater simplicity and of determining the average value of the gradient over the same dimensions as the sample.

Because of hysteresis in the magnet core there may be variations in B and dB/dz produced by a given current, depending on the previous cycling of the magnet and the magnetic characteristics of the core and pole-pieces. In a magnet used by the author the maximum variation in B (and therefore in dB/dz) encountered with the same current on the upward and downward part of the cycle $B = 0 \rightarrow 1.4\ \mathrm{T} \rightarrow 0$ is 1–2%: this is barely significant for most applications and is comparable to the accuracy with which the field and gradient are measured.

3.3.2 The vibrating sample magnetometer (VSM)

The principle of these instruments is similar to that of a spinner magnetometer, but the flux change in the pick-up coil system is produced by vibrating the sample rather than by rotation and induced as well as remanent magnetization can be measured. Although they are generally not sufficiently sensitive for the measurement of weak NRM, vibration magnetometers have been used in many other magnetic studies and are useful in rock-magnetic investigations for the measurement of J_i and hysteresis loops. The main advantage of a VSM over a magnetic balance is that the method of measuring J_i is independent of the inducing field B and no field gradient is required.

A VSM consists essentially of the sample vibration mechanism, a pick-up coil system in which an oscillatory voltage is generated, an electromagnet or coil for applying a field to the sample, and a signal processing circuit.

(a) *General design features*

If the flux ϕ linked with the pick-up coil, due to the magnetization of the sample, is a function of z, the position of the sample on the axis along which vibration takes place, then the instantaneous output voltage V generated in the coil is

$$V = \frac{d\phi}{dt} = \frac{d\phi}{dz}\frac{dz}{dt} \tag{3.21}$$

In general the waveform of the output will be complex, but if z varies sinusoidally V will contain a fundamental component at the vibration frequency. Provided the frequency and vibration amplitude remain constant the average output voltage is proportional to the strength of the vibrating magnetic moment. If ϕ is proportional to z over the amplitude range then $d\phi/dz$ is a constant, and the output signal will be sinusoidal if the sample is driven sinusoidally.

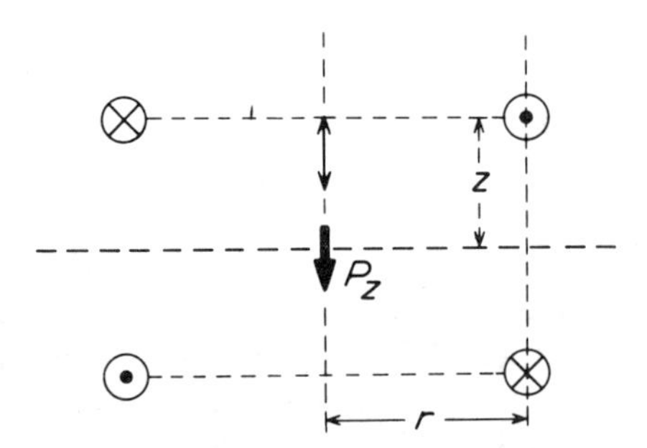

Fig. 3.12 Pick-up coils for vibrating sample magnetometer with magnetic moment parallel to vibration axis.

There are two modes of operation for measurements of induced magnetization according to whether the applied fields are parallel or perpendicular to the axis of vibration of the sample. If the inducing field is provided by a solenoid, sample vibration parallel to the field is convenient and a common arrangement is two coaxial coils connected in series opposition (Fig. 3.12). This system is insensitive to changes in the uniform external (inducing) field but the signals from each coil are additive for flux variations due to the vibrating sample. For systems in which the field is provided by an electromagnet the vibration axis is most conveniently perpendicular to the field. With some coil systems a complete determination of the direction of a permanent magnetic moment can be made, in principle, without removal of the sample from the holder, by rotating either the sample or part of the coil system through 90° about an appropriate axis.

An approximate calculation of the signal due to a vibrating axial dipole moment, P, on the axis and midway between two coils of small winding cross-section (Fig. 3.12) is easily carried out. The flux ϕ linked with one coil of N turns is given by

$$\phi = \frac{\mu_0 N P r^2}{2(r^2 + z^2)^{3/2}} \tag{3.22}$$

where r is the radius of the coil and z is the axial distance of the dipole from the coil plane. The winding cross-section of the coil is assumed to be small compared with r. The rate of change of flux with z is

$$\frac{d\phi}{dz} = -\frac{3\mu_0 N P r^2 z}{2(r^2 + z^2)^{5/2}} \tag{3.23}$$

and if the dipole is undergoing axial sinusoidal vibration of amplitude a and circular frequency ω

$$z = z_0 + a \cos \omega t \tag{3.24}$$

where z_0 is the mean distance of P from the coil plane. Then

$$\frac{dz}{dt} = a\omega \sin \omega t \tag{3.25}$$

Therefore the instantaneous induced voltage $V_v = d\phi/dt$ is

$$V_v = \frac{3a\omega\mu_0 N P r^2 z}{2(r^2 + z^2)^{5/2}} \sin \omega t \tag{3.26}$$

Note that V_v is not sinusoidal because z is varying according to Equation 3.24.

For maximum output with a coil pair the coils should be in the Helmholtz configuration and connected in series opposition, i.e. $z_0 = \frac{1}{2}r$. For multi-turn coils of extended cross-section the turns should be grouped as closely as possible around the Helmholtz position. Springford, Stockton and Wampler (1971) show that if the coil separation corresponds to $z_0 = (\sqrt{3}/2)r$ the output is least sensitive to sample positioning along z, with only a small (~20%) sacrifice in output relative to the Helmholtz setting.

An important feature arising from the series opposition connection of the coils is the reduction of spurious output voltages arising from variations in the applied field. Springford *et al.* (1971) consider other aspects of the above coil configuration and describe a vibration magnetometer in which it is used.

It is often more convenient to obtain a strong induced field (1–2 T) by using an electromagnet rather than a solenoid, and the pick-up coils are then required to respond to flux variations arising from sample vibration perpendicular to the direction of magnetization.

There is an extensive literature on the design of pick-up coils for this arrangement and practical systems fall into two classes, those with the coil axes lying in the x–z plane only and more complicated systems in which the coils are displaced along the y direction as well. In the following discussion the x and y axes are horizontal and parallel and perpendicular respectively to the applied field (and sample magnetization), and the z axis, the axis of vibration, is vertical.

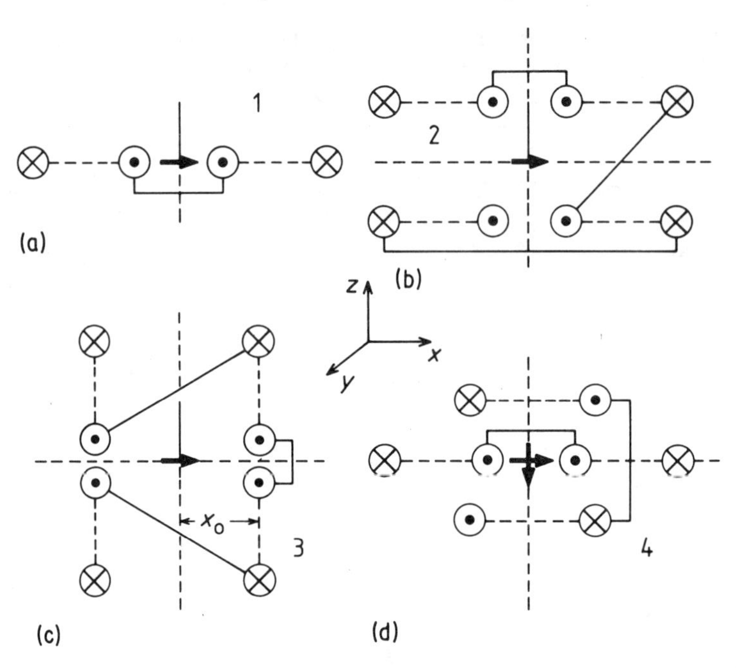

Fig. 3.13 Pick-up coil configurations for vibration magnetometers in which the induced moment is perpendicular to the vibration axis (z). (System 4 can also measure a parallel moment.) Dotted lines connect parts of the same coil, full lines show connections between coils.

Some systems in which the coils are confined to the x–z plane are sketched in Fig. 3.13. Although a single coil can be used it is, for a variety of reasons, less satisfactory than multicoil arrangements. A single coil lacks sensitivity, is susceptible to time variations in the applied field, and its output is rather sensitive to sample position. Foner (1959) describes the output characteristic of a single coil with its axis vertical and the sample vibrating close to its upper surface and displaced a short distance along the x-direction from the coil axis. A possible advantage of such a coil is that the magnet gap can be smaller than with multicoil systems, enabling higher magnetic fields to be obtained. A single-coil system is also discussed by Bragg and Seehra (1976).

The coil systems shown in Figs 3.13(a) and (b) sense the variation in the vertical component of the field due to the vibrating horizontal dipole P_x. The (b) configuration is essentially an extension of the (a), and can consist of either two pairs of coils centred at $\pm x$, as shown, or two extended cylindrical coils at these positions. Both arrangements, when connected as shown, are additive for the output of each coil due to the vibrating sample and (ideally) present zero area-turns to, and therefore zero output from, variations in the applied field.

The system shown in Fig. 3.13(c) senses the variation in the axial horizontal field of the dipole due to its vibration along z, and also presents zero area-turns to the applied field. Figure 3.13(d) is a configuration with which a total remanent magnetization vector can be determined with minimum sample manipulation. The components along z and x are measured simultaneously and the component along y is determined by rotating the sample or the x- and y-sensing coils through 90° about a vertical axis.

Various authors have examined the sensitivity of these different coil systems, but exact calculations are difficult. Mallinson (1966) and later Guy (1976) consider two-dimensional representatives of systems 1, 2 and 3, in which each coil consists of a pair of infinite straight wires parallel to the y-axis. From this analysis, based originally by Mallinson on the reciprocity principle, system 3 emerges as the most sensitive arrangements for a fixed magnet gap width and sample–coil separation x_0 measured along the x-axis. For each system the output is proportional to x_0^{-2}, and for systems 1 and 2 the ratio of the coil diameter to x_0 should be larger than is normally possible (> 2) to obtain a sensitivity comparable to 3. A general-principle derived by Mallinson and others (for an x-dipole vibrating along z) is the '45° rule' whereby maximum sensitivity is achieved if the return conductors of each pick-up loop are above or below the lines $x = y$, $-x = -y$ and $x = -y$, $-x = y$ drawn through the centre of the system. This can be achieved in system 3 but not in 1 or 2. The optimum design for system 3 is $d = \sqrt{3}x_0$, where d is the coil diameter, and for system 1 the inner coil edge, for a solenoidal coil, should extend up to the 45° line.

Foner (1959) and Bragg and Seehra (1976) have also investigated systems 1 and 2. The former author was the first to use the vibration principle for measurements in classical magnetism, although in fact Blackett and Sutton (1956) had earlier described a sample vibration magnetometer for measuring induced magnetization in rock samples. Foner pays some attention to the dependence of output signal on sample displacement from the coil centre position. A possible advantage of system 1 is that the output signal, for small (sinusoidal) vibration amplitude, is very nearly sinusoidal.

Coil systems in which the centres of the coils are not confined to the x–z plane have been described by Krause (1963) and Bowden (1972). Krause used the vibration principle for high-temperature measurement of the NRM of rock samples during thermal demagnetization (Section 11.2.4). An ingenious system of eight-plane trapezoidal coils arranged in an upper and lower group of four surrounded the sample (Fig. 3.14(a)). The plane of each coil of the upper group sloped upwards from the sample and the planes of the lower group downward at the same angle, 20° from the horizontal. By interchanging the coil connections in turn, it is possible to measure the three perpendicular components of the sample magnetization while simultaneously providing a compensated coil system to nullify the effect of ambient field variations.

A feature of systems with coil centres at (x, y), $z = 0$ or (x, y, z) is their generally broader 'saddle points', i.e. a smaller variation in signal output with sample displacement along x or y, compared with the systems previously discussed. It is this aspect with which Bowden (1972) is mainly concerned. His optimum design consists of eight coils mounted on the corners of a cube

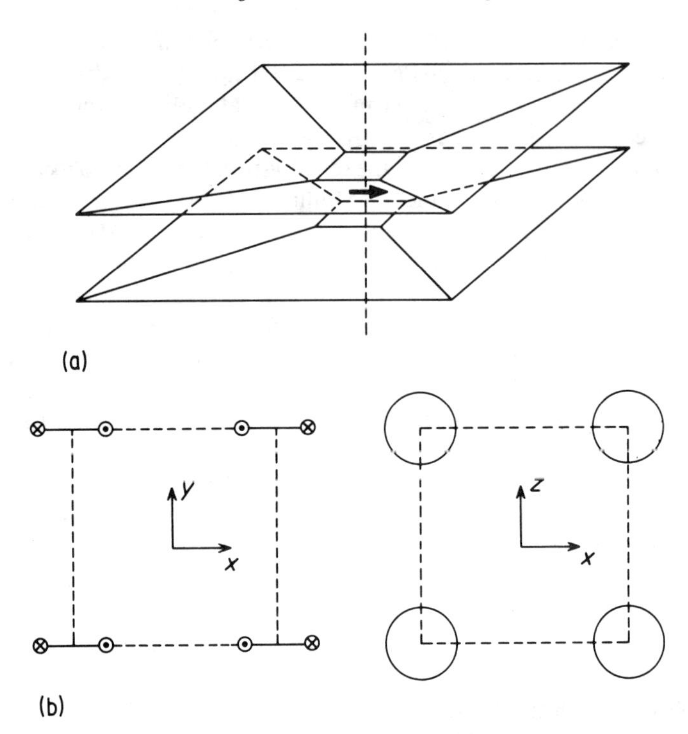

Fig. 3.14 (a) Trapezoidal pick-up coils used by Krause (1963); (b) the eight-coil system of Bowden (1972). The magnetic moment is directed along x and vibration is along z.

centred at $x = y = z = 0$ with their axes parallel to y (Fig. 3.14(b)). Although providing broad saddle points and good access volume, the typical sensitivity of this configuration falls well below that of systems 2 and 3 of Fig. 3.13, as does another system investigated by Bowden consisting of four coils in the x–y plane with their axes along z. In rock-magnetic studies the sample size is often somewhat larger than is used in the type of investigations in classical magnetism for which vibration magnetometers were developed, and misplacement of the sample is less important. In practice the advantage of higher sensitivity of the x–z configurations has led to their general use for rock-magnetic purposes.

As in the case for spinner magnetometers, the noise signals occurring in the output of a vibration magnetometer are usually above the ultimate noise, i.e. the random Johnson noise due to the resistance of the pick-up coils. The Johnson noise r.m.s. voltage is proportional to $R^{\frac{1}{2}}$, where R is the coil resistance, and in principle it is possible to design a coil system in which the signal/Johnson noise ratio is maximized, as can be done for spinner magnetometers (Section 9.3.1). In practice this is not easy to do rigorously for VSMs because of the difficulty of calculating coil outputs, and most coil systems are designed along the general lines indicated and the coils wound with many turns of fine wire in order to maximize the output voltage.

A potential source of noise is interaction between the applied field and pick-up coils, either vibration of the latter in the former or time variation in the field causing a spurious voltage output from the coils. Noise from the coil vibration, if it is excited by the sample vibration mechanism, is synchronous and must be avoided. The effect of variations in the applied field and ambient field is reduced by appropriate coil configurations, as already described, and if possible by improved stability of the current in the electromagnet or solenoid. This problem does not arise if a permanent magnet or superconducting solenoid is used. An advantage of systems 1 and 2 of Fig. 3.13 is that the coil axes are perpendicular to the applied field, and therefore even without their series-opposition connection the coils interact very little with the field. The required degree of balancing of area-turns in system 3 is more stringent. If necessary a trimming circuit can be used to fine tune the two coil systems for effective equality of area-turns (Fig. 3.15) and the same method can be adopted for multicoil systems.

Apart from coil design the two parameters which also enter into vibration magnetometer sensitivity are the vibration frequency and amplitude. In general the greater these are the greater the signal from a given magnetic moment, but there are practical limits to the vibration amplitude and frequency and there is usually a trade-off between the two. For instance, loud-speaker movements provide an amplitude of up to ~1.0 mm at frequencies of ~100 Hz, whereas mechanical vibrators, often based on a 'Scotch crank' (Fig. 3.16) to provide sinusoidal motion can easily provide a stroke of ~1.0 cm but generally at lower frequencies of ~10–40 Hz. Mechanical vibrators can carry a substantially heavier load than a loudspeaker movement, the amplitude is independent of the load, and there is no magnetic noise generation but there is the possibility of noise from mechanical vibration. Loudspeaker-type vibrators may be a source of electromagnetic noise and show load-dependent amplitude. More robust electromechanical vibrators are now available, working in the frequency range 30–100 Hz with stable amplitudes of 1–5 mm.

Too large an amplitude of vibration may result in increased distortion of the output waveform through the generation of higher harmonic signals. The coil output will only be even approximately sinusoidal when the amplitude varies sinusoidally and is small compared with a typical sample to pick-up coil distance in particular coil configurations. Since most pick-up coil systems must be small (1–2 cm) in order to be accommodated in the gap of an electromagnet, vibration amplitudes in such systems are typically ~2–6 mm.

The general output waveform of the pick-up coils is not sinusoidal, but consists of a fundamental component corresponding to the sinusoidal drive frequency plus harmonics of this frequency. The harmonic content is a function of the coil geometry, vibration amplitude and coil size relative to sample size.

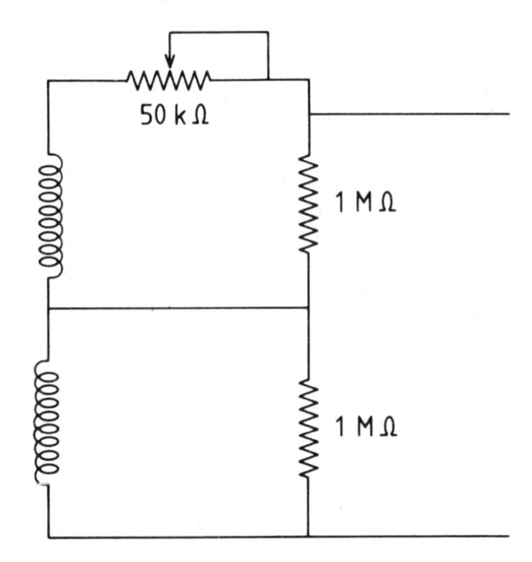

Fig. 3.15 Coil trimming circuit, based on that of Foner and McNiff (1968) for effectively equalizing area-turns of series-opposed pick-up coils. The 50 kΩ resistance is a ten-turn 'helipot' in series with the coil having the greater area-turns.

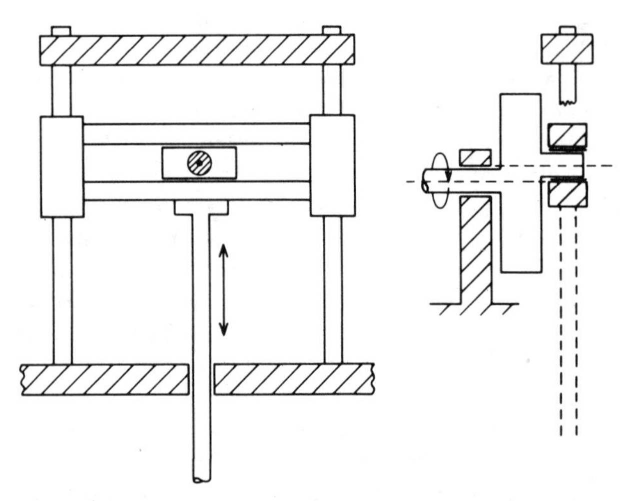

Fig. 3.16 Basic design of Scotch crank for converting rotational to oscillatory motion.

Although the output from some pick-up configurations varies somewhat with sample position this is not usually an important effect but it is desirable to ensure that motion of the sample along axes other than that of the vibration is minimal. Other important precautions are the avoidance of any material in the sample holder with significant electrical conductivity (to avoid generation of eddy currents) or with significant paramagnetic susceptibility and, as already mentioned, the avoidance of relative motion between the coils and magnetic field. The former two requirements are not normally a problem (or can be allowed for) and the latter can be achieved by appropriate mechanical design in rigidly securing the coils to the magnetic pole-faces or to the inner former of a solenoid.

Signal processing with a VSM is essentially similar to that used for spinner magnetometers (Fig. 9.17), although phase determination is not required unless NRM is to be measured. However, phase-sensitive detection is a convenient method of noise reduction in the output and is commonly employed. The reference signal can be obtained from a pick-up coil adjacent to a small magnet attached to the vibrator shaft, a photoelectric system (more suitable if there is a rotating shaft associated with the drive mechanism), or from a capacity transducer, one plate of which is fixed and the other vibrates with the drive shaft (a system employed in the commercial instrument marketed by Princeton Applied Research Corporation, New Jersey, USA). Since the signal from the pick-up coils contains unwanted harmonics it is usually necessary to employ some broad-band filtering before amplification and in some instruments the signal is first passed to a step-up transformer. This provides some filtering and amplification, and also isolates the pick-up coils from possible earth-loops.

Calibration of vibrating magnetometers is best carried out in one of two ways. An absolute determination using a known magnetic moment is achieved by vibrating an accurately made coil carrying a known current and placed in the sample holder. The coil should enclose the same volume as that of the samples. Another satisfactory method is to use a sample consisting of high-purity nickel or iron powder dispersed and sealed in a non-magnetic matrix. The sample can be magnetically saturated in the magnet field, providing a known test magnetic moment of magnitude determined by the value of J_s and the proportion of iron or nickel present. The values of J_s for iron and nickel are 218 and 54.5 A m^2 kg^{-1} respectively. A similar but less satisfactory method is to use a sample made up of paramagnetic salt and calculate its moment from the magnitude of the applied field. This method requires the field to be known reasonably accurately. All these methods can be used to test the linearity of the VSM output.

(*b*) *Practical instruments*

A simple VSM for rock-magnetic measurements was first descibed by Blackett and Sutton (1956). A disc of rock 2.3 cm in diameter and 0.5 cm thick was mechanically vibrated at 30 Hz with an amplitude of 1.2 mm. The vibration axis was parallel to the disc axis and to the applied field, provided by an electromagnet, and the output voltage was generated in two pick-up coils on either side of the disc and coaxial with it. The signal was amplified and read on a vibration galvanometer which with maximum gain gave a 1 mm deflection for a sample moment of $\sim 2 \times 10^{-7}$ A m^2. Variation of the calibration factor due to variations in vibration amplitude was monitored by means of a two-turn coil wound on the sample holder, which, when carrying a known current, provided a calibration signal. The coils could also be used to provide an absolute calibration of sample moment since the magnetic shell produced is closely coincident with the specimen, assumed to be uniformly magnetized.

Kobayashi and Fuller (1967) describe a 90 Hz system driven by an electric motor and mechanical converter, with a variable amplitude of 0.2–5.0 mm Pick-up coils of type 2 are used and samples of up to ~25 g can be accommodated. Phase-sensitive detection is used for signal processing, the reference signal being obtained from a coil placed near a magnet attached to the vibrator drive shaft. There is provision for slow rotation of the vibration sample about a vertical axis so that a vector NRM measurement ($> 10^{-8}$ A m^2 in 10 cm^3) in the x–y plane is possible. The coil system of Krause's (1963) instrument has already been described: his instrument was designed mainly for NRM measurement at elevated temperatures, and the internal radius of the coils was larger than normal in order to accommodate a 2.5 cm cubic sample and a furnace. With a 1.3 cm stroke at 40 Hz the minimum detectable magnetization was $\sim 2 \times 10^{-5}$ A m^2 kg^{-1}.

Creer, Hedley and O'Reilly (1967) constructed an air turbine-driven VSM. This drive system (working a Scotch crank) was employed to avoid the electromagnetic noise sometimes associated with motor drives. A 1 g sample was vibrated at 70 Hz with a 1.2 cm stroke, and the system was designed for use in solenoid with axial pick-up coils (Fig. 3.12). A modification of this instrument, using a motor drive at 11 Hz and 2.2 cm stroke is described by Hoye (1972). The frequency of the air turbine drive was somewhat difficult to stabilize and the turbine was also acoustically noisy.

In a later version of this VSM (Tucker, 1978) the axial pick-up coils are replaced by type 2 coils (Fig. 3.13), for use in the gap of a 17.5 cm electromagnet. Each coil is 1.7 cm long and 0.65 cm in diameter and carries 5000 turns, with a resistance of 850 Ω. The coils are fixed to the gap faces with their centres 5.0 cm apart: this separation is chosen to maintain a high maximum field in the gap (~ 1.0 T at 25 A) and also to allow room for a furnace and controlled atmosphere or vacuum around the sample. The measuring space also allows 2.5 cm rock cylinders to be measured at room temperature: for high-temperature work small (~ 1 g) samples are used. An amplitude of 1.5 cm and frequency of 11 Hz are used, and the noise level corresponds to a total moment of $\sim 1.0 \times 10^{-7}$ A m^2.

At the time of writing, Highmoor Electronics Ltd (Salford, England) market a vibrator and pick-up system using an electromechanical vibrator unit of 2.0 mm amplitude at ~80 Hz, with a sample of ~1 g. A type 3 pick-up coil system (Fig. 3.13(c)) is used: the coils have an internal and external diameter of 2.0 cm and 1.0 cm and are 0.5 cm wide, and each is wound with ~3500 turns of 48 SWG wire. The coils are centred on $x = \pm 0.7$ cm, $z = \pm 1.3$ cm. The amplitude of the vibrator is maintained constant to ~1 part in 10^3 by a simple feedback circuit (de Sa, 1980). The method relies on the fact that if a pick-up coil is placed around the vibrator shaft at the correct distance from an axial dipole fixed centrally in the shaft, then the vibrating dipole induces a voltage in the coil of twice the vibration frequency of magnitude proportional to the square of the stroke amplitude. This output is amplified, rectified and then passed via an electronic attenuator to control the input of the power amplifier driving the vibrator. The sensitivity of the magnetometer allows measurement of a total moment of $\sim 2 \times 10^{-6}$ A m^2, about twice the noise level.

Assuming that zero drift is absent or negligible, hysteresis (J_i–B) curves, or any required segment of them, can conveniently be plotted directly on an X–Y recorder receiving input from the VSM (Y) and from the applied magnetic field (X) via an analogue signal. Hall effect magnetometers are commonly used to place in the field near the sample. Commercial instruments also often provide access to a signal proportional to the magnetic field being measured and magnet power supplies can be obtained with programming facilities or can be adapted for this purpose.

Zijlstra (1970) describes a novel instrument for measuring J_i and hysteresis in single magnetic particles. The particle is mounted at the end of a horizontal reed in the gap of an electromagnet. An alternating magnetic field gradient, generated in small coils surrounding the particle, excites vibration of the magnetized particle and reed at the resonant frequency, the vibration amplitude being proportional to the magnetic moment of the particle.

3.3.3 Torque magnetometers and measurement of rotational hysteresis

Measurements of anisotropy of initial susceptibility of rocks, described in Section 2.8, are widely used for investigations of flow directions and fabrics in geological units. The observed anisotropy is due to alignment of non-equidimensional magnetic particles, or crystalline alignment of anisotropic uniaxial crystals. The former occurs where the magnetic particles are sufficiently strongly magnetized by the applied field (titanomagnetite series, pyrrhotite) such that the variation of demagnetizing factor with direction in a particle also significantly varies the induced magnetization. In sediments and some metamorphic rocks, low field anisotropy is mainly due to crystallographic alignment of haematite or pyrrhotite particles.

Measurements of susceptibility anisotropy in high fields can be used to investigate other magnetic properties of rocks and minerals. The magnetocrystalline anisotropy of cubic magnetic minerals (magnetite, titanomagnetites) is not seen in low fields but is observed in high fields, and this type of anisotropy is also observed in the haematite minerals in high fields. The phenomenon of exchange anisotropy associated with a two-phase system of hard and soft magnetizations in a mineral has been proposed by Stacey (1963) as the origin of a stable self-reversal mechanism, and this anisotropy can also be detected by high-field measurements. Another type of anisotropy that can be distinguished by high-field measurements is due to the 'stringing' together of particles in lines or planes, observed in rocks by Grabovsky and Brodskaya (1958) and by Stacey (1960a) in synthetic samples.

Closely associated with high-field anisotropy is the phenomenon of rotational hysteresis, the work done due to irreversible magnetization processes when a sample of magnetic material is rotated through 360° in a magnetic field (Vlasov *et al.*, 1967b; Day *et al.*, 1970). Rotational hysteresis is of interest as a physical process in rock magnetism, and also as a non-destructive technique of identification of magnetic minerals (Section 13.5) and for the detection of exchange anisotropy (Banerjee, 1966).

The basic principle of high-field anisotropy and rotational hysteresis measurements is the measurement of the torque exerted on a sample by an applied magnetic field due to the anisotropy as the sample is rotated to different azimuths about an axis perpendicular to the field. The torque T is given by $T = \mathrm{d}E/\mathrm{d}\theta$, where E is the enetgy of magnetization of the sample and θ is the direction of the applied field. T takes the general form

$$T = T_0 + T_1 \sin(\theta + \delta) + T_2 \sin 2\theta + T_4 \sin 4\theta + T_6 \sin 6\theta \quad (3.27)$$

where θ is the angle between the field direction and a reference direction fixed in the sample. T_0 is due to rotational hysteresis, resulting in a displacement from the zero axis of the torque curve, and the $\sin(\theta + \delta)$ term (of arbitrary phase) can arise in the measurement of rock samples if there is inhomogeneous distribution of susceptible material, non-uniformity of magnetic field, or asymmetries in the torque system. A $\sin\theta$ term can also be contributed if there is exchange anisotropy present. T_2, T_4, T_6 . . . terms arise from anisotropy of different types. Sin 2θ terms arise from shape anisotropy in strongly magnetic particles or from uniaxial anisotropy in weakly magnetic minerals, e.g. haematite, if the crystallographic axes are aligned. Crystalline alignment of cubic minerals (titanomagnetites) gives a dominant $\sin 4\theta$ term with minor contributions from $\sin 2\theta$ and $\sin 6\theta$ terms (Stacey, 1960a), depending on the degree of alignment of all three axes. Textural anisotropy gives rise to a curve containing $\sin 2\theta$ and $\sin 4\theta$ terms, with the ratio $T_4/T_2 = \sim 0.1$–0.2 (Stacey, 1960b, 1967). Anisotropy induced in isotropic material by stress, or superimposed on an already anisotropic material by stress, gives rise to a dominant $\sin 2\theta$ torque curve (Stacey, 1963; Banerjee, 1963).

For rotational hysteresis measurements torque curves are determined for the angular cycle $0° \rightarrow 360° \rightarrow 0°$. Hysteresis causes the 0°–360° and 360°–0 curves to be displaced from the axis of zero torque, and the area enclosed by the curves is proportional to twice the rotational hysteresis W_R, the irreversible work done during one cycle of the field. For the purposes of mineral identification torque curves are determined for different applied fields, and curves of W_R against B plotted. Examples of such curves are shown in Fig. 3.17.

(*a*) *Torque magnetometers*

The torque exerted on a rock sample in a saturating magnetic field can be approximately expressed by

$$T = \frac{m\rho J_s^{\,2}}{4\pi} \mu_0 (N_b - N_a) \sin 2\theta \tag{3.28}$$

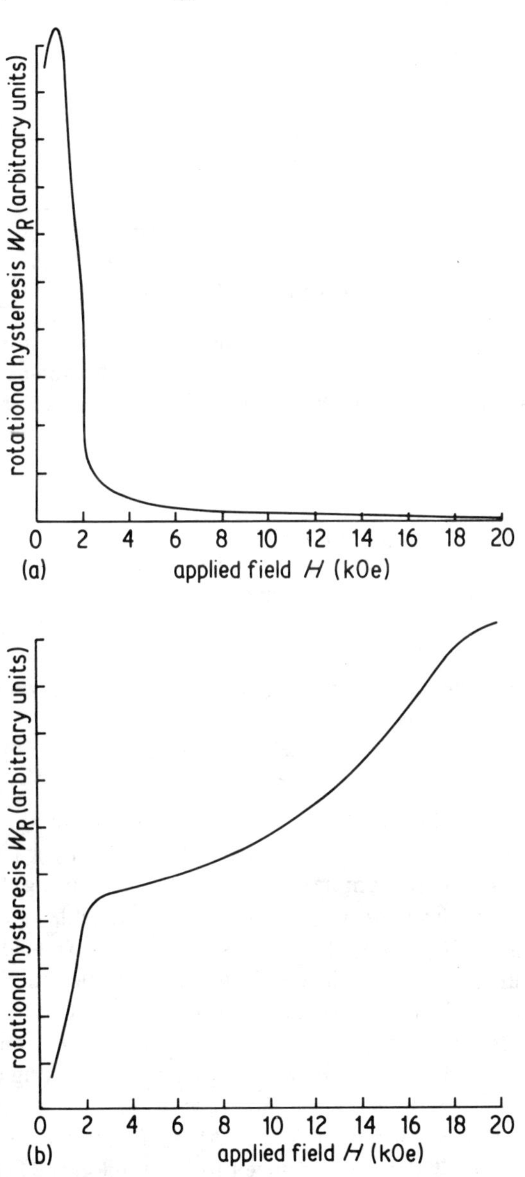

Fig. 3.17 (For caption and part (c) see overleaf)

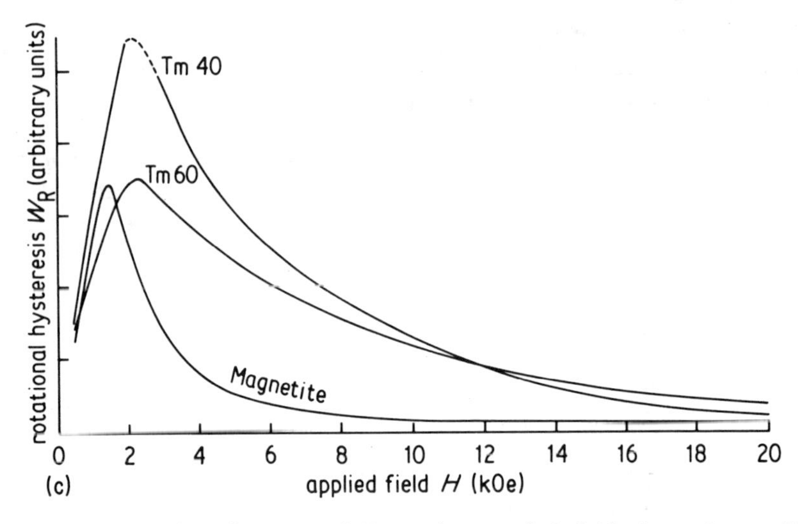

Fig. 3.17 Examples of curves of W_R against applied fields for polycrystalline materials: (a) maghemite; (b) haematite; (c) magnetite and titanomagnetites containing 40% (TM 40) and 60% (TM 60) ulvospinel. (From Day *et al.*, 1970)

where m, ρ, and J_s are the mass, density and saturation specific magnetization of the magnetic particles and N_a, N_b are the demagnetizing factors along the semi-major and semi-minor axes of the ellipse which is the cross-section of the equivalent ellipsoid (in the plane of measurement) representing the effective anisotropy of the sample. θ is the angle between the major axis and the field direction.

In terms of susceptibility differences the maximum torque is

$$T = \frac{mB^2}{2\mu_0}(\chi_a - \chi_b) \tag{3.29}$$

where B is the applied field and χ_a, χ_b are the maximum and minimum effective mass susceptibilities in the field B of the assembly of magnetic particles of mass m. Thus for a sediment sample of mass 10 g containing 1% by weight of haematite and showing 1% anisotropy in a bulk susceptibility of $2.0 \times 10^{-7}\,m^3\,kg^{-1}$, the maximum torque is $\sim 10^{-8}$ Nm in a field of 0.3 T. As an approximate guide, this results in a $\sim 2°$ deflection if the sample is suspended on a 25 cm length of 46 SWG phosphor-bronze wire. The torques resulting from igneous samples are normally much larger and thus extreme sensitivity is not required, and early equipment was very simple, consisting essentially of a torsional suspension supporting the sample and a graduated torsion head. An electromagnet supplied the magnetic field. Measurements were made either by restoring to zero the deflection produced by turning on the field, by rotating the torsion head, the sample being turned to different angular positions

(0–360°) relative to the field before switching on (Stacey, 1960b), or by slowly rotating the electromagnet about a vertical axis and recording the deflection by means of an optical lever (Vlasov and Kovalenko, 1964; Ozima and Kinoshita, 1964). The latter procedure is satisfactory provided the deflections are small.

More recent equipment is usually based on a modified galvanometer movement to which the sample is rigidly attached below the coil and suspended in the magnetic field. The torque is measured by the current required in the galvanometer coil to restore a deflection to zero. There are several advantages in this method, particularly when it is combined with an automatic feedback system.

The most important advantage is the elimination of the instability which can occur if the restoring torque exerted by the suspension is not in stable equilibrium with the torque produced by the sample. This arises through the non-linear nature of the magnetic torque, which causes, in certain regions of the torque curve, the magnetic torque to decrease faster than the restoring torque as the suspension is twisted to restore the deflection to zero, and zero deflection cannot be achieved. Harrison (1956) showed that this effect sets a lower limit on the torsional constant of the suspension (σ) that can be used (thus restricting sensitivity), given by $\sigma > (\mathrm{d}T/\mathrm{d}\theta)_{\mathrm{max}}$, where the right-hand term is the maximum slope of the torque curve. In a feedback system in which the zero angular deflection of the sample is maintained, the above instability effect is avoided. Other advantages of feedback are reduction of measurement time, easy adjustment of damping, extension of sensitivity range and the facility for direct recording of torque curves.

The basic circuit and components in a direct-recording automatic torque magnetometer are shown in Fig. 3.18. The most common detection device is a split-photocell receiving a light beam reflected from a mirror attached to the suspended system. After d.c. amplification the photocell output provides the appropriate feedback current in a coil attached to the system to null a deflection, through the torque exerted on the coil in a magnetic field. The voltage across a metering resistance in the feedback circuit provides the signal (Y) input for an X–Y recorder. A potentiometer attached to the shaft of the rotating magnet system can be used to provide a voltage for the X-input proportional to the angular position of the magnet relative to a fixed direction in the sample. If required, a digitizing circuit and tape punch or computer input can be included in the Y-circuit.

Descriptions of two such magnetometers for rock and mineral measurements are given by Banerjee and Stacey (1967) and, in more detail, by Fletcher, de Sa, O'Reilly and Banerjee (1968). Both use modified galvanometer movements for suspending the sample and providing the restoring torque. In the latter instrument the upper end of the phosphor-bronze suspension strip 0.013 cm × 0.0013 cm (also providing one lead to the coil) passes through a Perspex insulating plug which fits tightly into the Duralumin collar. This collar can be rotated manually and also by means of a fine screw about a

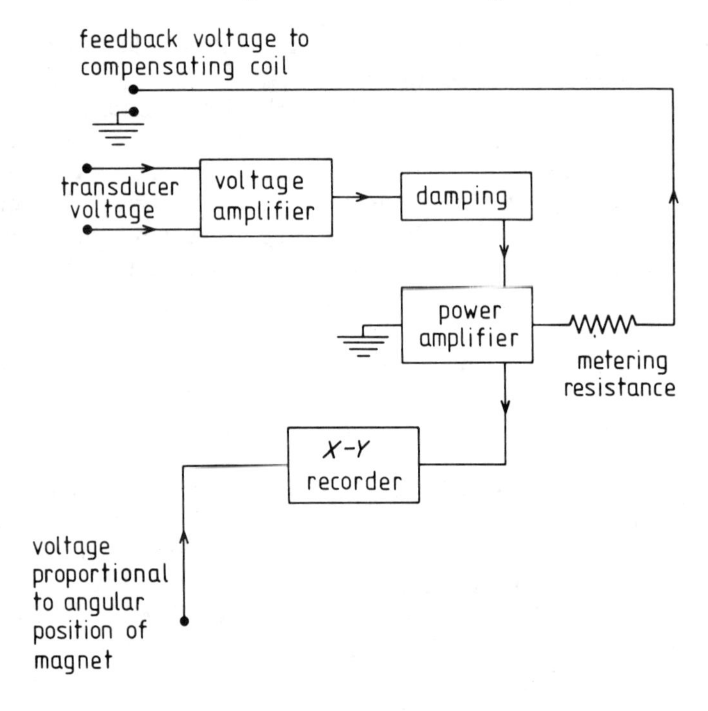

Fig. 3.18 Block diagram of torque magnetometer.

vertical axis for zero displacement. The second lead to the coil is a very fine copper wire which exerts no significant torque on the system. The coil consists of 100 cm^2 turns placed in the field ($\sim$50 mT) of the galvanometer permanent magnet. Below the coil and rigidly attached to it is a Perspex rod and below that a quartz rod terminating in the sample holder. The magnetometer was designed for measuring torque curves on single crystals and small ($\sim$0.1 g) rock samples at elevated temperatures and more details of this aspect of the instrument are given in Section 4.1.

Provision is made for direct recording of torque curves on an *X–Y* recorder, and also for digitized output of the torque readings at 5° intervals of the magnet rotation. The digitizer is based on a voltage-to-frequency converter described by de Sa (1965, 1966), and the torque curve is stored on punched tape for subsequent Fourier analysis.

Using a 17.5 cm magnet a maximum field of 1.2 T is available with the furnace in place (4.0 cm gap) and 2.1 T with the furnace removed and the gap reduced to $\sim$1.0 cm. The noise level of the feedback system is equivalent to a torque of 1.0×10^{-8} N m and the maximum restoring torque which can be applied is $\sim 1.0 \times 10^{-4}$ N m.

There are some difficulties associated with using a rotating magnetic field in torque magnetometers. A large and heavy electromagnet is necessary if a

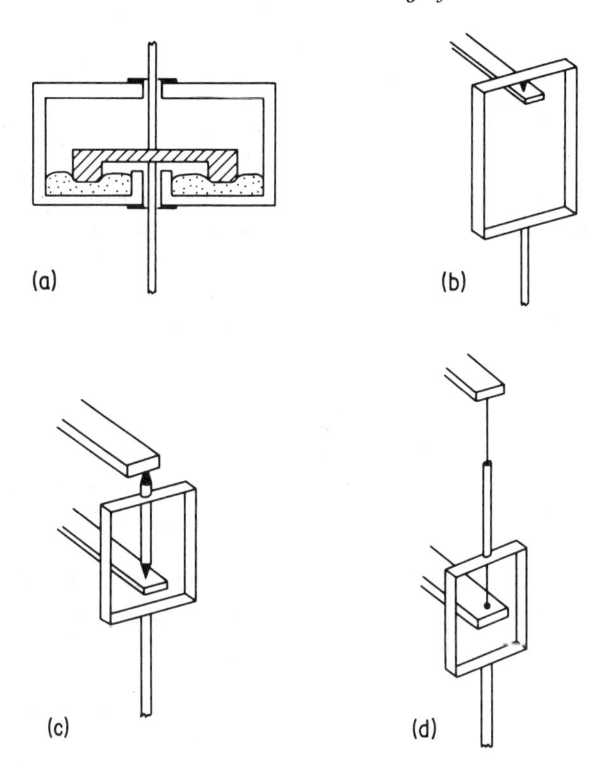

Fig. 3.19 Suspension systems for torque magnetometers: (a) mercury bearing; (b) and (c) single- and double-jewel bearing; (d) taut suspension.

strong (> 1.0 T) homogeneous magnetic field is to be obtained, requiring a substantial thrust bearing to support it and causing some problems in precise angular setting (if required) because of the large inertia of the system. It is also possible for the leakage field of the magnet to interact unfavourably with the feedback system of the magnetometer and with other equipment in its vicinity. These aspects of torque magnetometer operation led Wilson, Heron, de Sa and O'Reilly (1977) to design an instrument in which the magnet remains fixed and the torque head rotates. The latter consists of a coil supported by taut suspensions above and below (Fig. 3.19(d)), hanging in a magnetic field ($\sim$80 mT) provided by two permanent magnets attached to the inner wall of the vertical, tubular torque head housing. The lamp and split photocell are attached to the housing and rotate with it. The housing is mounted on a base-plate and rotates about a vertical axis on ball-bearings, using a synchronous or stepping motor. Analogue plots of torque curves can be obtained with an X–Y recorder, and there is also provision for on-line data processing using a microprocessor feeding into a teletype. The measurements range is approximately $5 \times 10^{-8} - 3 \times 10^{-4}$ N m, and sample masses up to 10 g may be used.

Other torque magnetometers designed for single-crystal and thin film measurements have been described by Penoyer (1959), Humphrey and Johnston (1963) and Abe and Chikazumi (1976).

Highmoor Electronics Ltd are currently marketing a torque head and associated electronic feedback system. The sample is suspended from a double-jewelled bearing system (Fig. 3.19(c)) and a sample mass of up to 30 g can be accommodated. Electrical connection to the feedback coil is by two very fine copper wires, and the controlling magnetic field is provided by a 'magnetron' magnet in which the gap field is 0.25 T. There is some advantage in using such a strong field: the leakage field from the main electromagnet is less likely to interact significantly with the feedback coil and only a small feedback current is required. The coil has a mean diameter of 1.5 cm and is wound with ~ 100 turns of 30 SWG wire, and the feedback current required to control the maximum usable torque ($\sim 10^{-4}$ N m) is ~ 4 m A. A capacity transducer is used to detect rotation of the head. It consists of a horizontal vane, attached to the suspended system, which moves between two pairs of fixed plates when the system rotates, the whole forming a capacitor in which the capacity depends on the position of the vane, i.e. the angular position of the suspended system. This variable capacity is used to vary the output of an amplifier, the output of which ultimately provides the feedback current. The useful range of measurable torques is 1.0×10^{-7}–3×10^{-4} N m.

Besides the suspension systems already described, other methods of supporting the sample have been used and may have advantages for particular instruments. In general, the single suspension suffers more from vibration-induced noise than other types and is less robust. Lateral motion of the sample is also possible if some field inhomogeneity is present. The dual, taut suspension system is robust (Fig. 3.19(d)) and allows the electrical connection to the coil without extra leads. Of the torsionless systems, which can, of course, only be used with a feedback system, the dual-jewelled bearing (Fig. 3.19(c)) is probably the most satisfactory but may be subject to some frictional forces. The magnitude of this force is dependent on the mass of the suspended system. The bearings can be modified from those used in electrical meters, and electrical connections are made by means of very thin wires or 'hair-springs'. The mercury bearing (Fig. 3.19(a)) is very friction-free if the surface of the mercury can be kept clean but in practice oxidation is difficult to prevent giving rise to variable surface effects. Aldenkamp, Marks and Zijlstra (1960) describe a torque transducer of high sensitivity which is used in a modified form to provide zero restoring force in a torque magnetometer used by Abe and Chikazumi (1976). Penoyer (1959) describes an air bearing, used in conjunction with a single-jewel upper support, to prevent transverse movement of the sample.

(*b*) *Torque magnetometer measurements*

Calibration of torque magnetometers is usually carried out by generating a known magnetic moment P_c in a small coil carrying a known current. The coil is placed in or symmetrically about the sample holder, with its plane parallel to the magnetic field B and the current is led in and out by very fine wires. In this position, the torque $T = BP_c = BNAi$, where i is the current in a coil of area-turns NA, but it is more satisfactory to rotate the magnet as for a torque curve measurement and record the maximum positive and negative torques observed for a given coil current. The linearity of the magnetometer response can be checked by using different currents and/or magnetic field. An alternative calibration method is to place a shaped sample of known saturation magnetization and demagnetization factors in the holder, and apply Equation 3.28. Penoyer (1959) used a nickel needle suspended perpendicularly to the field, for which $N_b - N_a = 0.50$. A less convenient method, also used by Penoyer, is to attach a torsion fibre of known torsional constant to the suspended system and twist the fibre through a measured angle, thus applying a known torque.

As with the vibration magnetometer magnetic contamination of the sample holder must be avoided, and it is preferable for it to be non-electrically conducting. The magnetic field must be reasonably homogeneous (maximum variation $< \sim 1\%$) over the volume of the samples used.

Interaction between the magnetic field and the feedback coil can be reduced if necessary by placing a mild steel base plate over the magnet, on which the torque and ancillary equipment can be mounted.

It is necessary to avoid anisotropy signals arising from the shape of samples, and to this end spherical samples, cylinders or discs, with a constant demagnetizing factor about one axis are ideal for measurements in one plane. Single crystals of natural or synthetic minerals can be formed into spheres by using a compressed air mill (Bond, 1951) or mechanical grinder (Lam, 1965). The degree of sphericity required can be estimated from Equation 3.28, using the approximation $N_b - N_a = 2(1 - b/a)/5$, valid when $N_b \simeq N_a$. Thus, assuming the non-sphericity takes the form of an ellipsoid with semi-major and semi-minor axes a and b, the corresponding maximum 2θ torque due to non-sphericity can be estimated and compared with the torque expected from other sources.

Porath, Stacey and Cheam (1966) have studied experimentally the optimum length/diameter ratio for cylindrical samples for high-field measurements, i.e. the value of l/d for which the difference between the demagnetizing factors parallel and perpendicular to the cylindrical axis is zero. Using iron and nickel cylinders they found a small dependence of optimum l/d with field strength, namely $l/d = 0.88(0)$ at $B = 1.2$ T, 0.89(3) at 1.7 T and, by extrapolation, 0.90(2) at infinite field strength.

The only modification required to a torque magnetometer if rotational hysteresis is to be measured is provision for rotating the magnetic field (or torque head) from 360° back to 0° after rotation from 0° to 360°. Since most applications of rotational hysteresis require its magnitude as a function of applied field strength it is also necessary to be able to vary the current in the electromagnet accordingly, but this facility will probably already be available.

If the required initial fields are less than the remanent field of the magnet (i.e. when the current through it is zero), typically 50 mT, lower fields can be obtained by passing the maximum current through the magnet in the opposite direction before measurement begins. This will leave a 'negative' remanent field in the gap, which can be brought to zero by a small reversed current, after which the whole range of 'positive' fields is available.

Solid or powdered samples can be used for rotational hysteresis measurements, and the limitations of sample shape in the plane of measurement are not so restrictive as for torque curve measurements. This is because the total W_R curve arises from the sum of the contributions of the contributing particles in the sample without being affected by the shape of the volume they occupy. It is necessary, as in the most rock-magnetic investigations, that in synthetic samples consisting of magnetic particles dispersed in a non-magnetic matrix the concentration is low enough to avoid magnetostatic interaction between particles. A typical range is 1–5 % by weight for titanomagnetite minerals, and somewhat higher, if required, for haematite and titanohaematites.

Since a complete range of W_R determinations in a sample involves two revolutions of the magnetic field for each of, say, 10–15 field values, the rate of rotation of the field is critical in determining the total measurement time. How fast this rotation can be is mainly governed by the response time of the feedback system, but may also depend on whether any vibrational noise arises with increasing rotation speed. Typical rates used are 1–3 min rev^{-1}, but the optimum rate for any instrument is a matter for experiment. Best results appear to be obtained if the sample is cycled through one revolution of the field prior to measurement, this being repeated for each change of field.

It is usual to plot rotational hysteresis on an X–Y plotter, from which W_R (in N m) can be obtained. The area enclosed by the curves is proportional to $\int_0^{2\pi} T(\theta)\mathrm{d}\theta \propto 2W_T$, and this area can be measured with a planimeter, or via the computer in computer-controlled equipment. The Y-axis calibration factor (torque cm^{-1} or torque m^{-1}) combined with the (fixed) length x corresponding to 360° can then be used to convert the area to the work done over two cycles of the field, and thus W_R for the sample determined. For comparison between samples it is convenient to normalize W_R to a fixed mass of magnetic material, if the amount in each sample is known, and plot W_R in N m g^{-1} against B.

3.3.4 Hysteresis measurements and loop tracing

The hysteresis curves and B–H loops associated with classical magnetic studies of ferromagnetics are not widely used in rock magnetism. In studies of hysteresis in rock samples the magnetizations normally encountered are so weak that the change in $B = \mu_0 H$ in air to $B = \mu_0 (H + M)$ in the presence of the sample is very small and a B–H loop of very small area results. It is more profitable to measure J or M, the induced magnetization per unit mass or unit volume, as a function of the applied field $B = \mu_0 H$, and this is the usual method of obtaining hysteresis data.

Hysteresis phenomena provide information on the domain structure of magnetic oxides and measurements at different temperatures and before and after heating can be used to monitor mineralogical changes in magnetic minerals.

The data may be obtained in the form of plots of induced magnetization J against applied field B, where B goes through the cycle $0 \rightarrow B_{max} \rightarrow 0 \rightarrow -B_{max} \rightarrow 0$. In general J will consist of induced and remanent components. For this purpose a vibrating sample magnetometer or magnetic balance can be used, and also a ballistic magnetometer of the type described by West and Dunlop (1971) (Section 9.5.3) although the usefulness of the latter instrument is somewhat limited by the low (coil-generated) maximum field of ~ 0.25 T. Some success has also been achieved in measuring B and H simultaneously by an a.c. method and displaying the B–H loop on an oscilloscope screen.

There are difficulties associated with measuring accurate hysteresis (or J–B) curves with a magnetic balance because of the way in which the magnetization is measured and also because of hysteresis in the electromagnet core. The latter feature implies that $B \neq 0$ when the magnet current is zero, and continuous monitoring of the magnet field is necessary for plotting curves from spot readings.

It can be seen from Equation 3.17 that $F_z \rightarrow 0$ as $dB/dz \rightarrow 0$. Therefore at or near the origin J becomes indeterminate and J_r cannot be directly measured.

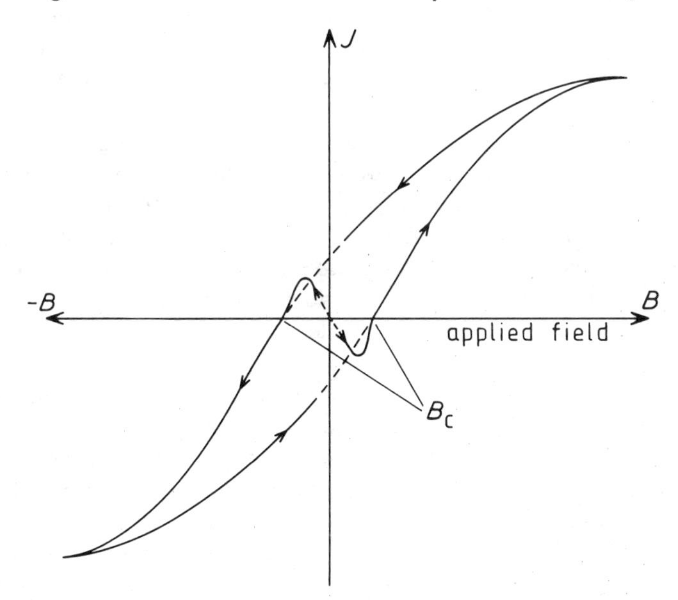

Fig. 3.20 Anomalous effect obtained near the origin when measuring J–B curves with a magnetic balance. Dotted lines show indeterminate region.

As B and dB/dz increase in the negative direction the effective magnetic moment of the sample is the vector sum of J_i induced in $-B$ and J_r along $+B$ and initially a negative signal will be obtained, corresponding to a positive value of J on the J–B curves. This signal becomes zero again when $J_i = J_r$ and the corresponding value of $-B$ is B_c, the coercive force measured by the balance (Fig. 3.20). As $-B$ is increased further the negative part of the curve is then measured. Since in many rocks the saturation remanence is very much less than the saturation magnetization, J_r and the J–B curve near the origin may be approximately obtained by interpolation through the anomalous region. J_r can be accurately determined with a remanence magnetometer.

With the vibration magnetometer the measurement of J is independent of the applied field, and complete J–B curves can be determined and plotted directly on an X–Y recorder, using a signal proportional to B for the X-axis.

It has always been an attractive idea to measure J and B simultaneously and display the results in the form of a J–B loop on an oscilloscope screen. The basic circuit consists of a coil or electromagnet providing an alternating exciting field B and a compensated pick-up coil placed in B and containing the sample. The output of the pick-up coil is proportional to dJ/dt in the sample, from which J can be obtained by electronic or numerical integration. With the former simultaneous signals proportional to instantaneous values of J and B can be displayed on the Y and X axes on an oscilloscope screen and the J–B loop displayed.

The main difficulties with the technique are associated with the strength of the applied field and in obtaining good sensitivity and signal/noise ratio. One of the earliest instruments designed for rock measurement was that of Bruckshaw and Rao (1950), and for simplicity the technique is analysed in terms of H, M and k. A coil former was wound with five coils, two of which, placed symmetrically about the centre, were used to provide the applied alternating field H and the other three, a central one and one outside each field coil, formed a compensated pick-up coil system to sense the induced magnetization in the sample. Insertion of a sample into the coil results in the instantaneous field B over the sample volume changing from $\mu_0 H$ to $\mu_0(H+M) = \mu_0 H + \mu_0 kH$, where M and k are the volume magnetization and incremental susceptibility of the sample. H is time-varying and the pick-up coils were adjusted for zero output in the absence of the sample. Thus, with the sample in place a voltage output appears in the coil proportional to the rate of change of kH, i.e. $k(dH/dt)$. After amplification the voltage across a resistance in the field coil circuit and the voltage from the pick-up coils are applied respectively to the X and Y planes of an oscilloscope. Assuming the exciting field to be of the form $H = H_0 \sin \omega t$, the instantaneous output signal is proportional to $kH_0 \cos \omega t$ and with constant susceptibility the displayed pattern is an ellipse. The analysis of the oscilloscope trace is illustrated in Fig. 3.21. Substracting the Y-signal without the sample from those with it in (Fig. 3.21(a)) gives values proportional to $kH_0 \cos \omega t$ from different values of

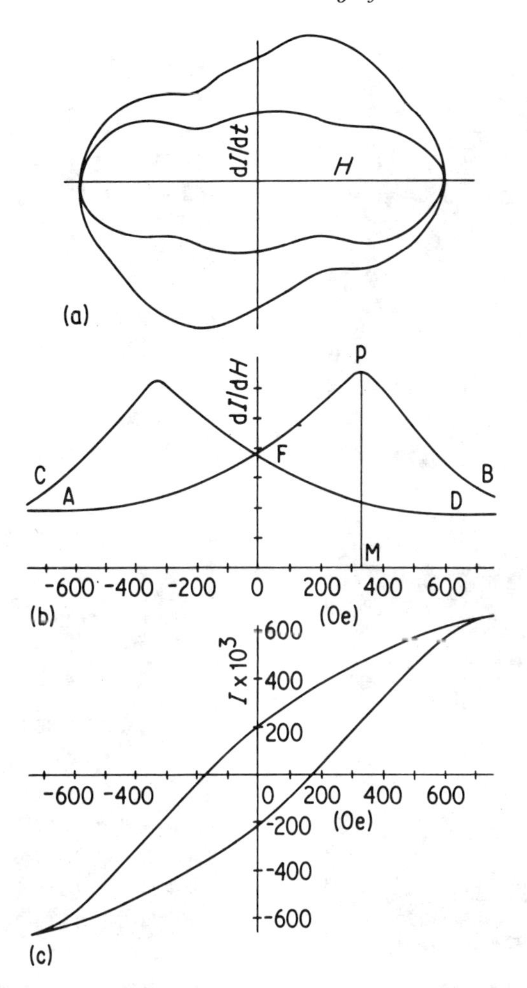

Fig. 3.21 Analysis of data obtained by the technique of Bruckshaw and Rao (1950): (a) output from pick-up coil with and without sample in place; (b) plot of quantity proportional to k against H (segments AB and DC are derived from the upper and lower parts of (a) respectively); (c) derived M–H loop. (From Bruckshaw and Rao, 1950, copyright of the Institute of Physics)

H. Since $\omega t = \sin^{-1}(H/H_0)$, these values of $kH_0 \cos \omega t$ could be converted to quantities proportional to k (Fig. 3.21(b)), which are then plotted against H. Since $k = \mathrm{d}M/\mathrm{d}H$, numerical integration of the curves of k against H provides values of M, and the corresponding M against H loop can be traced (Fig. 3.21(c)). Although the maximum applied peak field was only 75 mT (at 530 Hz) the equipment was used successfully to provide hysteresis data for some igneous rocks.

To overcome the limitation in the strength of the exciting field inherent in

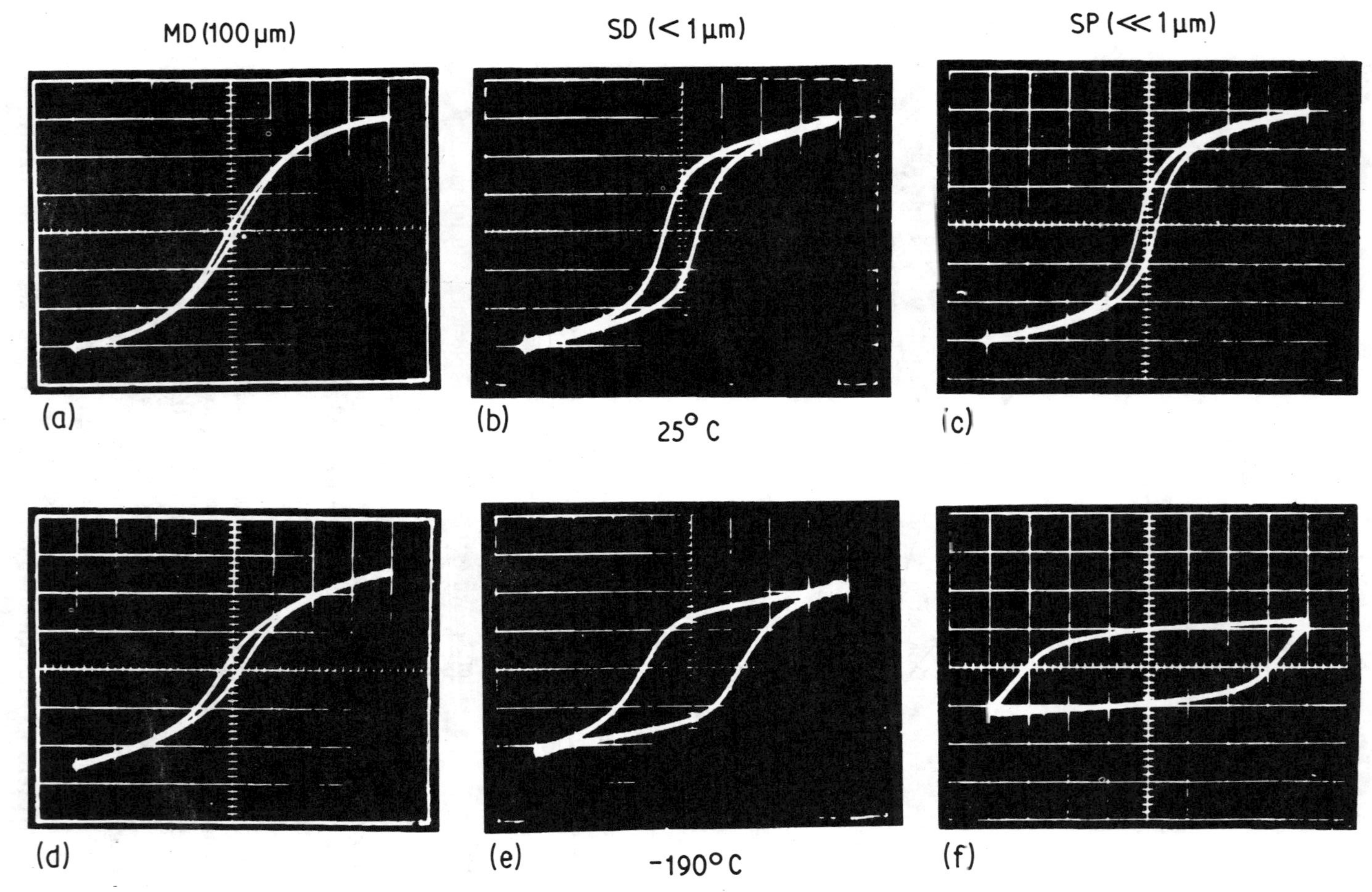

Fig. 3.22 Hysteresis loops obtained from basalt samples containing multidomain (MD), single domain (SD) and superparamagnetic (SP)

using coils, Deutsch (1956a) constructed equipment using an electromagnet. To avoid eddy currents in the magnet core it was made of high permeability transformer laminations, and provided a peak field of 0.24 T at 50 Hz in a 3.0 cm gap. The pick-up coil system consisted of a coil wound on one of the magnet pole-pieces close to the gap in series opposition with a second adjustable coil just outside and close to the gap, which was used to back off the signal due to $\mathrm{d}H/\mathrm{d}t$ only. The signals were displayed on an oscilloscope screen and analysed in the same way as described for Bruckshaw and Rao's instrument.

Provided that the magnet gap is reasonably large, the field in the gap is proportional to the current in the magnet coils. However, Stacey and Parry (1957) pointed out that it is assumed in Deutsch's instrument that there is no phase lag between the magnetic field in the magnet gap and the magnetizing current, from which the H-signal in the trace is obtained. Such a phase lag is likely to occur, and can lead to significant errors, for instance in determinations of coercive force. Stacey recommends the use of a search coil in the magnet gap, with electronic integration of the output voltage to provide a signal proportional to and in phase with the magnetizing field.

An improved loop tracer is described by Likhite, Radhakrishnamurty and Sahasrabudhe (1965) in which the most significant advance is electronic integration of the output signal, enabling hysteresis loops to be displayed directly on an oscilloscope screen. An electromagnet provides a maximum peak field of 0.36 T in a 1.2 cm gap of 5.0 × 2.5 cm cross-section. An L–R network corrects the phase lag between magnetic field and magnetizing current. A four-coil compensated pick-up coil, with the coils wound on top of each other, contains the cylindrical rock sample (2.5 cm in diameter) and slides into the pole-gap for a measurement.

The signal/noise ratio of the instrument is limited by the effect of uncompensated harmonics in the magnetizing field. In an applied peak field of 0.24 T, the signal/noise ratio is ~ 10 for a sample with a total moment of 1.0×10^{-3} A m^2, i.e. a specific magnetization of $\sim 6 \times 10^{-2}$ A m^2 kg^{-1} in a sample 1.2 cm high. Some examples of loops obtained with the equipment are shown in Fig. 3.22.

Investigations of hysteresis properties of rocks and minerals using loop tracing equipment are described by Deutsch (1956b), Likhite and Radhakrishnamurty (1966), Radhakrishnamurty and Sahasrabudhe (1967) and Radhakrishnamurty and Deutsch (1974). Measurements of hysteresis at high temperatures is described in the next chapter.

Chapter Four

High- and low-temperature measurements

These measurements are generally concerned with identification of magnetic minerals by their Curie temperatures or by other temperature-dependent transitions involving magnetic properties, and they are also an aid to determining the physical and magnetic state of magnetic minerals in rocks. The magnetism of paramagnetic, superparamagnetic and ferromagnetic minerals has distinctive thermal characteristics depending, for instance, on grain size, and in ferromagnetics on the existence of single and multidomain grains.

The types of measurement involved are variation with temperature of induced magnetization J_i, remanent magnetization J_r and initial susceptibility χ, acquisition of TRM and PTRM and the detection of distinctive low-temperature transitions that occur in certain minerals.

The highest temperature required is usually governed by the Curie or Néel temperatures of magnetic minerals, and is therefore not very demanding. Metallic iron has a Curie point of 780°C, and iron oxides and the titanium–iron minerals and pyrrhotite all have lower Curie temperatures. For low-temperature work, liquid nitrogen, at -196°C, is below the Morin transition temperature in haematite (~ -20°C) and the magnetic transition in magnetite (~ -140°C), and below the Curie temperatures of titanomagnetites and titanohaematites possessing high contents of ulvöspinel and ilmenite respectively. It is only necessary to go to liquid helium temperature (-269°C) for specialist investigations of basic phenomena in rock and mineral magnetization.

4.1 High-temperature measurements

4.1.1 Equipment

The equipment used for measurement of the variation of NRM (and IRM) with temperature and for experiments on TRM and PTRM is described in the sections on thermal demagnetization and magnetometers and will not be described further. Most of the instruments used for investigations of the variation of induced magnetization with temperature (J_i–T curves) are essentially those already described for applied field investigations in the previous chapter, modified where necessary for operation with the sample at elevated temperatures.

There are three potential problems associated with measurement of magnetic properties at high temperatures. They are:

(a) Increased noise of thermal and electrical origin arising from the furnace
(b) Decrease in sensitivity and/or maximum applied field, because of the increased sample–sensor distance or reduced sample size and increased magnet gap-width necessary to accommodate the furnace
(c) The desirability of heating the samples or a controlled atmosphere to inhibit mineral alteration at high temperature.

As far as (a) is concerned, electrical noise is reduced by using non-inductive windings on furnaces, as described in Section 11.2, and by the choice of either alternating or direct current, depending on whether the sensing device is sensitive to alternating or direct magnetic fields. Noise of thermal origin is reduced by suitable insulation round the furnace and/or water-cooled shields.

It is difficult to avoid loss of sensitivity when using a furnace and the loss can only be minimized by close attention to the design of the furnace and its insulation. If a vacuum or controlled atmosphere is used it is usually possible to place the sensor in the vacuum or inert gas, thus avoiding further loss of sensitivity. Alternatively, if a vacuum is required, it may be satisfactory to place the sample in an evacuated capsule, thus eliminating the need for a vacuum chamber around the instrument. In general, heating experiments are done on small samples to reduce thermal hysteresis, i.e. lag between sample temperature and thermocouple temperature, and to reduce thermal gradients in the sample.

4.1.2 Initial susceptibility

Variation of initial susceptibility with temperature can in principle be measured with any one of the types of instrument described in Section 2.6, after suitable modification. However, the most satisfactory method is probably that described by Stephenson and de Sa (1970). This instrument (Section 2.6.2) measures the change in inductance of a coil when the sample is inserted, and for high-temperature measurements a tubular non-inductively-wound furnace is placed in the coil with a smaller sample ($\sim$1.0 cm long and 0.4 cm in diameter) fitting inside an axial quartz tube inside the furnace. The furnace is thermally insulated from the coil by means of a tubular water-jacket. The sample may be heated in vacuum or controlled atmosphere or in a sealed capsule.

Temperature measurement is by a Pt/13% Pt–Rh thermocouple in contact with the sample. The noise level at the instrument output is equivalent

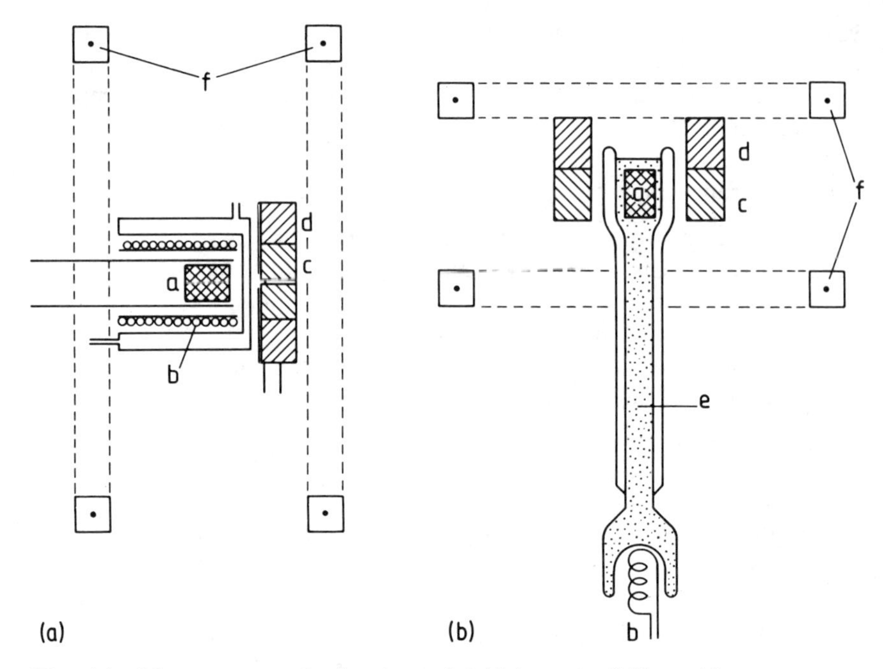

Fig. 4.1 Measurement of vibration of initial susceptibility with temperature. a, sample; b, heater; c, pick-up coil; d, compensating coil; e, heated liquid; f, field nulling coils. (After Radhakrishnamurthy and Likhite, 1970)

to a total susceptibility of about $1.0 \times 10^{-11}\,m^3$, i.e. approximately $4.0 \times 10^{-8}\,m^3\,kg^{-1}$ in a sample of the above size.

Radhakrishnamurty and Likhite (1970) use a modified version of their balanced transformer susceptibility meter (Section 2.6.2.2). The sample is surrounded by a d.c. platinum heating coil and water jacket, with the pick-up and compensating coils close to the sample at one end and separated from it by the water jacket (Fig. 4.1 (a)). In an alternative version for moderate temperatures (20–200°C) the sample is immersed in a suitable liquid in a vertical double-walled container heated at the base. The detector coils surround the sample outside the tube (Fig. 4.1(b)). The minimum detectable signal in each instrument corresponds to a susceptibility of $1.0 \times 10^{-7}\,m^3\,kg^{-1}$ in a standard 20 g sample.

4.1.3 Curie points and J_i–T curves

Curie point measurements fall into two main classes, those in which Curie points are determined from the decay of induced magnetization with temperature, and those in which a transition in some other magnetic or physical property, which also occurs at the Curie temperature, is observed.

Thermal decay curves of induced magnetization can be measured with magnetic balances, vibration magnetometers and ballistic magnetometers of the type described by West and Dunlop (1971). In the vertical balance used by the author (Section 3.3.1), the sample and suspension are enclosed in a quartz tube (1.5 cm in diameter) secured at its upper end to the underside of the torsion head baseplate by an O-ring seal. The lower end of the tube is closed by another O-ring seal through which leads are taken for the thermocouple, placed just below the sample. A small stainless steel container, the same size as the sample bucket, fits over the thermocouple tip and is filled with the sample material. This ensures a closer agreement between sample and thermocouple temperature. A non-inductively-wound furnace is wound directly on to the quartz tube (Fig. 3.10).

The furnace is 20 cm long and is wound with 20 turns of 27 SWG Nichrome wire ($R = 36\,\Omega$) sealed in place with fireclay cement to a total diameter of 2.3 cm. 700°C is achieved with a current of 2.6 A. Water-cooled copper plates fit over the magnet pole-pieces to keep them cool. It is not strictly necessary to wind the furnace non-inductively from the point of view of the magnetic field generated, but such a winding reduces vibration arising from the interaction between the alternating current (if used) and the applied field. An aluminium cover encloses the torsion head and rests on an O-ring seal let into a groove cut in the baseplate, thus completing an air-tight chamber for heating in vacuum or inert gas. Alternatively, an evacuated quartz capsule can be suspended in place of the sample bucket. Powdered samples or rock chips are contained in a quartz bucket suspended from a quartz fibre 0.5–1.0 mm in diameter, the upper end of which is outside the furnace and is attached by a hook to the fine copper wire suspension attached to the torsion arm. In the vertical balance described by Schwarz (1968b) the furnace is wound on a separate quartz tube and can be raised out of the magnet gap for low-temperature measurements, when a specially designed liquid nitrogen Dewar surrounds the sample. Curves of J_{i} (or a quantity proportional to J_{i}) against T are recorded directly on an X–Y recorder. This is a quick way of obtaining the data, but it is necessary to ensure that the signal is free of a contribution arising from zero drift and/or change of weight of the sample, e.g. loss of water or oxidation. This can be checked by a preliminary test, or by reducing the applied field to zero at intervals and checking the zero level. It may be advantageous to pre-heat some samples to ~100°C, to drive off water.

Because of the suspension arrangement in the horizontal (translation) balance it is difficult to enclose it in a vacuum chamber and, where necessary, samples are measured in evacuated capsules. The furnace and thermocouple used are essentially as for vertical balances. The advantage of the translation balance for heating experiments is that change of weight of the sample produces no deflection of the system. A simple balance for Curie point measurements described by Domen (1977a) employs a combination spiral and flat quartz spring to measure the force acting on the sample. The extension of the spring is monitored by a differential transformer, the output of which is passed to the *Y* axis of an *X–Y* recorder. The thermocouple is in contact with the sample and the leads taken up through the upper part of the sample holder and out to the *X* axis of the recorder with very fine connecting wires. The balance can be used with a conventional electromagnet to provide the field and field gradient and the use of a 'solenoidal' electromagnet is also described. This consists of a solenoid wound on a tubular soft-iron core with its axis vertical. The sample is thus magnetized in a vertical field, and at any point on the solenoid axis away from the centre only a vertical force will act on the sample thus avoiding the off-axis radial (outward) force acting in the normal electromagnet configuration. However, the maximum field and gradient available with the solenoid magnet falls short of those available in the conventional arrangement.

For high-temperature measurements with the vibrating sample magnetometer the furnace and vacuum arrangement is similar to that for the vertical balance. The pick-up coils are, of course, outside the furnace and therefore it is desirable to keep the diameter of the latter and its associated thermal insulation as small as possible if good sensitivity is to be maintained. In the instrument in the author's laboratory (Tucker, 1978), the vacuum arrangement is similar to that used for the vertical balance. A cylindrical aluminium chamber seated on an O-ring covers the vibrator mechanism, the drive shaft of which enters through a rotary vacuum seal. The vertical drive shaft and sample are enclosed in a quartz tube closed at the lower end and attached at the upper end to a brass flange which is bolted to the underside of the base plate carrying the vibrator unit via an O-ring seal. The furnace is 15 cm long and is wound non-inductively in grooves cut into a quartz tube of external diameter 2.3 cm and wall thickness 1.5 mm which fits closely over the vertical quartz vacuum tube. The windings are covered with Refrasil silica tape and the whole enclosed in a copper tube to which are soldered several longitudinal copper pipes connecting with reservoirs at top and bottom, through which cooling water is circulated. The diameter of the complete unit is 3.2 cm. To promote uniformity of temperature over the vibration amplitude range ($\sim$1.5 cm), the linear winding density is increased at each end of the furnace. The thermocouple leads are taken into the vacuum system via air-tight contacts and down the drive shaft, and the hot junction is in contact with the top of the platinum or copper sample holder. In spite of the leads being closely twisted together, a small alternating voltage corresponding to a temperature uncertainty of ± 2°C appears on the thermocouple voltage, arising from vibration of the leads in the applied magnetic field. A low pass filter is used across the DVM (digital voltmeter) with which the temperature is read which reduces the uncertainty to $< \sim 0.2$°C.

The Highmoor VSM is supplied with a tubular furnace which fits inside the standard pick-up coil configuration. A water jacket insulates the coils from the furnace.

The ballistic magnetometer used by Dunlop and West (Section 9.5.3) is fitted with a furnace consisting of a stainless steel tube carrying axial alternating current. A thin (0.7 cm) layer of magnesium oxide and an outer Pyrex water jacket provide the insulation. This type of furnace element promotes good temperature homogeneity over the measuring volume, but is of low electrical resistance and therefore a high current (~300 A at 3.6 V) is required to achieve 750°C. The ballistic instrument used by Schwarz and Whillans (1973) is fitted with a conventional tubular furnace thermally insulated from the pick-up coils by means of an asbestos layer and a water jacket.

An alternating method of determining Curie temperatures depends on the sudden decrease in magnetic permeability at the Curie point. One instrument of this type has already been described, in which the temperature variation of initial susceptibility is measured using a low-frequency inductance bridge. Another approach is to use the high frequency resistance of a coil to indicate the permeability change. If the sample is placed in the coil, the h.f. resistance will depend on the sample permeability relatively strongly, and it therefore offers a means of Curie point detection.

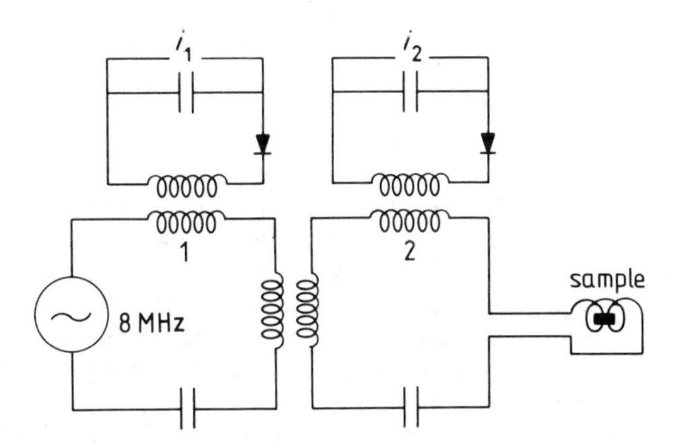

Fig. 4.2 Basic circuit of Petersen's (1967) method.

The basic circuit used by Petersen (1967) is shown in Fig. 4.2. The sensing coil containing the sample is wound with silver wire and both coil and sample are heated in the furnace. The two circuits are in resonance with the 8 MHz generator and if the current i_2 in circuit 2 is kept constant by varying the generator output, the current i_1 is proportional to the h.f. resistance of the sense coil. Although the method lacks sensitivity, it is potentially useful in being better able to distinguish two or more Curie points lying close together than a J_i–T measurement. Markert, Trissl and Zimmerman (1974) have further developed the principle in equipment for the measurement of susceptibility, Curie points and hysteresis losses.

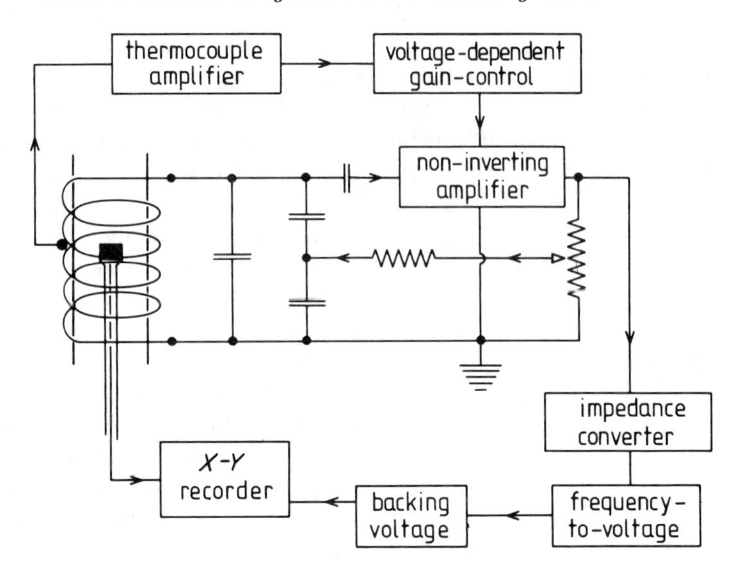

Fig. 4.3 Block diagram of marginal oscillator circuit. The high Q-value is obtained by a positive feedback circuit in parallel with the tuned circuit, of which the sense coil is a part. (Redrawn from de Sa, 1968, copyright of the Institute of Physics)

De Sa (1968) utilizes the change of sample permeability, and of the inductance of a coil containing the sample, to vary the resonant frequency of a 5 MHz high-Q marginal oscillator circuit (Fig. 4.3). To avoid spurious signals due to temperature variations in the coil during heating the variations are sensed by a thermocouple in contact with the coil windings and the thermocouple output used to control the gain of an amplifier in the output of the tuned circuit. A frequency-to-voltage converter is used to convert the output frequency variation to a direct voltage which, with the voltage from the sample thermocouple, is used to drive an $X-Y$ recorder.

The magnetic transition at the Curie point is accompanied by thermodynamic effects and Vollstädt (1968) has investigated these changes by means of differential thermal analysis (DTA). In this technique, thermal changes are monitored through the absorption or emission of heat (endothermic or exothermic reactions) during the heating of the sample. Although a well-defined endothermic peak is obtained at the Curie point of magnetite and a less well-defined peak with a haematite sample, results obtained with titanomagnetites are ambiguous (Fig. 4.4). The technique does not appear to have been developed further and is probably of limited application in rock magnetism.

Deutsch, Kristjansson and May (1971) describe a novel thermomagnetic balance suitable for igneous or strongly magnetic sediments. The sample, surrounded by a tubular furnace, is held in a quartz bucket at the upper end of

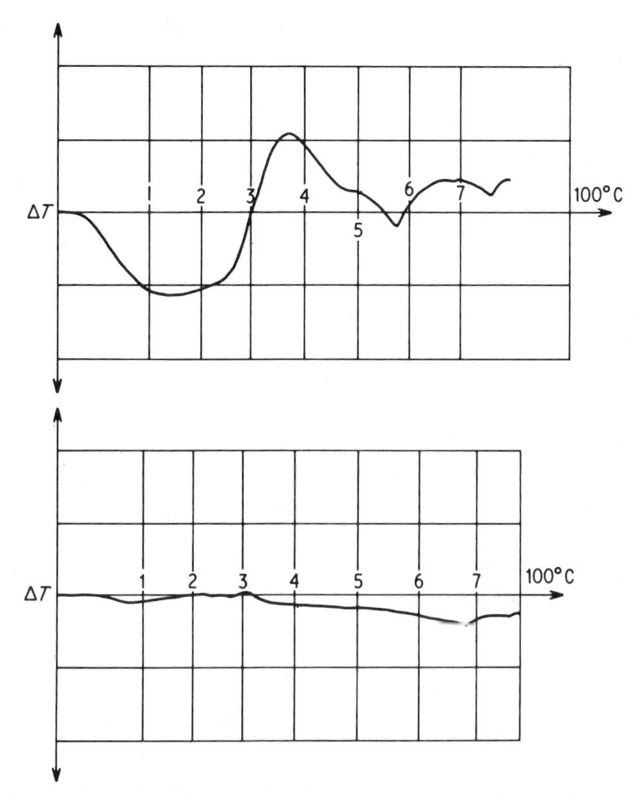

Fig. 4.4 Measurement of Curie points by differential thermal analysis. Upper trace, magnetite; lower trace, haematite. (From Vollstadt, 1968)

an oil-damped vertical pendulum of long period. The force exerted on the sample in an inhomogeneous magnetic field deflects the pendulum from the vertical, and the vertical position is restored by passing a current through a coil adjacent to a pair of astatic magnets attached to the lower end of the pendulum.

4.1.4 General comments

The problem of mineral alteration during heating, and of the formation of new magnetic minerals, is sometimes a severe one in thermal studies. Igneous rocks are usually the most susceptible to change, through oxidation, and it is not always possible to inhibit alteration by heating in either a vacuum or inert atmosphere. Another possible change is in the exsolution structure of different mineral phases. It may be possible to limit alteration by rapid heating: this usually necessitates the use of a small sample (< 1 g) if thermal hysteresis is to be avoided, and such samples, in the form of rock chips, are also less likely to break up through thermal stresses generated by rapid heating. The latter is not a problem where powdered samples are used but powdered material is probably more prone to oxidation than a solid sample.

It is usually possible to detect any changes in magnetic minerals, for instance during measurement of $J_i - T$ curves. Mineral alteration will be reflected in a difference between curves obtained during heating and cooling, and if repeat readings over, say, a 5-minute interval are taken at a constant temperature during a run, these readings will show a difference if alteration is occurring. The nature of the change may well be indicated by one or more different Curie points appearing in the cooling curve. Whether such Curie points occur in the heating curve depends on whether the new or altered mineral is formed at a temperature above or below its Curie point.

Since haematite is at the end of the oxidation chain, oxidation is not a problem in heating these rocks although physical alteration may occur, e.g. a change in grain size of the pigment form of haematite in red sediments. However, the possible breakdown of clay minerals in sediments and the formation of magnetite has already been mentioned in connection with thermal demagnetization, and whether this is seen in $J_i - T$ curves depends on the temperature of formation and whether or how fast the newly formed magnetite is oxidized to haematite.

The shape of J_i–T curves can also be informative. In igneous rocks, apart from discontinuities which may indicate and determine intermediate Curie points, the overall shape is usually indicative of the presence of one or more magnetic minerals. In a one-mineral system, assuming J_i is the saturation or near-saturation magnetization, an initially slow decrease with temperature is expected, with an increasingly rapid decrease as the Curie is approached. A more linear decrease of J_i with temperature or an 'exponential' decrease suggests the presence of other phases with progressively lower Curie temperatures.

In igneous rocks, the contribution to J_i of paramagnetic minerals can usually be neglected, but this may not be true of sediments of low iron oxide content, and J_i–T curves may be modified accordingly. J_i–T curves obtained from red sediments show variations between the 'standard' haematite-dominant form and those observed in rocks of very low magnetic iron oxide content, when the approximate dependence of J_i on the reciprocal of the absolute temperature, arising for instance from paramagnetism in clay minerals, is seen (Fig. 13.7). If the clay mineral content is low and quartz and/or carbonates are the dominant non-magnetic minerals the diamagnetism of the latter may be seen, i.e. a very small dependence of J_i on T. When measuring J_i in these weakly magnetic materials it is important to carry out a blank run with the empty sample holder over the temperature range to be used. With a quartz holder, ferromagnetic and paramagnetic contamination and the diamagnetism of the quartz can produce signals which vary with temperature in a complicated way.

The determination of accurate Curie points from J_i–T curves obtained from rock samples is not always straightforward. Because of the existence of paramagnetism above the Curie point, the presence of paramagnetic minerals in the rock and possibly the holder signal, the observed J_i does not fall to zero at the Curie temperature of the dominant mineral and there is often no sharp discontinuity there. A common procedure is to designate the Curie temperature as the intercept on the temperature axis of the tangent to the J_i–T curve at its steepest point.

The requirement often arises for a compact sample in which a small percentage by weight of a natural or synthetic mineral is dispersed in a non-magnetic matrix and which can be heated up to ~700°C. Among the matrix materials that can be used are potassium bromide (KBr), calcium fluoride (CaF_2) and aluminium oxide (Al_2O_3). After mixing the mineral powder with one of these materials, the sample is compacted into the required size and shape, usually cylindrical, by means of a screw press. Potassium bromide has the advantage of allowing the mineral to be easily recovered by solution in water. Aluminium oxide does not bind together very well under pressure unless very fine-grained. Silica gel was used by Hargraves and Young (1969), forming a consolidated sample of extracted mineral grains which had acquired a depositional magnetization while settling in the liquid gel. The gel solidifies on dehydration to form a solid, if somewhat fragile, sample which can be heated.

4.2 Low-temperature measurements

4.2.1 General

In comparison with investigations at high temperature, those at low temperature – at least down to liquid nitrogen temperatures – generally involve less interference with measuring instruments and are non-destructive, i.e. there is unlikely to be any mineral alteration.

Liquid nitrogen (−196°C) is a convenient coolant for investigations of magnetic transitions, and can be contained in rather simple Dewars or insulated containers which do not require much increase in the sample–sensor distance. A common procedure for some investigations is to cool the sample to −196°C and monitor magnetization changes as it warms to room temperature: in others it is necessary to monitor changes during cooling and warming.

In some rather specialized rock-magnetic studies, e.g. magnetic granulometry, a lower temperature is desirable for which liquid hydrogen (−253°C) or liquid helium (−269°C) are used. In view of the dangers and restrictions involved in using and transporting liquid hydrogen, helium is generally favoured and it requires a more complicated containment and handling system if excessive evaporation is to be avoided.

Copper–constantan is a suitable thermocouple for low-temperature use.

Resistance thermometers may also be used, suitable materials being platinum (which possesses a high temperature coefficient), iron, and (for $T < \sim 20$ K) carbon. Suitable materials for the sample holder and other parts at low temperature are Perspex (Lucite), glass, Pyrex, aluminium, copper and stainless steel. With a component of Perspex or glass there may be problems with cracking if it is initially at room temperature and is brought into direct contact with liquid nitrogen.

4.2.2 Low-temperature variation of NRM

The measurement of weak NRM ($< 10^{-5}$ A m^2 kg^{-1}) in a low-temperature environment (usually between liquid nitrogen temperature, -196°C, and room temperature) is not entirely straightforward. One of the most convenient methods employs a sensitive astatic magnetometer using measurement position (a) in Fig. 9.9, although such an instrument is not now commonly available. For more strongly magnetized material, an astatic fluxgate system might be useful. The author has used a cylindrical, double-walled Pyrex container 7.5 cm in diameter with a layer of expanded polystyrene ~ 1 cm thick between the walls and on the base. The sample is located at the centre of the central chamber by packing cotton wool round it, and it is advisable to surround the sample with a thin Perspex sheath, to prevent too-rapid cooling and possible break up by thermal shock. A Perspex (Lucite) lid with ventilation holes is placed on the container after the container has been filled with liquid nitrogen and the sample and container are cooled to -196°C. This retards heat loss (or gain) and helps to prevent cold air circulating around the magnetometer housing: this may be important in reducing zero drift of the magnetometer due, for instance, to Thomson currents circulating in a copper damping plate across which there are temperature differences.

Sample size is a compromise between the requirement of as large a total moment as possible for good signal/noise ratio and the need to avoid undue temperature gradients in the sample, which is easier to achieve in a small sample. The conventional 2.5 cm cylinder is probably too large for the latter requirement and ~ 1.0 cm cylinder is more satisfactory. If possible, the thermocouple is best placed in a hole drilled to the centre of the sample.

Even with a 1 cm cylinder there will be a significant temperature gradient across the sample during cooling or warming, resulting in a loss of definition of any transitional NRM change. It may not be feasible to record changes during cooling, but a decrease of NRM intensity between room temperature and a temperature lower than that of the transition being sought is probably indicative of the transition and, since the phenomenon is usually at least partially reversible, the intensity will increase again during warming.

The sample container can be made to fit on to the column which supports the normal sample holder for the astatic magnetometer, and can then be lowered from the magnet system in order to take zero readings. If the off-centre method of measurement is available a complete vector measurement of low-temperature NRM can be made. This is not normally necessary for the detection of magnetic transitions but may be an advantage for some investigations, for example low-temperature demagnetization.

For obvious reasons spinner magnetometers are not well suited for low temperature investigations without considerable modification. At the time of writing a low-temperature head is available for the Digico instrument. An enlarged ring fluxgate encloses a Dewar container and the sample (a standard 2.5 cm cylinder) is rotated at the lower end of a vertical shaft inside a tube which isolates the rock from the liquid nitrogen. The motor drive pulley, timing discs and slip-ring system for the thermocouple are at the upper end of the shaft. Only single-axis measurements are possible, and there is some loss of sensitivity (by a factor of 5–10) over the normal NRM magnetometer because of the reduced coupling between sample and sensor.

Measurements of the variation of NRM at low temperatures can also be carried out on the large access SCT (Superconducting Technology, Inc.) cryogenic magnetometer and, with small samples and careful Dewar design, with the 3.5 cm access CCL (Cryogenic Consultants Ltd) instrument. The Dewar providing the low-temperature enclosure can either be placed independently at the measuring position, or it can be attached with the sample to the insertion device. The total NRM vector is measured with a three-axis instrument, and if necessary a device such as that described for high-temperature measurements with the Digico magnetometer (Section 11.2.4) could be used for a total NRM determination with a two- or one-axis instrument.

For isothermal remanence and strong NRM measurements a fluxgate magnetometer can be used. Morris (1971) placed the sensor outside and close to one arm of a vertical U-tube containing liquid nitrogen. The sample is inside the tube at the same level as the sensor and is rotated at ~ 0.1 Hz by means of a vertical shaft. When the sample is at $-196°C$ the nitrogen is evaporated and changes in horizontal magnetization on warming monitored by the fluxgate. The temperature is recorded with a thermocouple at the centre of a dummy sample in the other arm of the U-tube. The minimum detectable magnetization in a 2.5 cm sample is $\sim 3 \times 10^{-5}$ A m^2 kg^{-1}.

4.2.3 Other measurements

The most convenient configuration for a low-temperature enclosure is a suitably insulated container with vertical access, and this is suitable for most instruments for rock-magnetic investigations including the VSM, torque magnetometer, susceptibility bridge and vertical magnetic balance. The horizontal balance is not, in general, suitable for low-temperature measurements.

Low-temperature variation of initial susceptibility has been measured in

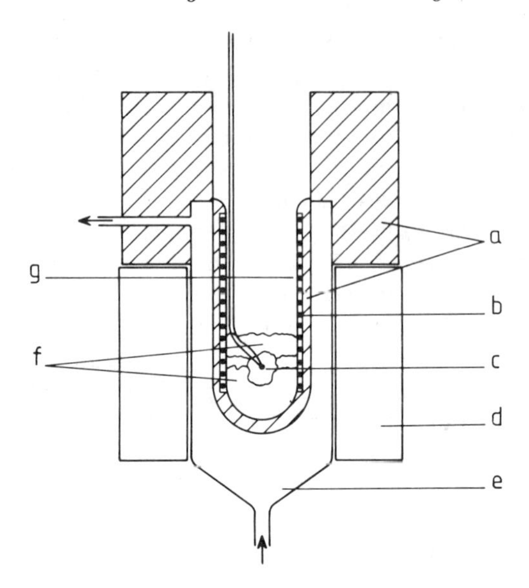

Fig. 4.5 Low temperature assembly for susceptibility bridge. a, polystyrene insulation; b, heating coil; c, sample and thermocouple; d, sense coil; e, water jacket; f, cotton-wool plugs; g, inner glass tube. (Redrawn from Stephenson and de Sa, 1970, copyright of the Institute of Physics)

small samples on the bridge of Stephenson and de Sa (1970) and essentially the same system (shown in Fig. 4.5) is available with the Highmoor susceptibility bridge. The water jacket prevents temperature variations in the pick-up coil, and the sample, inside a Plasticine sphere, is cooled by pouring liquid nitrogen into the inner tube. The cooling rate depends on the size of the Plasticine sphere. When $-196°C$ is reached the nitrogen is allowed to evaporate and warming begins. At $\sim -80°C$, when the natural warming rate becomes slow, it can be speeded up by passing a small current through the heating coil. The complete cooling and warming cycle takes 20–30 minutes using a small ($\sim$10–20 mg) sample.

Some general design information for controlled low-temperature enclosures is given by Creer *et al.* (1967). A system suitable for vertical magnetic balances and vibration consists of a vertical double-walled tube of thin ($\sim$0.1 mm) constantan (of low thermal conductivity), the space between the walls being connected to a vacuum system. A small heater coil ($\sim$10–20 W) is wound non-inductively on the outside of the inner wall at the sample position. For cooling, the constantan tube, with air between the walls, is immersed in liquid nitrogen contained in a glass-walled Dewar. For bringing the sample back to room temperature, the wall space is evacuated and the heater coil turned on, the heating rate being adjusted to avoid undue thermal lag between sample and thermocouple. In the low-temperature vibrating sample magnetometer described by Creer *et al.* (1967), the axial pick-up coils (Section 3.3.2, Fig. 3.12) are wound on the constantan tube and immersed in the liquid nitrogen. Although other sources of noise are likely to dominate it is worth noting that pick-up coil resistance (Johnson) noise will be reduced at low temperatures through reduction in both resistance and temperature.

Chapter Five

Controlled fields and field-free space

A common requirement in palaeomagnetic laboratories is a means of carrying out experiments or routine techniques in 'zero' magnetic field, the best-known examples being the removal of secondary magnetization by thermal and alternating field cleaning. Magnetometers for the measurement of NRM usually perform better with their sensors in a low ambient field (with single fluxgate detectors this is essential). Other investigations require a range of small controlled fields (0–100 μT), for example the acquisition of TRM, VRM (viscous remanent magnetization) and DRM (depositional remanent magnetization).

The type of coil or other system required depends mainly on two factors, the volume over which the field is to remain within the specified limits, and for 'zero' field the degree to which the field is in fact zero. Some experiments in controlled fields are carried out on a single rock sample (~ 10 cm^3) or a small quantity (~ 1 g) of rock powder and the required field uniformity is not difficult to achieve. For thermal cleaning there is a considerable saving of time if many samples (20–50) can be heated simultaneously, involving a volume of 'zero' field of ~ 1000 cm^3, and a larger volume is needed if many samples are to be stored in a low field.

The permissible upper limit to the magnitude of a field conventionally termed 'zero' field also depends on the nature of the investigation. The ± 50 nT field in which most alternating field cleaning can be carried out satisfactorily is not sufficiently low for some rocks undergoing thermal cleaning, when a maximum ambient field during cooling of ± 1 nT is often desirable. The maximum allowable field also determines the stability required in the power supply or automatic control system used with coil systems: since it is normally the geomagnetic field (~ 50 μT) which is nulled, the above residual fields correspond to a current stability of $\pm 0.1\%$ and $\pm 0.002\%$ respectively. For investigations in small applied fields $\pm 1\%$ stability is usually sufficient.

Controlled fields are usually produced by one of a variety of coil systems or by magnetic shielding with a high permeability enclosure. The former method is the most flexible for zero and low field applications since very low (~ 2 nT) and variable, reproducible low fields are not easy to obtain over large volumes with shields. The principles and design considerations of the two techniques are described in the following sections.

5.1 Coil systems

5.1.1 Solenoids and single coils

For single-sample or other small volume applications a solenoid may be used (Section 3.1), although for some purposes axial access only may be a disadvantage. To improve access a small central sector of the solenoid can be removed, thus forming a system of two coaxial finite solenoids separated by a narrow gap. The central field and its axial variation can be calculated from Equation 3.3 (using correct signs for θ_1 and θ_2) (Fig. 3.2(a)). Compared with a complete solenoid of the same overall dimensions there is a significant decrease in the axial field at the centre together with much inferior homogeneity along the coil axis. As an example, if there is a 2 cm gap in a solenoid 24 cm long of radius 3 cm the field at the centre decreases by approximately 17% for a given current and at $z = 1$ cm (i.e. at the inner end of one solenoid) there is a 6% increase in the field: in the complete solenoid $\Delta B \approx 0.1\%$ at this position. A more general treatment of the field between two coils of finite winding cross-section is given by Kroon (1968).

For larger volumes of uniform field solenoids are unsuitable and it is usual to employ a Helmholtz or other coil configuration. However, if uniformity requirements are not too severe the use of a single coil is worth considering. Assuming a line current of i A

$$B_z = \mu_0 N i r^2 / 2(r^2 + z^2)^{3/2} \tag{5.1}$$

and

$$B_0 = \mu_0 N i / 2r \tag{5.2}$$

where N is the number of turns, r the coil radius and B_0 and B_z the axial field at the centre and at a distance z from the coil plane. The field on the axis at a distance z from the centre is given with sufficient accuracy for $z \ll R$ by

$$B_z = B_0\left(1 - \frac{3z^2}{2r^2}\right) \tag{5.3}$$

Thus, B_z will be within 1% of B_0 for $z < \pm 0.08R$, equivalent to ~ 200 nT if the Earth's horizontal field (in mid-latitudes) is cancelled at the coil centre.

The variation of B_z in the plane of the coil can be obtained from a general expression given by Ference, Shaw and Stephenson (1940).

$$B_z = \frac{\mu_0 i r^2}{2(r^2 + z^2)^{3/2}}\left[1 + \frac{3y^2(r^2 - 4z^2)}{4(r^2 + z^2)^2} + \ldots\right] \tag{5.4}$$

where y is the radial distance from the coil axis. Thus in the coil plane ($z = 0$) when $y \ll r$

$$B_z = B_0\left(1 + \frac{3y^2}{4r^2}\right) \tag{5.5}$$

and the field increases by 1% at $y \approx \pm 0.12r$.

It is sometimes advantageous to use a square coil, for which the field on the axis and at the centre are given by

$$B_z = \frac{2\mu_0 Nia^2}{\pi(a^2+z^2)(2a^2+z^2)^{1/2}} \tag{5.6}$$

and

$$B_0 = \frac{\sqrt{2}\mu_0 Ni}{\pi a} \tag{5.7}$$

where the coil is of side $2a$. The axial and radial variation of B_z for $z, y \ll a$ are

$$B_z = B_0\left(1 - \frac{5z^2}{4a^2}\right) \tag{5.8}$$

and

$$B_z = B_0\left(1 + \frac{3y^2}{8a^2}\right) \tag{5.9}$$

where y is parallel to a coil side. Thus, for $B_z = 1\%$, $z = \pm 0.090a$ and $y = 0.16a$, a marginal improvement over the corresponding inscribed circular coil.

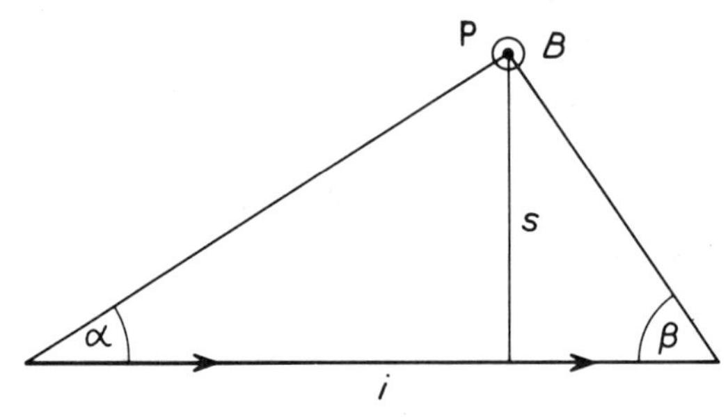

Fig. 5.1 Geometry for magnetic field at point P due to a line current of finite length.

If required, the axial (and radial) field at any point $P(x, y, z)$ can be derived by summing the contributions of each of the coil sides, using the expression for the field due to a straight current of finite length

$$B = \frac{\mu_0 i}{4\pi s}(\cos\alpha + \cos\beta) \tag{5.10}$$

where s is the perpendicular distance from the current to P and α and β are the angles shown in Fig. 5.1.

In the foregoing analyses it has been assumed that the winding cross-section is very small compared with the coil radius. For practical coils being considered here (i.e. providing a maximum field of $\sim 500\ \mu\text{T}$) this is usually the case, provided currents of up to ~ 1 A can be used. For reasonably small ratios of winding cross-section to radius the field homogeneity near the coil centre is essentially independent of the shape of the winding cross-section.

5.1.2 Helmholtz coils

(a) Theory

For larger volumes of uniform field one of a variety of coil configurations can be employed, the best known and simplest of which is the two-coil Helmholtz system.

The field on the axis of a circular coil at a distance z from the centre can be expressed in the form

$$B_z - B_0 + \frac{z^2}{2!}\left(\frac{\delta^2 B}{\delta z^2}\right)_0 + \frac{z^4}{4!}\left(\frac{\delta^4 B}{\delta z^4}\right)_0 + \ldots \qquad (5.11)$$

the terms of odd order vanishing because of the symmetry about the origin. The zero subscript denotes the value of the derivative at the origin, i.e. the coil centre.

Equation 5.11 also applies to a system of coaxial coils symmetrically placed about a central plane parallel to the planes of the coils and normal to their axes: the origin is taken at the centre of the system. Then a field of increasing uniformity can be obtained as successively higher derivatives of the field there are made zero. The first approximation to uniformity is when $\mathrm{d}^2 B/\mathrm{d}z^2 = 0$, achieved in the Helmholtz coil system which in its original form consists of two equal, coaxial circular coils of radius r separated by a distance $2z = r$. This result is easily obtained by differentiating Equation 5.1 twice and equating to zero. As with a single coil, B_z is maximum at the centre of the system ($z = 0$), but in the Helmholtz pair the maximum is 'flatter' and B_z becomes increasingly more uniform over a given distance from the centre in systems in which higher derivatives of the field are zero.

Differentiation of Equation 5.6 shows that the Helmholtz condition is satisfied for a pair of square coils when the separation $2z = 1.088a$, where $2a$ is the length of a side (Fig. 5.2).

For Helmholtz pairs in which there are N turns carrying a current i in each coil the field at the centre is given by

$$B_0 = (8.992 \times 10^{-7})\frac{Ni}{r} \qquad (5.12)$$

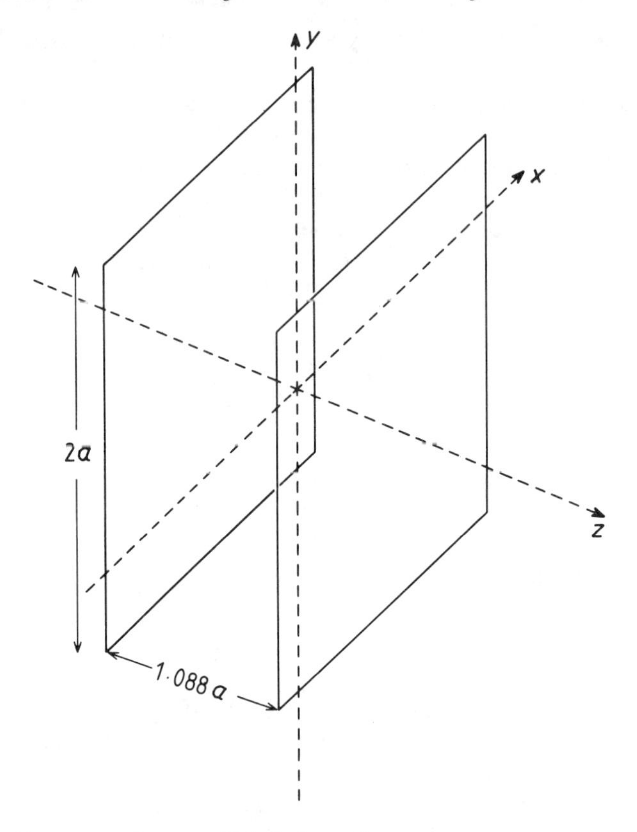

Fig. 5.2 Geometry of square Helmholtz coils.

and

$$B_0 = (8.148 \times 10^{-7}) \frac{Ni}{a} \tag{5.13}$$

for circular and square coils respectively. Thus, as a useful rule of thumb, the ampere-turns required to provide a central field B_0 μT in coils of radius r or side $2a$ metres is given by the product $B_0 r$ or $B_0 a$.

The fourth-order terms in Equation 5.11 can be used to determine the field uniformity near the origin with sufficient accuracy for present purposes. Parry (1967) gives for circular coils

$$B_z = B_0 \left[1 - \frac{0.144}{r^4} (8z^4 - 24z^2 r^2 + 3r^4) \right] \tag{5.14}$$

$$B_r = B_0 \left[0.576 \frac{zr_a}{r^4} (4z^2 - 3r^2) \right] \tag{5.15}$$

where B_r is the radial field at a radial distance r_a from the axis. There is, of course, cylindrical symmetry about the axis.

The corresponding expressions for square coils are

$$B_z = B_0\left[1 - \frac{1}{a^4}(0.783z^4 - 1.668z^2(x^2+y^2) - 1.95x^2y^2 + 0.458(x^4+y^4)\right] \tag{5.16}$$

$$B_x = B_0\left[\frac{zx}{a^4}(1.577z^2 - 0.167y^2 - 1.770x^2)\right] \tag{5.17}$$

B_x is the field along the x-axis at the point $P(x, y)$ (Fig. 5.2). B_y is obtained by interchanging x and y in Equation 5.17.

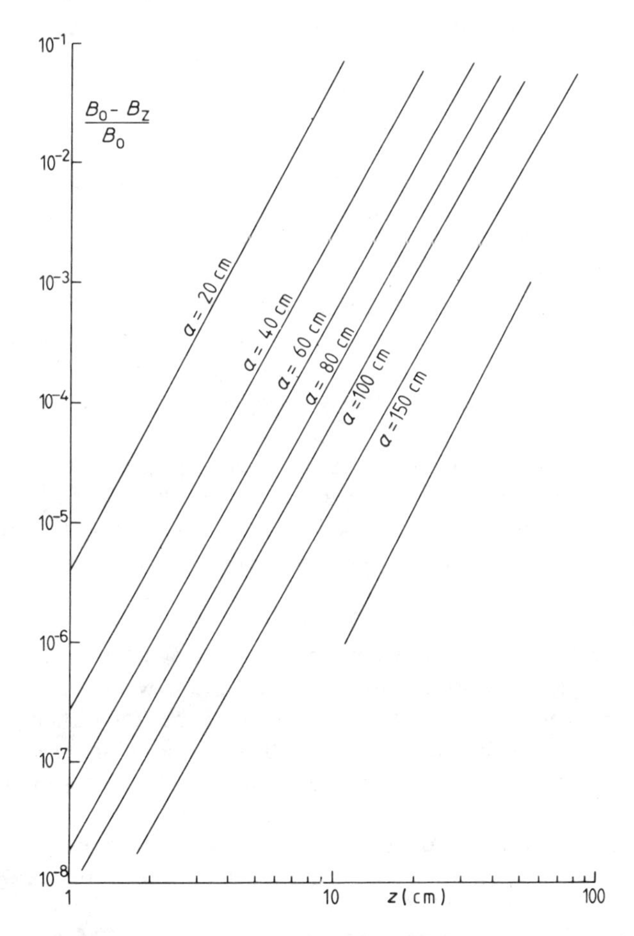

Fig. 5.3 Field homogeneity, expressed as $(B_0 - B_z)/B_0$, along the axis of square Helmholtz coils of side $2a$ for different axial distances z from the centre. The line at the right gives the radius of the spherical volume within which the given homogeneity is obtained in the coil system of McElhinny *et al.* (1970) (Section 5.1.3).

Figure 5.3 shows values of distance z from the centre of square coils plotted against the field homogeneity $(B_0 - B_z)/B_0$ for different coil sizes. Square coils show a $\sim 10\%$ increase compared with circular coils $(a = r)$ in the axial distance within which the field possesses a given homogeneity. However, perhaps a more realistic comparison is between circular and square coils of equal winding length, i.e. $2\pi r = 8a$, or $a = 0.78r$. The axial distance for a given homogeneity is then $\sim 15\%$ less in the square coil system compared with a circular pair.

Using Equations 5.14–5.17 it can be shown that the greatest field variation occurs along the z-axis. Thus, within a sphere of radius z centred at the origin, the uniformity of the total field is everywhere better than that existing along the axis: this is not true of the exscribed cubic volume of side $2z$, where the field variation between the centre and corners is typically two to three times greater than between $z = 0$ and $z = z$. Firester (1966) and Everett and Osemeikhian (1966) discuss off-axis homogeneity in circular and square systems.

It has been assumed in the foregoing that each coil consists of a line current. In practice the coils have a finite winding cross-section and only the windings closely adjacent to the mean radius, which is used to calculate the coil separation, will satisfy the Helmholtz condition. The distortion of the field will only be important in most applications if the winding width and depth are $> 0.1\ (r, a)$. The optimum winding cross-section was first discussed by Clerk-Maxwell (1892), who derived an optimum depth/width ratio of $(31/36)^{1/2} = 0.93$, and Franzen (1962) gives the same value, derived by a different method. Both analyses assume that the dimensions of the winding cross-section are small compared with r or a.

The simplest solution for a finite winding area is a single-layer winding on an inward-sloping flat surface of angle $\tan^{-1}(1/0.50)$ or $\tan^{-1}(1/0.54)$ for circular and square coils respectively, thus maintaining the Helmholtz criterion for each turn. Such a system is rarely of practical use.

(b) *Practical considerations*

Helmholtz coils are a convenient source of magnetic fields (generally $< 10^{-3}$ T) of good uniformity over a useful volume where applied fields of moderate uniformity are required. Coils 1 m in diameter are of ample size for many purposes, providing approximately 16 000 cm^3 of space in which the field is uniform to $\sim 1\%$. For more demanding situations, e.g. nulling of the Earth's field during thermal demagnetization 2 m square or circular coils provide a field uniform to $\sim 0.001\%$ in a volume of ~ 700 cm^3, approximately a 9.0 cm cube: this is the order of uniformity which is desirable if several rock samples which are susceptible to strong PTRM acquisition are to be thermally cleaned simultaneously. It corresponds to an ambient field of ± 0.5 nT or less over the samples, assuming that the Earth's field of $\sim 50\ \mu$T is cancelled at the centre of the coil system.

If the field uniformity is only required in a region close to the axis, it is possible to extend the length of such a region with a given coil pair by increasing the coil spacing above the Helmholtz value. Instead of a central plateau in the field there is a central low with a maximum on either side: if this 'ripple' amplitude is $\pm\delta B$ about the average field, the field remains within these limits over a greater distance than that in which it falls by $2\delta B$ from the central value with the Helmholtz spacing. Rudd and Craig (1968) show how the axial field varies for different (circular) coil spacings, and compute the length over which the field is within given limits. As an example, for $\delta B/B = \pm 0.5\%$ and $\pm 1\%$ the distances from the coil centre over which these uniformities are obtainable are increased from $0.30r$ and $0.38r$ to approximately $0.48r$ and $0.60r$ respectively, using coil spacings of 1.13 and 1.18 times the Helmholtz setting. The authors note that their curves can be used for square coils with only small error ($\sim 5\%$) if distances and positions are measured in units of the Helmholtz spacing. Daniels (1950) and Crownfield (1964) describe the use of this technique with solenoidal coil pairs of large winding cross-section.

The theoretical performance of Helmholtz systems is probably rarely achieved in situations where the objective is a high degree of homogeneity (better than 0.01%). This is mainly because the total field at any point in the coils is usually the sum of the Helmholtz coil field and a component of the local ambient field in which there is likely to be some gradient. Imperfections in coil geometry may also be a contributory factor.

In most laboratories the ambient geomagnetic field is significantly modified by local fields arising from iron and steel in the building structure and in equipment and fittings, and possibly by circuits carrying direct currents. Not only do these sources contribute magnetic fields and gradient at the working space but their magnitudes and directions may vary with time. For applied fields in the range 50–500 μT the gradients may not be important, but if zero field over an extended region is required they may be the limiting factor in the attainable field uniformity.

The gradients are of two forms, axial (e.g. dB_z/dz) and transverse (e.g. $dB_z/dy, dB_x/dz \ldots$). In a three component coil set, the former, dB_z/dz, dB_x/dx and dB_y/dy can be removed near the coil centre by a separating winding on each of two of the Helmholtz coil pairs, the axes of which are the y and z directions, each pair of windings being connected in series opposition (Fig. 5.4). Although convenient this is not the optimum coil separation for producing a uniform gradient at $z = 0$: this is given by $(d^3B/dz^3)_0 = 0$, for which the solution is $z = 0.866r$ for circular coils and $z = 0.946a$ for square coils, neglecting higher odd orders of the field. Transverse gradients are more difficult to deal with: for square coils Parry (1967) describes a system of currents flowing in the *same* direction in wires placed along opposite sides of the coils (Fig. 5.4). Although there are a total of nine first derivatives of the field, there are only five independent terms in the resulting tensor

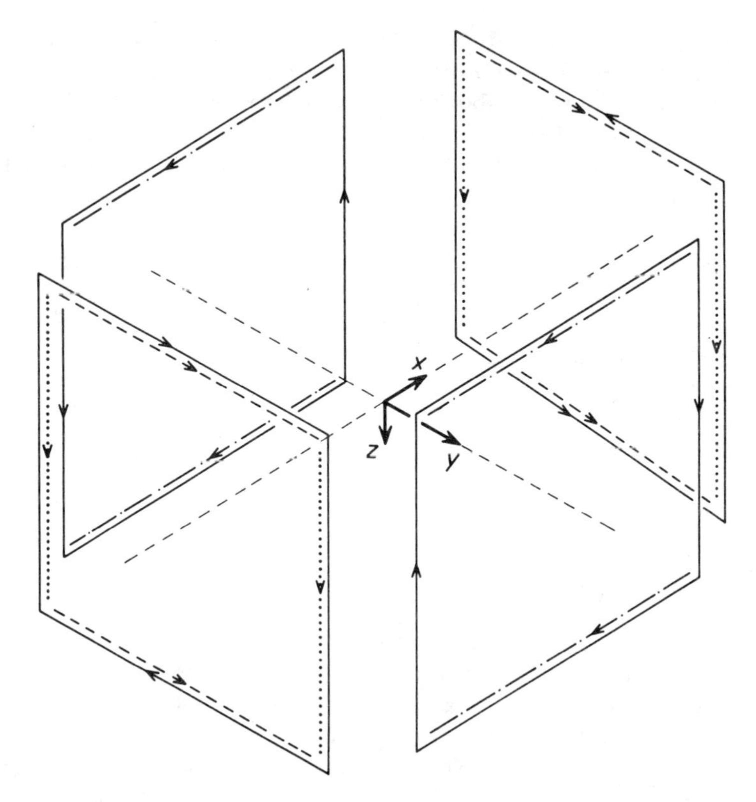

Fig. 5.4 System of square coils to cancel all first derivatives of the field at the centre. —— dZ/dz, dY/dy, dX/dx; – – – dY/dx, dX/dy; . . . . dX/dz, dZ/dx; –·–·– dZ/dy, dY/dz.

(since div B = curl B = 0), such as those enclosed by the dotted line

$$\begin{array}{ccc} \dfrac{\partial B_x}{\partial x} & \dfrac{\partial B_x}{\partial y} & \dfrac{\partial B_x}{\partial z} \\[2ex] \dfrac{\partial B_y}{\partial x} & \dfrac{\partial B_y}{\partial y} & \dfrac{\partial B_y}{\partial z} \\[2ex] \dfrac{\partial B_z}{\partial x} & \dfrac{\partial B_z}{\partial y} & \dfrac{\partial B_z}{\partial z} \end{array}$$

These field gradients can be cancelled at the centre of the coils by five coil systems, of which two are for the axial gradients described above. Figure 5.4 shows the windings required to cancel the remaining gradients. To avoid contributions to the central field from the external connections to the windings, they should ideally run outward from the coil corners along directions away from the centre of the coil system. If multiple windings are used the leads to each length of wire on the coil must lie along these directions for a reasonable distance from the coils.

The axial field gradient ($\partial B_z/\partial z$ or $\partial B_y/\partial y$) produced at the centre of a square Helmholtz pair of side $2a$ with the gradient windings on the same formers as the field cancellation windings is obtained by differentiating Equation 5.6 and is given by

$$\left(\frac{\partial B_z}{\partial z}\right) = 876\left(\frac{Ni}{a^2}\right) \text{nT m}^{-1} \tag{5.18}$$

At the optimum spacing for uniform gradient production ($z = 0.946a$), the numerical coefficient in Equation 5.18 is 658.

The transverse gradient produced at the coil centre by parallel wires on opposite sides of the square coils can be calculated from the expression for the

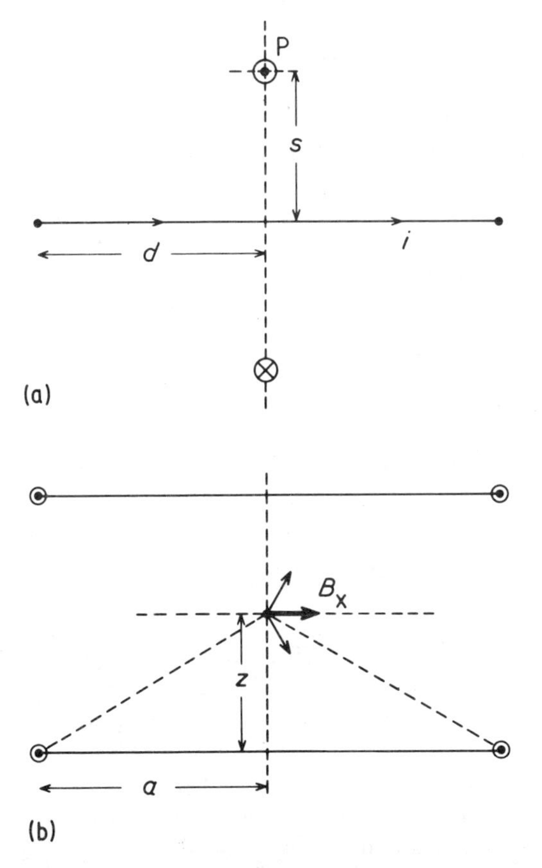

Fig. 5.5 (a) Magnetic field on the perpendicular bisector of a finite line current; (b) magnetic field due to four parallel wires. Current flow is into the paper and only the contribution of two of the wires is shown.

field produced at a point P on the perpendicular bisector of a finite line current i (Fig. 5.5(a)).

$$B = \frac{\mu_0 id}{2\pi s(d^2 + s^2)^{1/2}} \tag{5.19}$$

where $2d$ is the length of the line current and s is the perpendicular distance from the current to P. Calculating the field B_x due to four parallel wires (Fig. 5.5(b)) from Equation 5.19, differentiating with respect to s and substituting $s = 0.544d$ ($= 0.544a$) we obtain

$$\frac{\partial B_x}{dz} = 3040\left(\frac{Ni}{a^2}\right)\text{nT m}^{-1} \tag{5.20}$$

where N is the number of wires on each coil side and B_x is measured along the z axis.

The measurement of gradient components is time-consuming and their successful cancellation is clearly not simple. The achievement of highly uniform low intensity fields through the use of coil systems in a normal laboratory environment is a difficult undertaking, and a magnetically quiet, gradient-free site for such a facility is highly desirable.

Because of the inclination of and secular variation in the geomagnetic field, it is not convenient to null the Earth's field by using a single coil pair with its axis aligned parallel to the field. The best arrangement is three orthogonal coil pairs to separately cancel three perpendicular components of the field (one vertical and two horizontal). This gives flexibility in allowing for secular variation of the ambient field and in applying small fields in any desired direction. There is also a small improvement in field uniformity in the central region compared with that obtained with a single coil. This is because the residual field at any point in the central region of a coil pair is proportional to the field component that is being nulled. Although at some points in a given spherical region the vector sum of the three residual fields is greater than that in the same region if a single coil pair is used, the maximum excursion of the field in the region is decreased.

(c) *Design and construction*

For small coil systems (up to ~30 cm in diameter) it is usually simple and convenient to use circular coils with formers of wood or aluminium with a suitable channel cut for the winding. For larger coils, particularly two- and three-component systems, square coils are usually easier to construct (of aluminium channel or warp-resistant wood) and easier to fit together as a self-supporting semi-rigid framework. The large Helmholtz spacing of square coil pairs also gives marginally improved accessibility to the central region.

Helmholtz coil design depends on the use to which the coils are to be put, and on available power supplies. Coil size is determined by the degree of field uniformity required over the working volume, or where uniformity is not critical by the field required at the centre and the available power supply. In the former case, the expressions given above for the field variation in circular and square coils can be used to determine coil size (a or r). A common situation that then arises is the calculation of channel size and optimum wire gauge to use to give the required central field with a given power supply.

Equation 5.12 or 5.13 gives the total ampere-turns required and the maximum available current then fixes N, the number of turns per coil. The coil maximum resistance R_c is determined by the voltage of the power supply, allowing for any controlling resistances needed in the circuit. R_c and N are related by

$$R_c = 4\pi r N s \tag{5.21}$$

or

$$R_c = 16aNs \tag{5.22}$$

for circular (radius r) and square coils (side $2a$) pairs respectively (assuming series connection) where s is the resistance of the wire per unit length. Thus s can be determined and the corresponding gauge of wire found from wire tables (see Appendix 3) and also the cross-section of the winding channel required to accommodate the turns.

Since a major part of the cost of a large coil set may be in the wire it is desirable to minimize the weight of copper used. Following the above analysis, this can be achieved by decreasing N or the wire cross-section A if a larger power supply is available.

If the length of one turn on the winding is l, then

$$R_c = \frac{2Nl\rho}{A} \tag{5.23}$$

and

$$B_0 = C\frac{Ni}{l} \tag{5.24}$$

where ρ is the resistivity of the material of the wire (assumed to be copper), B_0 is the central field and C is a numerical constant, obtainable from Equations 5.12 or 5.13. The power input is $W = i^2 R_c$, and from Equations 5.23 and 5.24

$$W = \frac{2B_0{}^2 l^3 \rho}{C^2 NA} \tag{5.25}$$

If M is the mass of copper on the coil pair

$$M = 2lNAd \tag{5.26}$$

where d is the density of copper. Substituting NA from Equation 5.26 in Equation 5.25

$$W = \frac{4B_0{}^2 l^4 d\rho}{C^2 M} \tag{5.27}$$

Thus, from Equation 5.27, the power input required to maintain a constant field in a system of constant size is inversely proportional to the mass of copper used. Equation 5.25 shows, again for constant B and l, that W is inversely proportional to A and N. However, if A is reduced it is only necessary to increase the voltage in the same ratio. If N is reduced the current increases in the same ratio, at constant voltage, and the same current field is maintained.

It may be possible to use a particular power supply with a given pair of coils by connecting the coils in parallel rather than in series. To provide a given field, parallel-connected coils require the same power input as when series-connected, but take twice the current from the supply at half the voltage, assuming they are of equal resistance. Parallel connection of coil pairs in two- and three-component systems supplied by one power source may be advantageous for the same reason.

The construction of Helmholtz coil systems is straightforward, with aluminium channel or, for square systems, wood formers. Stable, non-warping wood such as teak or the less expensive ramin should be used. The cross-sectional area of the former and winding channel depends on the coil size and number of turns. It is important that the corners of the winding channel in square coils are well rounded, to avoid 'bunching' of the winding at the corner. Aluminium formers should have an insulating layer on the channel base and walls to avoid possible short circuits developing.

5.1.3 Multicoil systems

There is an extensive literature on the design of multicoil systems for the provision of magnetic fields of maximum uniformity over volumes which are as large as possible relative to the volume occupied by the coils. Although Helmholtz coils are suitable for the great majority of application in palaeo- and rock magnetism, other more complex coil systems in which fourth and higher order derivatives of the field are zero have found use, and these are described in this section.

A four-coil system for increasing the axial distance over which a given uniformity is obtained (relative to a Helmholtz pair) is due to Fanselau (1929). Two pairs of circular coaxial coils of radius r_1 and r_2 are symmetrically spaced about the origin and with their planes distances $2z_1$ and $2z_2$ apart (Fig. 5.6). The optimum setting is given by the ratios $z_1/r_1 = 1.1072$ (outer pair), $z_2/r_2 = 0.2781$, $r_2/r_1 = 1.3092$, and $z_2/r_1 = 0.3640$. The coils have the same number of turns N and are connected in series. The central field B_0 is given by

$$B_0 = (1.237 \times 10^{-6})\frac{Ni}{r} \tag{5.28}$$

The Fanselau system is useful where an elongated volume of good field uniformity near the coil axis is required. In a system with $r_2 = 50$ cm, the axial homogeneity factor $(B_0 - B)/B_0$ is $\sim 5 \times 10^{-5}$ at the planes of the inner coils,

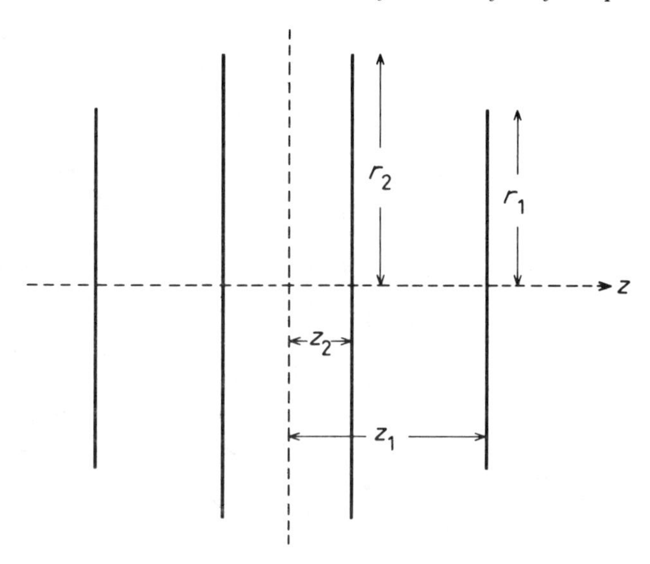

Fig. 5.6 Fanselau coil system.

i.e. over an axial length of $2z = \sim 28$ cm. This corresponds to a variation of 5 nT in a central field of 100 μT. To achieve this homogeneity over the same distance in a circular Helmholtz pair would require coils of diameter approximately 4.0 m.

In multicoil systems it is possible to manipulate five variables, coil size, coil separation and distance from the centre turns per coil and current in each coil, although the latter two are interchangeable and can be combined as ampere-turns per coil. By taking advantage of these variables, the volume of field of given homogeneity relative to the volume occupied by the coils can be increased.

A system of five equally-spaced square coils enclosing a cubic volume is described by Rubens (1945) (Fig. 5.7). The procedure for designing the system is as follows. Let the central coil consist of one ampere-turn, which produces a field of $B_0(z)$ at a point P a distance z along the coil axis from the centre. Then if the inside and outside coil pairs produce fields at P of $B_1(z)$ and $B_2(z)$ per ampere-turn, the total axial field at P is

$$B(z) = B_0(z) + B_1(z) + B_2(z) \tag{5.29}$$

α and β are the ratios of currents in the inside and outside coil pairs to the current in the central loop. If three values of z are chosen at which values of $B(z)$ are to be equal, Equation 5.29 provides two simultaneous linear equations which can be solved for α and β. Using values of z of 0, 0.075a and 0.125a, where $2a$ is the length of a coil side, then Equation 5.29 gives 0.405 for α and 1.92 for β, leading to the set of integers 71, 15, 37, 15, 71, which are proportional

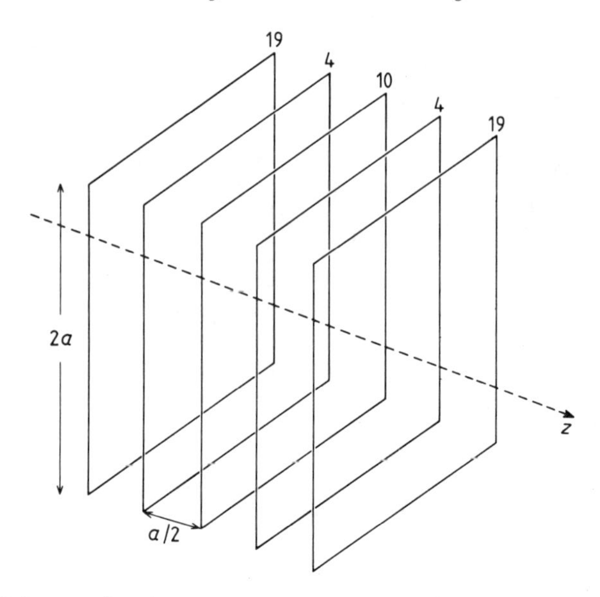

Fig. 5.7 Rubens coil system.

to the turns on each coil if they are series connected. The author gives details of a coil set in which the relative numbers of turns are modified slightly to produce better off-axis homogeneity: these revised numbers are 19, 4, 10, 4, 19 (Fig. 5.7).

The field at the centre of this modified system is

$$B_0 = (17.85 \times 10^{-6})\frac{i}{a} \tag{5.30}$$

where a coil side is $2a$ m.

The author does not give precise information on field homogeneity to better than 1 part in 1000. This homogeneity is achieved in the above system within a cylindrical volume, coaxial and concentric with the coil axis, of length $0.90a$ and radius $0.40a$, i.e. 45 cm long and 40 cm in diameter in a coil side 1 m. Equivalent homogeneity is obtained in a square Helmholtz system of approximately 2.5 m on a side.

It should not be forgotten that the homogeneity theoretically available in a coil system may not be achieved in practice because of constructional inaccuracies. A four-coil system which is shown to be rather tolerant of reasonable constructional errors and has improved access to the central region has been designed and extensively analysed by Alldred and Scollar (1967). It is geometrically similar to the Fanselau arrangement, but square coils are used, with unequal windings on the inner and outer pairs. With the notation of Fig. 5.8, the optimum ratios are $a_1/a_2 = 0.9555$, $z_1/z_2 = 3.6456$, $a_1/z_1 = 0.9094$

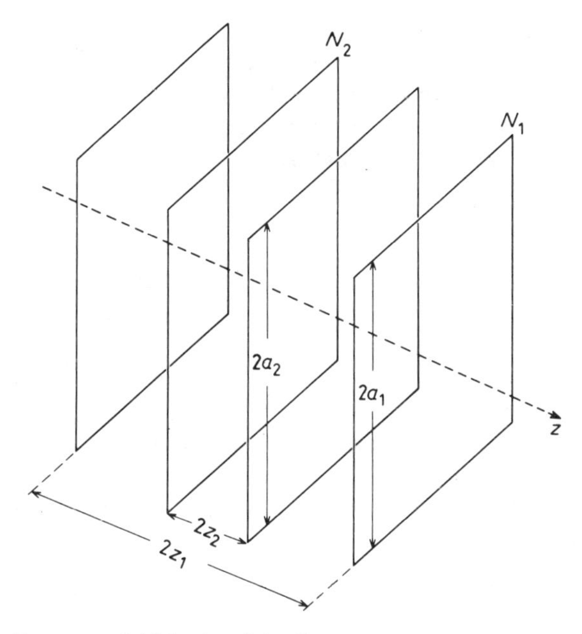

Fig. 5.8 Coil system of Alldred and Scollar (1967).

and $N_1/N_2 = 21/11$: the coils are connected in series. The field at the centre of the coils is

$$B_0 = (1.831 \times 10^{-6})\frac{N_2 i}{a_2} \tag{5.31}$$

A large three-component coil set, consisting of two of the above coil sets and a square Helmholtz pair is used in thermal demagnetization equipment by McElhinny, Luck and Edwards (1971). In the two four-coil systems, $2a_2 = 2.48$ m and 2.22 m respectively, and $N_2 = 220$ and 110. The authors give the radii of the spherical volumes at the centre of the coils within which homogeneities in the range 10^{-6}–10^{-3} are achieved, and these are shown in Fig. 5.3. It can be seen that there is a very substantial increase in the volume in Alldred–Scollar coils within which a given homogeneity is maintained, compared with a square Helmholtz set of side a_2. The improvement factor is approximately 2.6, measured as the distance from the centre (or the radius of the spherical volume) within which a given homogeneity is achieved: thus the factor by which the uniform field volume is increased is ~ 18.

The coil systems described in this section and Section 5.1.2 are quite adequate for most applications in palaeomagnetism and rock magnetism. However, they by no means exhaust the possible configurations that have been designed and analysed, but in general these more complex systems are inconvenient and their performance is unnecessarily superior for present purposes. Some references of interest are Clark (1938) and Everett and Osemeikhian (1966) (spherical coils), Lyddane and Ruark (1939) and Scott (1957), who describes a double Helmholtz system. McKeehan (1936) analyses systems of the Fanselau type and Hart (1967) gives useful data on magnetic fields due to line and distributed currents.

5.1.4 Coil control systems

There are two main purposes for which the coil systems described in this section are used in rock and palaeomagnetism: the provision of small (0.1–1 mT) fields of reasonable uniformity over a small volume and the achievement of 'zero' magnetic field ($< \sim 1$ nT) over an extended volume, in particular for thermal demagnetization. For the former purpose a power supply stable to $\pm 0.1\%$ provides magnetic fields of adequate stability and if the coil field is superimposed on a component of the geomagnetic field typical daily variations in the latter are not large enough to significantly affect the total field.

For the provision of good magnetic field-free space ($\pm$ 1 nT) the stability requirements are much more demanding. A stability of $\pm$ 1 nT in the coil field, cancelling the Earth's vertical field in mid-latitudes, implies a current stability of ~ 2 parts in 10^5. However, there is the further complication of the variation of the geomagnetic elements with time, causing the net field at the centre of the coil system to vary as well. It is not only the variation in intensity which may be important, but also in declination and inclination, since a 1′ change in declination of, say a 20 μT horizontal field results in an orthogonal field component of ~ 6 nT. The important feature of the time variations is the magnitude of the field change they produce during the period over which it is required to maintain the zero field. For thermal demagnetization this period is usually in the range 10–20 minutes, in which time geomagnetic field changes of 2–10 nT in intensity and 1′–5′ in direction are not uncommon, particularly during magnetic storms. It is therefore clear that for $\pm$ 1 nT stability or better a field compensation system is required which continually monitors and controls the central field.

For an intermediate stability of $\pm$ 10–100 nT, power supplies are available with stabilities in the range $\sim 10^{-4}$. Constant current is preferred to constant voltage as the coil current is then independent of any temperature change in the windings: for copper windings, the fractional current change at constant voltage for $T = \pm 0.1$°C is 4×10^{-4}.

The usual requirement in maintaining a field constant to 1 part in 10^3 or 10^4 is for the coil current to be monitored but not to be measured to this accuracy, since the current required is usually determined by direct measurement of the field and not by calculation from the coil geometry and ampere-turns. The above accuracy is not available in standard meters, but a simple potentiometer circuit can be used by which a current can be maintained to an accuracy of 1

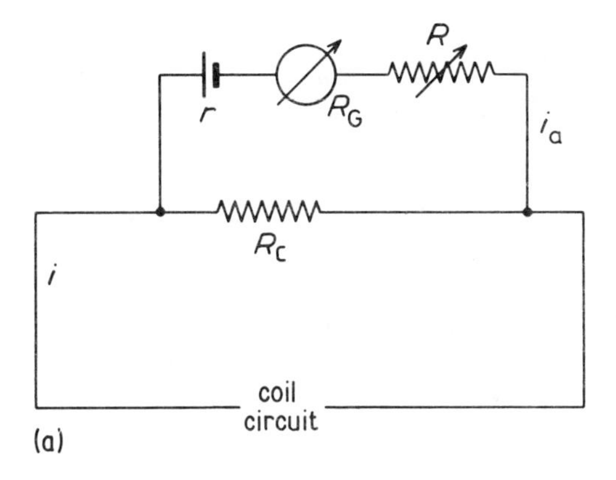

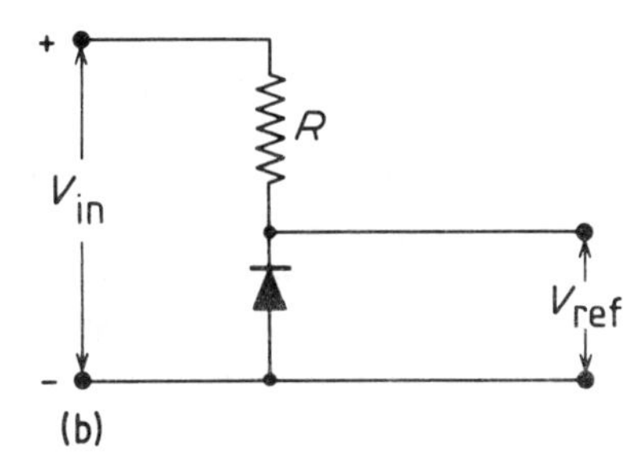

Fig. 5.9 (a) Control circuit for coil currents; (b) Zener diode circuit. V_{in} is the voltage source and V_{ref} is the stabilized output voltage. R is chosen such that the diode operates on the correct part of its current characteristic.

part in 10^3–10^5 (Fig. 5.9(a)). R_c is an adjustable resistance of constantan or manganin (low temperature coefficient of resistance) the voltage drop across which, when the required coil current is flowing, exactly opposes that of the reference voltage source of internal resistance r and no current flows in the galvanometer circuit. A current change Δi in the coil results in a change in the galvanometer current of Δi_a, where

$$\Delta i_a = \frac{\Delta i R_c}{R_c + r + R_g + R} \tag{5.32}$$

where R is a variable resistance for controlling the sensitivity. At maximum sensitivity ($R = 0$), $R_c/(R_c + r + R_g)$ is typically between 1/3 and 1/30 depending on the type of reference voltage source and meter used. Therefore, if $\Delta i = 0.1$ mA, $\Delta i_a = 3$–$30\ \mu$A, easily detectable on a galvanometer or micro-ammeter. Since the coil current is typically 100 mA, the above range of Δi_a corresponds to an accuracy of 1 part in 10^3, and there is no difficulty in improving this to 1 part in 10^{-4}–10^{-5} by using a sufficiently sensitive detector.

At the lower end of the accuracy range a mercury cell (Mallory) can be used for the reference voltage. For higher accuracy a standard (Weston) cell or a Zener diode are suitable. Figure 5.9(b) shows the circuit for the latter, which is capable of providing a low resistance reference voltage source of 1–2 V, stable to 1 part in 10^4–10^5. It can easily be arranged for the test circuit to be switched into each circuit of a three-component coil set, each with its appropriate value of R_c.

Because of the time variation of the geomagnetic field and therefore of the residual field at the coil centre, continuous compensation of the residual field variation is desirable in order to achieve optimum 'zero' field for periods longer than a few minutes. For some purposes manual compensation may be satisfactory but clearly an automatic system is preferable.

The principle of such a system is simple. A three-component fluxgate detects variations in the North, South and vertical component of the residual field in the coils and the three amplified outputs are used either to modify the coil currents in the appropriate sense via the power supplies, or to feed appropriate currents to secondary windings wound on the three coil sets. A slight complication arises from the fact that there is usually equipment in the region where the zero field is to be maintained, and it is necessary to place the magnetic field sensors outside the region. The observed field difference between the coil centre and the sensor position enables the field components at the latter position to be maintained at values such that the central field is zero.

The basic circuits used in the author's laboratory to control the field for thermal demagnetization are shown in Fig. 5.10. Each coil of the three-component system (square Helmholtz configuration, 2.4 m on a side) is supplied from a programmable constant-current power supply. A three-component fluxgate sensor is placed outside the furnace ~40 cm from the coil centre and the output from each is amplified and used, via a variable resistance R_2, for sensitivity adjustment, to control the output of the power supply to the appropriate coil. Each fluxgate sensor has an additional winding through which a small current is passed, derived from a voltage divider, to compensate the small field component at the sensor position (~5–20 nT) and ensure that the sensor output is zero when the corresponding central field component is zero. To set up the system, an independent field sensor is placed at the centre of the system and each component is adjusted as follows. The switches S_1 and S_2 are opened and the current through the coils adjusted from the power supply to bring the offset meter to zero. S_1 and S_2 are then closed and R_1 is adjusted until the central field component being adjusted is zero. This causes an offset current to flow and the offset meter is reset to zero using the power supply control. The central field is then reset to zero by adjusting R_1 again, and the above procedure repeated until the central field component is zero and the offset meter reads zero. The maximum variation in each field component which can be compensated is $\sim \pm 200$ nT, and over a period of 30 minutes the total field at the centre of the coils can be maintained to within ± 2 nT of zero.

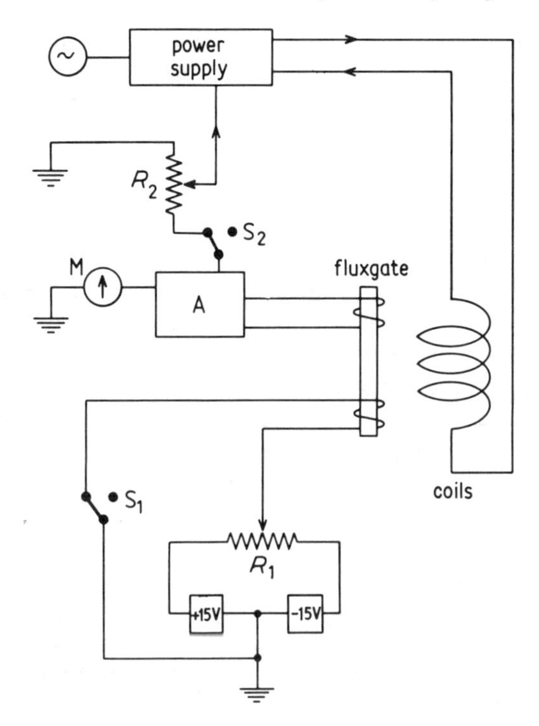

Fig. 5.10 Block diagram of circuit for controlled field-free space. The circuit for monitoring one field component is shown. A is the fluxgate output amplifier and M is the offset current meter.

McElhinny *et al.* (1971) use a similar system, obtaining a field variation no greater than ± 0.5 nT over the critical cooling period of thermal demagnetization.

Roy, Robertson and Keeping (1969) describe an array of five three-component sets of square Helmholtz coils 2.4 m on a side, placed at separate locations within the grounds of the Blackburn Magnetic Observatory, near Ottawa. Each coil carries three separate windings: primary, secondary and trimmer. The primary windings carry constant currents to cancel the average geomagnetic field at the site, the windings of each pair of the five sets cancelling the vertical, North and East field component being connected in series. For field-free space applications the East component primary windings carry no current, since their axis in each set is perpendicular to the magnetic meridian. One coil set contains a three-component fluxgate for monitoring field variations, the fluxgate outputs being used to control the currents in the secondary windings on the other four coil sets to compensate the field variations. The trimmer windings are used to compensate for small differences in the coil characteristics at each location. Under normal magnetic conditions the central field is maintained to within $\pm$ 1 nT of zero. Among the advantages of this 'master' and 'slave' coil system are the absence of monitoring fluxgates in the slave coils and the ability to produce controlled fields (0–100 μT) in any direction in the coil sets.

5.2 Magnetic shields

Magnetic shielding is a general term for the technique by which a magnetic field is prevented from entering an enclosure, or is greatly reduced inside it, through interaction of the field with the material of the enclosure walls. Both static and alternating magnetic fields can be attenuated in this way although different physical processes may be involved in each case. Attenuation of static and alternating fields is achieved by the partial trapping of the field in a layer of high magnetic permeability, whereas alternating field shielding can also be achieved through the generation of eddy currents in an electrically-conducting surface surrounding a region, the currents flowing in such a way as to generate a magnetic field opposing the external field and attenuating it as it permeates the conducting medium.

The most common application of magnetic shielding is as an alternative to coil systems for attenuating the ambient laboratory field to provide a volume of low and stable magnetic field. Shield sizes range from small containers for transporting rock specimens between instruments to room-sized shields providing a large volume of weak magnetic field in which one or more items of equipment can be housed. It is not usually feasible to achieve the very low residual fields ($\sim$1 nT) available in coil systems, but magnetic shields are often useful for providing a weak-field environment ($\sim$10–100 nT) suitable for a.f. demagnetization, some thermal demagnetization, rock storage etc. A potential advantage of magnetic shielding over coil systems is the capability of reducing the often severe field gradients occurring in research laboratories.

In the following sections the shielding factor S is defined as the ratio of the field immediately outside the shielded enclosure (B_0) to the field inside (B_i). In the literature S is sometimes expressed in decibels. The conversion factor is $S(\text{dB}) = 20 \log_{10}(B_0/B_i) = 20 \log_{10} S$.

5.2.1 Electromagnetic shields (alternating fields)

A sinusoidal electromagnetic disturbance entering the plane surface of an electrically conducting medium is attenuated to e^{-1} of its amplitude at a depth δ, the skin depth, given by

$$\delta = (2/\sigma\omega\mu_r\mu_0)^{\frac{1}{2}} \tag{5.33}$$

where σ is the electrical conductivity, $\omega = 2\pi f$ is the circular frequency of the disturbance and μ_r is the relative permeability of the medium. For copper and aluminium δ is approximately 1.0 cm at 50 Hz, 0.2 cm at 1 kHz and

7×10^{-3} cm at 1 MHz. For non-ferrous metals $\mu_r \approx 1$. Thus, the ratio of magnetic field strength on either side of a medium of thickness a in which the skin depth for the field is δ, is $\exp a/\delta$. As a rule of thumb, shielding factors of 10 and 100 are achieved in thicknesses (or depths) of 2.5 and 4.5 skin depths respectively. The actual shielding factor achieved in practical enclosures are not easy to calculate, owing partly to the presence of corners, openings and joints. From the figures quoted it can be seen that reasonable depths of conductor are efficient at high frequencies ($> \sim 1$ kHz), whereas for most applications in the palaeomagnetic laboratory the important frequency is that of the mains electricity supply, 50–60 Hz, and thick walls are necessary for significant shielding.

Theoretical shielding factors obtainable in spherical and long cylindrical enclosures through this process have been derived by King (1933). The expressions are complex, but greatly simplify if the thickness t of the wall is much less than the radius r of the sphere or cylinder. Under these conditions, King's values are

$$S = \left(1 + \frac{r^2 t^2}{C^4}\right)^{\frac{1}{2}} \tag{5.34}$$

for a sphere, and

$$S = \left(1 + \frac{r^2 t^2}{4C^4}\right)^{\frac{1}{2}} \tag{5.35}$$

for an infinitely long cylinder. C has dimensions of length and is given by

$$C = \left(\frac{10^7 \rho}{8\pi^2 f}\right)^{\frac{1}{2}} \tag{5.36}$$

where ρ is the resistivity of the wall material and f the frequency of the ambient field.

For the sphere and infinite cylinder the shielding factor is independent of the direction of the external field. For a finite open-ended cylinder S is a maximum when the field direction is perpendicular to the cylinder axis, and S_{max} is less than the value for the infinite cylinder. A closed, approximately equidimensional cylinder may be expected to give similar shielding to that of the equivalent sphere.

Equations 5.34 and 5.35 have been used to determine theoretical shielding factors at 50 Hz for a variety of aluminium cylinders, an aluminium box, and the cylindrical eddy current shield around the sample and pick-up coils on the PAR (Princeton Applied Research) spinner magnetometer. The agreement with experimentally determined S-values is reasonable, even for the spinner shield (approximated to a sphere) for which the condition $d \ll a$ does not obtain and the wall thickness (~ 2.5 cm) is different from the end face thickness (~ 5 cm). As expected, the shielding factor for axial fields for the PAR shield is considerably greater than for transverse fields. The shield is in the form of two hinged sections to allow access to the sample holder, and the eddy current paths appropriate to a transverse field will probably not traverse the small air gap between the two halves of the closed shield, because of its high resistivity, but will circulate separately in the two halves. The shield (internal and external diameter 12.5 cm and 17.5 cm, internal and external length 7.5 cm and 17.5 cm) has experimental longitudinal and transverse shielding factors of ~30 and ~5, at 50 Hz.

5.2.2 Magnetic shielding against static fields

The shielding of a region from a static magnetic field is achieved by enclosing the region with material of high permeability which forms a path of low reluctance along which the field preferentially passes. The shielding efficiency depends on many factors, the most important of which are the permeability of the wall material and its variation with the flux density it carries, the number of layers in the wall and the thickness of each, the remanent magnetism of the wall material and the degree to which the region is completely enclosed. Because of these and other variables, exact calculations of shielding factors (the ratio of the external and internal fields) for particular shield configurations and wall material are not possible. However, several theoretical investigations of geometrically simple shields (spherical and cylindrical) have been carried out, usually with some simplifying assumptions such as constant permeability and zero remanence of the wall material. These calculations are at least useful for estimating shielding factors of single-stage shields for various configurations and permeabilities.

It might appear at first sight that the optimum use of a given mass of shielding material would be a single layer of maximum thickness enclosing the required region, thus providing a path of minimum possible reluctance to the field. Rücker (1894) was the first to point out that this is not necessarily correct, and he showed, for a spherical shell of wall thickness t, radius r, and relative permeability μ_r, that if $t > 3r/2\mu_r$ it is advantageous to use two or more concentric shells separated by air or other low permeability material. Thus, with present day high-permeability materials ($\mu_r \approx 10^4$–10^5) this theoretical thickness is inconveniently small, but in practice the above criterion may not be reliable because of a decrease of permeability with wall thickness. At the time of Rücker's investigations the most common shielding material was soft iron of much lower permeability (~100) than is currently available.

(*a*) *Spherical shields*

Schweizer (1962) gives expressions for the shielding provided by a single and two-shell spherical shield in which $t \ll r$, and, in the case of the two-shell configuration, the separation of the shells is also much less than r. Assuming that $\mu_r t/r \gg 1$, the shielding factors S_1 and S_2 (one and two shells) are given by

$$S_1 \approx \frac{2\mu_r t}{3r} \tag{5.37}$$

and

$$S_2 \approx \frac{4\mu_r^2 t_1 t_2}{9r_1 r_2}\left[1-\left(\frac{r_2}{r_1}\right)^3\right] \tag{5.38}$$

where t_1, r_1 and t_2, r_2 refer to the outer and inner shells respectively: μ_r is assumed to be the same in each shell. It can be seen from Equation 5.38 that the shielding factor for the two-shell system is the product of the shielding factors for each shell considered individually, multiplied by a factor which arises from mutual interference of the two shells. Patton (1967) uses this principle to derive a general expression for multistage shielding. When a new shell is added internally a small fraction of the flux in the previous internal shell now passes through the new shell. The effect of this flux redistribution is to increase the field against which the new shell must shield, and the ideal combined shielding factor ($= S_1 S_2$) is reduced by the mutual interference factor. The general expression for multistage shielding then takes the form

$$S_m = S_1 S_2\left[1-\left(\frac{r_2}{r_1}\right)^3\right] S_3\left[1-\left(\frac{r_3}{r_2}\right)^3\right] \ldots S_n\left[1-\left(\frac{r_n}{r_{n-1}}\right)^3\right] \tag{5.39}$$

Mager (1970) and Thomas (1968) give a more exact expression for multiple shielding of spheres which consist of the terms on the right-hand side of Equation 5.39 plus the sum of the individual shielding factors. It can be seen that if the terms $[1-(r_n/r_{n-1})^3]$ are very small, i.e. very small separation of the shielded stages, the total shelding factor is additive rather than multiplicative. The greater the decoupling of the stages the greater the shielding factor.

(b) *Cylindrical shields*

Spherical shields are not usually a practical proposition because of the difficulty of fabrication, and cylindrical and cubical configurations are more commonly used. Because of the lesser symmetry of these shapes, shielding factors are more difficult to calculate and practical values are dependent for cylindrical shields on the length/radius ratio, whether the shield is closed, closed at one end or open-ended and the direction of the external field relative to the shield axis. The presence of the access opening is, of course, a feature which affects the shielding factors of all types of shield. Wills (1899) investigated theoretically the shielding produced by multishell cylinders of infinite length and constant permeability when B_0 is perpendicular to the shield axis. If μ_r is sufficiently high his formulae take the form (Wadey, 1956)

$$S_1 = \frac{\mu_r(1-q_1)}{4} \tag{5.40}$$

for a single shell, and

$$S_2 = \frac{\mu_r}{4}\left[(1-q_1 q_2)+(\mu_r/4)n_1 n_2 n_{12}\right] \tag{5.41}$$

for the double shell configuration. If a_1, b_1 and a_2, b_2 are the inner and outer radii of the inner and outer shells respectively, $q_1 = a_1^2/b_1^2$, $q_2 = a_2^2/b_2^2$, $n_1 = 1 - q_1$, $n_2 = 1 - q_2$ and $n_{12} = 1 - q_1 q_2$. Substituting for the shell thickness $t = b - a$ in Equation 5.40 and assuming that $t \ll a, b$, Equation 5.40 reduces to

$$S_1 = \frac{\mu_r t}{2a} \tag{5.42}$$

which is slightly less than the theoretical factor for a single shell sphere of the same radius as the cylinder (Equation 5.37). Wadey (1956) points out that Wills' formulae indicate that S depends only on μ_r and ratios of the shell dimensions. This should be true even if, as is usually the case, the permeability in each shell is different. Thus a given configuration can be scaled up or down and the same shielding factor maintained, provided the thickness and spacing of the shells are changed in the same proportion and permeability is not dependent on wall thickness, which it may be (Section 5.2.2 (e)).

The value of S at the centre of a finite cylindrical shield is not easy to calculate exactly. Among other factors involved is the state of magnetization of the shield material which will change as the end is approached and therefore alter μ_r. It is also necessary when describing shielding factors for cylinders to specify the direction of the field relative to the cylinder axis. If the field direction is parallel to the axis of an infinite cylindrical shield, $S = 1$ and there is no shielding effect. As the length of a long finite cylinder decreases the longitudinal shielding factor S_L increases and becomes comparable to the transverse shielding factor S_T. For a single, open cylindrical shell Mager (1968, 1970) gives the approximate expression

$$S_L \approx 4NS_{T(0)} + 1 \tag{5.43}$$

where $S_{T(0)}$ is the transverse shielding factor for the equivalent infinite cylinder and N is the demagnetizing factor of an ellipsoid with dimensional ratio $\beta = l/2r$, where l and $2r$ are the shield length and diameter respectively. Provided $t \ll 2r$ and $\mu_r \gg 1$, $S_{T(0)} \approx \mu_r t/2r \approx S_T$ for a finite cylinder, for $l/r > {\sim}4$. Equation 5.43 appears to be valid if $l/r > {\sim}10$ (see below). N for $\beta > 1$ is given by

$$N = (\beta^2 - 1)^{-1}\{\beta(\beta^2 - 1)^{-\frac{1}{2}} \ln[\beta + (\beta^2 - 1)^{\frac{1}{2}}] - 1\} \tag{5.44}$$

Values of N are given in Appendix 2. Equation 5.43 indicates the analogy between the effect of the magnetostatic demagnetizing field inside a magnetized cylinder and the field reduction inside the shield.

For short open cylinders ($l/a < {\sim}10$) S_L is substantially reduced by penetration of the external field into the cylinder. The effect of this penetration (Mager, 1970) is an effective longitudinal shielding factor due to the openings S_{op} where

$$S_{op} \approx 0.38\,(l/2a)^{-\frac{1}{2}} \exp\,(kl/2a) \tag{5.45}$$

where $k \approx 2.3$. The observed shielding factor S'_L is then given by

$$\frac{1}{S'_L} = \frac{1}{S_L} + \frac{1}{S_{op}} \tag{5.46}$$

where S_L is the shielding factor for the shell derived from Equation 5.43. There is a similar, but much smaller modifying factor for S_T. The effect of closing the shield ends is discussed in Section 5.2.2(e).

(c) *Cubic shields*

The cubic configuration is of some interest because of its convenient form for small shields and it is also the only practical form for large room-sized shields.

There is no exact solution for the shielding factor of a closed cube but it is likely to be comparable to that of a sphere of equivalent size. Either l can be substituted for r in Equation 5.37, where $2l$ is the cube side, or, perhaps better, the value of r for which the cube and the sphere have the same volume, i.e. $8l^3 = 4\pi r^3/3$ or $r = 1.24\,l$. From Equation 5.37 the shielding factor for $\mu_r \gg 1$ for a single cube of side $2l$ is then

$$S \approx \tfrac{2}{3}\left(\frac{\mu_r t}{1.24l}\right) \approx 0.54\left(\frac{\mu_r t}{l}\right) \tag{5.47}$$

Cohen (1967b) uses a coefficient of 0.36 in Equation 5, based on theory and experiment, and Patton and Fitch (1962) round it off to 1.

Since it evolves ratios only, the multistage expression for spherical shields (Equation 5.39) can also be used for cubical shields. Because of flux leakage at the corners and edges of cubical shells Equations 5.39 and 5.47 are likely to provide an upper limit to the shielding available: there is also some distortion of the internal field near the corners and edges, depending on the angle between the external field and the cube faces.

It is sometimes convenient, if expensive, to operate one or more complete pieces of equipment in a magnetically shielded environment, i.e. in a shield which is the size of a small room. Because of the large values of l in such shields very high values of S (based on Equation 5.39) are not easy to obtain, but by utilizing the effect of magnetic remanence in the shielding material theoretical S-values can be much improved (Section 5.2.2(e)).

(d) *Shielding against direct and alternating fields*

Magnetic shields against direct fields have the advantage of also shielding against alternating fields. If the latter are present there is additional eddy current shielding as well as the static shielding effect and the shielding factor rises with increasing frequency of the field, at least in the low frequency range.

The quantity governing the relative importance of the eddy current contribution to the shielding is the skin depth (δ) of the shield material and its relation to wall thickness t. For $t < \delta$ the eddy current shielding is weak, and S tends to the static value (assuming $S > {\sim}10$). However, it can be seen from

Equation 5.33 that S is inversely proportional to $(\sigma\mu_r)^{\frac{1}{2}}$ and is therefore much smaller in the shield material than in non-ferrous metals because of the high values of μ_r which more than offset the lower electrical conductivity of the shield. If $\sigma = 1 \times 10^6\ \Omega$ m, a typical value for shield material, then $\delta \approx 0.6$ mm and 0.2 mm for $\mu_r = 10^4$ and 10^5 at $f = 50$ Hz. Thus, if $t \approx 1$ mm a significant increase in the shielding factor is expected for alternating fields. It should be noted that the shield is subjected to a varying magnetic field and therefore the effective permeability is uncertain because of the dependence of μ_r on the field inside the wall material.

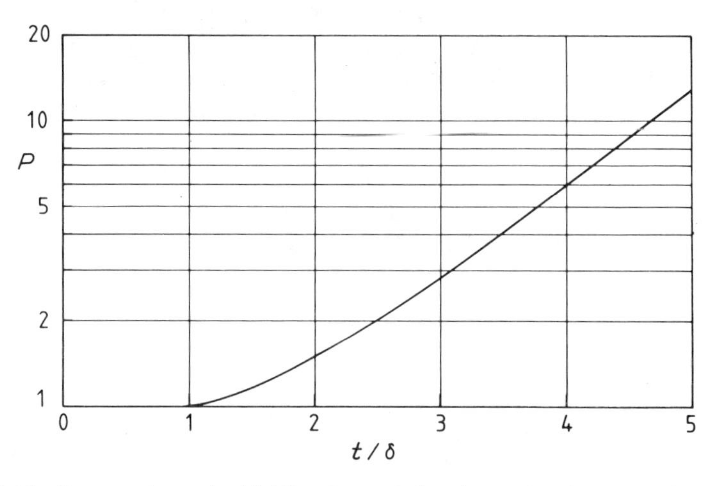

Fig. 5.11 Increase in static shielding factor S for alternating fields as a function of wall thickness/skin depth. P is the factor by which S is increased. (Redrawn from Mager, 1970)

The shielding factor increases with increasing frequency until the increase in f is offset by the decrease in μ_r which occurs at high frequencies. For the common shielding materials this occurs at 10–100 kHz. Figure 5.11 shows the increase in the static shielding factor expected when alternating fields are shielded, plotted as a function of wall thickness/skin depth for a single-layer shield (Mager, 1970). Although strictly applicable to a sphere or infinite cylinder, a useful estimate is also obtained for a cylinder long compared with its diameter. Further details of alternating field shielding are given by Mager (1970), including the phenomenon of resonance caused by a phase difference between the field penetrating into the ends of an open cylinder and the field entering through the shell. If at any point the phase difference is π a local increase in the shielding factor is observed.

(e) *Practical considerations*

The high-permeability shielding materials are specially annealed alloys of iron–nickel (Mumetal, Permalloy) or iron–cobalt (Permendur). Table 5.1

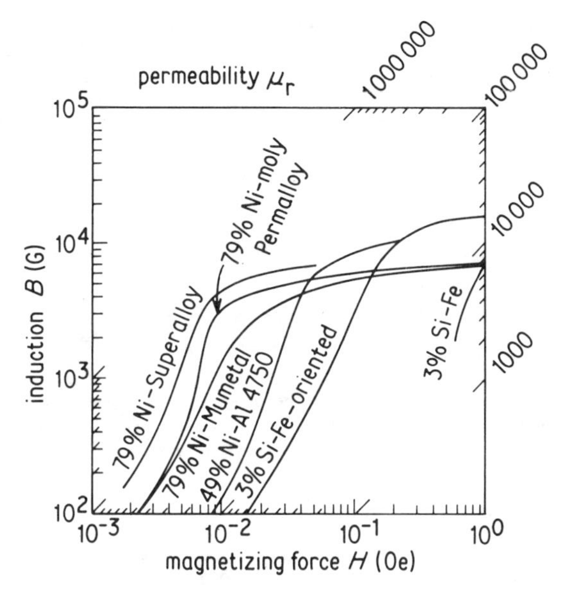

Fig. 5.12 Values of B, H and μ_r for some common shielding materials. $1\ \mathrm{G} = 10^{-4}\ \mathrm{T}$, $1\ \mathrm{Oe} = 79.6\ \mathrm{A\,m^{-1}}$. The permeability shown is relative permeability μ_r. (From Patton and Fitch, 1962)

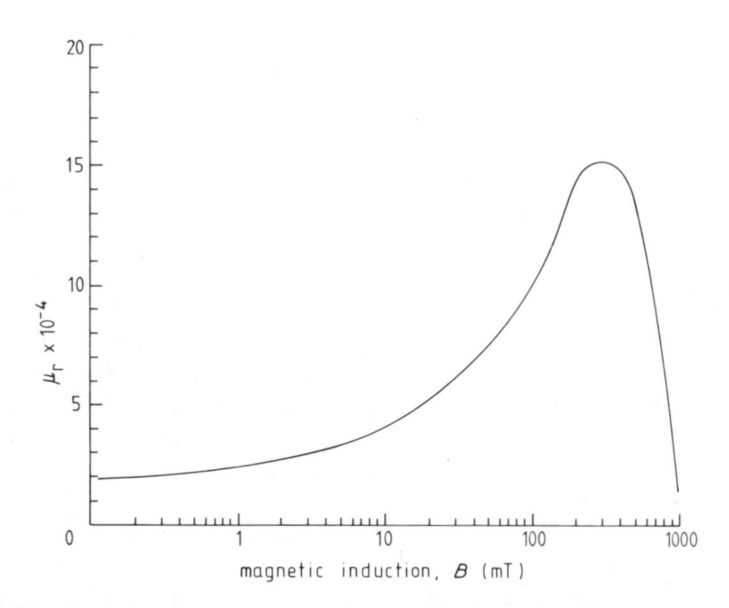

Fig. 5.13 Typical curves of variation of μ_r with the magnetic field B in the material, for Mumetal.

Table 5.1 Magnetic properties of high-permeability materials

Material	μ_r ($B = 2$ mT)	μ_r (max)	H_c (A m^{-1})	Resistivity ($10^{-8}\,\Omega$ m)
Iron	200	5 000	80	10
4% Silicon–iron*	1 500	30 000	12	47
Mumetal	20 000	100 000	4	62
Permendur	800	5 000	160	7
Permalloy (4-79)	20 000	100 000	4	55
Supermalloy	100 000	800 000	0.2	60

* Grain-oriented.
The values given are typical and μ_r may vary between samples and with sheet thickness (Data from *Handbook of Chemistry and Physics*, 53rd Edn (1974), Chemical Rubber Co., Cleveland, Ohio, USA.)

shows some of their properties which are of interest for magnetic shielding. Their permeabilities are sensitive to working during fabrication, and they become rather easily magnetized, for example through shock in the geomagnetic field, and therefore require careful handling. Figure 5.12 shows some representative values of B, H and μ_r.

The foregoing expressions for shielding factors are strictly theoretical and neglect various factors which are now considered. An important factor in the practical design of shields is the value of μ_r, and it should be known in each part of the shield. However, not only is the permeability of different samples of the same shielding material somewhat variable but μ_r also varies with wall thickness and, more important, with the magnetic induction in the wall. Figure 5.13 shows the variation of μ_r with B in the material for a typical high-permeability material such as Mumetal. It can be seen that in a multistage shield μ_r varies greatly in the outer and inner walls and that the possibility of saturation of the outer walls must be considered. μ_r can also be a function of sheet thickness, as it is in Mumetal and Permalloy. Patton (1967) quotes maximum values of μ_r of 1.2×10^5 and 1.5×10^5 for 0.015 cm and 0.035 cm thickness of Mumetal, and 1.5×10^5 and 3.0×10^5 for the same thicknesses of a Permalloy sample.

If μ_r is to be known with reasonable accuracy it is clearly necessary to know the value of B in the material (B_m). For a single shield on the outside of a multistage shield the magnetic flux through the shield material is equated to the same flux returning through the air, a fraction f of which passes through the inside of the shield and $(1-f)$ outside. When $S \gg 1$ the interior flux is approximately equal and opposite to the flux there in the absence of the shield. Following Patton (1967) consider an external field B_0 perpendicular to a face of a cubic shell of side $2l$ and wall thickness t. Then, from the above condition

$$fB_m 8lt = B_0 4l^2 \qquad (5.48)$$

or

$$B_m = \frac{B_0 l}{2ft} \qquad (5.49)$$

A reasonable and not very critical assumption is $f = \frac{1}{2}$, leading to

$$B_m \approx \frac{B_0 l}{t} \qquad (5.50)$$

Because B will be very low in the space inside the outer layer of a multistage shield, the working permeability for the inner shells will usually by close to the initial permeability of the wall material. Patton also shows that a useful approximate relation is $B_m/\mu_r = B_i$ (or $H_m = H_i$) where B_i is the field (in air) inside a shell. This expression enables a rough estimate of the shielding factor to be obtained by first calculating B_m from Equation 5.50 and then using $B_i = B_m/\mu_r$, the appropriate value of μ_r being obtained from a μ_r–B_m plot (Fig. 5.13).

In practice the above results are usually modified by the presence of remanent magnetism in the shield material. Equation 5.50 then becomes

$$B_m + B_r \approx \frac{B_0 l}{t} \qquad (5.51)$$

where B_r is the remanent field in the material. An important feature of Equation 5.51 is that B_m and B_r are vectors, and B_r can take values from zero to $> B_m$. *A* consequence of some interest is when $B_r = B_0 l/t$, whence $B_m = B_i = 0$, and the internal field is zero, i.e. $S = \infty$. This is the underlying principle of the achievement of very high shielding factors by adjustment of the remanent magnetism of the shield.

The above condition is valid for a single-stage shield. In a multistage shield B_r of the inside stage is controlled and B_0 is the field acting on this stage. B_0 is given by the field external to the whole shield times the ratio of inner stage shielding factor to the total shielding factor.

It can be seen that although shield remanence is a factor that can be exploited to improve S, it increases the difficulty of designing shields for a given shielding factor. A related factor which does not appear in the shielding expressions is the coercive force of the shielding material. This is usually of the order of 1–10 μT and a shield ceases to be effective if the external field falls substantially below this value. Thus, coercive force is another reason why theoretically high values of S are misleading.

In the design of magnetic shields it is necessary to take into account the magnitude of the external field B_0 which it is required to shield against. In palaeomagnetism the most common requirement is shielding against fields comparable to the geomagnetic field i.e. ~ 100 μT, and such an external field is assumed in the following section.

It should be clear from the foregoing discussion that shield design is not an exact science. However, it is possible to state some principles which give a lead for optimizing the use of shielding material for particular purposes and from which shielding factors can be estimated. Since cylindrical shields are the most convenient configuration for many applications features of their design are considered first.

Equation 5.42 can be used to estimate the transverse shielding factor S_T for an open single shell, provided the length $l > \sim 4r$. For typical values of t and r of 0.1 cm and 7 cm respectively, $B_m = 14$ mT and $\mu_r \approx 50\,000$ (Mumetal), giving $S_T \approx 350$. However, tests by the author on a variety of Mumetal cylinders in an external field of 100 μT suggest that estimates are often high, by as much as a factor of 2. Equations 5.43 and 5.46 appear to be useful as a rough guide to S_L. In open Mumetal cylinders in which l/r varied from 4.0 to 11.0, r from 3.5 cm to 6.4 cm and $t = 1.0$ and 1.3 mm, S_L varied between 30 and 250 and S_T between 150 and 300, both factors improving as l/r increased and r decreased. Part of the discrepancies observed may be due to remanence in the shields. Demagnetizing the shields is not entirely straightforward, because the skin depth at 50 Hz is comparable to the wall thickness and therefore the demagnetizing field is not uniform in the material. However, provided the peak field is substantially greater than the coercive force of the wall material, demagnetizing will occur throughout the wall. It is, of course, necessary to separate the shields in a multistage shield for demagnetizing and demagnetize each one separately. For wall thickness of ~ 1 mm a peak field of 500 μT at 50 Hz should be satisfactory. One way of detecting whether a remanent field is present inside the shield is to measure the interior field with the shield aligned along some axis relative to the external field, and then remeasured with the shield turned through 180°. A significant difference between the two residual fields indicates a remanent field contribution to the shielded external field. It is, of course, possible to use the remanent field to provide very high S-values for a fixed shield and a constant external field direction.

Although expressions have already been given for the shielding factor (S_T) for double and multiwalled shields, in practice they again only give a very rough estimate and appear to overestimate the shielding factors available. The design of multishell shields is a compromise between the need for wide shell separation for good decoupling (Equation 5.39) and the resulting increase in the size of the outer shell and cost of the shield. In practice, wall separations of ~ 0.1–0.2 of the inner shell radius are used in cylindrical shields, with wall thicknesses of 1–2 mm. Thomas (1968) gives a detailed analysis, including the effect of wall thickness. Tests by the present author indicate that shielding factors are approximately additive for double-walled shields, at least for S_T. S_L also increases if an outer shell is added to a single shell, but in a rather unpredictable fashion depending on the value of l/r of the inner shell.

For open cylindrical shields of small diameter (~ 10–20 cm) and $l/r > 4$ maximum shielding against low fields is essentially achieved with two shells. S_L may be increased somewhat with a third shell. For larger diameter shields, when S for each stage decreases, more stages may be advantageous if a high shielding factor is required.

The effects on S of closing the ends of a single-stage cylindrical shield have been analysed theoretically mainly in the German literature and are quoted by Mager (1970). The results obtained appear to the present author to be physically implausible since they imply a reduction of S_L relative to S_T by a factor of $1 + (r/l)$. This factor is often small, and experiments show that the effect on S_L of closing the shield ends is small for single-stage shields and is rather critically dependent on the contact with the shield and the remanent state of the ends. The effect on S_T is negligible for $l/r > \sim 4$ in single and multistage shields: however, there is a marked increase in S_L in multistage shields if one or both ends are closed even with poor magnetic contact.

The axial homogeneity of the longitudinal field within an open cylindrical shield has qualitative similarities with that of the field inside a finite solenoid, i.e. greater uniformity of interior field is achieved as l/r increases. However, there is also the transverse field to consider in a shield and, as might be expected, the homogeneity of this field also improves as l/r increases and the transverse field is generally more uniform along the axis than the longitudinal field. If a single-stage shield is converted to a multistage one by adding external shells the homogeneity of the residual field is also improved by a small amount.

Cubical shields have found use in palaeomagnetism in the form of small boxes for transport and storage of cut rock specimens, and large room-sized shields in which complete items of equipment operate. A Mumetal box with a well-fitting lid $\sim$10–20 cm on a side with $\sim$1.0 mm walls should provide a shielding factor of $\sim$100 or better against the Earth's field, which is adequate, for instance, for inhibiting the acquisition of VRM in red sediments after thermal cleaning and before NRM measurement.

The design and performance of room-sized cubic shields has been described by Cohen (1967a, 1970), Patton and Fitch (1962) and Patton (1967). Two important factors are the availability of the required shielding material in large sheets, and the cost and how it depends on the wall thickness of a shield.

The main points of the design procedure of Patton and Fitch (1962) for a two-stage shield are as follows. Mumetal was the best available material, and a shielding factor of 500 was required in a cube of maximum side length 2.76 m. The inner cube side is 2.44 m, a compromise between maximum working space and shielding efficiency. The two-stage form of Equation 5.39 is then applied to find the product of the individual stage shielding factors, $S_1 S_2$. A graph of S against t, the wall thickness, is now plotted for each shield using Equation 5.47, the appropriate value of μ_r being obtained for each t-value from Equation 5.50 and Fig. 5.13. The values of μ_r used for the inner shield take account of the field acting on the inner shield for different t- (and S-) values of the outer shield. From these graphs a further graph can be constructed relating the values of t_1 and t_2 to obtain the required product $S_1 S_2$.

If the ratio of the outer to inner cube side is n, the total volume of Mumetal used in the shield is proportional to $t_1 + n^2 t_2$, where subscripts 1 and 2 refer to the inner and outer cubes respectively. Assuming that the cost of the Mumetal increases with its volume, the graph relating t_1 and t_2 for the given values of $S_1 S_2$ is used to plot $(t_1 + n^2 t^2)$ against t_1 and t_2. This graph shows a marked minimum in $(t_1 + n^2 t_2)$ for a restricted range of values of t_1 and t_2 and therefore the thicknesses of the two shields can be chosen. For the shield under discussion $t_1 = 1.5$ mm and $t_2 = 0.75$ mm, giving a theoretical shielding factor of ~700.

The performance of the shield was somewhat better than predicted, providing a central residual field of ~35 mT instead of the expected ~70 nT in an external field of 50 μT. This may be due to slight shield remanence or different values of μ_r from those used. The residual field can be reduced further by magnetizing the shield in the appropriate direction by means of a Helmholtz coil system at the shield centre providing fields of the order of 10–100 μT at the walls. The typical field gradient in the shield is 6 nT m^{-1}. The stability of the residual field is normally ~5 nT and a heavy truck passing within 10 m of the shield produced a field change of only ~11 nT.

The uncertainties associated with the performance of magnetic shields is illustrated by Cohen (1967a), who constructed a two-stage cubic shield of Permalloy of similar dimensions to that just described, with $t_1 = t_2 = 1.5$ mm. The measured shielding factor was only 43, well below the most conservative theoretical estimate (~100), but no mention is made of the possible influence of remanence, nor does the variation of μ_r with B_m appear to have been considered.

Cohen increased the shielding factor by 'shaking' the shield. This is the process whereby the application of an alternative magnetic field to a permeable material results in an apparent increase in permeability, arising from the continuous motion of the domains and domain walls. Using a 60 Hz field of ~ 100 μT provided by a Helmholz coil system external to the shield, Cohen reports an increase in S for static and low-frequency fields by a factor of ~ 8. (Cohen 1967a, b).

Patton and Fitch (1962) point out that high-permeability walls are not essential to produce very low fields in a shield. By adjusting the remanence of the walls any desired low field can be obtained inside, for instance in a room 'papered' with mild steel sheet. However, external field variations would only be attenuated by a small amount.

If an enclosure is to be shielded against strong magnetic fields (~0.01–0.1 T) the same principles apply as for weak fields, except that saturation of one or more shells of a multistage shield must be avoided to achieve a good shielding factor. This usually necessitates the outer shell being of sufficient thickness, derived from Equation 5.50, such that B_m is below the saturation value, and/or the use of materials with higher saturation flux density than Mumetal or Permalloy (~ 0.7 T), such as the cobalt–iron alloys (e.g. Permendur, B_{sat} ~ 2.5 T).

In shield construction it is obviously important to avoid any high reluctance regions such as might occur at joints and lids in small shields and at joints and the door in room-sized shields. Small shields are usually welded, giving good joint continuity, but large shields are made from smaller sections of shielding material which require joining together. Patton and Fitch (1962) recommend an 8 cm-wide strip of the shield material along the length of the junction, pressed firmly against the joining sheets by frequently-spaced brass screws screwed into a wooden support behind the join. The internal and external doors overlap the door openings by 5 cm all round, and the space between the shield walls is evacuated with a vacuum cleaner. The modest pressure drop ($\sim$0.1 bar) provides a sufficient force to firmly press the doors against the overlapping walls.

A comparatively recent development in magnetic shielding is the availability of thin tapes and flexible sheets of interwoven ribbon of shielding material for 'wrap around' shields ('Metshield', 'Metglas', 'Telshield'). These materials have maximum permeabilities ($\sim$10 000–50 000) somewhat less than conventional alloys, with comparable saturation flux density and electrical resistivity. Typical thicknesses used are 0.05–0.2 mm. Because the material is very thin a single layer saturates in rather low external fields ($\sim$0.5 mT), and the shielding achieved against static fields is also low. The use of several layers partially overcomes these disadvantages. Better shielding factors are achieved against low frequency magnetic fields, which is the most useful application of these materials. For instance, published data for Metshield shows S_T values of $\sim$20 at 60 Hz for a single-layer cylindrical shield 5.0 cm internal diameter and 18 cm long with $B_0 = 0.2$ mT, and $\sim$50 with $B_0 = 0.1$ mT for a two-layer shield of the same dimensions with 5 mm separation of the layer.

An advantage of these shielding materials is that they can be shaped for different purposes and reworked many times without degradation of their magnetic properties, in contrast to the careful working and handling required for the conventional alloys.

The use of a shield or coil system for providing a low field environment depends on the particular conditions required and the nature of the field to be shielded. For a small volume of stable low field (relative to the Earth's field) and for protection against alternating fields and reduction of field gradients, a magnetic shield is convenient and portable and will attenuate external field variations. An equivalent coil system does not shield against alternating fields and requires a feedback system to counteract field changes, probably resulting in a greater cost than for a shield.

A coil system is preferred where a large volume of 'zero' field is required and where weak alternating fields are not a problem. The main advantages of three-component coils over shields are their flexibility in providing controlled fields in any desired direction and greater accessibility of the working space. A shielded room provides the latter but not the former and the expense of shielded rooms can be excessive. The double-walled 2.4 m room described by

Patton and Fitch cost $10 000 (in 1962). In some circumstances a combination of coils and shields is effective. A single-stage cylindrical shield inside a coil system may provide sufficient attenuation of external field variations to obviate the need for feedback controls of the coils and will efficiently attenuate alternating fields.

Some interesting papers on the development of static and dynamic shielding are those of Rücker (1894), Wills (1899), King (1933) and Gustafson (1938). More recently, in addition to the literature referred to in the text, there is a useful survey in a journal special issue (*IEEE Trans. Electromagn. Compat.* (1968), **10**, 1–175.

5.2.3 Superconducting shields

The ultimate in magnetic shielding is achieved by utilizing the phenomenon of superconductivity. A closed superconducting shell traps within it the magnetic field present at the onset of the superconducting state. Subsequent changes in the external field cause persistent, non-decaying currents to circulate in the shell which generate magnetic fields which exactly cancel the field changes. The effect arises from the zero electrical resistance of the shell and consequently 'zero' skin depth, providing a theoretically infinite shielding factor for fields of frequencies $\sim 10^{10}$ Hz down to zero. The only proviso is that the critical field of the superconductor is not exceeded, causing the superconducting state to be lost. Another property of a superconducting medium is the tendency for magnetic fields to be excluded from it as it cools through the critical temperature – the Meissner effect. This effect is not perfect, i.e. absolute zero field is not achieved in the medium, but it contributes to the shielding of direct fields and other techniques are used in practice to obtain extremely low residual fields in a superconducting enclosure.

The important feature of superconducting shields, and ones in which they differ from conventional shields, is that they trap the magnetic field present in the shield as it cools through the critical temperature, and therefore these shields cannot be used by themselves to produce a zero-field environment. The normal procedure to achieve this is to carry out the cooling in a magnetic field which is as near zero as possible, usually provided by a large coil system. If necessary, additional reduction of the interior field can be achieved. If the shield is rotated during cooling, eddy currents generate a magnet field which opposes any field in the shield perpendicular to the rotation axis and these currents persist when the superconducting state is achieved (Vant-Hull and Mercereau, 1963; Deaver and Goree, 1967). In a cylindrical shield rotated about its axis at one revolution per second the residual transverse field can be reduced by a factor of 10^2–10^3 by this method. The axial field can be reduced by using a superconducting solenoid carrying the required persistent current (Deaver and Goree, 1967).

Once the required field has been trapped in the shield it remains extremely constant even if the external field changes. The stability is such that any variations are less than those detectable by currently available field magnetometers ($\sim 10^{-2}$ nT). For external field changes very high shielding factors are obtainable. For a cylindrical shield open at both ends the transverse and longitudinal shielding factors for uniform external fields are (Goree and Fuller, 1976)

$$S_T = 36^{x/2r} \tag{5.52}$$

and

$$S_L = 31^{x/r} \tag{5.53}$$

where x is the distance in from the open end and r is the cylinder radius. Thus, at the centre of a cylinder of length six times its diameter, $S_T \approx 4.7 \times 10^4$ and $S_L \approx 8.9 \times 10^8$. With superconducting shields $S_T/S_L < 1$ in contrast to a single-stage conventional shield, for which $S_T/S_L > 1$. From Equations 5.52 and 5.53 it is clear that S_T and S_L decrease rapidly as the ends are approached, but they indicate the high shielding factors that can be achieved.

In many applications the facility of a wide range of trapped magnetic fields is a useful one. Fields of 0–1.0 T can be trapped and maintained with a stability several orders of magnitude better than a servo-controlled system. This property can be utilized for measurements in rock magnetism with the cryogenic rock magnetometer of low and high field susceptibility, and for investigations of the acquisition of thermoremanent and viscous magnetization.

The high capital cost of superconducting shields and the additional running costs associated with liquid helium and its loss usually make the use of these shields prohibitively expensive for routine use in the palaeomagnetic laboratory. The extremely high stability and shielding factors are also only rarely, if ever, necessary. Their main use in rock and palaeomagnetism is in association with cryogenic magnetometers (Section 9.4).

Chapter Six

Miscellaneous techniques

6.1 Separation methods

There is an extensive literature on mineral separation and only an outline of the different techniques is given here. The emphasis is on those methods applicable to separation or concentration of the common magnetic minerals from different types of rock.

Separation techniques for magnetic minerals can be divided into two categories, those which depend on the difference in magnetic properties between magnetic and other minerals, and those depending on other physical properties. The choice of method may depend on whether it is required to preserve the magnetic state of the minerals, for instance if particle NRM is to be investigated, or whether, as for mineral identification, this is of no consequence.

In general, good separation becomes increasingly difficult as the grain size of the rock decreases. It is usually not too difficult to extract the comparatively abundant magnetic mineral in many igneous rocks, whereas the fine grain size and often low magnetic mineral content of lake sediments and some red siltstones poses a more difficult problem.

6.1.1 Non-magnetic methods

These techniques are essentially those used in geological studies, the most common method involving the use of heavy liquids in which particles float or sink according to whether their density is less or greater than that of the liquid. In the simplest method the crushed or otherwise disaggregated rock is thoroughly mixed into the heavy liquid and the particles allowed to settle or rise to the surface according to their density. An improved method is to use a steep-sided conical funnel to hold the liquid and rock particles, with a stop-cock or pinch clip at its lower end. The heavy fraction can then be removed by opening the stop-cock and allowing the particles to flow through, with a portion of the liquid, into a filter paper in a second funnel from which, after washing, it can be recovered. For good separation, especially of material of fine sandstone grade or smaller, it is essential that the individual grains are

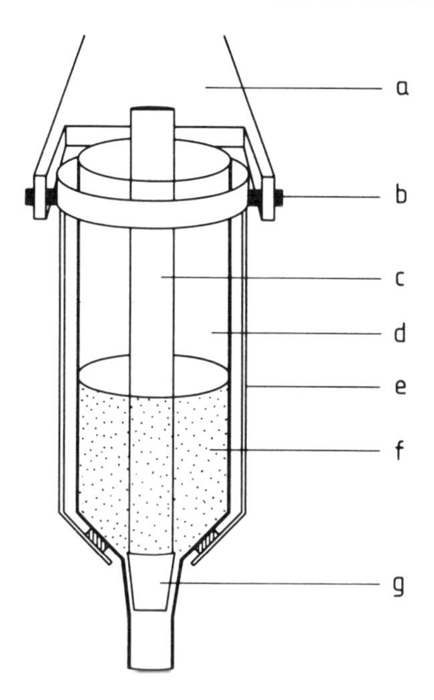

Fig. 6.1 Slow-speed centrifuge for heavy mineral separation in sediments. a, rotating arm; b, swing axis; c, glass tube; d, heavy liquid container; e, brass frame; f, heavy liquid; g, ground glass joint.

dispersed in the liquid and wetted by it. This can usually be achieved by thorough stirring, or by ultrasonic agitation.

In the author's experience, this method is not very satisfactory for sandstones or siltstones of typical size 20–200 μm, for example when extracting the sparse ($\sim$1–3%) polycrystalline haematite from redbed samples. Prior removal of carbonates by treatment with dilute hydrochloric acid improves dispersion of the grains and the use of a slow-speed centrifuge (providing 10–20 g) greatly increases the rapidity of separation. Figure 6.1 shows a diagram of the author's apparatus for separating specularite from red sediments. Bromoform (specific gravity 2.89) is a suitable liquid to use. If the sediment is well dispersed, good separation is achieved in 5–10 minutes after which the light fraction is stirred again and a second and, if necessary, third run is carried out. The heavy fraction is removed by briefly lifting the central tube and allowing the material to flow out into a suitable container. Although less successful with finer grained material, e.g. lake and sea-bottom sediments, some concentration of heavy minerals can be achieved with this type of equipment, particularly if higher centrifuging speeds are used.

Liquids in common use for mineral separation are bromoform, $CHBr_3$ (specific gravity 2.87 at 20°C), tetrabromoethane, $C_2H_2Br_3$ (2.96), mercury

Table 6.1 Density of minerals (relative density or $\times 10^3 \text{ kg m}^{-3}$)

Mineral	Density	Mineral	Density
Magnetite (Fe_3O_4)	5.2	Rutile (TiO_2)	4.3
Haematite (α-Fe_2O_3)	5.3	Calcite ($CaCO_3$)	2.7
Maghemite (γ-Fe_2O_3)	4.6–4.9	Magnesite ($MgCO_3$)	2.8–3.1
Ilmenite ($FeTiO_3$)	4.8	Dolomite ($MgCa(CO_3)_2$)	2.9
Pyrrhotite (FeS_{1+x})	4.5–4.6	Feldspars	2.6–2.8
Troilite (FeS)	4.8	Pyroxenes	3.0–3.7
Pyrite (FeS_2)	5.0	Olivines	3.2–4.4
Ulvöspinel (Fe_2TiO_4)	4.8	Biotite	2.7–3.1
Goethite (α-FeOOH)	4.3	Mica	2.7–3.1
Lepidocrocite (γ-FeOOH)	4.0	Quartz	2.6
Siderite ($FeCO_3$)	3.9		

potassium iodide (Thoulet's solution, 3.20), and thallium formate-mallonate, $CH_2COOTl_2 + HCOOTl$ (Clerici's solution, 4.25). Table 6.1 lists the densities of magnetic and other common minerals. The reader is referred to the books cited below for details of the preparation and use of these liquids, their admixture with other liquids to obtain intermediate densities and precautions to be observed when using them. It is usual to make a preliminary separation with bromoform, with which the opaque (including all the magnetic) minerals can be separated from quartz, feldspars, carbonates, pyroxenes, etc. Further separation of the heavy fraction can then be achieved with smaller quantities of the more expensive, higher-density liquids. For some purposes the concentration of magnetic minerals achieved with bromoform alone is sufficient. Among the texts in which heavy liquid techniques are described are Twenhofel and Tyler (1941), Krumbein and Pettijohn (1938), Tickel (1965), and Allman and Lawrence (1972), the last mentioned containing much useful information.

A direct but somewhat tedious method of separation is by hand-picking of grains from a powdered rock sample dispersed on a plate under a low-power microscope. Provided the particles of interest can be identified the method can be useful for extracting a small amount of not too fine-grained material ($\sim$100–1000 μm), for example for X-ray analysis. A simple aid in the technique is a glass tube tapered to a small hole at one end with the other end connected to a vacuum line (a water pump is quite sufficient). A finger held over a hole in the side of the tube forms a simple valve by which a particle sucked on to the end of the tube can be released into a container: alternatively, if the tube ends in a larger hole, particles can be drawn into the vacuum line and trapped in an appropriate container. Further details are given in Allman and Lawrence (1972).

Two other methods of separation which are of some interest but not in common use are based on differences in the electrical conductivity and dielectric constant of minerals.

In the former technique, mineral particles dispersed on a metal plate are subjected to an electric field of an electrostatically charged rod or plate held over them. Positive and negative charges on the particles are attracted to or repelled from the rod according to its polarity, those repelled leaking away to the metal plate if the particle conductivity is sufficiently high. These particles are then attracted to and held by the rod, whereas the poorly conducting minerals do not allow the repelled charges to leak away and the net attractive force is very small. The magnetic minerals, including haematite, have high conductivity relative to others and this method is potentially useful for their extraction. Some aspects of the technique are described by Crook (1910), and also by Twenhofel and Tyler (1941) and Krumbein and Pettijohn (1938).

Electrostatic forces are also present when mineral particles are subjected to an electric field in a medium whose dielectric constant K_m differs from that of the particles K_p. If $K_p > K_m$, the particles tend to move into the field, whereas when $K_p < K_m$ they are repelled from it. The dielectric constant of most minerals lies in the range 5–80, with the oxide and sulphide minerals such as magnetite, haematite, ilmenite and pyrrhotite in the upper part of the range. Various organic liquids and mixtures of them provide a range of values of K_m. The technique has only been reported for small-scale use: an electrode system of two needles with their points ~ 0.1 cm apart and a field of $\sim 10^6$ V m^{-1} across the gap was used by Berg (1936), and the method has also been described by Hatfield and Holman (1924) and in Krumbein and Pettijohn (1938) and Tickel (1965).

6.1.2 Magnetic methods

These techniques separate minerals according to their magnetic susceptibility or saturation magnetization. If a particle is in a region of magnetic field B and field gradient $\mathrm{d}B/\mathrm{d}x$, a force F acts on it, given by

$$F = \frac{m\chi B}{\mu_0}\frac{\mathrm{d}B}{\mathrm{d}x} \tag{6.1}$$

or

$$F = \frac{mJ_s}{\mu_0}\frac{\mathrm{d}B}{\mathrm{d}x} \tag{6.2}$$

where m is the particle mass and χ and J_s its specific susceptibility in the field B and its saturation specific magnetization. Equation 6.1 is the relevant expression for paramagnetic minerals and those which are not saturated in B and Equation 6.2 applies where saturation is achieved. As B and $\mathrm{d}B/\mathrm{d}x$ increase, particles of weaker susceptibility can be extracted. Since most common minerals are ferrimagnetic or paramagnetic, magnetic separators can be designed to separate not only the minerals commonly considered to be magnetic but many others as well. It can be seen from Equations 6.1 and 6.2 that the forces acting on paramagnetic and diamagnetic minerals will be opposite in direction because of the negative property of diamagnetism. Thus there is a small separating action between paramagnetic particles, which move towards regions of increased B, and diamagnetic quartz, calcite and dolomite particles which tend to move out of the field.

In rock magnetism it is usually required to extract magnetic minerals only, a procedure which is facilitated by their high effective susceptibility relative to other minerals. The exception is haematite, the susceptibility of which is comparable to some paramagnetic minerals (Table 2.1).

A simple but often effective method of magnetic separation is to cover one end of a strong bar magnet with a thin polythene or other plastic sheath and move it about just above a layer of rock particles dispersed on a flat surface. Alternatively, the sheathed magnet can be used to stir a suspension of particles in water. The particles adhering to the magnet can be removed by withdrawing it from the sheath, from which the particles will drop or can be brushed off. This technique is often effective with igneous material, from which magnetites and titanomagnetites can be extracted, and also composite particles possessing a sufficiently high proportion of magnetic mineral.

The magnetic analogue of hand picking from dispersed rock particles is the use of a magnetized needle. Particles which possess appropriate susceptibility can be detected by their attraction to the needle point and then removed from the matrix minerals. Removal of particles from the needle is facilitated if it is in the form of a small electromagnet. A short length (~ 5 cm) of iron or mild steel wire is tapered to a point at one end and forms the core of a small solenoid at the other. Switching off the current in the solenoid then reduces the needle magnetization to its low remanent value and the particles will drop off or can be easily removed. Varying the current in the solenoid also enables a certain selectivity of extraction to be achieved.

For larger quantities of material, separation is best achieved by allowing the crushed rock to settle under gravity in a column of water in a vertical glass tube. If an external magnet or electromagnet provides a suitable field and gradient over a region of the column magnetic particles will be held there on the wall of the tube. Figure 6.2 shows pole-piece configurations for these separators. For fine-grained lake sediments P. Readman (private communication) continuously circulates a suspension of the sediment in water past the region of magnetic field until no further material is extracted.

It can be seen that in magnetic separators the particles are often held by a magnetic force acting against the gravitational force on the particle. Therefore, for particles to be held in air

$$F = \frac{m\chi B}{\mu_0}\frac{\mathrm{d}B}{\mathrm{d}x} \geqslant mg \tag{6.3}$$

where g is the acceleration due to gravity. The holding force is independent of the particle mass, provided B and $\mathrm{d}B/\mathrm{d}x$ do not vary significantly over the particle volume. In water, or any other liquid, the gravitational force is less in the ratio $(\rho_\mathrm{m} - \rho_\mathrm{l})/\rho_\mathrm{m}$, where ρ_m, ρ_l are the densities of the mineral and liquid respectively.

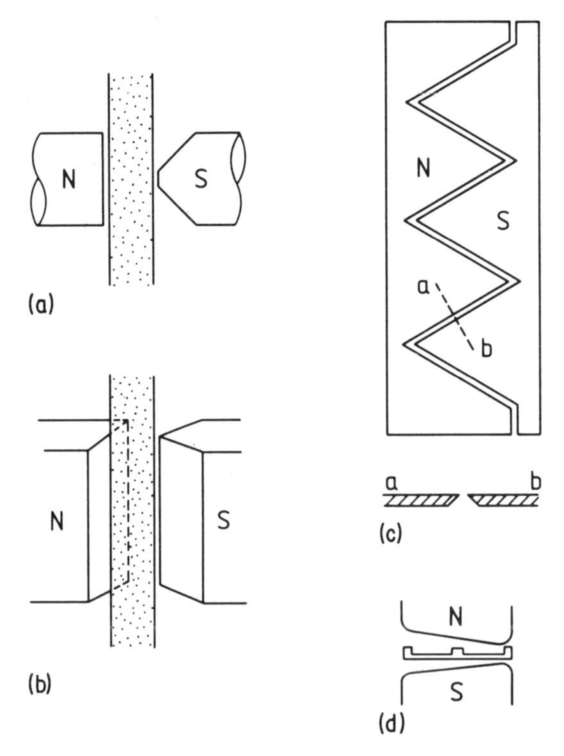

Fig. 6.2 Examples of pole-piece configurations for magnetic separators; (b) is probably more satisfactory than (a) in that there is a larger region of field and gradient; (c) is used vertically with a narrow rectangular chamber held against it down which the sediment suspension flows; (d) is a cross-section of the pole-gap in the Frantz separator.

Rearranging terms in Equations 6.3 leads to $(B/\mu_0)\mathrm{d}B/\mathrm{d}x > g/\chi$. For a paramagnetic mineral of specific susceptibility $\sim 50 \times 10^{-8}\,\mathrm{m^3\,kg^{-1}}$ to be held against gravity, $B(\mathrm{d}B/\mathrm{d}x) \approx 20\,\mathrm{T^2\,m^{-1}}$, e.g. $B = 0.5\,\mathrm{T}$, $\mathrm{d}B/\mathrm{d}x = 40\,\mathrm{T\,m^{-1}}$. These high values show that normally only the strongly magnetic minerals will be held, and that these only require a modest field and field gradient to extract them because of their much enhanced susceptibility.

The Frantz 'isodynamic' separator is designed to separate minerals with a wide range of susceptibilities, but it is not suited to the strongly magnetic iron minerals which are often first removed by another method. Its greatest use is in the geological laboratory for the separation of different paramagnetic minerals from crushed rock. The separator consists of a shallow brass channel about 20 cm long placed between the poles of an electromagnet of the same length (Fig. 6.2). The channel and magnet can be tipped about both longitudinal and transverse axes. In operation, the channel is angled so that material from a hopper at its upper end moves slowly down the length of the channel in the magnetic field. Because of the transverse slope of the channel, all particles tend to fall under gravity towards its lower edge: this tendency is opposed by a force towards the upper side exerted on paramagnetic (and ferrimagnetic) particles by the magnetic field and gradient. According to whether the gravitational or magnetic force on a particle dominates, which depends on its susceptibility and the field and gradient values, it will move down the channel near the lower or upper edge. The lower end of the channel has a central divider which directs the two groups of particles into separate containers. Precautions to be observed to obtain the best results are given in Allman and Lawrence (1972) and an application of the separator to alluvial mineral separation is described by Flintner (1959).

Lovlie (1972) has adapted the Frantz separator for the extraction of magnetic minerals from sea-bottom sediments. The brass channel is replaced by a narrow glass tube and the magnet and tube are placed in a nearly vertical position. The tube is filled with water and a suspension of the sediment allowed to flow slowly down it. Magnetite, haematite and some paramagnetic minerals are held against the section of tube wall in the region of strongest field, while non-attracted minerals fall under gravity towards the opposite wall owing to the slightly off-vertical setting of the tube.

After extraction of the magnetic fraction from a rock sample, it may be desirable to further separate particles of different magnetic properties from the extract. In the case of titanomagnetite particles extracted from igneous rocks this may be possible by thermomagnetic separation. This technique is based on the change in Curie temperature with composition of titanomagnetites. If the mineral extract is heated to a temperature T on a plate placed below and close to one pole of a magnet, those grains with $T_c < T$ are attracted to the magnet while those with $T_c > T$ will be paramagnetic and remain on the plate. By decreasing T in steps from the maximum Curie point, present grains of different composition can, in principle, be extracted.

Figure 6.3 shows experimental equipment used by Parry (1957). At each temperature material will be attracted by the magnet to the upper part of the tube provided

$$\frac{mJ_B}{\mu_0}\frac{dB}{dz} > mg \qquad (6.4)$$

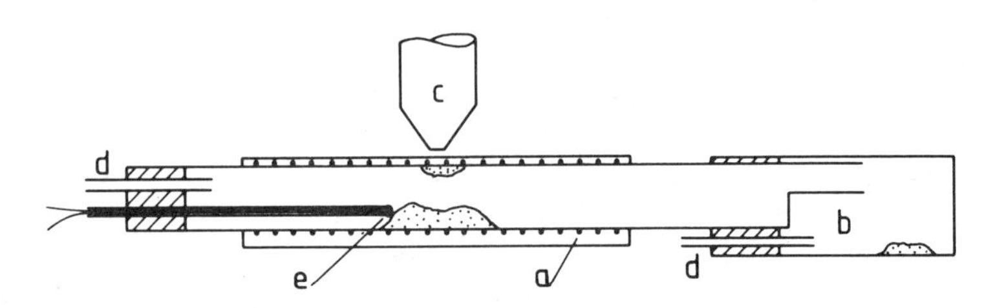

Fig. 6.3 Thermomagnetic separator. a, furnace; b, trap for extracted material; c, magnet; d, inlet and outlet for nitrogen; e, thermocouple.

where m is the mass of a grain, J_B is its magnetization per unit mass in the magnetic field B and dB/dz is the field gradient, assumed constant over the grain. The attracted material is collected by moving the electromagnet along the furnace until the material is in the specimen tube and switching the magnet off. A similar separator is described by Nagata (1961).

For successful extraction, thermomagnetic separators require properties in the particles which may not commonly occur. The desirable properties include purity of composition, absence of composite grains (i.e. of ferrimagnetic and paramagnetic materials or containing particles of differing Curie points) and J_i–T characteristics such that dJ_i/dT near T_c is large. It is also often difficult to avoid mineral alteration during heating.

The J_i–T characteristic determines the 'resolving power' of the technique, i.e. the range of Curie points in the particles lifted at a given temperature. It can be seen that the steeper the J_i–T curve is near T_c the better the resolving power.

Thermomagnetic separation has not proved to be a very satisfactory technique in practice and is little used. Parry (1957) and Nagata (1961) give some examples of its use.

Equation 6.4 can also be applied to the separation of particles possessing different magnetic properties at room temperature. Blackett (1956) applied the technique to distinguish and extract different minerals from a magnetic separate and to estimate their coercive force and saturated remanence. The analysis given here differs slightly from Blackett, who considered the particles to be in the Gauss B position relative to the magnet. If a small magnet is held with its axis vertical over the separate, a grain will just be lifted if (Equation 6.4)

$$g = \frac{J_B}{\mu_0}\frac{dB}{dz} \tag{6.5}$$

If the magnet can be considered as a dipole of moment P, $B = \mu_0 P/2z^3$ and $dB/dz = 3\mu_0 P/2\pi z^4$, where z is the distance of the magnet centre over the grain. As z decreases, B and dB/dz increase until Equation 6.5 is satisfied for a grain at a distance z_1. If the magnetization curve for the grain is approximated by the relation

$$J_{1B} = \frac{\chi B}{\mu_0} \tag{6.6}$$

where χ is a constant susceptibility, then

$$\chi = \frac{4\pi^2 z_1{}^7 g}{3\mu_0 P^2} \tag{6.7}$$

Now if the particle is initially magnetized to saturation and possesses a saturated remanence J_r, its total magnetization as the dipole distance decreases will be approximately of the form

$$J_{2B} = J_r + \frac{\chi B}{\mu_0} \tag{6.8}$$

and

$$\chi = \frac{\mu_0 J_r}{B_c} \tag{6.9}$$

where B_c is the coercive force measured in the field. The particle now lifts when $z = z_2$, when Equation 6.5 is satisfied for the value of J_{2B} given by Equation 6.8. Equation 6.5 now becomes, using Equations 6.6 and 6.8

$$g = \left[J_r + \frac{\mu_0 \chi}{2\pi} \left(\frac{P}{z_2^3} \right) \right] \frac{3\mu_0 P}{2z_2^4} \tag{6.10}$$

Inserting the expression for χ from Equation 6.7 we get

$$J_r = \frac{2\pi g}{3\mu_0 P z_2^3} (z_2^7 - z_1^7) \tag{6.11}$$

and

$$B_c = \frac{\mu_0 P}{2\pi z_1^3} \left(\frac{z_2^4}{z_1^4} - 1 \right) \tag{6.12}$$

Blackett gives results which show reasonable agreement for values of J_r and B_c for individual iron, magnetite and haematite particles and values obtained from measurements on the bulk materials.

Housley, Grant and Abdel-Garwad (1972) use a similar technique for selectively separating magnetic particles from lunar fines. A lower limit can be placed on the amount of iron metal present in particles that are lifted in a known field and field gradient. The variable field and gradient are supplied by a vertical permanent magnet with the upper pole-piece adjustable in height above a horizontal plate containing the particles.

6.2 Colloid techniques

The use of a finely divided ferro- or ferrimagnetic powder for mapping the magnetic structure at the surface of a magnetized sample (Bitter patterns) is well established. The most common application is to studies of domain structure in ferromagnetics, the particles being attracted to and marking as dark lines and areas visible under a microscope the regions of high magnetic field and field gradient associated with the discontinuities of magnetization at domain walls.

In the original form of the technique as developed by Bitter (1931), fine iron powder was sprinkled on to the surface under examination but in later experiments (Bitter, 1932) a suspension of maghemite particles in ethyl acetate was used. The technique was not very successful initially because of the large size of the particles and the lack of careful surface preparation which it is now known is essential to obtain good results. True colloidal particles of magnetite, i.e. in the approximate range 0.001–0.1 μm (10–1000 Å), were first prepared by McKeehan and Elmore (1934) and used in a soap solution as a dispersant. Methods of preparing colloidal magnetite are given in Appendix 4. Colloid suspensions generally consist of isolated particles together with aggregates of 10–100 particles: magnetically there will effectively be superparamagnetic, single and multidomain grains present.

The formation of a colloid pattern requires that the energy associated with the magnetic attractive force at a boundary exceeds the energy of thermal fluctuation of the particles. The relevant theory is summarized by Carey and Isaac (1966). Elmore (1938b) describes some of the magnetic properties of colloidal magnetite.

One of the difficulties of the colloid technique is the achievement of a flat, strain-free surface on the sample. Conventional mechanical polishing techniques leave a thin (~ 1000 Å), partially melted and distorted surface layer (the Beilby layer) and a colloid pattern on such a surface may only record distortions of magnetization due to localized, stress-induced anisotropy rather than the true domain structure. In rock samples and mineral crystals there is also the possibility of thermal alteration of the surface layer. Various techniques have been developed for removal of surface stress, on the stressed layer and also for 'etching' or increasing the relief between different phases in a particle. These techniques are thermal annealing, electropolishing, thermochemical etching, and ionic etching. Soffel and Petersen (1971) describe the latter technique in which a mechanically polished sample, placed behind a cathode containing a circular hole, is bombarded with positive ions produced by an electric discharge in air at 10^{-2} torr pressure.

The temperature of the rock surface is prevented from rising too high ($\sim 50°$ C) by periodic interruption of the bombardment and by restricting the discharge voltage. For the polishing of magnetite and titanomagnetite in a rock matrix, the authors used a 3.5 kV discharge at a pressure of 2×10^{-2} torr, with a '20 minutes on – 20 minutes off' bombardment programme. The maximum temperature rise was 40° C, a necessary limitation because of evidence of oxidation of the surface at higher temperatures. A typical time required for polishing is 10–20 hours.

A method of thermochemical etching for separated magnetite grains is that used by Hanss (1964). Slabs about 2 mm^2 in area are cut parallel to the crystal structure and placed in a crucible containing a mixture of boron and lead oxide in the ratio of 1:10 by weight. The crucible is heated to $\sim 650°$ C for 10 minutes and the resulting 'glass' removed from the magnetite grains by solution in 3N nitric acid.

Electrolytic polishing (electropolishing) can be used for magnetite (and possibly pyrrhotite) crystals, and is described by Soffel (1963). The mineral crystal is held in a metal clamp and forms the anode of the electrolytic cell, a piece of sheet iron 5 cm^2 in area forming the cathode. The electrolyte is a 5:1 mixture by volume of anhydrous acetic acid and perchloric acid and a current density of 20–100 A cm^{-2} is required. For further details and for precautions to be observed in using the above electrolyte the reader is referred to Jacquet (1956).

Thermal annealing of a strained surface is a comparatively simple technique, but involves the possibility of chemical alteration. Soffel (1965) heats mechanically pre-polished samples to 800° C for ~60 hours in a vacuum of 10^{-5} torr, followed by slow cooling. Vlasov and Bogdanov (1964) obtained good results with magnetite particles by first polishing with chromium and magnesium oxide powder to a mirror finish and then heating to 900° C for 4 hours.

The colloid technique has not found wide use in rock magnetism and has largely been confined to the study of magnetite and titanomagnetites, the only common terrestrial magnetic minerals (with haematite) in which the surface fields and gradients are sufficiently strong for pattern formation. Surface preparation of single magnetite crystallites is possible by thermal annealing (Vlasov and Bogdanov, 1964; Hanss, 1964), and electropolishing or ion etching. The electrical conductivity of titanomagnetite particles in a rock matrix in which magnetite or titanomagnetite is dispersed is not high enough for electropolishing, and ion etching or annealing may be used. An advantage of the former technique is that the surface temperature can be kept low (~50° C), thus reducing the possibility of thermal alteration.

Some of the results obtained in domain studies in rock magnetism are given by Vlasov and Bogdanov (1964), Soffel (1968, 1971, 1978, 1981), and Soffel and Petersen (1971). As an example, Soffel (1971) demonstrated the dependence on grain size of domain structure in titanomagnetite containing 45 mol % Fe_3O_4, and determined the critical radius of grains for transition to the single-domain state and also the type of domain structure occurring just above single-domain size. The relation between the remanent magnetism of some iron meteorites and their domain structure is described by DuBois (1965). Blackman and Gustard (1962) and Eaton and Morrish (1971) have investigated domain patterns on crystals of haematite.

Apart from studies of domain structure, the colloid technique may also have potential for monitoring changes in domain configuration resulting from the application of a magnetic field (both direct and during a.f. demagnetization) and mechanical stress (Soffel, 1966) and from changes in temperature (Soffel, 1977). There is also the possibility of detecting particles which are carrying the natural remanence in a rock and particles which are capable of carrying remanence if the sample is first given a saturated IRM. Informative results may be obtained from these latter applications without the sophisticated polishing techniques required for domain observations.

6.3 Piezomagnetism and shock effects

Piezomagnetism is a general term for the effects of stress (uniaxial and hydrostatic) on the magnetic properties of rocks and minerals and is of some interest in both rock-magnetic and palaeomagnetic studies. In the former there are reversible and irreversible effects of stress on induced and isothermal magnetization (Kume, 1962; Girdler, 1963), susceptibility (isotropic and anisotropic) (Stacey, 1960b) and the acquisition of TRM by certain rocks (Kern, 1961). In the practical sphere, stress-induced changes in rock magnetization produced by the build-up of stress prior to an earthquake (the 'seismomagnetic' effect) has been proposed as a possible prediction technique by Stacey (1964), and Johnston and Stacey (1969) have also examined the evidence for a similar effect associated with volcanic eruptions. A useful selection of papers on these topics was given at the International Association of Geomagnetism and Aeronomy (IAGA) meeting in Seattle in 1977, and published in 1978 (*J. Geomagn. Geoelectr.*, **30**, 477–607).

In the field of palaeomagnetism the remanent magnetism acquired by some rocks when stress is applied and released in the presence of an ambient field has been considered as a possible source of secondary NRM. This is piezoremanent magnetization (PRM). Also of possible interest is the deviation of TRM direction from the ambient field direction when acquired by magnetically anisotropic rocks under stress (Stott and Stacey, 1960; Kern, 1961). A purely mechanical effect of stress on NRM is the possible shallowing or decrease of inclination in a depositionally magnetized sediment due to compaction by overlying beds.

The magnetic effects of shock have mainly been investigated in connection with remanent magnetic properties and only irreversible effects can generally be studied. The two main areas of interest are the modification of an existing NRM by shock, and the generation of a remanent magnetization when a rock is shocked in the presence of a magnetic field (shock remanent magnetization, SRM). Although of no great practical interest in terrestrial palaeomagnetism, shock effects are of potential importance as a factor in the interpretation of remanent magnetism on the Moon, where the widespread occurrence of meteorite impacts provides the shock source.

An important possiblity currently being investigated theoretically (Srnka, 1977) and in the laboratory (Martelli and Newton, 1977) is the acquisition of a shock remanent magnetism by rocks in a local transient magnetic field generated as a result of the impact.

Natural and laboratory-induced shock effects on the NRM of terrestrial rocks have been investigated by Hargraves and Perkins (1969), Shapiro and Ivanov (1967), Cisowski and Fuller (1978) and Halls (1979). Experimental work on samples from the Lonar crater in India and on lunar samples and their analogues is described by Cisowski, Dunn, Fuller, Rose and Wasilewski (1974), Cisowski, Fuller, Wu, Rose and Wasilewski (1975) and Cisowski, Dunn, Fuller, Wu, Rose and Wasilewski (1976).

No special equipment is required for the study of irreversible effects of pressure on magnetic properties since the samples are not under stress during measurement and conventional instruments can be used. Modified equipment is required when reversible changes are of interest, and measurements are carried out with the samples under pressure.

Pressure effects on the NRM of rocks have been measured by a variety of methods. The application of pressure during measurement usually precludes continuous sample rotation, and astatic and ballistic magnetometers have been adapted for NRM measurements.

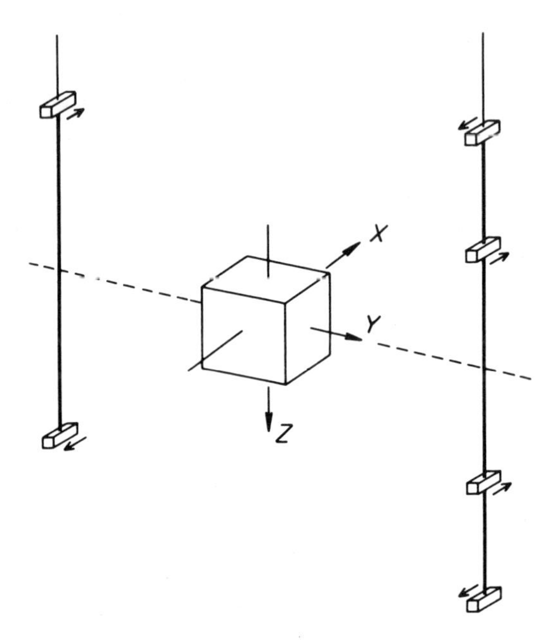

Fig. 6.4 Astatic system of Hodych (1967). The small arrows show relative polarity of magnets.

Hodych (1967) describes an ingenious combination of astatic and parastatic magnet systems which allows simultaneous measurement of two orthogonal components of NRM of a sample under uniaxial stress. The two systems are suspended with their centres on the same horizontal line and their magnets aligned parallel (Fig. 6.4). If the geometry of the systems and their astaticisms are perfect, the two magnet systems do not interact (i.e. a deflection of one produces no magnetic torque on the other). If a magnetized sample is placed midway between the systems and on their line of centres the astatic system measures the vertical NRM component (Z) and the parastatic system the horizontal component parallel to the line of centres (Y). As constructed, with the magnet systems astatic to about 1% and the instrument set up in a city laboratory, changes in magnetization of about 2×10^{-5} A m^2 kg^{-1} could be detected in a 2.5 cm cubic sample, the limiting noise being due to stray magnetic fields. This limit could be much improved by increased astaticism and operation in a magnetically quiet location. Uniaxial stress of up to ~400 bars along the vertical axis of the sample was applied with a fixed, non-magnetic lever system. Valeyev (1975) reports some results obtained with rocks under uniaxial pressure and a conventional astatic magnetometer.

Pozzi, Godefroy and Legoff (1970) and Pozzi (1977) use a parastatic magnetometer of the type described in Section 9.2.1(e), mounted close to and mechanically isolated from a bronze and aluminium hydraulic press mounted on vertical guides. The sample can be rotated about a vertical axis to different azimuths while under uniaxial compression, and the horizontal component of NRM is measured with the sample opposite the central magnet of the system. The vertical component is measured by raising the sample above the level of the central magnet – a procedure similar to the off-centre method described for the astatic magnetometer (Section 9.2.3(b)). The standard sample is a cylinder 3.5 cm in diameter and 4.0 cm long, which can be subjected to a maximum uniaxial stress of ~3500 bars. A change in magnetization of $\sim 2 \times 10^{-7}$ A m^2 kg^{-1} can be detected, and there is also provision for making measurements in an applied field of up to 1 mT.

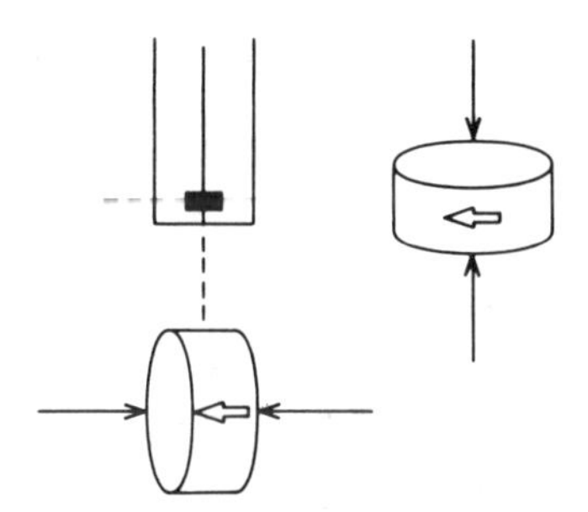

Fig. 6.5 Measurement of stress effects with astatic magnetometer (Ohnaka and Kinoshita, 1968). Full arrows show uniaxial stress, open arrows the sample NRM vector.

Ohnaka and Kinoshita (1968) measured changes in NRM parallel and perpendicular to the applied stress by placing the stressed sample in two different positions relative to the magnet system (Fig. 6.5). The disc-shaped sample was compressed in a 'pincer'-type press allowing the sample to be placed within 5.0 cm of the sensing magnet.

In their investigations of the seismomagnetic effect in igneous rocks, Bezuglaya and Skovorodkin (1972) measured changes in specific magnetization with pressure at elevated temperatures (20–200° C), the magnetization being a TRM imparted in the laboratory. The sample is confined between pistons and surrounded by a furnace, outside which are placed three orthogonal fluxgate sensors to detect changes in the three components of TRM: the sensitivity is given as $\sim 2 \times 10^{-5}$ A m^2 kg^{-1}, but the sample size is not stated.

The cryogenic magnetometer is ideally suited for continuously monitoring changes in NRM with stress, and Reval, Day and Fuller (1977) and Henyey,

Pike and Palmer (1978) describe techniques. The former use a largely non-magnetic tubular hydraulic press of plastic and aluminium to axially compress the sample at the measuring position in a three-axis, 6.4 cm access SCT magnetometer. Stresses of up to 2 kbar can be applied. Henyey *et al.* use a beryllium–copper press in which oil is pumped along alternative paths to provide uniaxial or hydrostatic pressures of up to 6 kbar and 1 kbar respectively. The press has a comparatively high remanent moment ($\sim 10^{-6}$ A m^2) which is, however, essentially stress-independent.

Measurements of induced magnetization in rocks under pressure are usually less demanding from the point of view of equipment sensitivity. In the ballistic method of Nagata and Kinoshita (1965) the sample, under uniaxial pressure, is surrounded by a pair of coaxial coils attached to a sliding frame enabling the coils to be moved such that first one coil surrounds the sample and then the other. A large solenoid for producing axial applied fields surrounds the sample and pick-up coils. For investigations of piezomagnetic changes in IRM, induced magnetization, coercive force and TRM in basalt samples, Schmidbauer and Petersen (1968) surround the axially compressed cylindrical sample (3 cm in diameter, 11 cm long) with a tubular furnace and solenoid giving axial applied fields of up to 0.3 T. Transverse fields are provided by a permanent magnet. Two fluxgate sensors just outside the solenoid are used to detect changes in magnetization of the sample.

Martin, Habermann and Wyss (1978) use a beryllium–copper press encased in a Mumetal shield, with two pairs of fluxgate detectors: one pair measures axial and radial changes in the NRM of the sample during loading and unloading, and the other monitors changes in the ambient field so that they can be subtracted from the sample fluxgate signals (Fig. 6.6).

Schult (1968b) has measured the effect of hydrostatic pressure (up to 50 kbar) on the Curie points of magnetite and titanomagnetite. Curie point measurement is by a high-frequency method (Section 4.1.3) and the pressure is achieved by using a form of press known as the 'belt', described by Hall (1960).

By mounting a sample compressed in a small press attached to the shaft of a vibrating sample magnetometer, Guertin and Foner (1974) succeeded in measuring changes in susceptibility and induced magnetization in 0.2 cm^3 samples in fields of up to 70 T provided by a superconducting solenoid. The beryllium–copper press, weighing 40 g is of the type described by Wohlleben and Maple (1971) and used for their measurement of changes in induced magnetization under pressure by the Faraday method. The sample is compressed (up to $\sim$10 kbar) between two axial pistons contained in a cylindrical case by means of an hydraulic press, and the pistons are then locked in position with lock-nuts and the pressure maintained when the hydraulic press is released. Two problems associated with the technique are magnetic contamination of the press and eddy currents induced in it in any gradient in the applied field.

Studies of the changes in magnetocrystalline anisotropy and magneto-

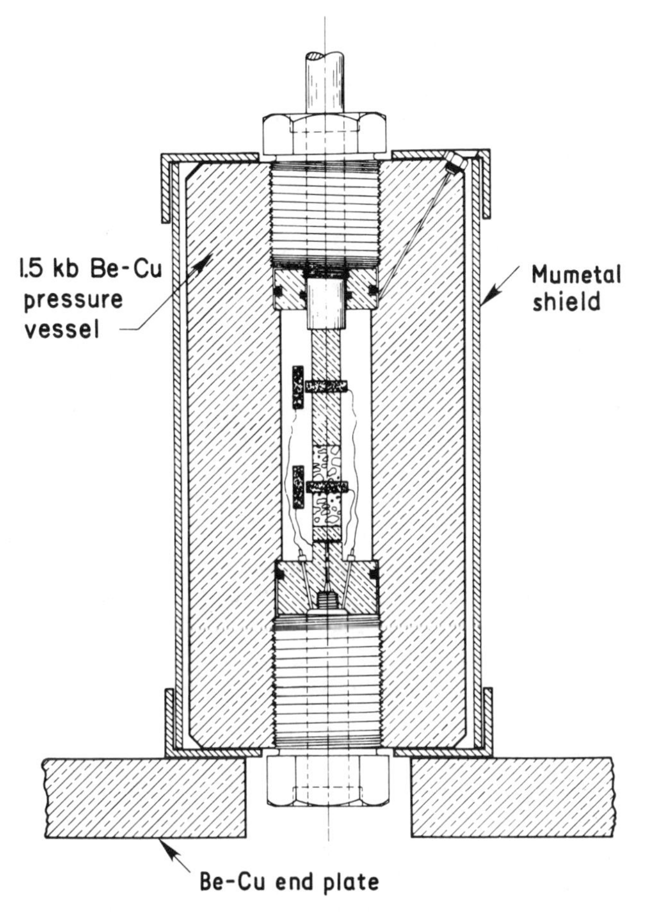

Fig. 6.6 Scale drawing of 1.5 kbar beryllium–copper pressure vessel, base plug, sample configuration, and fluxgate magnetometer array. One pair of magnetometers is positioned around the cylindrical rock sample. A second pair of magnetometers is positioned around a beryllium–copper dummy sample and monitors changes in the background field within the pressure vessel. (From Martin *et al.*, 1978)

striction in samples under hydrostatic pressure are reported by Kinoshita and Nagata (1967). Measurements were made using a kerosine-pumped press in the gap of an electromagnet, tensional stress and magnetostriction being detected with strain gauges attached to the sample.

Only irreversible effects of shock on remanent and induced magnetic properties are normally measurable, and these effects are measured with conventional instrumentation. It is beyond the scope of this book to describe the techniques used for applying shock, and they are usually only available in specialist laboratories. References to work done in this field in rock magnetism have already been given.

6.4 Viscous remanent magnetization (VRM)

The characteristics of viscous remanent magnetization are described in Section 7.2 and some aspects of its measurement and applications are considered here.

Since investigations of VRM acquisition normally require samples in which the natural VRM component, acquired in $\sim 50\ \mu T$ over $\sim 10^6$ years, has been removed we first consider this aspect of VRM measurement. Removal of VRM is not a straightforward procedure because it is unlikely that it is carried by a distinct group of grains with a coercivity spectrum or blocking temperature distribution which does not overlap with that of other magnetizations in the rock. Also, since the maximum coercivity or blocking temperature of grains carrying VRM is not usually known, there is no way of deciding up to which alternating field or temperature to carry out a conventional demagnetizing procedure. Alternating field demagnetization is a suspect procedure since there are well-documented examples of VRM showing anomalous hardness (Biquand and Prévot, 1971). Russian workers have favoured prolonged heating ($\sim$ 10 days) at 100°C in zero field (Trukhin, 1966), but this treatment is based on results obtained from a group of sedimentary rocks and is unlikely to be either an exact or universally applicable procedure. However, moderate heating is certainly expected to remove some of the VRM in all rocks. On the basis of theoretical calculations, Dunlop and Stirling (1977) state that VRM acquired by haematite and low and high titanium titanomagnetites ($x > 0.6$) should be erased by heating to 375°C, 250°C and 80°C respectively. Some earlier work in this field is that of Wilson and Smith (1968), commented on by Dunlop (1969). It may be satisfactory simply to allow the VRM to decay by maintaining the samples in zero field for as long as possible. This was the treatment adopted by Creer (1957) in his studies of the acquisition of VRM in the Keuper Marls of south-west England: the 'aging' time was 15 months. The time used in this technique will, of course, always fall short of the acquisition time of the natural VRM by several orders of magnitude. Bol'shakov and Faustov (1976) have investigated the temperature required to remove VRM in some Permian red clays; both theory and experiment indicated 200°C for 8 hours.

For the same reasons that there are difficulties in removing only the viscous magnetization from a rock it is usually only possible to estimate its magnitude. It may happen that in some rocks there is evidence from demagnetization of a distinct VRM component, e.g. a sharp discontinuity in the curve of NRM intensity versus alternating field or temperature, at low values of field or temperature.

Subtraction of the NRM vector after a cleaning step from the vector before cleaning should confirm the presence of VRM if the erased vector has a direction near that of the local geomagnetic field. However, if the viscosity coefficient is such that significant VRM is built up in a sample between removal from the outcrop and measurement this VRM will generally be randomly directed. It will also usually be removed at a low alternating field or temperature (although the possibility of anomalous a.f. hardness should not be forgotton) and the vector subtraction technique may only be useful at higher fields or temperatures.

Trukhin (1967) notes that some Russian workers have estimated the VRM component (presumably in sediments) by redeposition of crushed samples in a magnetic field of the same magnitude as the geomagnetic field. Viscously magnetized particles are presumed not to contribute to the resulting depositional magnetization and the VRM component is derived from the difference between the DRM and initial NRM intensities. Apart from the doubtful validity of the basic assumption that particles carrying VRM do not contribute to the DRM, the many uncertainties associated with the comparison of laboratory DRM and the natural DRM process make this technique highly suspect.

The logarithmic time dependence of the magnitude of VRM implies an initial relatively rapid acquisition of VRM in an applied field and equally rapid initial loss on removal of the field. Thus, for measurement of the initial rate of acquisition the time required for a measurement of VRM intensity and whether it can be carried out in the applied field assumes some importance. The most satisfactory technique is to measure the acquisition of VRM continuously in the presence of the field. This can be done in principle with an astatic (or parastatic) or cryogenic magnetometer. If sufficient sensitivity is available a ballistic instrument may also be used in an applied field, although a continuous record will not be obtained but a series of spot readings separated by time intervals, the length of which depends on the time constant of the magnetometer. It is not possible to apply the external field during measurement with a spinner magnetometer, and the spinner instrument is only useful for measurements in which VRM is allowed to decay for a period before measurement (Gose and Carnes, 1973) or if the viscosity coefficient is very low, i.e. significant VRM is only acquired over periods of several hours or longer. It is usually possible to reduce the measurement time of VRM relative to that required for a total NRM vector determination by ensuring that the VRM is acquired along an axis in the sample such that a single reading measures it.

In principle the measurement of VRM with an astatic magnetometer is rather simple. The sample is placed in the appropriate measurement position and the external field applied in such a direction that the resulting VRM causes deflection of the magnet system. The deflection (or output of a feedback circuit) corrected for any signals due to NRM and susceptibility then provides a continuous record of the growth of VRM. Possible sources of difficulty are drift of the magnetometer zero and the effect on the magnetometer of the applied field. If the field is applied with a large coil system surrounding the magnetometer, the field over the magnet system will be essentially uniform and should not cause too great a deflection if the field is not too large and the magnet well astaticized. The magnetometer can be re-zeroed before putting the sample in position and the sensitivity should then be checked, since the applied field usually lowers the sensitivity because of the finite astaticism.

Many VRM investigations are carried out in applied fields comparable to the geomagnetic field, and with a three-component coil set normally used to null the latter it is sometimes convenient to obtain the applied field by switching off the current in the appropriate coils. If stronger fields are required a reversing switch in a coil circuit can be used to obtain double the field normally cancelled by the coil.

If a horizontal applied field causes too great a deflection of the magnet system and the astatic magnetometer can be used with the off-centre measuring method or the mid-magnet sample position (Fig. 9.9 (d)), it may be advantageous to use a vertical external field. This usually exerts a smaller deflecting torque on the magnet system and the VRM can be measured directly with the sample in either of the off-centre positions. There is some reduction in the sensitivity with the off-centre technique because a given vertical magnetic moment produces a smaller deflection ($\propto z^{-4}$) than the same moment when horizontally directed ($\propto z^{-3}$), for the same distance, z, from the magnet system.

Zero drift of the magnetometer can be checked by rapidly removing the sample to a site as near as possible to the magnetometer where an identical magnetic field has been set up, and replacing it when the zero reading has been recorded. If the coil system around the magnetometer is large enough it may be possible to remove the sample to a position where it produces no significant deflection of the magnet system yet remains in essentially the same applied field. The procedure in which the applied field is a component of the geomagnetic field is clearly useful in this respect. However, in a square Helmholtz coil pair of side 2 m the field in a cubic volume approximately 60 cm (80 cm) on a side is within 1% (2%) of the central value, sufficiently close for most purposes. If the measuring position is a few centimeters from the sensing magnet the deflection produced by the sample will be very small when the sample is 30–40 cm distant.

The cryogenic magnetometer is also very convenient for VRM investigations providing an ambient magnetic field can be established at the sample measuring position. The field normally present is the weak residual field inside the shielding before the superconducting state is achieved and which is 'trapped' when the shield becomes superconducting. Thus, if a stronger field is required and can be set up at the appropriate position before cooling, it will be trapped at superconducting temperatures with the very high stability associated with superconducting shielding systems.

In principle the field can be set up in any orientation in the measuring space by coils suitably placed outside the superconducting shield. However, axial fields are convenient and easily provided by an axial solenoid. Some instruments which have this facility are also provided with a heater for raising the temperature of the shield to above the critical value. The trapped field is then lost and a different field can be established using the solenoid, and subsequently trapped. In some of the SCT instruments, fields of up to 10 mT can be trapped in this way.

Under optimum conditions the zero of cryogenic magnetometers is almost free of drift, an important feature for VRM measurements. If necessary, the zero can be checked as with the astatic magnetometer, by temporarily removing the sample into an equivalent applied field, in the upper part of the shield or in a coil at the mouth of the magnetometer. Since a typical time constant for the cryogenic instrument is only 1 s it is often feasible to make VRM measurements by using the field in a coil at the entrance to the magnetometer, rapidly lowering the sample to the measuring position and then raising it again into the coil.

Investigations of the rate of decay of VRM or IRM in zero ambient fields are much less demanding than acquisition studies and are carried out with any magnetometer of time constant comparable to or less than the intervals at which critical VRM decay is to be measured, and in which samples can be measured in near-zero magnetic field, i.e. in a field very small compared with that in which the VRM is acquired. The initial rate of decay of VRM in the great majority of rocks is the same as the initial rate of acquisition, i.e. the viscosity coefficient $S = \mathrm{d}J_{\mathrm{VRM}}/\mathrm{d}(\log t)$ is the same for acquisition and decay, for t not too large, and thus decay measurements may more easily provide data on S than the more difficult acquisition measurements. Many spinner magnetometers are convenient for decay measurements, with those providing continuous read-out of magnetization components having some advantage for frequent initial decay readings over the Digico instrument in which the minimum sampling time is 5 s (Urrutia-Fucugauchi, 1981).

Apart from its importance as a secondary magnetization, VRM is of interest in its own right as a physical phenomenon and various attempts have also been made to utilize it for geophysical purposes. Of most interest perhaps is the attempt to use VRM as an aid to dating rocks, by comparing the magnitude of a present VRM component in a rock with that acquired in the laboratory in a known time in the geomagnetic field.

Assuming that the growth is logarithmic with time during the period since the rock was deposited, the age of the rock can be estimated. The simplest case is where the rocks are younger than about 700 000 years when the last reversal of the geomagnetic field occurred. However, Trukhin (1967) had little success in attempting to date some Quaternary sediments around 100 000 years old. Heating the samples to 100°C to remove the natural VRM prior to measuring the laboratory-induced VRM (over a period of six months), Trukhin found that the viscosity coefficient (S) was different from that obtained after first allowing the natural VRM to partially decay in zero field, an alternative method of isolating the VRM component. Changes of $\sim$10°C in the ambient temperature during VRM acquisition were also found to have a significant effect on S. Apart from the uncertainties associated with extrapolating from laboratory to geological time scales, other complicating factors include variations in the strength of the geomagnetic field and the occurrence of 'events' such as the Laschamp (~20 000 years ago) when the field either reversed or underwent a large excursion from the normal direction.

Application of the technique to older rocks involves the effect of multiple reversals of the geomagnetic field, possible geological alteration of the primary NRM (e.g. by heating due to burial) and greater difficulty in accurately isolating the present-day component of VRM. In his study of VRM in the Keuper Marls from south-west England, Creer (1957) recognized that some of the VRM presently in the rocks would be due to the field when last reversed, and that an estimate of the time to the last reversal based on the magnitude of the present VRM and observed viscosity coefficient would be a lower limit. The acquisition of VRM (J_v) by the marls, as determined in the laboratory, followed the law

$$J_v = SB \log (t/t_1) \tag{6.13}$$

where B is the ambient field, taken to be a constant, 50 μT, and t and t_1 are the time to the last field reversal and the 'aging' time before the present VRM in the rocks was measured. The latter was 15 months, the time between collection of the samples and measurement of VRM for the purposes of the dating test; during this period the samples were in zero field. The time t obtained was ~100 000 years with an upper experimental error limit of 250 000 years. The VRM component in the samples was isolated by vector subtraction of the observed 'stable' direction of NRM in the marls from an apparently reliable primary Triassic direction obtained from other rocks.

A different technique was employed by Heller and Markert (1973) in their attempt to correlate the age of Hadrian's Wall in the north of England with the magnitude of the VRM component in the NRM of some dolerite blocks forming part of the structure. The method is based on the expression derived by Néel (1955) which introduced the concept of a thermal fluctuation field B_f related to the growth of VRM by

$$J_v(t) = J_{(0)} + \chi S'(Q + \log t) \tag{6.14}$$

where

$$S' (Q + \log t) = B_f \tag{6.15}$$

$J_v(t) - J_0$ is the VRM acquired in time t after the application of an external field B_{ex} to a sample of initial state J_0, χ is an irreversible susceptibility and S' is a viscosity factor related to S. Q is a constant of the form $\log t_0$. The authors show that an estimate of B_f is given by the difference between B_{ex} and the peak alternating field required to demagnetize the VRM component in the rocks. B_{ex} is the geomagnetic field, again presumed constant. S' is derived from the laboratory growth rate of VRM, and Q and χ are also derived experimentally. In spite of some uncertainty concerning the significance and magnitude of Q

and the particular susceptibility involved, reasonable agreement was obtained between the magnetic age derived from some samples (1600–2400 years) and the true age of Hadrian's Wall (1600–1900 years). Another relevant factor casting some doubt on the validity of the technique is the varying hardness exhibited in different rocks by even short-period VRM against a.f. demagnetization (Biquand and Prévot, 1971). This phenomenon is at present imperfectly understood in igneous rocks in which the VRM is carried by multidomain titanomagnetite particles, as in the Hadrian's Wall dolerites.

The possibility of using VRM as a means of orienting borehole cores has been investigated by Fuller (1969) but no successful results are reported.

6.5 Anhysteretic remanent magnetization

If a rock sample or an assembly of magnetic particles is subjected to an alternating magnetic field superimposed on a small direct field and the alternating field is decreased to zero, the samples will in general acquire an 'anhysteretic' remanent magnetization (ARM). Anhysteretic remanence is a type of magnetization that is mainly of interest in the laboratory as a source of spurious remanence in alternating field demagnetization and as the electromagnetic analogue of thermoremanence. However, to the extent that lightning strikes are sometimes considered to consist of both direct and alternating current components, ARM also occurs in nature. The 'ideal' magnetization encountered in the Russian literature is another term for ARM.

ARM is acquired parallel to the direct field, and equipment for imparting or investigating ARM consists essentially of coils to provide direct and alternating fields of the required strength and direction. The direct field is commonly in the range 50–500 μT and the peak alternating field up to 200 mT. For a given magnitude of direct and peak alternating field maximum intensity of ARM occurs when the two fields are parallel.

Two applications of ARM are concerned with investigations of ancient magnetic field strengths (palaeointensities). In one the alternating field demagnetization curve (or coercivity spectrum) of ARM is determined before and after heating a rock sample in the Thellier technique to test for thermal alteration of the remanence-carrying minerals (Shaw, 1974). In the other application (Stephenson and Collinson, 1974) ARM is used as an analogue of TRM for palaeointensity investigations of lunar samples, and the coercivity spectra of the NRM and saturation ARM of a sample are compared (Section 14.4).

The most efficient and most common technique of a.f. demagnetization is to tumble the sample about two or three mutually perpendicular axes during application of the field. Thus, if a.f. demagnetization curves of ARM are to be compared with each other or with those of NRM it is desirable that the ARM acquisition procedure follows as closely as possible the demagnetization procedure, i.e. the sample should be tumbled during ARM acquisition, preferably in an identical way as is used during demagnetization. This requirement implies the provision of a direct field constant in direction relative to axes in the tumbled sample during ARM acquisition.

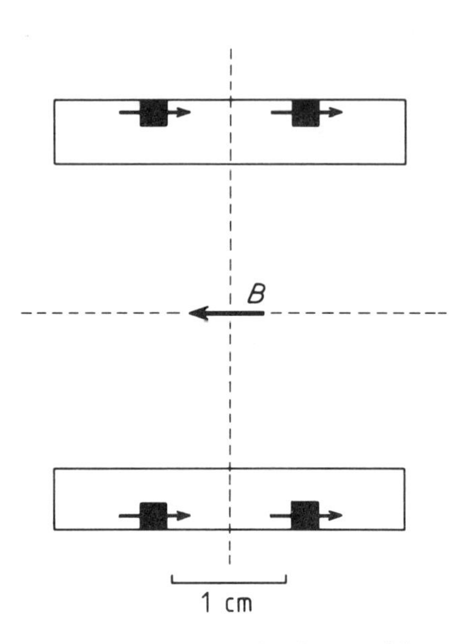

Fig. 6.7 Cross-section of Stephenson's device for providing applied magnetic field during tumbling.

In the device developed by A. Stephenson (personal communication) for investigations of lunar and other small samples the direct field is provided by four small permanent magnets placed in the wall of a tubular holder containing and tumbling with the sample (Fig. 6.7). Each magnet is a 2 mm cube of Alcomax IV magnetized parallel with the tube axis with a magnetic moment of $\sim 2.0 \times 10^{-3}$ A m^2. The central field is 180 μT and it is uniform to $\sim \pm 10\%$ over a volume of ~ 1.0 cm^3 at the centre. Even with very hard permanent magnet materials there may be some demagnetization of the magnets by the applied alternating field, and it is necessary to measure or compute the central field after the magnets have been subjected to the maximum proposed alternating field to be used in imparting ARM.

An alternative approach is that of J. Shaw (personal communication) and Barton (1978) who used a small coil system surrounding the sample to provide the direct field. Both use a form of the Rubens square coil design (Section 5.1.3) to obtain maximum field uniformity over the sample volume. The two-axis tumbling system of Shaw is adapted so that current can be passed to the coil from the outer rotation via brushes bearing on brass plates attached to the coil former on the inner rotator. To overcome rotational remanent magnetization the rotation of the tumbler is reversed every two revolutions so that wires can be connected directly to the outer rotating section, the wires then twisting and untwisting as the tumbler operates. The central winding and inner and outer pairs of windings are as in Fig. 5.7 and a current of 50 mA through the coil provides a central field of 50 μT. A choke (8 H) in the supply circuit suppresses the induced alternating current arising from the ambient alternating field.

Barton (1978) designed a coil to accommodate two 2.1 cm cubic samples along the axis. The coils are 3.6 cm on a side with a 1.0 cm separation and contain 9, 5 and 16 turns on the central and inner and outer windings respectively. Field uniformity is $\pm 3\%$ over 96% of the sample volume.

6.6 Depositional remanent magnetization (DRM)

This process, in which magnetized particles are partially aligned by an ambient magnetic field during sedimentary deposition, attracted some early research activity in connection with the NRM of varved clays and their possible use as recorders of secular variation (King, 1955; Griffiths, King, Rees and Wright, 1960). Redeposition in the laboratory of disaggregated varves and other sediments showed that DRM is affected by a variety of variables and perturbing influences and that quantitative, reproducible results are not easily obtained. Among these influences are water currents, the relative size of magnetic and matrix particles and the rate and mode of drying out of the sediment. The phenomenon of the inclination error, whereby the DRM of laboratory-deposited sediments often exhibits a lower inclination than that of the ambient field in which the DRM is acquired, also attracted some attention because of its possible bearing on the reliability of pole-positions derived from depositionally magnetized sediments.

More recently the profitable investigations of secular variation and field reversals in the NRM of lake-bottom and oceanic sediments has revived interest in the DRM process (Opdyke, 1972; Creer, Thompson, Molyneux and MacKereth, 1972; Kent, 1973; Verosub, 1977b). Collinson (1974) used laboratory DRM to derive information on the magnetization of the pigment and specularite form of haematite in sediments.

The DRM process is easily demonstrated in the laboratory by allowing a disaggregated rock sample to settle in water in the Earth's or any low magnetic field and measuring the resulting NRM of the sediment before or after drying out. Quantitative investigations usually require more complex equipment, including an ambient field which is variable in strength and direction, variable depth of deposition, facilities for deposition in flowing water and controlled conditions during drying out and sampling.

Some of the earliest investigations were carried out on very fine-grained clays with very long natural settling times of the constituent particles when allowed to settle in water. Johnson, Murphy and Torreson (1948) decreased this time by containing the suspension in a Buchner funnel and drawing the water through with the aid of atmospheric pressure above. By continuing the suction after deposition was complete the drying out and compaction of the sediment was facilitated, although it seems possible that some disturbance of the sediment and its DRM would be encouraged.

King (1955) describes a system for deposition under gravity on a larger scale, in order to obtain several samples from one run and to avoid wall effects. The sediment, in the form of a slurry, is introduced into a feed tank and the resulting suspension passes over a weir into the settling tank (30 cm deep) under the influence of circulating currents induced by a rotating paddle at the base of the feed tank. Return flow is achieved through a tube from the bottom of the feed tank leading to a point about half-way down the settling tank. An adjustable Helmholtz coil pair surrounds the tank and provides an aligning magnetic field, variable in strength and direction, over the lower half of the settling tank. When deposition is complete the water is drained off and the sediment, in a removable tray, is lifted, excess water allowed to drain off and the sediment allowed to dry out until sufficiently compact for sampling and 'blocking in' of the DRM. Modification of the sediment tray enables deposition on a sloping surface to be investigated.

A modified version of this system, described by Blow and Hamilton (1978), is shown in Fig. 6.8. The authors also describe an alternative to continuous deposition of sediment, in which the total mass of sediment to be deposited is introduced in dispersed form into the tank, thoroughly stirred and allowed to settle in the ambient field. This method was also used by Vlasov, Kovalenko and Tropin (1961).

If the experimental requirement is to produce a DRM in a few grams of sediment without attempting to approach the original environmental conditions, very simple equipment can be used. In the author's experiments on thermal demagnetization of the NRM of specularite and pigment extracted from red sandstones (Collinson, 1974), the requirement was a DRM (as strong as possible) in $\sim$0.1 g samples of specularite and $\sim$2 g samples of the pigment–quartz separate. The arrangement is shown in Fig. 6.9. The vertical tube is pressed down on the rim of the quartz sediment tray by means of the brass clamp and the tube filled with water or acetone (the latter drying out of the sediment more rapidly). A 1.0 mT horizontal aligning field is used, and the specularite or pigment sample dropped into the liquid column over a period of about a minute. When deposition is complete the liquid is drained off through the side tube and any remaining liquid allowed to leak slowly through the seal by slackening off the brass clamp. The quartz tray is then removed and the sediment allowed to dry out prior to measurement of the DRM with an astatic or cryogenic magnetometer.

A largely unknown factor in DRM investigations is the extent to which DRM is disturbed during the final drying out and compaction process, although some work has recently been done in this field (Stober and Thompson, 1979; Henshaw and Merrill, 1979). This uncertainty, together with

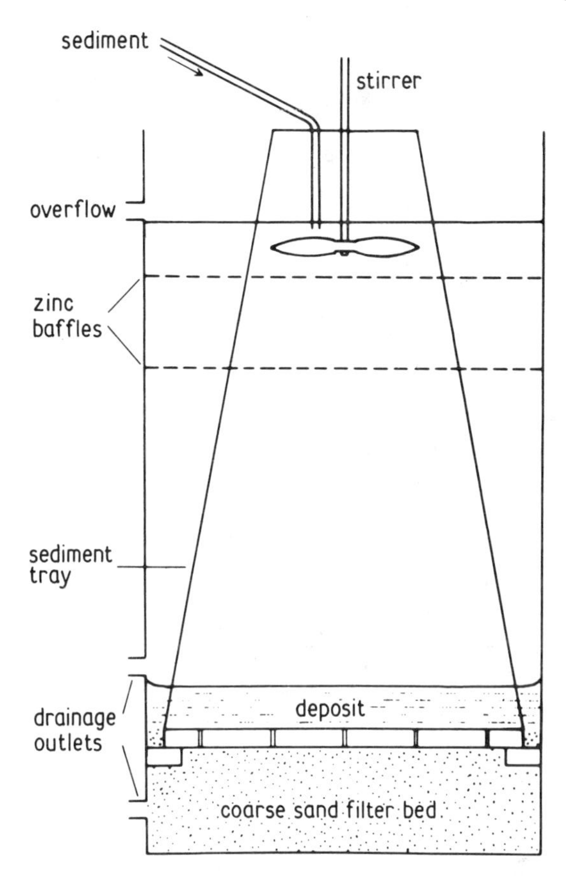

Fig. 6.8 Sedimentation tank for DRM experiments. A suspension of the sediment is fed continuously into the top of the tank by gravity feed from another tank. The depositional tank is approximately 50 cm high and 22 cm across. (From Blow and Hamilton, 1978)

a developing interest in post-depositional remanent magnetization (PDRM), i.e. the occurrence of the DRM process in a wet sediment after deposition, led to the desirability of measuring DRM in situ without drying and sampling the sediment. Since the conditions under which sedimentation occurs is not now important, deposition is usually carried out by shaking up the disaggregated material in water and allowing the slurry to settle out. Good dispersion may be facilitated by mechanical agitation (e.g. using a kitchen food mixer and/or by ultrasonic agitation). Otofuji and Sasajima (1981) simulate the compaction process by centrifuging the sediment in the aligning field.

In his investigations of PDRM in synthetic sediments containing $\sim 2\%$ by weight of magnetite, Tucker (1980) was able to measure the comparatively strong remanence with a fluxgate gradiometer. Since the applied field and

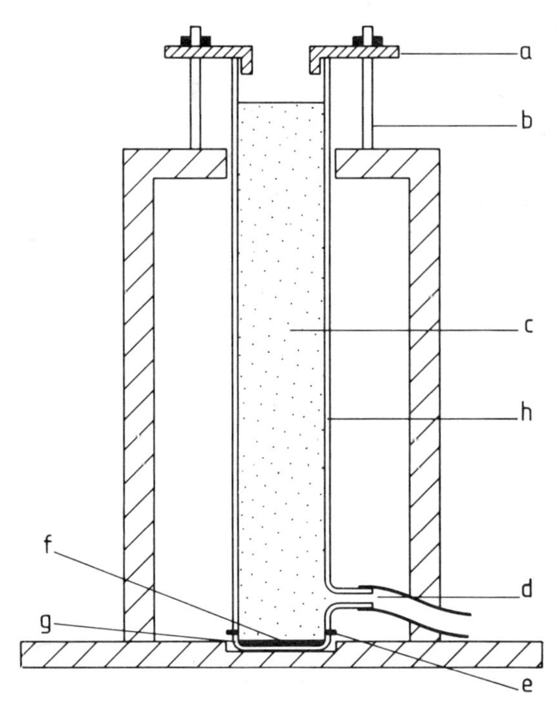

Fig. 6.9 Deposition column for DRM investigations. a, clamping plate; b, threaded rod; c, depositing sediment; d, outlet tube; e, seal; f, sediment; g, quartz tray; h, glass tube.

resulting DRM were horizontal, measurements in one plane only were required, with one of the fluxgates' probes close to the sediment and the other further away and oriented to form an astatic pair with the sensing probe.

Vlasov, Zvegintsev and Pavlov (1964) used a ballistic method for wet DRM measurement. The equipment consists essentially of a vertical water-filled deposition tube 1.6 m long surrounded by a jacket carrying ethylene glycol solution. A suspension of kaolin and the magnetic particles is fed into the upper end of the tube and the sediment collects on a filter paper in a container at the lower end. Holes in the bottom of the container allow the water to drain through, assisted if necessary by reduced pressure in the lower chamber. Deposition experiments can be carried out at different temperatures by heating the circulating ethylene glycol solution or cooling it with liquid nitrogen vapour in a heat exchanger. DRM is established by a vertical field produced by a large Helmholtz coil system, ensuring a reasonably uniform field over the depth of settling. The measurement of DRM intensity is by means of a pair of coils coaxial with and outside the deposition tube, a ballistic signal being generated by moving the coils such that the sample effectively passes from the centre of one coil to the centre of the other.

A more sensitive method of measuring wet sediments is provided by the cryogenic magnetometer, with which it is also possible to investigate the acquisition of DRM as a slurry settles out. Barton, McElhinny and Edwards (1980) examined the acquisition of DRM by dilute and concentrated slurries of organic muds contained in tubes 30 cm long which could be lowered through the measuring region of a two-axis cryogenic magnetometer. By bringing different levels in the tubes into the sensor region, the magnetization of the sediment at different stages of the settling process can be measured. The procedure is similar to that used in the measurement of the NRM of long cores of sediment, and the method of calibration and other details are described in Section 9.5.2.

Chapter Seven

A brief review of palaeomagnetism

The purpose of this chapter is to give a short account of the origins of remanent magnetism in rocks and the way in which it is interpreted, to describe some applications of the remanent magnetism of rocks and to define some of the terms used in the following chapters.

The basic phenomenon on which palaeomagnetism rests is the acquisition by rocks at or near the time of their formation of a permanent magnetization derived from and parallel to the then existing magnetic field at the site. Thus, the rock preserves a record of the ancient ambient field direction, and indirectly its strength, at the site at the time of formation of the rock.

This original magnetism is termed the primary magnetization. Between the time of formation of the rock and the present, other, secondary magnetizations may be acquired and the vector sum of the primary and secondary magnetizations as they exist at present in the rock is called the natural remanent magnetization (NRM). It is also common usage to refer to either the primary or secondary component as an NRM.

7.1 Primary magnetization

In the majority of rocks the NRM is carried by members of the iron–titanium oxide series of minerals. The primary NRM can be acquired by one of three basic processes.

Igneous rocks, e.g. volcanic lavas and intrusives, which have been at high temperatures during their formation become magnetized initially by thermo-remanent magnetization (TRM) when the constituent magnetic minerals cool through their Curie or blocking temperatures in the ambient geomagnetic field. Though weak ($\sim 50\ \mu T$) the field is sufficient to align the remanent magnetization along the field direction and the TRM generally has a high stability (i.e. resistance to demagnetizing influences), enabling it to persist unchanged in the rock for millions or billions of years. In the expected ambient field range (0–100 μT) TRM intensity is usually proportional to field strength, which is the basis of determining the ancient field intensity. TRM has been extensively investigated both theoretically and in the laboratory, and among many references are Néel (1955), Verhoogen (1959), Everitt (1962b, c), Stacey (1962, 1963) and Day (1977).

Sedimentary rocks can acquire a primary NRM through two processes, namely depositional (or detrital) and chemical remanent magnetization (DRM and CRM).

As a sediment, containing permanently magnetized oxide grains, is deposited in water the grains tend to orient themselves in a position of minimum energy, i.e. with their magnetic axes aligned along the ambient field direction. Although there may be some disorientation when the particles reach the bottom and also during subsequent drying out and consolidation, a statistical alignment parallel to the field direction remains: this is DRM. With magnetic grains sufficiently small relative to interstitial volumes, alignment may take place in water trapped in the interstices of the sediment after deposition – this is referred to as post-depositional remanent magnetization (PDRM).

In laboratory deposition of natural and artificial sediments it has sometimes been observed that the inclination of the DRM (i.e. the dip of the magnetic vector below the horizontal plane) is less than that of the applied field, whereas the direction of the horizontal component of the field (declination) is accurately recorded. The former is the 'inclination' error, a possible cause of which is the presence of elongated magnetic particles which tend to lie with their axes parallel to the deposition surface. In nature, consolidation during burial may also be a contributory cause. However, the inclination error has only rarely been definitely established in the NRM of sediments. Griffiths *et al.* (1960), Irving and Major (1964), Collinson (1974) and Lovlie (1974) have investigated DRM in the laboratory, and Opdyke (1961) and Elston and Purucker (1979) refer to field observations. The theory of DRM has been investigated by Nagata (1961), Stacey (1962), Collinson (1965a) and King and Rees (1966).

It should be noted that DRM is a process whereby a sedimentary rock, rather than its constituent magnetic minerals, becomes magnetized. These minerals have already acquired a thermoremanent or chemical magnetization prior to deposition.

Chemical magnetization (CRM) is another process that can occur in sediments. In TRM the permanent magnetization of the particles is acquired as they cool through their Curie point or blocking temperature below which temperature magnetostatic alignment forces dominate over the randomizing effect of thermal fluctuations. In sufficiently small magnetic particles thermal fluctuations dominate at normal earth-surface temperatures but magnetostatic forces overcome thermal fluctuations if the particle volume grows through a critical value. In the presence of an ambient magnetic field the particles acquire a stable remanence parallel to the field direction. This is chemical (or, as it is sometimes termed, 'crystallization') remanent magnetization.

CRM can be acquired either by precipitation of a magnetic mineral out of solution or by the alteration of one mineral to another. Examples of these processes occur in red sediments, in the precipitation of the reddish haematite pigment (with possibly an intermediate hydrated stage) from circulating iron-bearing solutions, and the in-situ oxidation of magnetite to haematite. Experimental work indicates that when CRM is acquired through alteration of a previously magnetized material, the new material is usually magnetized parallel to the external field and has no memory of the magnetization of the original mineral.

The stability of CRM is similar to that of TRM. Laboratory and field investigations of CRM are described by Haigh (1958), Kobayashi (1959), Porath (1968), Hedley (1968), Larson and Walker (1975) and Roy and Park (1972). Theoretical aspects are discussed by Stacey (1963) and Banerjee and Stacey (1974).

It is not uncommon for a rock to acquire more than one type of primary magnetization. A sediment may initially possess a DRM carried by magnetite which subsequently oxidizes to haematite with a resulting CRM: a weathered igneous rock may contain a component of CRM carried by a weathering product or a late-stage exsolution product.

7.2 Secondary magnetization

Time intervals of up to $\sim 10^9$ years may elapse between the formation of a rock and the present and in this period additional, secondary components of magnetization may be acquired.

Viscous remanent magnetization (VRM) is acquired if there are magnetic grains in a rock in which thermal fluctuations cause irreversible changes in domain alignment or domain wall positions. In zero ambient field no net magnetic moment results from these changes, but even a weak magnetic field can bias the process such that a magnetization builds up parallel to the field direction. This is viscous magnetization and its intensity is proportional to the (weak) ambient field strength and generally to the logarithm of the time of exposure to the field, on which the magnetic hardness also depends. Viscous magnetization is common as a secondary magnetization in all types of rock, and the theory of VRM has been investigated by Street and Woolley (1949), Néel (1955), Stacey (1963) and Dunlop (1973c). Observations of VRM in terrestrial and lunar rocks include those of Creer (1957), Bol'shakov (1975), Biquand and Prévot (1971) and Dunlop (1973c).

Rocks which have been subject to burial by overlying deposits or are adjacent to an intrusive undergo a rise in temperature. If this exceeds the blocking temperature of any magnetic grains present they will acquire a partial thermoremanence (PTRM) when the rock subsequently cools, i.e. a TRM acquired below the Curie point: 'total TRM' is a term sometimes used to denote that the Curie point was exceeded during the acquisition process. Since the rocks may be at the elevated temperature for a long time, a 'viscous' PTRM can also be acquired, which is more stable than a PTRM acquired after heating to the same temperature for a short time. Examples of viscous PTRM have been investigated by Chamalaun and Creer (1964), Briden (1965), and Irving and Opdyke (1965).

At exposures where thunderstorms are frequent rocks may be encountered in which a lightning strike has produced a magnetization in the immediately surrounding area. Approximating the effect of a strike on a horizontal surface to an infinite vertical line current of i A, the (circular) horizontal magnetic field at a distance r is $(\mu_0 i/2\pi r)$ T. Thus, at $r = 2$ m and $i = 10^4$ A, the field is 1 mT and rocks inside the circular area may possess a significant magnetization. Such a magnetization, acquired when a large field acts for a short time, is called an isothermal remanent magnetization (IRM). The anomalously high intensity of NRM of rocks very near the strike point is often an indicator of this secondary NRM. Examples of lightning-induced IRM in a basalt are given by Cox (1961) and in a sandstone by Purucker (1974).

The observed NRM of rocks may consist of any combination of primary or secondary components between the extreme cases where one or other is entirely dominant. Since it is the primary NRM that is of interest for palaeomagnetic interpretation, it is first necessary to remove any secondary NRM, as far as possible leaving the primary NRM intact: this is the important procedure of partial demagnetization, sometimes referred to as 'magnetic cleaning'.

Magnetic cleaning can be successfully employed because the magnetic hardness, or stability, of secondary magnetization is generally less than that of the primary NRM. Cleaning by means of a decreasing alternating magnetic field is based on the classical technique used in magnetic studies whereby demagnetization of a ferromagnetic material is achieved by subjecting it to hysteresis cycles of decreasing magnitude. The effect of this treatment is to randomize domain directions leaving the material, in the ideal case, with no net magnetic moment. In thermal cleaning, rock samples are taken to successively higher temperatures and cooled in zero field thus releasing and, on cooling, randomizing the magnetization of grains with successively higher blocking temperatures. Primary magnetization is associated with high coercivity and high blocking temperatures and thus its separation from magnetically softer secondary NRM is in principle possible. Partial or complete removal of secondary NRM in some rocks can be achieved by subjecting them to high pressure or low temperature, but these are not routine techniques.

In sediments a quite different technique, chemical cleaning, is sometimes employed, in which the mineral phase carrying the secondary magnetization is removed. In some redbeds the pigment form of haematite on the matrix grains and in the rock interstices carries the secondary NRM. This pigment can sometimes be removed by allowing acid to percolate through the rock and dissolve it out, leaving the black, crystalline haematite particles, often carrying a harder, primary NRM, essentially intact.

Secondary magnetization can be carried by particles of low coercivity or low blocking temperature originally present in the rock, or by a new mineral produced in the rock by, for instance, weathering or diagenetic alteration. In the latter situation it is possible for the new mineral to acquire an NRM which is more stable than the primary magnetization. Thus, the most stable NRM revealed in a rock by magnetic cleaning is not necessarily the primary NRM.

7.3 Palaeomagnetic surveys and interpretation

A palaeomagnetic survey of a geological unit comprises the collection of a representative group of rock samples suitably marked so that their orientation in the exposure can be reproduced, the measurement of their NRM directions and intensity, and the testing for and, if necessary, removal of secondary magnetizations. A mean direction of primary NRM for the rock unit is then computed, together with its associated error at any desired level of probability.

On the assumption of a geocentric axial dipolar geomagnetic field during geological time, the ancient latitude λ of the collection site is simply related to the ancient inclination of the geomagnetic field there (i.e. the inclination, or dip I of the NRM above or below the horizontal plane) by the relation.

$$\tan I = 2 \tan \lambda \tag{7.1}$$

The declination, D, of the NRM (i.e. the direction of its horizontal component) gives the direction along which the ancient magnetic pole lies. Thus a knowledge of D and I allows, in principle, the establishment of the ancient magnetic and coincident geographic pole position at a particular geological time. This is the starting point for the interpretation of palaeomagnetic data, which through the evidence it has provided for polar wandering, continental drift and plate tectonics has proved so fruitful in furthering our understanding of the development of the Earth's surface.

7.4 Other aspects of palaeomagnetism

Although the major impact of palaeomagnetism has been in the field of continental drift and polar wandering, palaeomagnetic research has contributed significantly to other fields in terrestrial and planetary sciences. Perhaps the most interesting of these applications is concerned with reversals of the Earth's magnetic field, a phenomenon discovered quite early in palaeomagnetic studies through the observations of 'normal' and 'reversed' directions of NRM in time-separated regions of a rock exposure. The existence of field reversals is of some interest to geologists as a potential aid to stratigraphy, and also to those concerned with the properties of the core 'dynamo' generating the geomagnetic field, but more recently reversals have played a major role in revealing the development of ocean basins by sea-floor spreading. The magnetic anomaly pattern, produced by normal and reversed

TRM of basalt cooling in the Earth's field and subsequently spreading on the ocean floor, has been confirmed and greatly extended back in geological time by measurement of dated reversals in rocks and by reversal sequences in cores drilled through ocean floor sediments.

Evidence of the past behaviour of the geomagnetic field on a shorter time scale is also available from palaeomagnetic studies. Secular variation over the last few thousand years can be estimated directly from the variation in NRM directions with depth in glacial varves and lake sediment cores, and indirectly in older material by investigating the scatter of NRM directions about the mean in lava flows. The intensity of the ancient geomagnetic field can also be determined, in principle, from the NRM intensity of suitable rocks.

An interesting and profitable interdisciplinary field of study is palaeomagnetism and archaeology. Artefacts such as pottery and ovens contain iron oxides (often haematite) which have been fired above the Curie point and therefore are capable of recording the ambient field through a TRM. The most profitable research has been concerned with documenting the geomagnetic field intensity and direction over the last few thousand years. An advantageous feature of these archaeomagnetic studies is that the archaeologist can often provide an accurate date with which to associate a field direction and intensity derived from an object, and, with the accumulation of reliable data, it then becomes possible to date some archaeological items by matching the field direction and intensity derived from them to master curves of field direction and intensity variation with time.

Among geological applications of remanent magnetism in rocks are the detection and measurement of relative rotation between blocks or landmasses, the timing of such rotations and of folding, the dating of intrusions and the detection of overturned strata.

Finally, palaeomagnetic studies on extraterrestrial materials have been extensively pursued during the last 15 years. Meteorite magnetism has not been a particularly rewarding study, but investigations of the strength of the magnetic field in which the chondrites and achondrites acquired their NRM may eventually provide important evidence about the meteorite parent bodies. The discovery of stable components of NRM in the samples returned from the Moon by the *Apollo* and *Luna* missions was one of the most interesting results of lunar sample research. Research is limited in scope because of the absence of oriented, bedrock samples and severe restrictions in the availability of samples. However, evidence has been obtained for an ancient lunar magnetic field of internal origin which, if a result of dynamo action in a molten core, clearly has important implications for lunar history and structure.

This necessarily brief and incomplete account of palaeomagnetic studies indicates the wide range of applications of the discipline, to which more will undoubtedly be added in the future. For comprehensive coverage of palaeomagnetism applied to the history of the Earth's surface features, the reader is referred to books by McElhinny (1973), Tarling (1971, 1983) and

Irving (1964). Archaeomagnetic techniques are described in a book by Aitken (1974) and have been reviewed by Tarling (1975). Papers on the magnetism of extraterrestrial materials include those by Herndon, Rowe, Larson and Watson (1972) (a review of Russian studies), Brecher and Arrhenius (1974) and Nagata (1979). Lunar palaeomagnetism is reviewed by Fuller (1974) and some more recent work is reported by Stephenson, Collinson and Runcorn (1976), Pearce *et al.* (1976), Fuller, Meshkov, Cisowski and Hale (1979) and Runcorn, Collinson and Stephenson (1981).

Chapter Eight

The sampling of rocks for palaeomagnetism

8.1 General

The aim of a palaeomagnetic survey of geological unit is to obtain the best value for the mean direction of the geomagnetic field over the time interval in which the rocks acquired their primary magnetization. The factors which have to be taken into account to achieve this are the nature of the Earth's magnetic field and its variations, the time–depth relationship in a geological structure and the process by which the rock acquired its remanent magnetism.

The relevant characteristics of the geomagnetic field are its secular variation and the longer term apparent secular movement of the magnetic pole. The former is observed as a variation about a mean value of the angular elements and intensity of the field at any point on the Earth's surface, and has an amplitude of $\sim 20°$ and a period of $\sim 10^3$ years. The long-term movement is a secular change of the field at a site at an average rate of the order of 0.2°/million years. Neither change proceeds at a uniform rate, and polar wandering or plate movement, which causes the long-term changes, may be zero for long or short periods.

The occurrence of reversals of magnetization, although they are of intrinsic interest and may even assist in interpretation of the NRM, will not generally effect the outcome of the survey because it is the mean axis of magnetization of the unit which is required: the time interval over which a reversal takes place ($\sim 10^3$–10^4 years) is very small compared with a typical interval covered by a survey.

The time intervals represented in different vertical thicknesses of a geological structure are very variable and normally unknown. Thus, a uniform spatial sampling through the outcrop depth will in all probability not be a uniform sampling in time. In practice, also, a geological unit may only be accessible in separate, isolated exposures with sections of the vertical thickness absent or inaccessible for sampling.

The procedure for sampling of igneous and sedimentary rocks is partly governed by the different way in which they acquire their primary remanent magnetization. A single lava flow cools through the Curie point of its constituent magnetic mineral(s) in a time which is short compared with the secular variation period ($\sim 10^3$ years) and will therefore record a 'spot' reading of the geomagnetic field. A dyke or sill, intruded below the surface, will take longer to cool and under appropriate conditions may record a substantial part of a secular variation cycle in the TRM directions measured from different parts, depending on the distribution of the Curie point isotherms in space and time.

Samples of sediments are unlikely to record a spot reading of the field in the above sense. Whether they acquire their NRM by depositional or chemical magnetization, a few centimetres thickness is likely to represent a period which is typically of the order of 10^3–10^5 years, largely averaging out secular variation. However, this can be considered as a spot reading relative to the time interval covered by the formation and to the time scale of polar wandering and plate movement.

In a sediment which has been magnetized chemically, the formation of a new mineral – and thus the acquisition of the primary NRM – can take place irregularly and in an unknown time, perhaps 10^7 years or more after deposition. Thus, the depth–time and magnetization–time relationships through the strata may bear little similarity to each other.

The minimum requirement for sampling a geological unit for palaeomagnetic interpretation is the averaging out of secular variation. (In this context 'palaeomagnetic interpretation' implies interpretation of derived pole positions in terms of polar wandering and plate movement, as distinct from, for example, investigations of lake sediments in which secular variation itself is of interest). The time interval represented by the geological unit will preferably be one that can be considered as a 'point' in geological time, covering a period in which any apparent polar movement is unlikely to be large, i.e. $> \sim 10°$. Therefore, a sampling unit will ideally be one covering a period of ~ 1–20 million years. Typical examples are a group of lava flows or intrusives, or a sedimentary layer 50–500 m thick.

Sampling schemes are devised in the light of the foregoing considerations. Rock samples are collected from sites distributed through the structure in a way which the investigator considers will afford the best mean field direction. Among the factors considered are lithological and petrological variations, changes of grain size and colour in sediments, evidence of relative movement between different parts of the exposure, presence or absence of weathering (a source of secondary NRM) and suitable mechanical properties for removal or drilling of samples and the cutting of specimens.

The usual definition of a collecting 'site' is one at which the previously described spot reading of the field is obtained. In a lava flow or intrusive, a site will cover a horizontal or vertical area of typically 0.5 m^2, in a sediment a 2 m length by 10–20 cm depth along a bedding horizon, or an area of ~ 1.0 m^2 of an exposed bed.

It can be seen that there are three levels of sampling, namely sites, samples and specimens. The term 'sample' usually refers to a hand sample or drilled core from which the several specimens measured on the magnetometer are taken.

It is unfortunately true that only after measurements of NRM and magnetic cleaning are completed is it really possible to judge how many sites should be sampled, and how many samples and specimens obtained from each site to obtain a mean NRM direction of reasonable precision. Multiple sampling is necessary not only because of the nature of the geomagnetic field, as already described, but also because of other sources of scatter that may be present in the NRM directions at the different sampling levels. In each specimen from a sample there are random measurement errors and there may be scatter arising from the magnetization process, for example statistical deviation of the direction of depositional NRM from the ambient field direction. Sample mean directions may have orientation errors associated with them and also errors arising from differing anisotropy among samples. Systematic differences between site mean directions can arise from undetected relative displacement between geological blocks, varying inclination and/or compaction error in DRM and errors in reading the local bedding planes in sediments. Also incorporated in the final mean NRM direction from all sites will be errors arising from differences in the completeness with which secondary magnetizations are removed from each specimen.

The minimum of sampling is one specimen from one sample from each of three to five sites, but this would essentially constitute only a preliminary survey of the structure. In practice, it is desirable to have information on within-sample and between-sample scatter of NRM vectors and, if the overall (random) scatter of directions in the geological unit is large, additional data are required to reduce the error in the mean direction of NRM. Whether more specimens, samples or sites are best used for this purpose ideally depends on a knowledge of the dispersion at different sampling levels, which is not normally available. However, within-sample scatter is usually less than that between samples, and three to five specimens should adequately define a sample mean while four to eight samples define a site mean, depending on the dispersion of the sample means. It is clear that little will be gained by using more data at the level at which there is usually the least scatter and it may be worthwhile to collect more samples per site than given above, for use if results from some of them suggest it would be profitable. The number of sites used is typically five to thirty depending on the areal extent and accessibility of the formation, or the number of flows or intrusives.

There is sometimes a choice of material to sample in a particular unit, e.g. different grain sizes and colour, weathered and unweathered and compact and fractured material. Within limits it is desirable to obtain samples of each different type of rock present except where it is likely that unsatisfactory samples or data will be obtained. Weathering is a well-known source of secondary magnetization in both igneous and sedimentary rocks and sampling of fresh, unweathered rock is always a good policy. For this reason, quarries, road cuttings and similar exposures are preferred areas for sampling, if available. In natural outcrops weathering may penetrate to below the depth of sampling, or a preliminary chipping off of a rock surface may reveal only a small depth of weathering. In general, the coarser grained or more porous the rock the greater the depth of weathering.

In variegated red bed formations it has been found that 'red' (i.e. red–brown–purple) rocks generally provide the best data, whereas drab or green samples often show anomalous NRM directions and/or a proneness for VRM acquisition.

8.2 Sampling techniques

The sampling of geological and archaeological material for palaeomagnetic studies involves both removal of samples from their natural environment and marking them in such a way that their original field orientation is known. This is necessary for subsequent conversion of the sample magnetization directions to a common reference system.

There are two techniques employed for the collection of rock samples from the outcrop, namely removal of blocks (hand samples) and field drilling of cores. Also described in this section are techniques for sampling lake sediments, friable and unconsolidated materials and archaeological material.

8.2.1 Rock sampling

The collection of hand samples, typically 1000–5000 cm^3 in volume, is cheap, rapid and convenient at many exposures and requires little in the way of special equipment. Among the disadvantages are that more sample weight is often collected than may be required and, more important, the choice of samples may be partly governed by which blocks are removable and of convenient size rather than by what is desirable from the standpoint of sample distribution and lithology.

A procedure that is often satisfactory in hand sampling is to first break off the selected sample, trimming it to size if necessary, and then replace it as precisely as possible in its original position for orientation. This will avoid possible wasted time through first orienting the sample and then having it break unsuitably on removal from the outcrop. If accuracy of orientation is important the sample should be marked before removal.

The drilling of cores in the field with a portable drill has gained popularity in recent years. The power source is usually a small petrol engine such as is used for portable chainsaws, modified to take the tubular drill bit and water feed for cooling. Electric drills have also been used, supplied by an oil or petrol-driven generator.

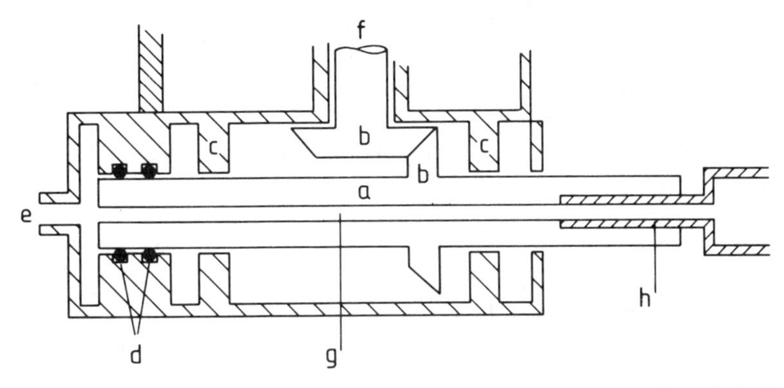

Fig. 8.1 Diagram of adapter for supplying coolant to drill. a, drill shaft; b, bevel gears; c, bearings; d, O-ring seals; e, coolant inlet; f, drive shaft; g, central hole; h, drill shank.

Cutting speeds are high with petrol motors, between 2000 and 4000 rev min^{-1}, and somewhat lower with electric drills. Lower speeds may be advantageous as far as drill life is concerned (cores per drill bit) but this is offset by longer cutting times and associated operator fatigue. Drill bits are usually of steel or stainless steel with diamond-impregnated tips, and the shank has an axial hole through which the coolant is fed to the cutting surface via a sleeve on the motor drive shaft (Fig. 8.1). Magnetic contamination from particles of steel rubbing off on to the core surface is possible and can be important in weakly magnetized rocks, although it is not difficult to recognize and remove in bad cases. Provided severe local heating has not occurred, any magnetization of the steel will be comparatively soft and will usually be removed by magnetic cleaning. Steel particles could conceivably be a source of magnetite during the thermal cleaning of sediments (Section 11.2).

With high drill speeds, it is sometimes difficult to get the coring process started because of the failure of the drill to bite at the desired place and the tendency of the drill to 'walk' over the rock surface. If necessary, the cut can be started by using a special drill bit fitted with a central tungsten-carbide tip. Some workers use a drill stand as an aid to locating the drill and cutting straight cores.

Water or other coolant (a small admixture of soluble oil or detergent facilitates drilling) is fed under pressure to the drill from a container that can be pressurized (e.g. a garden insecticide spray), or in some instances it may be sufficient to place the water container 2–3 m above the drilling site (if the outcrop allows it) and use the pressure of the head of water. Depending on the rock type 10–20 cm of core 2.5 cm in diameter can be drilled with a litre of water, the drilling time being 3–15 cm/min for sediments and 1–5 cm/min for basalts. A typical core length is 15 cm.

Field drilling requires more elaborate equipment than hand sampling but is more economical of sample weight and there is usually little restriction on material that can be sampled. A water supply is required and a considerable weight of water must be carried to collection sites at which it is not available and which are not accessible by road. Block sampling may be less rapid in terms of time spent in the field (although the total sample preparation time is comparable in the two collection methods), and field drilling has the advantage of less time involved in orientation.

Portable drilling systems have been described by Graham and Keiller (1960), Brown and Khan (1963), Doell and Cox (1965), Helsley (1967a) and McElhinny and Néel (1967). A larger drill for obtaining cores up to 6 m in length is described by Opdyke (1967).

A question that is often raised is whether the remanent magnetism of rocks is affected by the shock and vibration associated with block sampling or field drilling and the cutting of specimens. There appears to be normally very little effect on the carriers of a stable primary NRM, but a mechanically induced secondary magnetization of variable and sometimes high coercivity may be acquired by less stable particles. In rare cases where magnetic particles of very low relaxation times are present, the mechanical energy imparted may be sufficient to allow their realignment along the ambient (Earth's) field. Such rocks are already likely to have acquired a low stability VRM in the geomagnetic field with a time constant of minutes or seconds, and the measurement of a stable NRM component in such rocks may be difficult anyway. Any primary NRM of sufficient stability to provide useful palaeomagnetic data is unlikely to be affected by the mechanics of sample collection and preparation, but possible contamination from tools and local heating due to cutting with insufficient coolant should always be borne in mind.

A brief account of observations which suggest the presence of drilling-induced remanence (DIR) is given by Kuster (1969). A detailed study of DIR in two magnetite-bearing rocks is reported by Burmester (1977). Two components of DIR were detected, a soft VRM acquired in the field of the steel drill bit and a harder viscous component which persisted in a.f. demagnetizing fields in excess of 90 mT. The latter magnetization appears to be a VRM carried by magnetite particles in the surface of the specimen. This VRM is magnetically hardened through stresses induced in the particles during the cutting process.

Since it is not usually feasible to drill and cut rock samples in zero magnetic field it is likely that DIR may occasionally occur, most probably in unstable magnetite-bearing rocks. There is also the possibility of it contributing a small systematic error to the mean NRM direction of a group of specimens. The above author shows that acid etching of the specimen surface may be successful in removing the stressed material and also suggests thermal demagnetization of the surface particles with a gas flame.

Sallomy and Briden (1975) describe evidence of DIR in samples of Lower Jurassic sediments taken from a borehole core from Wales and in field-drilled cores from Yorkshire. DIR appears to be aligned with the direction of drilling (i.e. the borehole axis) rather than the local geomagnetic field, and steel contamination of samples is not the cause. The effect is more often seen in harder rocks in which drilling is slow and more heat is generated at the drill tip, leading the authors to suggest a partial TRM in the immediately adjacent material as the cause. However, the reason for the alignment of the DIR along the core axis is not clear, since the measured magnetic field at the drill tip is only of the order of 1 μT. Grain reorientation by vibration or by percolating drilling fluid also appear to be unlikely causes.

8.2.2 Lake sediments

Recent interest in the magnetism of lake and sea-bottom sediments has led to the use of devices for the remote acquisition of long cores of these materials. Details of the construction of sea-bottom coring equipment is outside the scope of this book and is not considered here.

Several laboratories are using a coring device based on the design of MacKereth (1958, 1969) for the acquisition of lake sediment cores up to 6 m in length in water depths of up to 100 m. Full details are given in the 1958 paper,

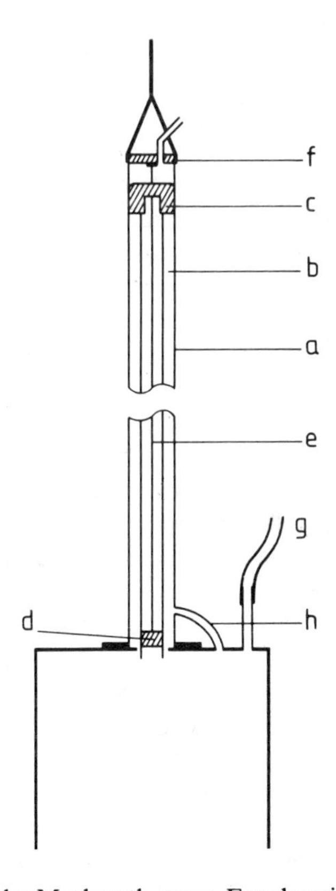

Fig. 8.2 Basic parts of the Mackereth corer. For description see text.

and a much simplified diagram of the corer is shown in Fig. 8.2. The outer stainless steel tube (a) ($\sim$8 cm in diameter, 6 m long) is attached by means of a flange at its lower end to a vertical steel drum 45 cm in diameter and 1.2 m high, open at its lower end. The Duralumin core tube (b), 3.5 cm in diameter, 0.5 cm wall, is attached to a piston (c) at its upper end and contains a piston (d) at its lower end. The piston is attached securely to a thin tube (e) passing up inside (b), through a pressure seal in piston (c) to a plug (f) securely attached to the top of (a) and closing it off. To operate the corer it is lowered to the lake bed until the lower end of the steel drum is embedded in the sediment, a small tension being maintained in the cable to keep the corer axis vertical. Water is then pumped out of the drum via a plastic tube attached to outlet (g), whereupon the hydrostatic pressure exerted by the overlying water forces the drum into the sediment, anchoring it firmly. Compressed air is admitted above piston (c) and forces the core tube down into the sediment. When the piston passes the entrance to the bypass tube (h), the coring is completed and compressed air passes into the drum, forcing it and the core tube out of the sediment. When the whole of the drum is out of the sediment and full of air its buoyancy completes the extraction of the core tube and the equipment rises to the surface. The core tube is taken out of (a) and the core forced out with a hydraulic pump into a suitable half-section tubular container or into a V-section trough made by two lengths of wood joined at right angles. In some versions of the corer the core tube is plastic and the core is then in a suitable form for continuous measurement of NRM (Section 9.5.2).

In shallow lakes another type of piston corer is used, based on a principle used by Livingstone (1955) and in a modified form by Vallentyne (1955). The corer is lowered to the lake bed on the end of rigid rods which hold it in place while the core tube is mechanically forced into the sediment with the aid of another rod to a depth of 1–3 m. Deeper cores can be obtained by repeated sampling in the same hole. The Kullenberg gravity corer relies on a heavy weight (100–500 kg) to force the coring tube into the sediment, the tube and weight falling freely through the last few meters before impacting the bottom (Wilson, 1941; Emery and Broussard, 1954). In a modification described by Jenkin and Mortimer (1938) the weight can slide up and down the upper portion of the coring tube. Using a cable from the surface the weight can then be used to hammer the tube into the sediment. Jenkin and Mortimer (1941) and Hvorsley and Stetson (1946) have described other types of gravity corer and the latter authors give a useful review of techniques.

Although one of the primary objects of lake-sediment studies is to examine geomagnetic field variations down a core, it is desirable to link these with an absolute declination, i.e. to know the North direction in the core when it was in place. This can be done by attaching a compass and camera to some fixed point at the top of the corer and photographing the compass before extracting the core. It is also necessary to ensure that the core tube does not rotate relatively to the rest of the corer during extraction from the lake bed. At the time of writing other methods of core orientation are being investigated, and also methods of recording any departure from the vertical of the core.

8.2.3 Sampling for archaeomagnetism

In archaeomagnetic investigations the aims are different from those in palaeomagnetic studies of the geological column. The time span involved is restricted to about 10 000 years and the major interest is in using the direction and intensity of NRM in dated archaeological material to establish the secular variation of the geomagnetic field over different areas of the Earth. The determination of a reliable and accurate variation curve also affords the possibility of providing contributory evidence of the age of poorly dated and undated artefacts, by fitting the NRM direction and/or intensity found in them with the appropriate region of the variation curves.

The most useful archaeological material for magnetic studies is any object which has been heated above the Curie point of the magnetic mineral in it (usually haematite, but sometimes magnetite), and which has remained in place and unaltered in attitude since the last heating took place. Thus hearths, ovens and furnaces are common sources of samples, although palaeointensity determinations can be carried out on unoriented material such as pottery pieces and bricks.

There is often a restricted choice of material for sampling and it may be necessary to utilize fragile or friable material. For secular variation investigations it is important to determine NRM directions as accurately as possible and it is usual to orient the samples rather more carefully than for a palaeomagnetic survey. This is also desirable where only a few samples are available from a particular site.

Archaeomagnetism was extensively developed by E. and O. Thellier, who used simple and accurate sampling and orienting techniques (Thellier, 1936).

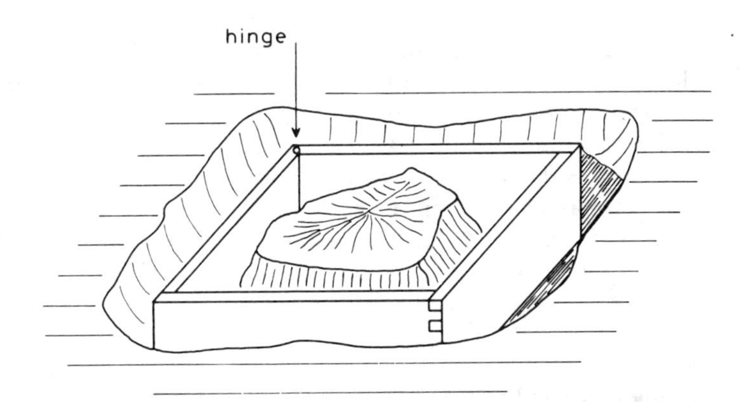

Fig. 8.3 Method of sampling archaeological material, due to Thellier.

For friable material a sample of it 5–10 cm across is exposed by removing the surrounding material and a hinged wooden or aluminium frame is placed around it (Fig. 8.3). The frame is then filled with plaster of Paris, which is allowed to harden. The plaster surface can be left rough, or, better, it can be smoothed during hardening to form a plane surface level with the top of the frame which is then used for the orientation marks. When hardened, the frame is opened and removed and the sample removed by cutting it away underneath the exposure. This method is also suitable for mechanically weak geological samples, and it is often possible to arrange for the plaster surface to be horizontal, by wedging the frame in position prior to filling with plaster. If the material sampled is expected to be weakly magnetized ($< 5 \times 10^{-6}$ A m^2 kg^{-1}) and is to be measured without removal from the plaster block it is advisable to test samples of the hardened plaster for remanent magnetism prior to its use in the field.

The above technique is capable of providing quite large samples where required and these were favoured in early archaeomagnetic studies. A modification of the method is for smaller samples which can be measured, for instance, on the Digico archaeomagnetic magnetometer (Section 9.5.1). An annular hole with its axis vertical is excavated around the sample, the hole being of diameter and depth such that an open-ended plastic tube 5.0 cm long and 5.0 cm in diameter with its axis vertical can be placed over the upper part of the sample, leaving a few millimetres between sample and tube. The top of the tube is levelled with a spirit-level and plaster of Paris poured in to fill the tube, and the top surface levelled off or a plastic disc fixed to the tube. After orientation the lower part of the sample is cut away and trimmed so that the underside of the tube can be filled with plaster and levelled off flush with the tube end, or another plastic disc fixed to the tube. The integrity of very friable material is best preserved by completely filling the remaining space in the tube with plaster. The samples are measured without removal of the plastic case but again it is a useful precaution to test samples of plaster and plastic tube for remanent magnetism prior to sample collection. A method described by Homonko (1978) for sampling flowstone may also be of use. Small plastic discs are glued to parts of the exposure which are likely to break off reasonably easily, the discs being set horizontal before the glue sets. An orientation mark is then placed on the disc before breaking the sample off. For measurement of NRM the sample and disc are held in a suitable cylindrical container. Some friable material may benefit from painting the surface with or dipping the sample in a hardening polyester resin either in the field or back in the laboratory.

8.2.4 Soft sediments, muds, etc.

It is sometimes required to take samples of sediments with a high moisture content which require permanent support to maintain the sample shape. Such materials include some varved clays, cores of lake and sea-bottom sediments, sand and soil samples, cave sediments and unconsolidated muds.

The basic method of obtaining samples suitable for NRM measurement is to press suitable containers into the material, mark orientation data on them, extract sample and container and seal with a lid.

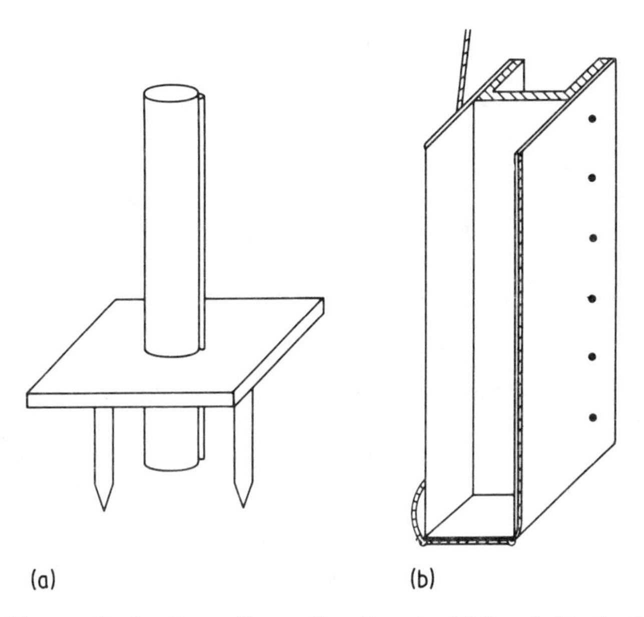

Fig. 8.4 Two methods of sampling soft sediments: (a) is suitable for horizontal surfaces and (b) for vertical exposures.

Homonko (1978) describes two methods of obtaining extended sections for long core measurement of soft material, used by the author to sample recent laminated clays. In one, a length of aluminium square-section channel has detachable steel blades fitted to two sides and the base (Fig. 8.4(b)) and is pushed or tapped into a vertical section until the base of the channel is against the face. A steel wire, attached to the box shown in the diagram, is then pulled up to cut the sediment at the back and the sample can be removed from the exposure. The cross-section of the box is chosen so that when the corners of the square-section core are shaved off the core can be slid into a length of plastic tubing (5 cm in internal diameter) with the aid of a strip of polythene sheet placed along the back of the box before the core is taken. Before being put in the tube the core surface is consolidated by painting with an epoxy-based resin. Both ends of the tube have plastic discs glued on prior to transport. In the laboratory the remaining space in the tube is filled with plaster of Paris. A modification of this method has been developed for use in a horizontal exposure of sediment. A hole is bored out to the required depth with an auger-type (post-hole) drill. The back of the box has an additional plate which is pressed against the box when it is pushed down the hole. A wedge is then forced down between plate and box and the blades are forced into the sediment. Core extraction is as before.

The second method is suitable for situations where a horizontal area of sediment is available for sampling. A plastic tube with one end sharpened is pushed or hammered into the sediment and is prevented from rotating by a longitudinal extension on the tube wall. During coring, this extension slides down a key-way cut into a plate which is pinned into the ground as shown (Fig. 8.4(a)). A bull's-eye spirit level set in a plug which fits into the upper end of the tube is used to keep the tube vertical during coring. After marking, the core can usually be extracted by opening up the hole somewhat by pushing and pulling the upper part of the tube and then pulling it out. In some material this coring method may cause some compaction of the core.

8.2.5 Sample preparation in the laboratory

After making any additional reference lines required on the sample with waterproof (Indian) ink, hand samples are cored with a stainless steel or phosphor-bronze drill bit with a diamond-impregnated tip, mounted on an ordinary workshop vertical drill press. As with the field drill the drive shaft is modified to take the bored-out drill shank and sleeve for supplying coolant to the cutting surface. Cutting speeds are usually in the range 2000–4000 rev min^{-1}, and cores up to 15 cm long and 2.5 cm in diameter are easily cut.

It is usually convenient to core sediments perpendicular to the bedding plane and igneous and other irregularly shaped rocks in their field position (although the strike or other reference azimuth is, of course, immaterial). The sample can be supported under the drill on the tapered ends of vertical bolts passing through a horizontal steel or brass plate (the plate ~15 mm thick with bolts ~15 mm in diameter on a ~4 cm grid). The bolts are screwed up or down to contact and support the underside of the sample when the upper surface is in the desired attitude. The sample and support are contained in a metal box with a detachable front for easy access when setting up the sample, and it is fitted with a drain hole at the bottom. The coolant is pumped through the drill-bit from the top of a settling tank below the drill and passes back down through the drain hole into the bottom of the tank.

To avoid jamming and possible damage to the drill and core, it is not advisable to drill right through a sample but to stop short of the underside and subsequently break the core off as described for field drilling.

Before cutting a core into discs or cylinders a fiducial line is marked along its length corresponding to the scratch mark on field-drilled cores on the strike direction in hand samples. A simple method of ensuring that the sequence and attitude of the specimens cut from a core are known after cutting is to mark a line at the top of the core starting at the fiducial line and diverging from it down the length of the core.

The cores are cut into cylinders or discs with diamond-edged cutting wheels (usually 15–25 cm in diameter) of stainless steel or phosphor-bronze. A tungsten carbide wheel is a possible and cheaper alternative but is usually suspect as a source of magnetic contamination. When weakly magnetic material is being cut it should be tested for contamination by measuring the NRM of a homogeneous, weakly magnetized rock sample before and after several shallow cuts have been made in it with the wheel.

Cutting speeds should be fairly high ($\sim$2000 rev min^{-1} with a 20 cm wheel) and there must be an ample flow of coolant at the cutting surface. A simple and rapid cutting procedure is to clamp 3–5 cores lying parallel on a horizontal brass carriage running on horizontal rails so that the wheel cuts through the cores successively as the carriage is traversed under the wheel. Any projecting pieces left by the cut samples breaking off can be removed by holding lightly against the edge of the wheel or by abrading with a paste of aluminium oxide on a glass plate.

Several commercial cutting machines are available which can easily be modified for the rapid cutting of rock cores.

8.3 Orientation methods

Before samples are removed from the outcrop their attitude is recorded, so that their NRM can be related to a common geographical reference system for subsequent analysis of the derived ancient geomagnetic field direction.

8.3.1 Hand samples

If the samples are from a well-bedded sediment with clean bedding surfaces exposed, or if they have reasonably flat surfaces of $\sim$20 cm^2 or more in area, the direct marking of strike and dip on the upper surface is the simplest method. The strike is a horizontal line on the surface and can be marked with the aid of a spirit level and either waterproof ink or a sharp-pointed brass scribe. The dip of the surface, the line of maximum slope (and therefore perpendicular to the strike), is measured with an inclinometer and marked in the same way: there will be no ambiguity if it is always marked on the 'down dip' side of the strike line. The strike direction is also recorded (see Section 8.3.3) and its sense indicated on the strike line by an arrowhead.

For samples with irregular surfaces a simple device may be used consisting of a small Perspex table in the shape of an isosceles triangle fitted at each corner with short (2–3 cm) brass legs of equal length. A convenient size for the table is a 4–5 cm base, 6–7 cm for the longer sides and $\sim$1 cm thick (Fig. 8.5(a)). The legs should be tapered to a point and they are placed on the rock surface in such a way that the short side is horizontal. The strike and dip,

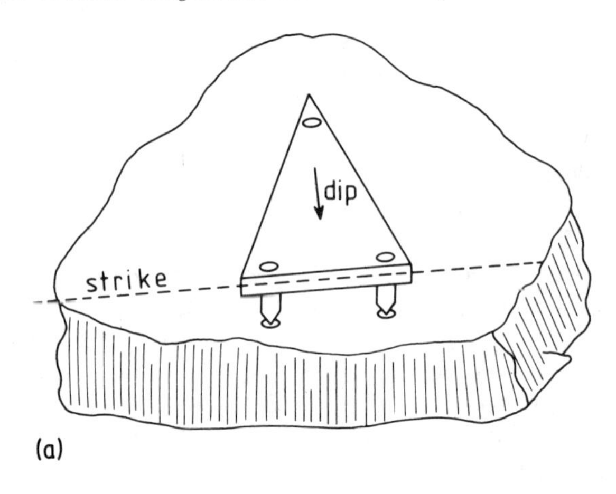

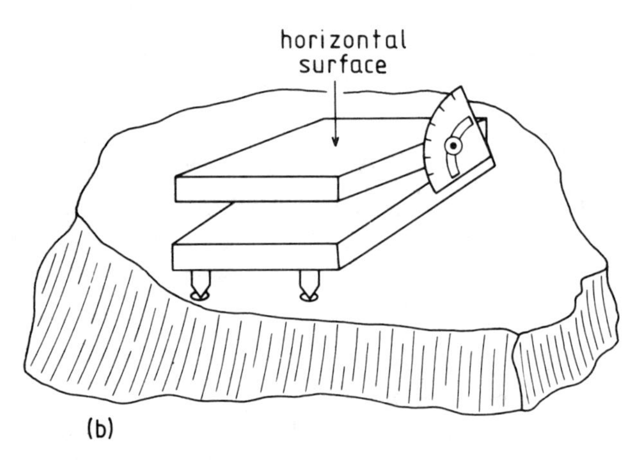

Fig. 8.5 Orientation methods for hand samples: (a) triangular Perspex table recording 'dip' and 'strike' which can be reproduced in the laboratory; (b) method of reading the 'dip' in situ.

as measured on the table surface, are then recorded as for bedding planes, and the position of each leg on the rock surface marked by a small drop of white ink or a scratch mark. A modification of this method is the inclusion of a built-in spirit level parallel to the short side of the table, which is satisfactory provided the table is not dipping so steeply that the spirit level becomes inoperative. Alternatively, the table can be fitted with a plate hinged to the short side, which is adjusted until horizontal, and the tilt of the lower plate read with a fixed inclinometer (Fig. 8.5(b)). Another method is the use of a right-angled triangular shaped table with two built-in spirit levels parallel to the equal sides:

on suitable samples this table can be adjusted on the surface until horizontal, and the direction along one side recorded.

The overall accuracy of the above methods depends mainly on the time which the operator is prepared to spend, and would normally be expected to be 2–4°. Assuming spirit levels and inclinometers are functioning correctly there are unlikely to be any systematic errors involved (but see Section 8.3.3).

The strike direction, or any other required direction, can be measured with a magnetic compass, theodolite, gyrocompass or sun compass, the latter method being discussed in Section 8.3.3.

The simplest method, of adequate accuracy, is to use a prismatic compass or Brunton compass in which a compass and inclinometer are combined. If the surface or orienting plate is horizontal or nearly horizontal the compass can be placed on it and the strike direction read off relative to magnetic North. If the strike line or direction is on a sloping surface, it is usually possible to sight along it with sufficient accuracy.

In an environment of sedimentary rocks with no igneous material (or significant quantities of iron or steel) in the immediate vicinity, the strike direction can be satisfactorily determined by this method. However, if igneous or other rocks with sufficiently strong induced or remanent magnetization are collected, or if such rocks are near the collection site, a significant deviation (up to several tens of degrees) of the local geomagnetic declination from the regional value is possible: the deviation may also be critically dependent on the sample position. The use of a magnetic compass is then unsatisfactory, and although it is possible to use a gyrocompass or theodolite sighting these methods involve costly and bulky equipment and the instrumental accuracy is somewhat in excess of that normally required. At some sites it may be possible to use a magnetic compass to determine the angle between the measured strike direction and the bearing of a topographic feature, from which the true strike direction can be determined by consulting a map (Bidgood and Harland, 1959). However, where the compass deviation is locally very variable it is necessary to ensure that both readings can be taken with the compass in the same position.

Whether a magnetic compass is giving false readings because of local magnetic anomalies can usually be determined by simple tests. In severe cases the indicated bearing of a local feature will vary significantly as observed from points a few metres apart. The compass needle will probably also be seen to rotate as a local exposure of rock is approached. (This situation may even be useful as a guide to the magnetic polarity of different rock exposures if samples with normal and reversed NRM are being sought.) Weaker local magnetic anomalies can be detected by comparing the observed bearing of a topographic feature with the true bearing derived from a map, or by taking a sight on the Sun or on the shadow cast by a vertical rod if tables of the Sun's position are available. The use of the Sun compass is described in Section 8.3.3.

A theodolite or gyrocompass is somewhat cumbersome and inconvenient to take into the field, particularly if the sampling sites are remote and inaccessible by road. The latter, essentially a motor-driven gyroscope whose rotation axis aligns itself parallel to the Earth's axis of rotation, requires batteries to power it and is expensive, as is a theodolite. The use of the instruments requires some skill and knowledge and will not be described here. For further information the reader is referred to manuals of surveying.

8.3.2 Field-drilled cores

The most convenient method is to use a device of which one version is shown in Fig. 8.6. The slotted brass or aluminium tube is pushed down over the drilled core while it is still in place, and the tube and hinged table adjusted until the latter is horizontal. An inclinometer attached to the table measures the dip of the plane perpendicular to the core axis, and a linc parallel to the table hinge axis defines the strike of this plane. A line is scratched down the length of the core with a brass scribe guided by the slot: this line serves to locate the strike axis on the diametral plane of the core. After recording the strike direction the brass tube is removed and the core broken off by leverage with a brass or stainless steel rod pushed down to the base of the core in the annular cut left by the drill. Some variations of the basic device are described in the references given at the end of Section 8.2.1.

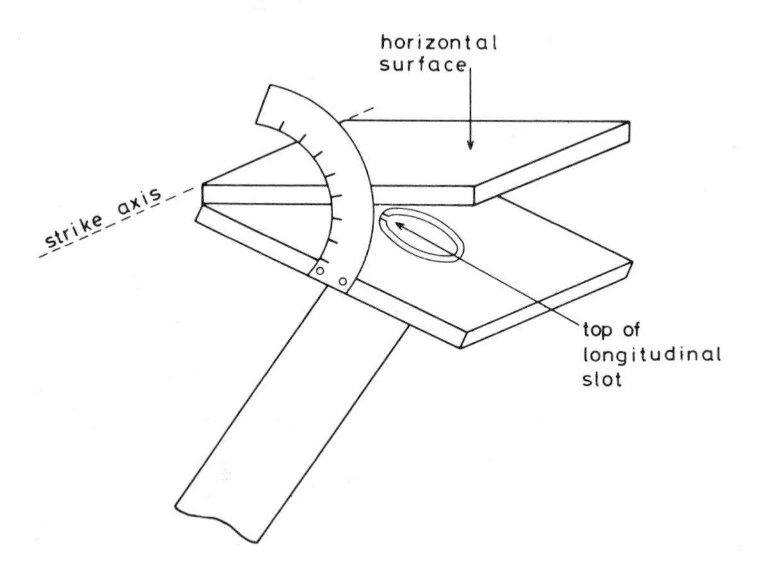

Fig. 8.6 Orientation device for field-drilled cores.

For the purpose of interpreting directions of NRM in sediments it is usual to assume, in the absence of evidence to the contrary, that they were deposited on level surfaces within 2–3° of the horizontal. Thus the bedding plane in the samples is the reference surface for determining the direction of NRM relative to the ancient horizontal plane, if the primary magnetization was acquired before the beds were tilted. In samples in which the bedding is poorly defined it is necessary to estimate the regional dip of the beds from field observations in the collecting area, or by consulting a geological map of the area: the NRM directions are then referred to this plane.

Lack of knowledge of the exact tilt of the bedding is an important source of error in sedimentary palaeomagnetism. Some deposits, particularly fine-grained beds, can become contorted under the pressure of overlying rocks and in different samples the observed bedding may depart significantly from the true value. However, there will usually be visual evidence for this, and scatter among the values of strike and dip recorded in different samples at different sites may also be a pointer. Other sources of error are less easy to detect, for example relative rotation of different parts of a formation.

8.3.3 The Sun compass

In its simplest form a Sun compass consists of a square plate with a white surface on which is marked a graduated circle with a thin rod inserted accurately perpendicular to the plate at the centre of the circle. Two spirit levels are fixed to the plate to enable it to be set horizontally. Assuming the Sun is shining and the site is not in shadow, the vertical rod will cast a shadow whose azimuth relative to a reference direction on the plate (along one of the sides) can be read off on the graduated circle. By calculating the Sun's azimuth and thus that of the shadow bearing the reference direction can be determined.

Figure 8.7 shows the relevant geometry. The angle H is the 'local hour angle', the angle between the meridians containing the Sun and the site. For a given time of day the day-to-day variation of H is not significant, but during any one day H takes values between 0° and 360°. The angle D is the Sun's declination, the angle between the Earth's equatorial plane and the Sun–Earth line. D does not vary significantly (for the present purpose) during one day but undergoes a cyclic variation of amplitude between 23.5° N and 23.5° S with a period of twelve months.

The following analysis essentially follows that of Creer and Sanver (1967) for a Sun compass on which the circle is graduated in a counter-clockwise direction. In solving the spherical triangle PCS, we use the cosine rule for h (Fig. 8.7) and note the general identity $\sin(90 - x) = \cos x$

$$\cos h = \sin L \sin D = \cos L \cos D \cos H \tag{8.1}$$

The sine rule then gives β, the Sun's azimuth, from

$$\sin \beta = \cos D \sin H / \sin h \tag{8.2}$$

or the cosine rule

$$\cos \beta = -(\sin D - \cos h \sin L) / \sin h \cos L \tag{8.3}$$

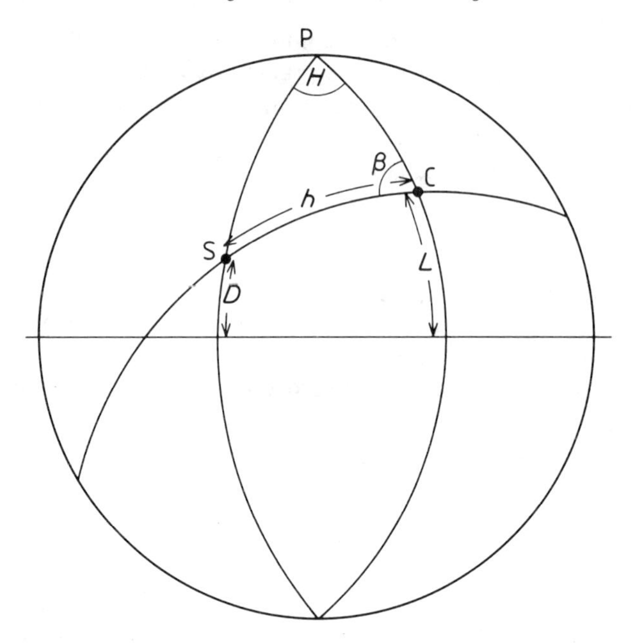

Fig. 8.7 Geometry for sun compass calculations. P is the pole, C is the collection site and S is the point where the Sun–Earth line intersects the Earth's surface; L is the site latitude.

Table 8.1 Sign convention for L, D and H

Latitude (L)	North, +ve; South, −ve
Declination (D)	North, +ve; South, −ve
Hour angle (H)	$< 180^\circ$, +ve; $> 180^\circ$, use $360^\circ - H$, −ve

Table 8.2 Quadrant in which shadow azimuth occurs

Sign of Equation 8.4 numerator	Sign of Equation 8.4 denominator	Azimuth of shadow and corresponding quadrant corresponding quadrant	
−	−	$\beta + 180^\circ$	(SW)
−	+	$-\beta$	(NW)
+	+	β	(NE)
+	−	$180^\circ - \beta$	(SE)

h can be eliminated by dividing Equation 8.2 by Equation 8.3 and substituting for $\cos h$ from Equation 8.1. Then

$$\tan\beta = -\sin H / |\cos L \tan D - \sin L \cos H| \tag{8.4}$$

The angles H and D are obtained from tables in the *Air Almanac* or *Nautical Almanac* or can be computed. The tables give the Greenwich Hour Angle (G) at intervals during the day of Greenwich Mean Time, from which H is found by adding the E longitude of the collection site to G. It is thus necessary to note the local time (with an accuracy of two minutes or less) when the shadow angle is being read, and then to convert this to GMT to find G. The declination D is read directly from the appropriate table. Both H and D are independent of L, the site latitude. The quadrant in which β lies may be obtained with the aid of Tables 8.1 and 8.2.

In low latitudes, at times around midday, the sun is near the zenith and the shadow cast by the vertical pin will be very short, resulting in some loss of accuracy. It may be worthwhile using a longer pin on such occasions, although with increasing length it becomes more difficult to ensure that the pin is perfectly straight.

Embleton and Edwards (1973) describe the construction of a combined compass, inclinometer and sun compass and provide working drawings, and Stone (1967c) has developed a sun compass with which the North direction can be mainly determined in the field. A simple device which is satisfactory at some outcrops is described by Verosub (1977a). A rectangular brass plate is inscribed near one end with lines radiating from a point at 10° angular intervals. A vertical rod is screwed into the plate at the origin of the lines. One edge of the plate is oriented with a magnetic compass at a point away from the outcrop, and the angular position of the shadow of the rod noted. The plate is then quickly transferred to the sample site and oriented so that the shadow falls in the same position, when the plate edge is again at the known orientation.

Chapter Nine

Measurement of the NRM of rocks

9.1 Magnetization of rock samples

9.1.1 Introduction

The magnetization M (magnetic moment per unit volume) of a rock sample can be expressed as the vector sum of the remanent magnetization M_r and the induced magnetization M_i in the ambient geomagnetic (or laboratory) field H

$$\mathbf{M} = \mathbf{M}_r + \mathbf{M}_i = \mathbf{M}_r + k\mathbf{H} \tag{9.1}$$

where k is the volume susceptibility, assumed here to be isotropic. In palaeomagnetism it is the intensity and direction of M_r that is of major interest and some aspects of its measurement are considered in this section.

The observed remanent magnetic moment of a rock sample is the vector sum of the remanent magnetic moments of all the magnetized particles contained within it. Because these particles vary in size, intensity of magnetization, number per unit volume and alignment of their moments, some degree of inhomogeneity of NRM is a common occurrence. It may be severe in coarse-grained rocks in which a very few large grains contribute the bulk of the observed remanence in a $\sim 10\ \text{cm}^3$ sample. Another form of inhomogeneity of NRM occurs in those sediments in which discontinuous deposition of the magnetic minerals has resulted in a markedly banded structure. It may be noted here that it is possible for inhomogeneity of NRM intensity to be associated with homogeneity of direction throughout a sample and vice versa. The above-mentioned banded samples are an example of the former case if the NRM direction in each band is the same, whereas other sediments may have uniform NRM intensity but variations of direction within a sample due, for instance, to the effect of water currents (if the NRM is depositional) or secular variation of the geomagnetic field during the period in which the sample acquired its magnetization. In the routine measurement of NRM it is usually necessary to adopt procedures which reduce errors arising from inhomogeneity and this topic is discussed further in Section 9.1.3.

Other features which bear on the measurement of NRM are the size and shape of samples and inhomogeneity and anisotropy of susceptibility. However, since most NRM measurements are carried out in zero field, susceptibility effects are not usually important.

The magnetic moment of a rock sample is not measured directly by a magnetometer but is derived from the intensity and direction of the magnetic field arising from, and in the vicinity of, the sample. The optimum sample is one for which the field configuration around it can be determined with a minimum number of readings, which in general implies a sample whose magnetic moment can be represented by a dipole and a minimal multipole contribution.

The ideal rock sample for a determination of NRM is therefore a uniformly magnetized sphere of zero susceptibility. Outside the sample its magnetization can be represented by a dipole at its centre of moment $\frac{4}{3}\pi r^3 M$ where r is the sample radius and M its volume magnetization. Such a sample would require a minimum of two readings for a complete determination of its NRM. In practice, spherical samples are not practicable for routine measurements and inhomogeneity should be assumed in the measuring procedures even though it may not be significant in many samples.

The preferred sample shape is that for which the external field at any point departs as little as possible from that of a dipole of equivalent magnetic moment at the sample centre, and depends minimally on the orientation of the sample magnetization relative to reference axes within the sample. A cube or an approximately equidimensional cylinder are convenient sample shapes which approach the above requirements: the latter shape, being somewhat simpler to prepare, is now generally favoured.

Some early measurements were done on thin discs of rock, particularly in sediments, for reasons which include the possible detection of secular variation in adjacent discs from a core and ease of measurement by Blackett's method with the then commonly used astatic magnetometer (Section 9.2.4(b)). The field on the axis of a uniformly magnetized thin disc of volume v and volume magnetization M_z and M_H perpendicular and parallel to the disc plane respectively can be expressed in the form

$$B_z = \frac{\mu_0 M_z v}{2\pi z_0^3} \tag{9.2}$$

$$B_H = \frac{\mu_0 M_H v}{4\pi z_0^3} \tag{9.3}$$

where B_z and B_H are respectively the axial field and the field parallel to the disc plane at a distance z along the axis from the disc centre, and $z_0 = (z^2 + r^2)^{1/2}$ where r is the disc radius. Thus, the field at any axial point is equal to that of the equivalent dipole displaced an extra axial distance $(z_0 - z)$ from the point.

There are two possible approaches to the measurement of non-spherical samples. On the assumption of uniform magnetization the field due to a cubic or cylindrical sample can be computed for the distance from the sample at which the sensor is placed, or, in the case of spinner-type magnetometers the flux linkage with the pick-up coil system can be determined. In principle this allows for the non-dipole nature of the equivalent moment, but the problem of the dependence of the field on the orientation of the magnetization relative to the sample axes remains, since a prior knowledge of this orientation would be required to compute the field intensity and direction at any point.

The alternative approach is to assume that the sensor–sample configuration in the magnetometer and the sample size and shape are such that the assumption of an equivalent dipole is a valid one, and that the variation of external field with NRM orientation in the sample can be neglected. The extent to which this approach is a valid one is now examined with respect to measurement of both NRM directions and intensity.

9.1.2 Magnetic field of non-spherical samples

It may be first noted that it is not necessary to consider demagnetization effects due to the shape of the sample. If the volume intensity of magnetization is M, the demagnetizing field H_d is given by

$$H_d = NM \tag{9.4}$$

where N is the demagnetizing factor appropriate to the sample shape and axis of magnetization. The value of N for a sphere is 1/3 and the effective value of N is similar for cubes and equidimensional cylinders. It is rare for a rock magnetization to exceed 10^2 A m^{-1} and thus H_d has a maximum value of ~ 30 A m^{-1}. The incremental susceptibility of these rocks is typically such that the back-induced magnetization is between two and three orders of magnitude less than the remanent magnetization, and less by a larger factor in typical sediments.

(a) Cubic samples

The calculation of the field at any point near a uniformly magnetized cube is complex and has been investigated in connection with NRM measurements by Helbig (1965). However, in the restricted case of magnetization parallel to a cube edge, the resulting axial field (Gauss A position) and the field at points along the two axes perpendicular to the axis of magnetization and to the other cube faces (Gauss B positions) is simpler to compute. This can be done rigorously by differentiation of the general expression for the potential of a magnetized cube at external points (Hellbardt, 1958) or, often with sufficient accuracy, by summing the fields due to square current loops uniformly distributed on the cube surface with their planes perpendicular to the magnetization axis, to which the cube magnetization is equivalent. Because of the symmetry of the cube the magnitude of the fields at points along the cube axes, although differing at small distances from those due to the equivalent central dipole, are still in the same ratio as those of the dipole field, i.e. Gauss A: Gauss B = 2 : 1 at points equidistant from the cube centre (Sharma, 1968).

This implies that the direction of NRM of a cubic sample can be accurately determined by measuring the fields due to the resolved components of magnetization along axes perpendicular to the cube faces, at points on these axes; the axial or transverse field can be measured, for instance by an astatic magnetometer (sample in one of the positions shown in Fig. 9.9) or a fluxgate sensor. The computed intensity will only be correct if the instrument is calibrated for cubic samples rather than for a dipole, or if an appropriate correction factor is applied to a dipole calibration.

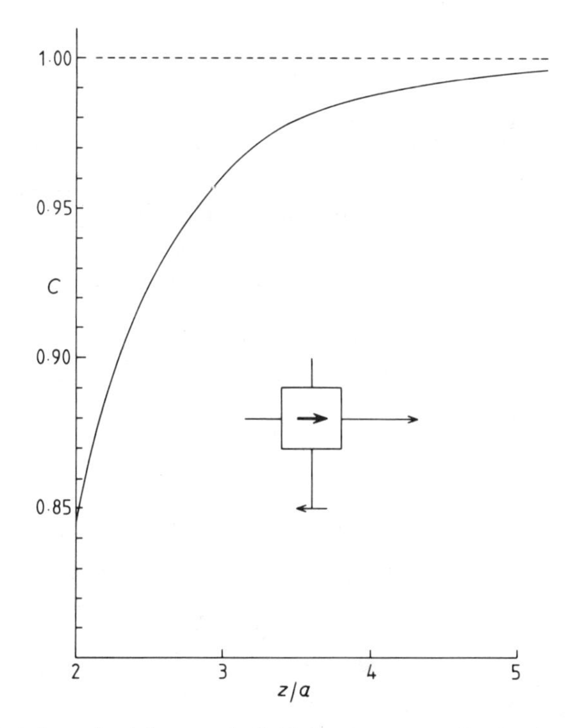

Fig. 9.1 Variation of axial magnetic field due to a magnetized cube compared with field of equivalent centred dipole. Values of C plotted against z/a or x/a.

The variation of the axial field with distance from the cube centre for a uniformly magnetized cube is shown in Fig. 9.1, in which the factor C is plotted against z/a or x/a, where $2a$ is the cube edge. The numerical factor C relates the equivalent centred dipole field to the field of a cube by

$$B_z = C\left(\frac{\mu_0}{2\pi}\frac{P}{z^3}\right) \tag{9.5}$$

$$B_x = C\left(\frac{\mu_0}{4\pi}\frac{P}{x^3}\right) \tag{9.6}$$

for Gauss A and B positions respectively: P is the dipole moment, directed parallel to a cube edge. The factor C is calculated from the modified expression of Hellbardt (1958)

$$C = \frac{1}{4}\left(\frac{z}{a}\right)^3 \tan^{-1}\sqrt{\left\{\left[1+\left(\frac{z}{a}-1\right)^2\right]^2-1\right\}} - \tan^{-1}\sqrt{\left\{\left[1+\left(\frac{z}{a}+1\right)^2\right]^2-1\right\}} \tag{9.7}$$

in which z and x, the distance to the cube centre, are interchangeable.

It can be seen from Fig. 9.1 that at $z = 3a$, i.e. at a distance equal to the cube side from a cube face, the field is within 4% of that of the equivalent centred dipole, and < 1.5% at $z > 4a$. These field values reflect the rapid decay with distance of the fields arising from the multipole contributions to the cube magnetization.

In a spinner magnetometer, in which the rotating sample induces an alternating voltage in a pick-up coil array, the direction of NRM in a cubic sample will be accurately determined, but the accuracy of the intensity measurement depends in general on a theoretical or experimental determination of the flux linkage between sample and coil(s).

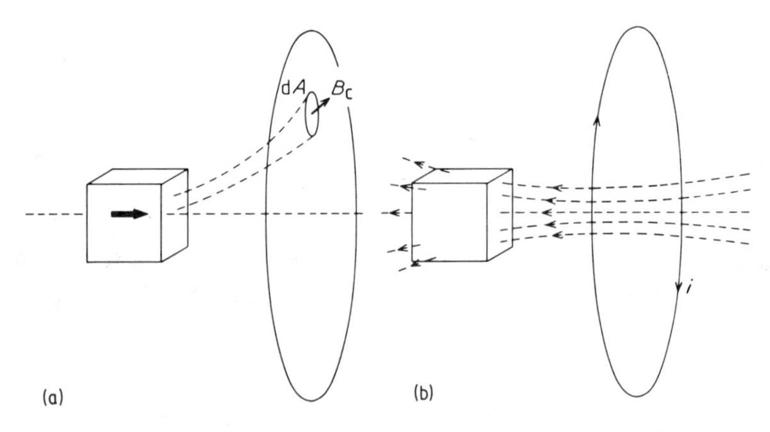

Fig. 9.2 (a) Flux linked with coil due to magnetized cube; (b) field at cubic sample due to the coil.

The flux linkage with a pick-up coil due to the magnetized cube (Fig. 9.2) is $\int_A B_c.dA$, where B_c is the field due to the cube at points in the coil plane and dA is an element of the coil area A. Parry (1967) points out that this flux linkage is also given by the integral over the cube volume of the field parallel to the cube magnetization due to unit current in the coil multiplied by the volume magnetization of the cube. The flux linkage due to a quadrupole or higher multipoles is equal to the produce of the appropriate field gradient over the sample and the multipole strength. Thus, it is not only necessary to know (or measure) the flux linkage for a particular sample-coil configuration but also whether, for a fixed direction of the sample magnetization relative to the coil, the linkage depends significantly on the orientation of the magnetization relative to the cube axes. In the simple pick-up system shown in Fig. 9.2, it can be seen from the above theorem that there will be some dependence of flux linkage on NRM orientation within the cube. The effect of different orientations is to slightly alter the multipole contributions to the total cube moment, and thus the fundamental and higher harmonic content of the induced voltage. This will generally be only a small effect and in routine measurements will produce changes in measured intensity much smaller than is typically observed between specimens and samples.

From the above concept of flux linkage it is apparent that if the pick-up coil system produces a uniform field over the sample, the flux linkage will be independent of sample shape (to the same extent that the field is 'uniform') and moreover the output voltage of the coil due to the spinning sample will be purely sinusoidal and independent of the shape of the sample and of any inhomogeneity of the NRM: the magnitude of the voltage will be equal to that generated by a dipole of equivalent moment to that of the sample. This property of certain coil systems has been stressed by Thellier (1967) and used by him in ballistic and spinner magnetometers for the measurement of large, irregularly shaped archaeological samples.

(*b*) *Cylindrical samples*

The magnetic field due to a magnetized cylinder is of more practical interest in palaeomagnetism because of the wider use of rock cylinders, a consequence of this shape being simpler to cut both from field drilled cores and hand samples. Because of the lower symmetry of the cylinder compared with the cube there is a greater potential for systematic errors to occur in NRM measurements, and it is of some interest to investigate such sources of error and also to enquire whether there is an optimum length/diameter ratio of the cylinder which minimizes them.

A potential source of error is illustrated in Fig. 9.3. Because of the cylinder geometry, the field at a point along or on a line perpendicular to the magnetization axis will depend on the orientation of the NRM within the cylinder. Thus the fields measured by a sensor at a point due to equal orthogonal components of the sample magnetization will not in general be

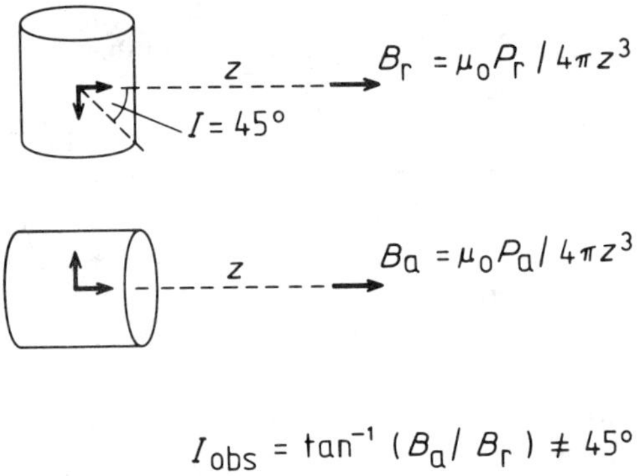

$B_r = \mu_0 P_r / 4\pi z^3$

$I = 45°$

$B_a = \mu_0 P_a / 4\pi z^3$

$I_{obs} = \tan^{-1}(B_a / B_r) \neq 45°$

Fig. 9.3 Possible source of error in measurement of NRM of rock cylinders. The sample has a true inclination of 45° and the fields B_r and B_d are measured at a distance z, assuming the dipole relationship $B_r \neq B_a$ and $P_r \neq P_a$ (where P_r, P_a are the observed moment components), because of the different distribution of the sample magnetization relative to the sensor.

equal. In a common measuring configuration in which two of the orthogonal components are in the plane of the cylinder perpendicular to its axis and are also the components of the declination of the NRM, a possible error can arise in the measured inclination in the sample. The effect has been investigated for the Gauss A position by Sharma (1965, 1968) and Larochelle and Pearce (1969).

The field at points on the axis of an axially magnetized cylinder can be obtained by integrating the field due to a thin disc (Equation 9.2), and is given by

$$B_z = \frac{\mu_0}{2} M_z \left[\frac{z + l}{(r^2 + (z + l)^2)^{1/2}} - \frac{z - l}{(r^2 + (z - l)^2)^{1/2}} \right] \tag{9.8}$$

where M_z is the volume magnetization of the cylinder of length $2l$ and diameter $2r$, and z is the axial distance from the centre of the cylinder. If the magnetization of the cylinder is represented by an equivalent centred dipole of moment P

$$B_z = \frac{\mu_0 P}{2\pi z^3} = \frac{\mu_0 M_z \pi r^2 2l}{2\pi z^3} = \frac{\mu_0 M_z}{2} \left(\frac{2r^2 l}{z^3} \right)$$

and as before the field due to the cylinder can be expressed in terms of the field due to the equivalent dipole by

$$B_z = C_{\mathrm{AA}} \left(\frac{\mu_0 P}{2\pi z^3} \right) \tag{9.9}$$

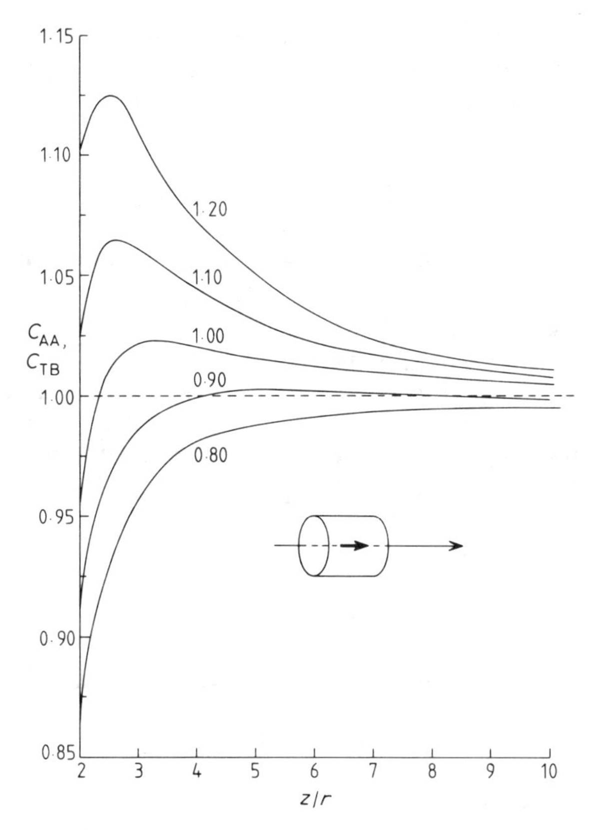

Fig. 9.4 Values of $C_{AA} = C_{TB}$ plotted against z/r for different height/diameter ratios of cylindrical samples.

where

$$C_{AA} = \left[\frac{z+l}{(r^2 + (z+l)^2)^{1/2}} - \frac{z-l}{(r^2 + (z-l)^2)^{1/2}} \right] \Big/ 2r^2 l/z^3 \tag{9.10}$$

In Fig. 9.4 the factor C_{AA} is plotted against z/r for values of l/r between 0.80 and 1.20. It can be seen that the shape providing the best dipole approximation for $z > 2r$ is $l/r \approx 1.0$ when the maximum departure of the axial field from that due to the equivalent dipole at the cylinder centre is $\pm \sim 3\%$.

At expression for the field along an axis in the diametral mid-plane of a cylinder due to uniform magnetization parallel to this axis (transverse magnetization) has been developed by M. Westphal and is quoted by Larochelle and Pearce. The expression takes the form

$$B_x = \frac{\mu_0 r^2 l M}{x^3} [1 + (0.75 - \gamma^2)\delta^2 + (0.702 - 2.814\gamma^2 + 1.125\gamma^4)\delta^4 + (0.684 - 5.476\gamma^2 + 6.56\gamma^4 - 1.25\gamma^6)\delta^6 \ldots] \tag{9.11}$$

where $\gamma = l/r$, $\delta = r/x$ and M_x is the magnetization per unit volume. The numerator of the quantity outside the brackets is simply $\mu_0 P/2\pi$, where P is the total moment of the cylinder, and thus B_x can again be expressed in terms of a factor C_{TA}, where

$$B_x = C_{TA}\left(\frac{\mu_0}{2\pi}\frac{P}{x^3}\right) \tag{9.12}$$

C_{TA} being the bracketed expression in Equation 9.11. In Fig. 9.5 C_{TA} is plotted against z/r for length/diameter ratios between 0.70 and 1.10. For $l/r \approx 0.80$ the dipole approximation gives B_x correct to within $\pm 2\%$ for $z/r > 2$. An alternative derivation of C_{TA} is given by Sharma (1965).

The corresponding coefficients C_{AB} and C_{TB} for the field at Gauss B positions for an axially and transversely magnetized cylinder will give an identical expression to that for the Gauss A field, except for a factor of 2. Therefore $C_{TB} = C_{AA}$ for all values of z/r and l/r (Fig. 9.4).

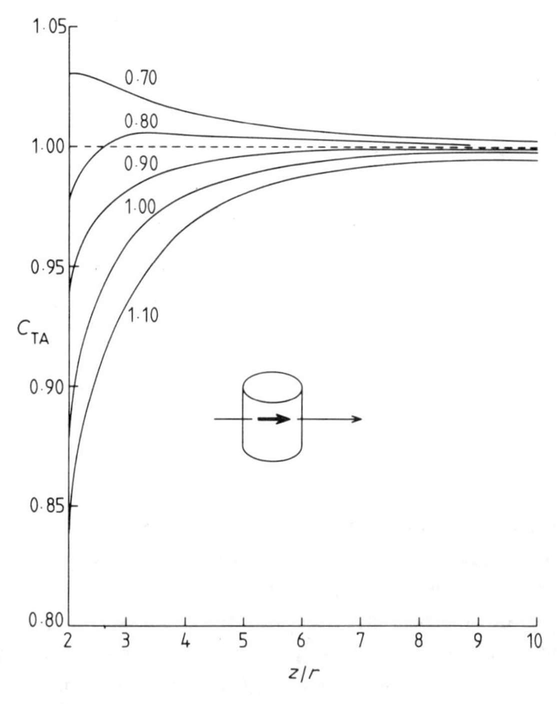

Fig. 9.5 Values of C_{TA} plotted against z/r.

Because of the different symmetry involved in the plane of the cylinder, in general, $C_{AB} \neq C_{TA}$. The Gauss B field and C_{AB} for an axially magnetized cylinder can be determined by considering the cylinder as consisting of a

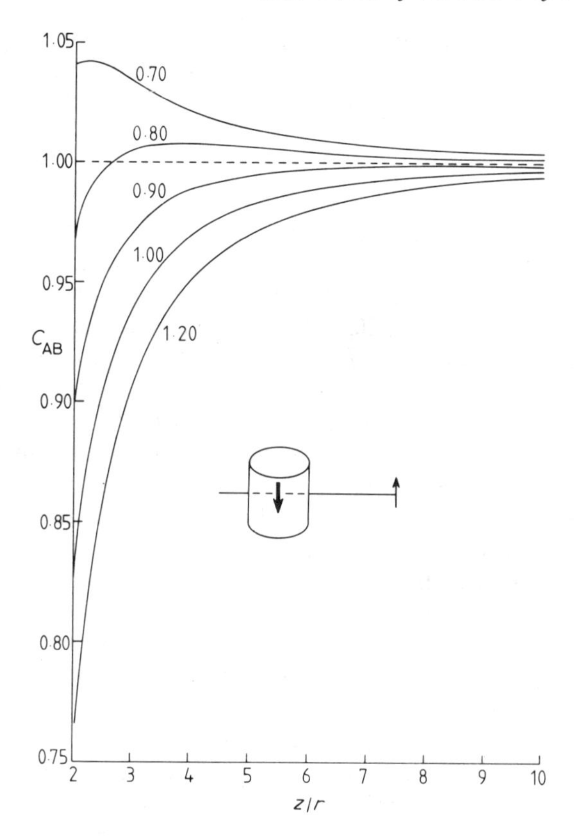

Fig. 9.6 Values of C_{AB} plotted against z/r.

number of thin magnetized rectangular slabs of height equal to that of the cylinder ($2l$) and of width varying between zero and $2r$ (Sharma, 1966). The present author is indebted to I. D. C. Gurney for carrying out the necessary computations and for checking the result by an independent method based on the magnetic analogue of the gravitational potential due to a disc of uniform density (Jeffreys and Jeffreys, 1956). Values of C_{AB} for cylinders of different shape are plotted against z/r in Fig. 9.6. It can be seen that the shape of the curves of C_{AB} and C_{TA} (Fig. 9.5) are similar for the same value of l/r, but the deviation from the equivalent dipole field is relatively more severe for C_{AB} at low values of z/r. In both cases the magnetization of cylinders with $l/r \approx 0.80$ is well represented by the equivalent dipole even when z/r is small.

It is apparent from the plots of C_{AA}, C_{AB}, C_{TA} and C_{TB} that the optimum value for l/r is $0.80 < l/r < 0.90$ if the cylinder magnetization is to be approximated by the equivalent dipole at its centre. The maximum difference between the cylinder and dipole fields at points along the axial and radial directions is 4% at $z/r > 3$ and $< 2\%$ for $z/r > 4$. In methods of measuring NRM in which the inclination is determined from the ratio of field components due to magnetizations along and perpendicular to the cylinder axis, the above field deviations correspond to maximum errors in the measured inclination of $\sim \pm 1°$ and $\sim \pm 0.6°$ respectively. These errors are comparable to, or less than, typical random errors of NRM measurement, but they are systematic. Larochelle and Pearce (1969) use a similar approach to assess measurement errors arising from cylinder shape when using a parastatic magnetometer. They show analytically that for a given distance from the sample to the central magnet (Gauss A measuring position) it is possible to choose an optimum value of l/r such that $C_{AA} = 1.00$, a result qualitatively confirmed by Figs 9.4 and 9.5. For a 2.5 cm diameter sample and $z/r = 4.7$ ($z = 6.0$ cm, the minimum sample–magnet separation) the optimum value of $l/r = 0.88$. The authors also show that errors in both direction and intensity of NRM arising from the shape of cylindrical samples can be made negligibly small for any $z > z_{min}$ by choosing the value of l/r near 0.88.

It may be noted that $0.80 < l/r < 0.90$ is a rather marked optimum value in that C_{AA} increases and C_{TA} decreases as l/r increases through values near unity, and the same is true of the coefficients C_{AB} and C_{TB}, appropriate to field measurements in the Gauss B position.

From the point of view of possible errors due to sample shape, there is little to choose between the Gauss A and B measuring positions for the sample when using a static sensor, i.e. one that responds to field (e.g. astatic or parastatic magnetometer, or spinner magnetometer with fluxgate sensor).

In spinner magnetometers which use pick-up coils for sensing the sample magnetization it is rotation of the axial and transverse fields due to the sample, depending on C_{AA} and C_{TA}, which gives rise to the induced voltage in the coils. To obtain good coupling between sample and coils, z/r is small (2–3) and inspection of Figs 9.4 and 9.5 shows that the optimum value of l/r is near 0.90.

Since declination is measured in the plane of the cylindrical sample, by symmetry no error in the measured declination arising from shape is expected but only in inclination and intensity. In samples obtained from a site where the core (and sample) axes are more or less randomly distributed in space, any error in the measured inclination due to shape will also be random when the NRM directions are referred to a common reference plane. In some sampling procedures small systematic errors can arise, for instance when cores taken from a sedimentary formation are all drilled with their axes perpendicular to the bedding plane.

The variation in the field coefficient C_{AA}, C_{AB} and C_{TB} with z/r and l/r shows that a significant reduction in measurement errors arising from the height/diameter ratio of cylindrical samples is achieved by using a value of l/r of 0.80–0.90, i.e. a height of 2.0–2.3 cm with the common sample diameter of 2.5(4) cm. In most magnetometers errors in the measured inclination and intensity are then expected to be comparable to or less than random errors of measurement.

9.1.3 Inhomogeneity of magnetization

Inhomogeneity of remanent magnetization in rocks is both expected and observed, and it is necessary to enquire whether it can cause significant errors in NRM measurement.

A favourable feature of the types of inhomogeneity described earlier in this chapter is that it generally varies in a random manner between samples. Although there may be errors in the measured NRM of each sample, any such errors are expected to be randomly distributed within a group of samples and to produce only an increase in the scatter of sample directions but no systematic error in the mean direction. The increased scatter will only be important if it seriously increases the circle of confidence around the mean.

To a first (and satisfactory) approximation, inhomogeneity of magnetization can be represented by a displacement of the equivalent sample dipole from the centre of the sample; this is also equivalent to a centred dipole plus a quadrupole. Thus, provided the measurement procedure is such that signals due to quadrupole components can be removed from the data used to calculate intensity and direction of NRM, the dipole magnetization can be satisfactorily determined.

Magnetometers respond to either the static field due to a magnetized sample placed near the sensor or to a varying field arising from sample rotation or translation. With continuous sample rotation, or if the sample is rotated to different azimuths about an appropriate axis in the former case or before translation in the latter, the magnetometer response to a dipole is of the form $A \sin \theta$, where θ is the instantaneous angular position of the measured NRM component relative to the sensor, whereas the quadrupole component results in a $\sin 2\theta$ response. Therefore in the static case, if a reading at an azimuth θ is $(d_d + d_q)$ corresponding to the dipole and quadrupole contribution, then at an azimuth $(\theta + \pi)$ the reading will be $-d_d + d_q$. Differencing the two readings then eliminates the quadrupole signal and allows the dipole signal to be measured.

This procedure illustrates the principle of minimizing the effects of inhomogeneity of NRM in measurements with astatic and ballistic (including cryogenic) magnetometers. In spinner-type instruments the quadrupole contribution to the output waveform is a sinusoidal signal of twice the rotation frequency: this is blocked in the signal processing circuits by using suitable filters or more commonly by phase sensitive detection.

It is also the usual practice, particularly where the sample–sensor configuration is unsymmetrical, to take readings with as many aspects as possible of the sample presented to the sensor. For example, the measurement procedure with an astatic or parastatic magnetometer or with the single fluxgate sensor of a Schonstedt spinner should include the presentation of each face of a cubic sample, or of a cubic holder of a cylindrical sample, to the sensing magnet or fluxgate.

Several studies of inhomogeneity of remanence and its effects are reported in the literature. Collinson, Creer, Irving and Runcorn (1957), Irving, Molyneux and Runcorn (1966) and Creer (1967) describe the measurement of inhomogeneous disc-shaped samples with the astatic magnetometer and the detection, determination and removal of the quadrupole contribution to the readings. Using samples of Torridonian sandstone Irving *et al.* demonstrated the dependence of the field due to the quadrupole component on the inverse fourth power of the sample–magnet system distance and also, as expected, the random distribution of azimuths of the equivalent off-centre dipole position in different rock discs. The displacement of the equivalent dipole from the sample centre due to inhomogeneity was found to be typically 0.2–0.4 cm.

Doell and Cox (1967a) investigated the effect of inhomogeneity of NRM on their spinner magnetometer (single coil pick-up) using artificial cylindrical samples made up of 10 discs of magnetized and non-magnetized material. The sample NRM direction was either parallel, perpendicular or at 45° to the cylinder axis. By varying the number and position of the non-magnetic discs, a wide range of inhomogeneity could be produced. With the six-spin measurement procedure, which minimizes inhomogeneity effects, the maximum departure of the measured from the true NRM was $\sim 2°$, whereas using a two-spin procedure, the deviation was sometimes $> 5°$. The maximum discrepancy in measured NRM intensity was $\sim \pm 5\%$.

The effect of NRM inhomogeneity on the accuracy of direction and intensity measurement may be expected to depend on the type of magnetometer used, and Collinson (1977) describes measurements in which the response of different magnetometers to four samples of known inhomogeneity was tested. The samples consisted of one or two thin discs and one or two small volumes of magnetized material (α-Fe_2O_3) placed in 2.5 cm Perspex cylinders: in two of the samples there were two magnetized regions, the NRM in each region differing in direction by $\sim 40°$. Figure 9.7 shows the results of the measurements on astatic, spinner and cryogenic magnetometers showing the range of errors to be expected from the rather severe inhomogeneity presented by the samples. It was not possible to investigate intensity errors because of the viscous decay of the sample magnetizations over the long (~ 2 year) period that the measurements took: the NRM was a saturated IRM, stabilized by partial a.f. demagnetization.

9.1.4 Other considerations

The great majority of routine palaeomagnetic measurements are now done on cylindrical samples 2.5 cm in diameter and ~ 2.0–2.3 cm high. This size is a reasonable compromise between the increased signal and often improved homogeneity achieved with a larger sample and the desirability of keeping the magnetometer sensor and the sample–sensor distance within reasonable limits. It is clear that for a given degree of inhomogeneity and sample–sensor

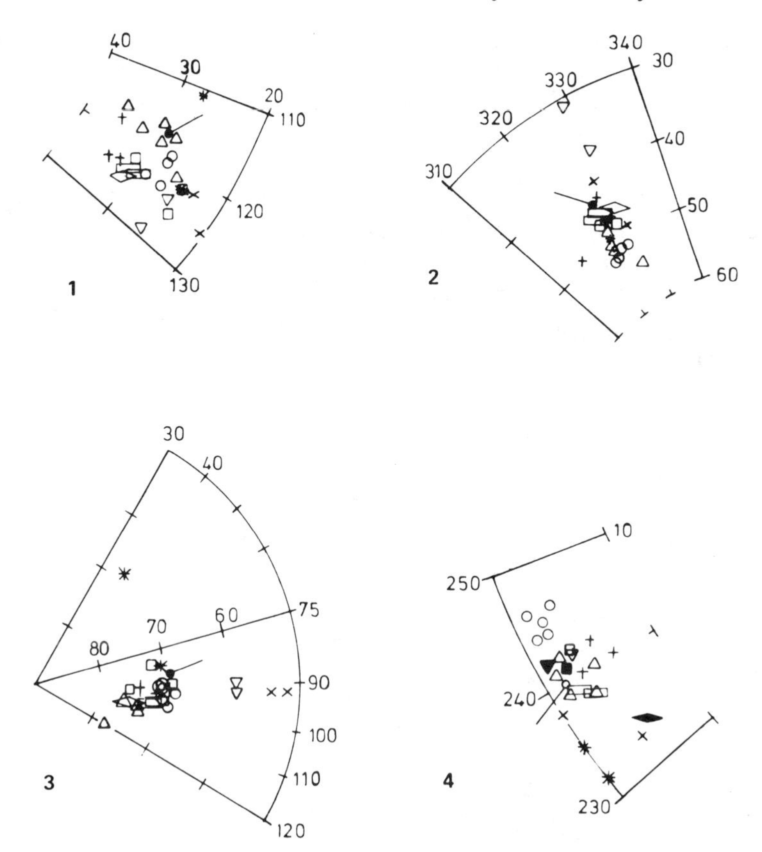

Fig. 9.7 Directions of *NRM* in four inhomogeneous samples as measured by different magnetometers. Cross, astatic magnetometer (off-centre); plus, astatic magnetometer (direct method); asterisk, astatic magnetometer, (position (d), Fig. 9.9 with sample rotation); diamond, astatic magnetometer (position (d), Fig. 9.9, sample not rotated); open square, PAR spinner; inverted triangle, Schonstedt spinner; triangle, Digico spinner; rectangle, UGF-JR3 spinner; open circle, SCT cryogenic magnetometer; full circle with line through, 'true' direction, as determined from original magnetizing direction of samples. In sample 4 all the inclinations are negative except the three full symbols.

distance (or pick-up coil size) a small sample is desirable from the standpoint of reducing the effect of the inhomogeneity.

Some workers have measured large (100–1000 cm^3) samples cut or chipped to a roughly equidimensional shape and many archaeological samples are conveniently measured in this form. If there is no marked departure from an equidimensional shape there is no reason to expect significant errors in the measured NRM direction arising from shape provided a suitable sample–sensor distance is maintained. If marked inhomogeneity of NRM is present it is also important to pay some attention to this latter factor so that the quadrupole and higher moment fields are as low in proportion to the dipole field as when smaller samples are used. In the simplest case of a single sensor, e.g. a fluxgate (Fromm, 1967), this can be achieved if the larger samples subtend approximately the same solid angle at the sensor as the smaller samples. If a dimension of the sample increases by a factor n, this condition implies an increase in the sample–sensor distance of n also. Since the field at the sensor is proportional to $M/z^3 \propto n^3/n^3$ there is no loss of sensitivity, expressed as field at sensor per unit magnetization of the sample. The effect on magnetometer sensitivity of sample size is discussed further in connection with the different types of magnetometer.

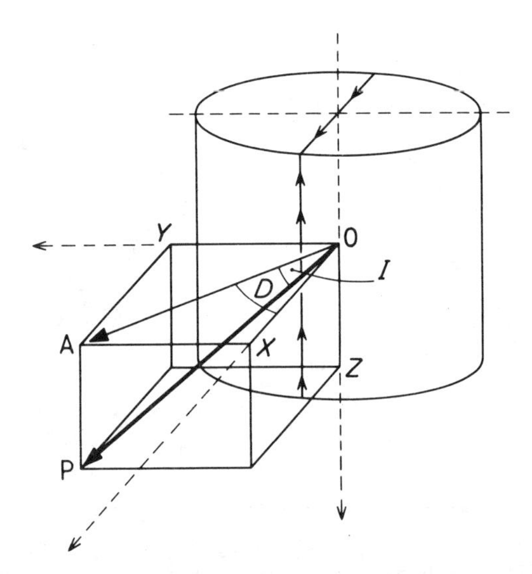

Fig. 9.8 Reference axes, declination D and inclination I, and magnetization components X, Y and Z for cylindrical samples.

The information obtained from a measurement of NRM on a magnetometer is the magnitude of the magnetic moment of a rock sample and the direction of the magnetization vector relative to a reference system in the sample. For the most common sample geometry, an approximately equidimensional cylinder, the reference system is an axis in the plane perpendicular to the cylinder axis and the plane itself. The reference axis is defined by a 'fiducial' direction line marked on the upper plane surface of the cylinder which usually represents a known direction from the field orientation data. The magnetization vector direction is then defined by two angles, declination D and inclination I. The former is the angle between the fiducial direction and the direction of the component of the sample moment in the plane of the cylinder and the latter is the angle between the total magnetization and the reference plane. The inclination is conventionally positive if the vector is directed below the reference plane (the upper surface of the cylinder) and negative if above (Fig. 9.8). The upper surface of the sample is usually taken to be that which is towards the top of the core or hand-sample when in position in the outcrop.

9.1.5 Historical development of rock magnetometers

Although not realized by them at the time, the Chinese around the 1st century AD may be considered to be the first to have measured the direction of NRM in a rock sample. They constructed primitive compasses by shaping a piece of natural magnetite (lodestone) to a convex surface and placing it on a smooth board, noting that any such piece of lodestone always rotated to a particular orientation. The axis in the lodestone parallel to the local magnetic meridian was, of course, the axis of the horizontal component of the lodestone's magnetization.

The deviation of the magnetic compass near strongly magnetized rock outcrops was observed from the 17th century onwards, for instance by Humboldt (1797), who conjectured that such rocks had been struck by lightning. Although this was probably the case with Humboldt's observation, it is not a necessary phenomenon for a compass deviation to be observed, since many igneous rocks possess an NRM of sufficient intensity to strongly perturb the local magnetic field. Indeed, the magnetic polarity of a rock outcrop, if not its direction of magnetization, may be determined by using the compass as a simple magnetometer.

During the 19th century there were isolated measurements of the NRM of rock samples and these were carried out by observing the deflections produced when the rocks were brought near to a compass or simple suspended magnet system, either a single magnet or an astatic pair.

Melloni (1853) investigated the NRM of lavas from Vesuvius with the aid of an astatic system of two long magnets. The samples were brought near to one end of one of the magnets, so that the magnetic field of the sample interacted essentially with one pole of the system only. The author observed that if the sample was too close to the magnet the induced magnetization in the sample adjacent to the magnet became comparable to or greater than the NRM, and it was later realized that measurements should be made in a way which allowed for the effects of induced magnetization arising from either the magnet or the geomagnetic field. Folgheraiter (1894) used a sensitive compass to measure the NRM of bricks and pottery and of rocks in situ and in the form of hand samples, and Brunhes (1901) also studying natural and archaeological material used a 'Mascart declinometer', which appears to have been essentially a compass mounted at the centre of a graduated rule. The compass needle carried a small mirror, enabling deflections to be read with an optical lever. During 1915–1935, Mercanton carried out extensive measurements on lavas from different parts of the world using an astatic magnetometer, and confirmed the existence of normally and reversely magnetized rocks (Mercanton, 1926, 1932). In his classic paper describing investigations of the NRM of Mt Etna lavas, Chevallier (1925) describes the use of a single horizontal magnet suspended by a weak torsion fibre, with an auxiliary magnet placed nearby to cancel the horizontal component of the geomagnetic field. This improved the sensitivity by replacing the strong torque on the magnet due to the Earth's field with the weak torsional control of the suspension. A 1 m optical lever was used to record deflections.

These early magnetometers suffered from lack of sensitivity and a non-stable zero due to natural and man-made ambient magnetic field variations. However, by using comparatively large samples (100–200 cm^3) close to the magnetometer magnetizations of $\sim 10^{-3}$–10^{-4} A m^2 kg^{-1} were successfully measured. The use of these large samples close to the magnetometer accentuated the effects of inhomogeneity of NRM, and several of the above authors comment on this and the necessity for reducing its effects on the accuracy of measurement.

During the 1920s, Chevallier initiated the development of two new types of magnetometer, the ballistic and resonant instruments. The former type of magnetometer was further developed by Thellier in France and by Japanese workers and became widely used. The resonance principle, in which an amplified resonant response is established in a detecting system by causing the magnetic field due to the sample to apply a periodic force or torque to the detector at the resonant frequency, proved subsequently not to be a profitable approach.

The first spinner magnetometer specifically designed for palaeomagnetic measurements was constructed by Johnson and McNish (1938) and the technique was subsequently steadily developed, aided by the rapid post-war advances in electronic techniques. During the same period astatic and parastatic magnetometers were being increasingly used, the latter type first being used for NRM measurements in the early 1930s by Thellier (1933) and the former coming into more general use after the classic paper by Blackett (1952) on the design features of such magnetometers. Both systems benefited from the development of new permanent magnet materials of high coercivity and high saturation remanence, and during the early period of systematic palaeomagnetic studies (1955–1965), the astatic (or parastatic) and spinner magnetometers were favoured for routine measurements.

The development in the late 1960s of the on-line computer-linked spinner magnetometer (Molyneux, 1971) and of the superconducting rock magnetometer were further landmarks in the progress to more rapid measurement of weakly magnetized samples. With the latter instrument the limit of useful sensitivity has probably been reached and under low noise conditions the signal/noise ratio enables the most weakly magnetized samples to be satisfactorily measured in minimum time.

The development of fast and sensitive rock magnetometers in the past

25 years has taken much of the tedium out of palaeomagnetic research, requiring as it does a large number of NRM measurements to obtain reliable data. To one who remembers long hours spent in a darkened room recording deflections of a light spot, followed by further exertions with slide rule, logarithms, trigonometrical tables and stereographic nets, the advent of the transistor, integrated circuit, computer and microprocessor is a welcome development. However, it is not without its dangers, in that there is sometimes a touching but not always justified faith in the results which emerge from electronic black boxes and which are then convincingly printed out. The student in the past was perhaps more in contact with the realities and imperfections of his magnetometer and may have had a better idea of the uncertainties associated with his results through a better understanding of the way in which they were obtained.

9.2 Magnetostatic instruments

These instruments measure the magnetic field of a rock sample by the deflection produced of an array of magnets supported on a vertical torsional suspension, or by direct field measurement with, for example, a fluxgate detector. The magnet array or other detection configuration is designed to be insensitive to the ambient magnetic field and to its time variations, the former property enabling high sensitivity to be realized and the latter ensuring stability of the instrument zero. The system responds to the field of rock sample appropriately placed near it. Design and adjustment of the magnet system is directed towards combining high sensitivity and a reasonable time constant with good zero stability. The design varies according to whether an optical lever is used to record deflections (direct reading) or an electronic feedback system.

This section is concerned with magnet detectors: the use of fluxgate detectors is described in Section 9.2.2.

9.2.1 The astatic magnetometer

(*a*) *General*

A single suspended magnet with its axis horizontal, free to deflect about a vertical axis, can only be used for strongly magnetic samples which produce deflections greater than those due to ambient field variations, and the two-magnet, or astatic system is the simplest type of sensitive instrument (Fig. 9.9). By closely equalizing the magnetic moments and setting them antiparallel the net moment of the system can be made extremely small, and therefore the torque about a vertical axis exerted on the system by an ambient uniform horizontal field is correspondingly weak. High sensitivity is then achieved by using a weak torsional suspension. Although there are variations in the way

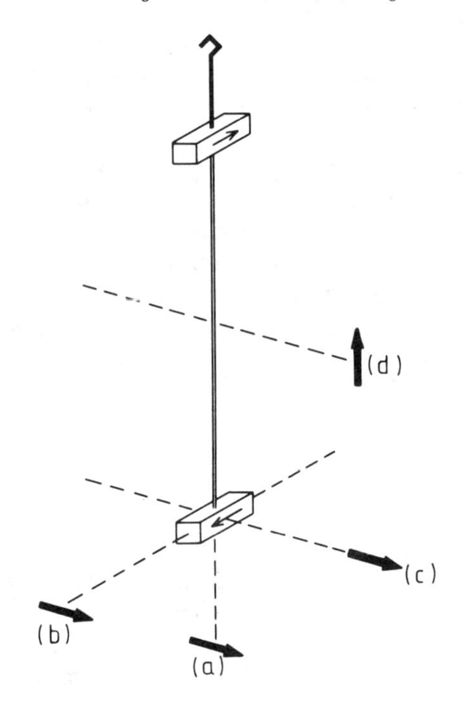

Fig. 9.9 Astatic magnet system and sample measuring positions. Arrows show component of sample magnetic moment measured in each position.

astatic systems are utilized, the above is the basic principle of all such instruments.

Much thought has gone into the design of astatic systems, both for maximum sensitivity and maximum signal/noise ratio. Because the fundamental (Brownian) noise of a system is usually exceeded by noise from other sources it is usual in palaeomagnetic instruments to design systems for maximum sensitivity for a given time constant.

The basic relationships governing the performance of astatic systems are

$$PB = \delta\theta \tag{9.13}$$

and

$$T = 2\pi\sqrt{\left(\frac{\alpha I_0}{\delta}\right)} \tag{9.14}$$

where P is the moment of the lower magnet, B is the horizontal field due to the sample at and perpendicular to P (sample position (a) in Fig. 9.9), θ is the angular deflection of the system (assumed small), δ is the torsional constant of the suspension, T is the period of torsional oscillation of the system (assuming no magnetic torque is acting) and I_0 is the moment of inertia of one of the magnets about the vertical axis. αI_0 is the moment of inertia of the whole system, α being necessarily > 2. In the following analysis the effect of the sample field on the upper magnet is neglected.

Eliminating δ between Equations 9.13 and 9.14, we get an expression for the sensitivity

$$\frac{\theta}{B} = \frac{1}{4\pi^2} T^2 \frac{P}{\alpha I_0} \tag{9.15}$$

Thus for a given T, P/I_0 should be maximized and α minimized, the latter implying, as might be expected, a minimum possible contribution to the total moment of inertia of the stem, mirror, magnet supports, etc. Since $P \propto l^3$ and $I_0 \propto l^5$, where l is a typical dimension of a magnet, $P/I_0 \propto l^{-2}$, and thus Equation 9.15 leads to the requirement of small magnets for high sensitivity if α can be made only slightly greater than 2.

(*b*) *Magnet systems*

The requirement of maximum P and minimum I_0 is to a certain extent mutually exclusive for many magnetic materials since for a given magnet mass maximum P requires a fineness ratio β (= length/breadth) as large as possible, because of the reduced demagnetization factor, whereas minimum moment of inertia requires β as small as possible. However, by expressing P/I_0 in terms of ρ, the density of the magnet material and its magnetization/unit mass (J_β) for a given β-value, an optimum value of β can be derived to provide maximum P/I_0. The magnet size is then chosen such that the required T is obtained when the system is hung on the weakest available suspension. Several analyses for optimizing magnetometer performance have been given (Blackett, 1952; Collinson and Creer, 1960; Roy, 1963; Deutsch, Roy and Murthy, 1967).

'Classical' magnetic materials (e.g. Alcomax, Ticonal), which show a significant increase in J_β with increasing β give optimum performance with β in the range 3–5. Ceramic and sintered magnets (e.g. Magnadur, Columax) and the recently-developed rare-earth magnet materials such as samarium–cobalt which possess very high coercive force and thus only small dependence of J_β on β, give the best results with $\beta < 1$ and are now favoured for the most sensitive instruments. Details of two typical magnet systems using the above types of magnetic material are given in Table 9.1.

Glass or aluminium tubing is suitable for the magnet system stem. It should be of sufficient internal diameter to take a small spigot at each end on which the magnets are mounted (Fig. 9.10). Before mounting the magnets they should be machined to the same mass to within 1% or less, and if not pre-saturated should be remagnetized in a strong field ($\sim$1 T), followed by a small ($\sim$0.05 T) back field to stabilize the moment.

The length L of the stem is not critical and is a compromise between the need for a short stem for good mechanical stability and low α of the system and the slightly improved sensitivity arising from a long stem. The lower limit is usually governed by the permissible loss of sensitivity arising from the

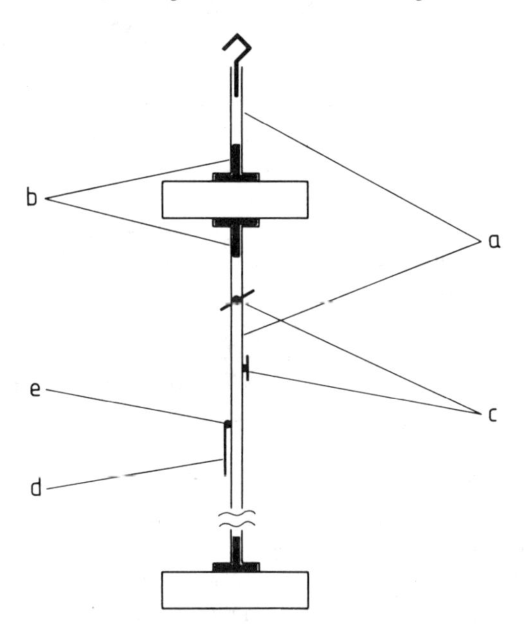

Fig. 9.10 Construction of astatic system. a, glass or aluminium tube; b, aluminium spigots; c, trimming magnets; d, mirror; e, spot of adhesive.

opposing torque on the upper magnet in measuring positions (a) and (b), Fig. 9.9. For a dipole in position (a) the sensitivity decreases by a factor $f = [z_l/(z_l + L)]^3$ compared with that when $L \to \infty$, where z_l is the distance between the dipole and the lower magnet. For a typical minimum value of z_l of 2.5 cm, $f = 0.13$ and 0.037 for $L = 2.5$ cm and 5.0 cm respectively. Typical values of L lie in the range 6–12 cm.

It is convenient to have the magnet system detachable from the suspension, and hooks of thin copper wire are convenient (Fig. 9.10): it is important that the hooks engage positively and cannot twist relative to each other.

Phosphor-bronze strip is commonly used for the suspension. The torsional constant of a strip of length l, breadth b and thickness d is given by

$$\delta = nbd^3/3l$$

where n is the shear modulus, which for phosphor-bronze is $\sim 4.0 \times 10^{10}$ N m^{-2}. A quartz fibre suspension may also be used, but it is difficult to produce such a fibre to give an accurate predetermined value of δ since for a circular cross section $\delta \propto r^4$, although some adjustment is possible by altering l. The drawing of quartz fibres is described by Braddick (1963) and Kirk and Craig (1948).

In direct reading instruments and some using electronic feedback, the deflection is measured with an optical lever. A small mirror ($\sim$6 mm $\times$ 6 mm) cut from a surface-silvered or surface-aluminized microscope cover slip is attached to the magnet system stem with a small spot of adhesive near one edge: this avoids undue distortion of the mirror when the adhesive hardens.

(c) *Astaticizing the magnet system*

The purpose of astaticizing is to reduce as much as possible the torque exerted on the system by the ambient horizontal magnetic field. This allows high sensitivity, controlled only by the torsional constant of the suspension, and a reduction in zero drift arising from time changes in the field. The total magnetic torque T_m acting on the magnet system is

$$T_m = B\,\Delta P + P\,\Delta B \tag{9.16}$$

P and ΔP are the moment of one magnet and the net moment of the system respectively, B is the average horizontal field strength perpendicular to ΔP and ΔB is the difference in the horizontal field at the upper and lower magnets perpendicular to their moments. Since there is likely to be a vertical gradient in the horizontal field, the magnet system can only be astaticized for the particular field configuration in which it is placed, and astaticizing ideally achieves the situation in which the vector sum of the torques in Equation 9.16 equals zero.

The astaticism S can be expressed as $P/\Delta P$, where ΔP is the effective net moment of the two magnets, the ambient field being assumed uniform. Igneous and sedimentary rocks will typically give fields of 10 and 10^{-2} nT respectively at the lower magnet (position (a), Fig. 9.9): thus, to obtain significant deflections from such rocks and allow the sensitivity to be controlled mainly by the suspension, the effective ambient field acting on the system must be comparable to or preferably less than the above fields. Assuming a horizontal Earth's field of 20 μT, an astaticism of, say, 500 can be thought of as reducing the ambient field to 40 nT. For magnetometers for measuring sediments, effective reduction of the ambient field is achieved by higher astaticism ($\sim$ 5000) and reduction of the ambient field by a suitable coil system. Note that as high an astaticism as possible is still desirable in the latter case, to minimize the effects of time changes in the ambient field. Thus, values of S of 300–500 are suitable for medium sensitivity instruments (i.e. for most igneous rocks) in average laboratory conditions, while for more sensitive magnetometers values of around 5000 should be aimed for.

During the astaticizing procedure it is necessary to rotate the lower magnet relatively to the upper one, and it is convenient to fix the spigot holding the former in the stem with shellac on other low-melting point material. A touch of the stem at the spigot with the tip of a small soldering iron will then enable the magnet to be rotated and fixed in the new position when the shellac cools.

After determining the polarity of each magnet with a small compass needle or other magnet of known polarity, the system is hung on a torsionless suspension (e.g. a single nylon or silk thread) at the proposed position of the magnet system and the period of oscillation determined with the magnet

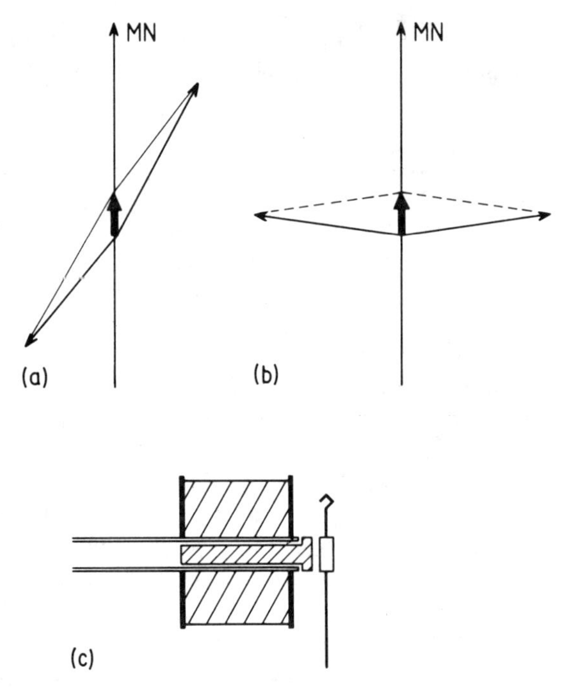

Fig. 9.11 Astaticizing of magnet system: (a) setting of magnet system when one magnet is stronger than the other; (b) setting when magnetic moments are closely equal but not antiparallel; (c) small demagnetizing coil. A soft iron or mild steel core increases the available field.

approximately parallel (T_1) and antiparallel (T_2). The equilibrium position of the system with the latter configuration will determine the stronger magnet, which will set along or near to the meridian (Fig. 9.11(a)) and the initial astaticism is given by $S = P/\Delta P = T_2^2/2T_1^2$. With the magnets antiparallel the stronger magnet is now demagnetized in steps (see below) and the oscillation period checked after each step. As demagnetization proceeds the period will increase and, as equality of the magnet moments is approached, the system will set at an increasing angle to the meridian as imperfect antiparallelism becomes more important than ΔP (Fig. 9.11(b)).

The lower magnet is now rotated through a small angle in the way previously described, and an improvement in astaticism is indicated by the system setting in or near the meridian with again an increase in the period. The demagnetization and magnet rotation procedures are repeated until the highest possible period is achieved and maintained over a period of a few hours.

Too much demagnetization will be indicated by the magnet system setting in the reversed direction (i.e. the stronger magnet becoming the weaker one), and can be remedied by demagnetizing the now stronger magnet: with repeated over-demagnetization it may be necessary to remagnetize the magnets and start the procedure again. Too much rotation of the lower magnet is indicated by the system setting to the West (East) of the meridian if it was originally setting to the East (West).

Typical values of T_1 and final T_2 in the Earth's horizontal field (20 μT) for magnet systems described here might be 0.5 and 15 s respectively, giving $S \approx 450$ which is probably as high a value as can be maintained over a reasonable period of time.

With magnetic materials such as Alcomax and Ticonal demagnetization can be carried out by the decreasing alternating field technique, using a small coil at one end of which the appropriate magnet can be placed and then removed (Fig. 9.11(c)). This technique may not be possible for the newer materials because of their high coercivity and in the author's experience the best method with these magnets is mechanical abrasion with fine carborundum paper.

Achievement of higher astaticisms ($\sim$3–5000) required for very sensitive magnetometers presents something of a problem because of the impossibility of reproducing changes in the ambient Earth's field for test purposes. However, since it is necessary to apply a reasonably uniform nulling field to reduce the external magnetic torque on the magnet system, it is likely that this field will be similar to the Earth's field. Thus, a common procedure is to astaticize the system against changes in the applied nulling field, following a method first described by Blackett (1952).

In this technique, the deflection of the magnet system (now hanging on its phosphor-bronze suspension) resulting from a small change in the horizontal nulling field is minimized by adjusting small 'trimming' magnets attached to the magnet system stem (Fig. 9.8). These magnets are small ($\sim$5 mm) lengths of hard steel wire of moment p, which can be rotated about a horizontal axis perpendicular to their length. Thus a small variable moment of $\pm p \sin \psi$ ($\psi = \pm 90°$) can be obtained by setting at the appropriate angle ψ, the angle between the trimmer axis and the stem of the magnet system. If, as is usual, the axes of the nulling coils are parallel and perpendicular to the axis of the magnets it is convenient to use trimmers also aligned parallel and perpendicular to the magnet axis, using the appropriate coils to apply the test field changes. If the trimmers are fixed to the stem by a drop of shellac they can easily be loosened and re-set by touching with the tip of a small soldering iron. It may save time if after initially attaching a trimming magnet the sense of the deflection produced by a field change is tested with the trimmer set at $+90°$ and $-90°$. If the trimming moment required is within the range available the deflections will be of opposite signs at the two extreme angular settings.

It is not possible to specify how far the astaticizing process needs to be taken, since this depends on local magnetic conditions and the relationship between the Earth's field and applied field stability. However, it should not be difficult to reduce the deflections obtained from a given field change without trimming magnets by a factor of 100 after their addition and adjustment. The system can then be tested for zero drift relative to sensitivity and the process carried further if thought necessary. The practical limit is governed by the magnitude of the changes in the geomagnetic field and the similarity between the geomagnetic and applied fields.

(*d*) *Magnetometer construction and performance*

The requirements for a low-sensitivity instrument (for standard rock samples down to $\sim 10^{-4}$ $\mathrm{Am^2\,kg^{-1}}$) are comparatively simple. Figure 9.12 shows the essentials and Table 9.1 gives further details. By using a folded light path the operator can sit at the instrument to manipulate the sample and read the deflections. A typical distance from the fixed mirror to the scale is 0.5–1 m, and if the latter is the normal 50 cm length it is desirable for it to have the appropriate radius of curvature to ensure linearity between deflections of spot and magnet: if a flat scale is used, the scale reading is too large by an amount $d(\tan\theta - \theta)$ where d is the distance from the scale to the mirror on the magnet system and θ is the angular deflection of the light beam. With $d = 1$ m, this amounts to 1.1, 2.7 and 5.3 mm for true deflections of 15, 20 and 25 cm respectively. Linearity is not affected by the angle between the incident beam and fixed mirror, and it is only necessary to ensure that the reflected beam is perpendicular to the scale.

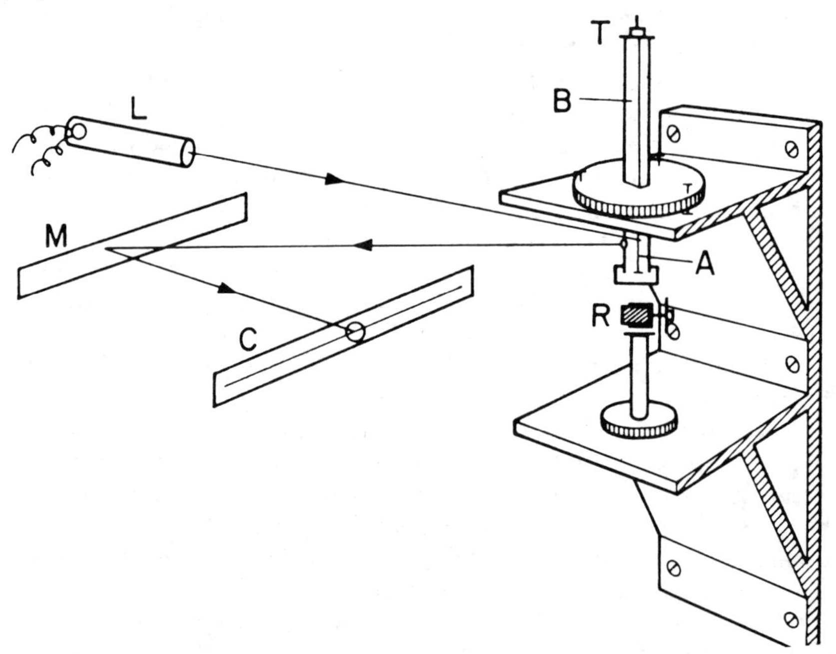

Fig. 9.12 Low-sensitivity astatic magnetometer. A, magnet system; B, suspension; C, scale; L, lamp; M, mirror; R, sample and holder; T, torsion head. (From Collinson and Girdler, 1967)

Table 9.1 Details of low-sensitivity astatic magnetometer

Magnets*	Alcomax IV, $0.60 \times 0.20 \times 0.20$ cm
Magnetic moment (P)	9.0×10^{-3} A m^2
Moment of inertia (αI_0)	1.3×10^{-9} kg m^2
Astaticism	~ 500
Suspension (phospor-bronze strip)	$0.013 \times 0.0013 \times 24.0$ cm
Period	~ 2 s
Sensitivity†	9.0 nT cm^{-1} at 1 m
Noise level	$\sim \pm 1.0$ mm on scale
NRM equivalent of noise‡	$\sim 1.0 \times 10^{-5}$ A m^2 kg^{-1}

* Longitudinally magnetized.
† Expressed as field required perpendicular to lower magnet to produce 1.0 cm deflection at the scale at 1 m distance.
‡ Magnetization of 2.5 cm sample which gives noise deflection when placed 3.0 cm below lower magnet.

The magnet system and suspension are housed in a glass tube with a window holding a lens for the light beam, and a threaded plastic ring at the lower end holding a copper plate ($\sim$3–4 mm thick) for electromagnetic damping of the magnet system. Electrolytic copper should be used, since commercial copper often contains sufficient ferromagnetic impurity to exert a torque on the lower magnet.

Magnet system vibration may be significant in a city laboratory, particularly in a second or higher floor room. There may also be severe magnetic gradients near the walls of a reinforced concrete or steel frame building and a site near the middle of the room is then preferable with the instrument mounted on a substantial pillar. A preliminary magnetic survey of a room will aid in choosing the best place, and will also give information about the effects of lifts. Before choosing a site it is advisable to roughly astaticize the magnet system and set it up in a temporary fashion and monitor it for a few days for vibration and zero drift.

To realize the full capability of more sensitive instruments for measuring down to $\sim 5 \times 10^{-7}$ A m^2 kg^{-1}, the requirements are more demanding. These include a magnetically and mechanically quiet environment and space enough to allow 3–5 m separation between the magnetometer and control desk. If available, an out-of-city site will fulfil the first two requirements: in a city laboratory continuously monitored field-free space may be required. Direct-reading high-sensitivity magnetometers usually have a period of 10–20 seconds, which serves to filter out shorter-period vibrations of the magnet system about its vertical axis. However, because of the difficulty of dynamically balancing the system, other vibration modes can be induced by microseisms arising from wind, railways, traffic, etc. At a quiet site it is usually sufficient to mount the magnetometer on a concrete block let into and isolated from the floor of the laboratory. At a more disturbed site a better anti-vibration mounting may be required. Braddick (1963) describes the latter, some of which could be modified for magnetometer use.

As previously mentioned, a set of Helmholtz or other coils are needed to cancel the ambient field over the magnet system: additional advantages of these are the elimination of induced magnetization in the sample, and use for applying fields for measuring low field susceptibility. A three-component Helmholtz set is satisfactory and a side of 2 m or more will give adequate uniformity of field. For astaticizing, it is convenient to arrange for the axis of one of the horizontal field coil sets to coincide with that of the magnet's system magnets. A stabilized power supply ($1{:}10^3$ or better) is necessary for the coils, and constant current is preferable to constant voltage to reduce zero drift due to change of coil resistance with temperature. In a copper-wired coil system producing 20 μT, $\pm 1\,^\circ$C temperature change will correspond to a 70 nT field decrease with a constant voltage supply.

To avoid disturbance of the magnetometer by the operator it is desirable to place the controls and reading scale at 3–5 m from the magnet system. Sample manipulation can be controlled by a pulley or other mechanical drive system, and the additional sensitivity arising from the increased optical path length may be useful provided there is sufficiently high signal/noise ratio. However, the longer path length will reduce the quality of the image on the reading scale through loss of illumination and loss of 'sharpness' through increased magnification. With the limited mirror size on the magnet system the requirements for a good image are a 30–60 W light source with a small filament, a fine wire object of adjustable position and a lens of the correct focal length. Because the thin mirror on the magnet system is unlikely to be flat, it is not possible to use a lens of standard focal length to obtain a sharp image. The author has found the most satisfactory method is to acquire a set of discarded spectacle lenses and find the correct one by trial and error: not only are such lenses of a variety of focal lengths, but many also vary along different diameters.

Figure 9.13 is a diagram of the high-sensitivity astatic magnetometer in the author's laboratory. A useful accessory shown is the zero coil. This is a coil placed above and on the vertical axis of the system and is used to apply a small, variable field gradient to control the position of the image on the scale. Coarse adjustment of the image position is by means of the torsion head. Another useful feature is the provision of an appropriately placed coil with which small field gradients can be produced at the magnet system. This provides a convenient method of adjusting the sensitivity of a direct-reading magnetometer over wide limits. Table 9.2 gives further details of the magnetometer.

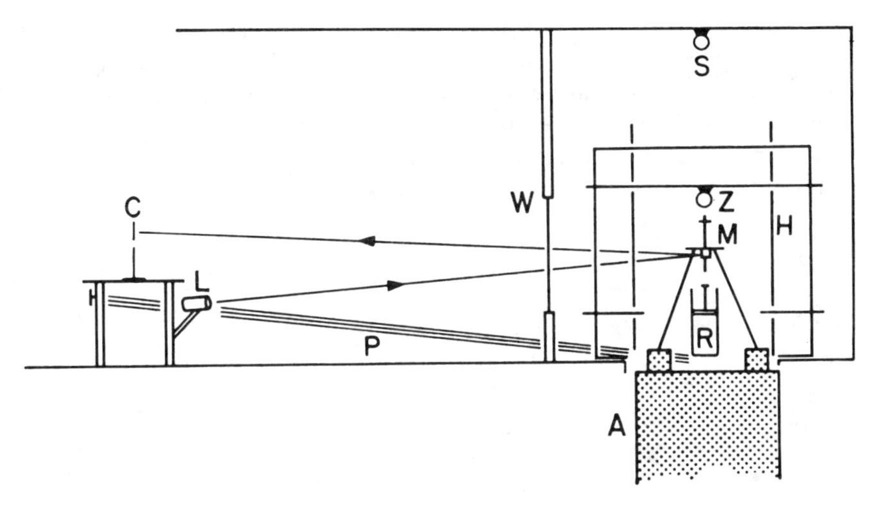

Fig. 9.13 High-sensitivity astatic magnetometer. A, concrete block isolated from floor; C, scale; H, Helmholtz coil system; L, lamp; M, magnet system; P, pulley system for sample manipulation; R, sample manipulation system; S, sensitivity coil; W, window in magnetometer room; Z, zero coil.

Table 9.2 Details of high-sensitivity astatic magnetometer

Magnets*	Samarium cobalt 0.20 × 0.20 × 0.60 cm
Magnetic moment *P*	1.0×10^{-2} A m^2
Moment of inertia (αI_0)	$\sim 2.5 \times 10^{-10}$ kg m^2
Astaticism	~3000
Suspension (phosphor-bronze strip)	0.005 × 0.0003 × 16.0 cm
Period	10 s
Sensitivity†	2.0×10^{-2} nT cm^{-1} at 5 m
Noise level	$\sim \pm 1.0$ mm
NRM equivalent of noise‡	$\sim 2.0 \times 10^{-8}$ A m^2 kg^{-1}
Brownian deflection	~ 0.01 cm

* Transversely magnetized.
† ‡ Defined as in Table 9.1.

9.2.2 Parastatic systems

In magnetically noisy environments where there are variable direct field gradients the astatic system may be insufficiently stable against field variations because of time-varying derivatives of the ambient field. In such situations the parastatic (or biastatic) system may be used (Fig. 9.14), consisting of a central magnet of moment P and antiparallel magnets of moment $P/2$ above and below it. Such a system is insensitive to changes in the magnitude of a uniform vertical gradient of the horizontal field and zero drift from this source can be much reduced. These magnetometers were pioneered by Thellier (1933) and Pozzi and Thellier (1963) and their performance has benefited from the improvement in magnetic materials in the same way as astatic instruments.

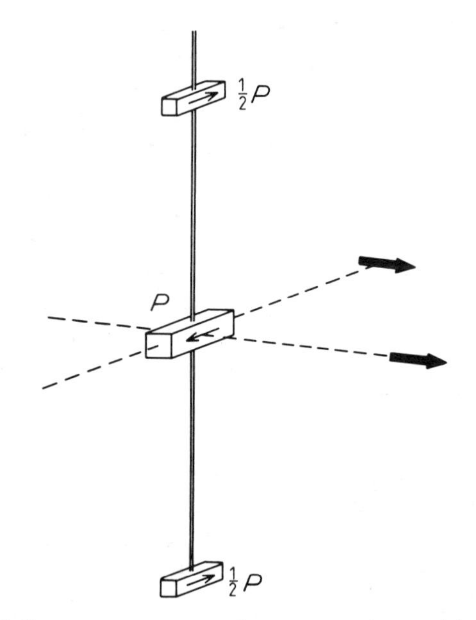

Fig. 9.14 Parastatic magnet system and two measuring positions.

The design of parastatic systems follows broadly the same line as for astatic systems, i.e. the magnets should have maximum magnetization with minimum moment of inertia and be small in size. The sample measuring position is usually on the horizontal axis through the central magnet perpendicular to the latter's magnetization and the field due to the component of NRM parallel to this axis exerts a deflecting torque on the magnet system. Analysis shows that if a reasonable minimum distance from sample to magnet is adopted, the sensitivity of the parastatic system is comparable to that of an astatic system of the same total mass of magnets, using measuring position (a) (Fig. 9.9). There may be an improvement in the performance of the parastatic system (sensitivity for a given time constant) because the combined moment of inertia of the outer magnets will be less than the equivalent single magnet of equal mass. There will be a small additional moment of inertia contributed by the longer stem and extra magnet mountings. The above result applies to a parastatic magnet system in which the central magnet is equal to one of the astatic pair and the other two magnets are formed by cutting the second magnet in half or when they are the same shape as the central one.

The torque acting on a parastatic magnet system of length $2l$ in a horizontal magnetic field with uniform vertical gradient $\mathrm{d}B/\mathrm{d}z$ is given by

$$L = PB - \left(B + l\frac{\mathrm{d}B}{\mathrm{d}z}\right)\frac{P}{n} - \left(B - l\frac{\mathrm{d}B}{\mathrm{d}z}\right)\frac{P}{m} \tag{9.17}$$

$$= PB - PB\left[\frac{1}{n}\left(1 + \frac{l}{B}\frac{\mathrm{d}B}{\mathrm{d}z}\right) + \frac{1}{m}\left(1 - \frac{l}{B}\frac{\mathrm{d}B}{\mathrm{d}z}\right)\right] \tag{9.18}$$

where B is the horizontal field perpendicular to the central magnet of moment P, and P/n and P/m are the magnetic moments of the outer magnets (Fig. 9.14). The magnetic moments of the outer magnets are assumed to be anti-parallel to the central moment. In the ideal case, assuming uniform $\mathrm{d}B/\mathrm{d}z$, $m = n = 2$ exactly and $L = 0$, and the system does not respond to changes in $\mathrm{d}B/\mathrm{d}z$. In practice, the effective astaticism that can be achieved depends on the uniformity of the field gradient and on how closely the moments of the outer magnets can be made equal to each other and to half the control magnet. The equalization cannot be achieved completely by the method described for astatic systems, and it is necessary to measure the moments as accurately as possible. Although the magnet masses can be accurately compared, they may not indicate the relative magnetic moments with the same accuracy because of interior flaws, inhomogeneity of magnetization, etc. Direct measurement of moments in the range 10^{-3}–10^{-1} A m^2 is not easy, being too strong for most rock magnetometers, but indirect methods are possible, e.g. timing the oscillations of the magnets in a known magnetic field. It is probable that an accuracy of $\pm 0.5\%$ in the relative magnetic moment is a realistic value that can be achieved with careful measurements. The response to changes in the uniform field can be minimized in the same way as for astatic magnet systems (Section 9.2.1). This is done by adjusting two small, perpendicular trimming magnets attached to the stem supporting the magnets until minimum deflection is obtained when an applied uniform field is varied.

The astaticism obtainable in practice cannot be determined with any precision because of the above uncertainty of magnetic moments and the possibility of L being near zero because of a fortuitous configuration in which $n, m \neq 2$, but are such that the bracketed terms in Equation 9.18 sum to 1. If B increases upwards and $(l/B)(\mathrm{d}B/\mathrm{d}z) = \Delta B/B = 0.001$, where ΔB is the change in B between the central and outer magnets, then $L = 0$ if n and m are 2.01 and 1.99 respectively, irrespective of whether the upper or lower magnet is the stronger. Although this arrangement is not 'perfectly' astatic, it is very nearly so for small changes in $\Delta B/B$. However, we may consider a simple case for an approximate comparison of an astatic and parastatic system in a field gradient $\mathrm{d}B/\mathrm{d}z$, with $n = 2.000$, $m = 2.004$ and $\Delta B/B = 10^{-3}$. For the parastatic system, putting Equation 9.18 in the form $L = PB(1 - x)$, we get $x = 0.9990$, and the torque is reduced by a factor of ~1000, compared with the torque on the central magnet alone, i.e. the astaticism S is 1000 and is independent of the magnitude of dB/dz. The astaticism that can be obtained in an astatic system without trimming magnets, by the method described in Section 9.2.1, is 200–300, and this would only exist in the particular field in which the astaticizing procedure was carried out. This also applies to a system in which a higher 'instantaneous' astaticism has been achieved by the use of trimming magnets.

If a dipole is at a horizontal distance z from the central magnet of a parastatic magnet system of total length $2l$, the deflecting torque L acting on the magnet system is given by

$$L = \frac{\mu_0 P_z P_m}{4\pi}\left[\frac{2}{z^3} - \frac{\cos^3\theta\,(3\cos^2\theta - 1)}{z^3}\right] \tag{9.19}$$

where the dipole moment P_z is perpendicular to the moment P_m of the central magnet, and $\theta = \tan^{-1} l/z$. When multiplied out the first term of Equation 9.19 is the torque on the central magnet and the second term the torque on the outer magnets. For $\cos\theta < 1/\sqrt{3}$ $(\theta > 55°, z < 0.71l)$ the second term becomes positive, i.e. the torque on the outer and inner magnets is in the same sense and there is small improvement in sensitivity contributed by the outer magnets instead of the decrease when $z > 0.71l$. Thus, the optimum value of l is such that $l = z_{min}/0.71$ where z_{min} is the smallest distance from the central magnet that a rock sample can be measured. A typical value of this distance is ~4 cm for a 2.5 cm cylindrical sample, and therefore $l \approx 6.0$ cm. Although the above value of l is derived for a dipole it will also be approximately correct for cylindrical rock samples of optimum diameter/height ratio (Section 9.1). Larochelle and Christie (1967) show that when $z = 0.5l$, the torque L reaches its maximum value of $2.046 P_z P_m/z^3$. Their value of z_{min} is 6.0 cm, leading to a total length of the magnet system of 24 cm, based on the above optimum value of z/l. The cylindrical magnets are of Platinax II, the central magnet being 3.2 mm in diameter and 4.6 mm long ($P = 17.4 \times 10^{-3}$ A m^2) and the outer ones 2.5 mm in diameter and 3.7 mm long. Using an automatic measuring procedure and a spot-following device to read deflections the magnetometer is capable of measuring a specific magnetization of ~1×10^{-6} A m^2 kg^{-1} in a 22 cm^3 sample with $\pm 1°$ accuracy in a measurement time of about six minutes, and ~6×10^{-7} A m^2 kg^{-1} with $\pm 5°$ accuracy.

This performance is achieved on the fifth floor of an eight-floor laboratory building where there is not only substantial magnetic noise but also mechanical vibration. The effect of the latter is reduced by mounting the magnetometer on 500 kg of cinder blocks supported on rubber pads and suspending the magnet system on a 60 cm-long suspension inside the tubular magnetometer casing, the upper end of which is attached to a gimbal system.

Pozzi (1967) and E. Thellier (private communication) describe an instrument consisting of three spherical magnets ($\sim$3 mm in diameter) of separation 15 cm. The magnets are mounted in cubical Perspex holders and magnetized in a direction perpendicular to an axial hole in each of them, partially demagnetized if necessary to bring their moments (as measured on a magnetometer) accurately to a 2:1 ratio and then mounted in their holders in a framework with the magnet holes collinear. A quartz rod is passed through the holes and fixed in place, ensuring that the two outer magnets are accurately antiparallel to the third. The tube enclosing the magnet system is evacuated to lessen drift arising from convection currents. With a 4 cm cubical sample, specific magnetizations of the order of 2×10^{-8} A m^2 kg^{-1} can be detected with a time constant of around 45 s. A feedback system now attached to this instrument reduces the time constant.

9.2.3 Application of feedback

The potential advantages of applying negative feedback to an astatic magnetometer are reduction of the magnetometer time constant while maintaining or improving maximum sensitivity, easy adjustment of sensitivity and a more convenient signal read-out. Since the period is controlled electronically rather than by the geometry and mass of the magnet system, the requirement of a small magnet system for high sensitivity with a given period is relaxed, and within limits larger magnets can be used and therefore greater sensitivity achieved. However, this may be of little advantage if the signal/noise ratio is reduced, for instance by electronic noise or mechanically-excited vibration of the magnet system of period near to the chosen one.

In the commonly used control system, the light beam reflected off the magnet system falls on a split photocell (photo-voltaic or photo-resistive) which forms part of a circuit from which the output voltage is zero when each half of the photo-detector is equally illuminated. Any deflection of the magnet system results in unequal illumination of the detector and a resultant output voltage which, after processing, is passed to a coil near or surrounding one of the magnets of the magnet system: the coil axis is horizontal and perpendicular to the magnet moment when the system is in the zero position. With suitable circuit design the current in the coil required to maintain the magnet system undeflected or the voltage across a suitable resistance in the coil circuit is proportional to the deflection obtained in the absence of feedback.

In the magnetometer of Petherbridge, de Sa and Creer (1972) the signal from the split photocell is amplified and fed to the control coil (a small pseudo-Helmholtz pair surrounding the upper magnet) via an impedance converter and loop-gain control unit. The latter is essentially a variable resistance to which the response of the magnetometer (output voltage per unit torque on the magnet system) and the time constant is proportional. Thus the time constant increases with sensitivity, which is a favourable feature for reducing the effects of noise.

Roy, Sanders and Reynolds (1972) use two adjacent photo-resistors as the detector and the control coil is a small solenoid placed near the upper magnet. The out-of-balance current from the detector is passed directly to the solenoid via different resistance–capacity networks according to the sensitivity and damping required. de Sa and Widdowson (1974) describe a system for modifying the loop frequency response of the feedback circuit using operational amplifiers, and outline a method for predicting the closed loop stability.

A direct-reading astatic magnetometer of high sensitivity normally has a time constant of 10–20 s and acts as an efficient rejection filter of noise in the frequency range 0.5–5.0 Hz. A potential advantage of the application of negative feedback is reduction of the response time to ~1 s or less, with consequent sacrifice of the long time constant filtering effect. The overall result is that low-frequency noise may be present in the output signal and an associated reduction in the signal/noise ratio of the magnetometer occurs. The most common source of low-frequency noise is ground vibration, and it is the excitation of oscillations of the magnet system about the vertical axis by these vibrations which is the main source of output noise. These oscillations are excited if there is dynamic unbalance of the magnet system resulting in dynamic coupling between rotation of the system about the suspension axis and disturbing moments about the horizontal axis, such as ground vibration produces.

Making some simple assumptions about the form of dynamic unbalance of the magnet system and postulating reasonable values for the asymmetry of the magnet configuration, de Sa, Widdowson and Collinson (1974) were able to predict the fundamental frequencies of excited oscillations of the system, which were subsequently verified experimentally (Fig. 9.15). The predicted first and second mode frequencies were near 1.2 Hz and 2.7–5.0 Hz for magnet systems either longitudinally or transversely magnetized. Similar frequencies were observed by Roy *et al.* (1972) in the output from their magnetometer.

It is not feasible to construct a magnet system with sufficient geometrical accuracy to achieve perfect dynamic balance, and de Sa and Widdowson (1974) use two high *Q* twin T rejection filters centred on the dominant noise frequencies to reduce the vibrational noise arising from the unbalance while maintaining the rapid response time (~10 ms). They also describe two methods of improving the signal/noise ratio of the magnetometer, namely by using a recorder or output meter of 1–2 s time constant or a phase-sensitive detector. The latter technique is advantageous if the random zero drift of the magnetometer is significant. The sample is rotated below the lower magnet of the magnet system (position (a), Fig. 9.9) at ~10 Hz and a reference signal, obtained photoelectrically from the same rotation system, is used with the output signal in a conventional phase-sensitive detector circuit.

Whether the application of feedback to an astatic or parastatic magnetometer achieves a significant improvement in performance depends critically on

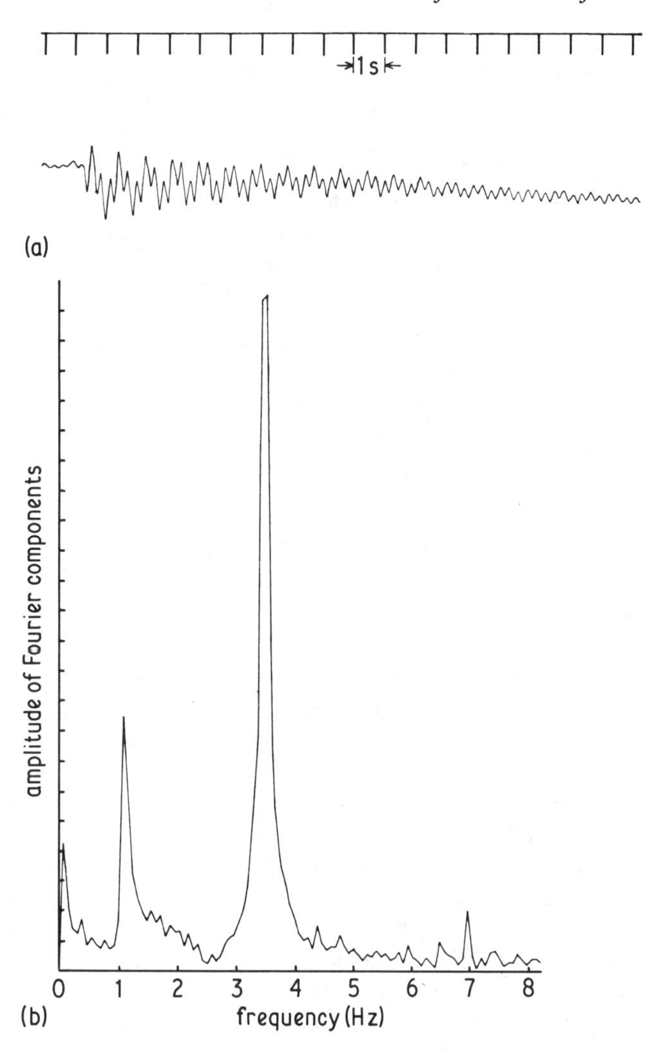

Fig. 9.15 (a) Response under damped feedback magnetometer to mechanical impulse, with open loop; (b) Fourier components of trace shown in (a). (From de Sa *et al.*, 1974)

the different noise of the modified instrument. It has not proved to be generally possible to take advantage of both potential sources of improved performance, i.e. much reduced time constant and increased sensitivity, and some sacrifice of the former is usually considered worthwhile to achieve the latter. Roy *et al.* (1972) claim an increase in sensitivity by a factor of 10 and a reduction of period from 13 s to ~5 s compared with the direct-reading instrument, a very significant improvement in performance. (Increased sensitivity is used here in the sense that more weakly magnetized samples can be measured with equivalent accuracy.) In the de Sa and Widdowson magnetometer the sensitivity is comparable with other direct-reading instruments, but the time taken to measure the NRM of a sample is reduced by a factor of about three. The authors draw attention to the fact that if the magnetometer time is constant in the order of 1 s or less, then it is likely that the time required to measure a sample will be governed by the time taken to manipulate the sample to different measuring positions and that full advantage of a very short response time cannot be realized.

9.2.4 Calibration and measurement of NRM

(*a*) *Calibration*

Magnetometers do not determine magnetic moments or specific magnetizations directly and they are not amenable to a theoretical calculation of their sensitivity. It is therefore necessary to determine their sensitivity experimentally in terms of a calibration factor to convert the deflection or voltage output produced by a rock sample at a given distance from the active magnet to the magnetic moment, or, more commonly, the volume or specific magnetization of the sample. This requires the use of an accurately known magnetic moment comparable in magnitude to the moments encountered in rocks.

Also necessary, but much simpler, is the calibration of direction. When the magnetometer deflection is zero, the appropriate component of magnetization in the sample is parallel to the magnetic axis of the active magnet (the magnet which is sensing the field of the sample) and this axis must be parallel to a fixed and known direction across the sample holder. This direction is most conveniently established by using a long, thin magnet (of piano wire or thin steel rod) mounted in a non-magnetic 'sample'. This can be rotated about a vertical axis in the holder until the magnetometer deflection is zero, and the holder then marked parallel to the test magnet axis, and/or the azimuth scale set to 0° (see also Section 9.2.4(b)).

The most satisfactory method for calibrating astatic and parastatic magnetometers depends on the nature of the measurements to be made and the sample geometry. Allowance must also usually be made for the effect of the upper magnet in astatic and the outer two magnets in parastatic instruments (positions (a), (b), (c), Fig. 9.9).

For 'dipole' samples (those which are small compared with the distance from the magnetometer at which they are measured) a small, accurately wound single-layer coil ~ 5 mm in diameter consisting of a few turns of fine (24–30 SWG) wire may be used. If a reliable milliameter is used to measure the current through the coil, this provides an absolute calibration through the relationship

$$B = \frac{\mu_0 P_c}{4\pi z^3} = \frac{\mu_0 N r_c{}^2 i}{4z^3} \tag{9.20}$$

where B and P_c are the field and magnetic moment produced by a current of i A in a coil of N turns of radius r_c (metres), and z is the distance from the centre of the coil to the centre of the lower magnet, where B is measured. B is measured in the plane of the coil (position (a), Fig. 9.9) or along the axis of the coil (position (c) and for parastatic systems). In the latter two cases the field is a factor of two greater than that given by Equation 9.20.

The standard coil can be used in two ways. The coil can be considered as a standard magnetic moment, and if measurements are made at different z-values the sensitivity can be expressed as the deflection/unit dipole moment for each z-value: in this way the effect of the non-active magnets will automatically be allowed for. Alternatively, the sensitivity can be expressed as the horizontal field required at the active magnet (perpendicular to P) to produce 1 cm deflection at the scale (or unit output in feedback systems). Expressed in this way the sensitivity will be independent of z, and the effect of the non-active magnets can be allowed for when the readings are processed.

The calibration coil can also be used in position (d) (Fig. 9.9), when the corresponding expression for B is

$$B = \frac{3\mu_0 N r_c{}^2 i l z}{2(l^2 + z^2)^{5/2}} \tag{9.21}$$

where $2l$ is the separation of the magnets. In Equation 9.21 B is the effective field at the magnet system such that the torque on the system is given by BP where P is the magnetic moment of one of the magnets.

If cubical or equidimensional cylindrical samples are to be measured, the assumption of the equivalent centred dipole is satisfactory if $z/r > \sim 3$, where r is a typical dimension of a sample of optimum shape (see Section 9.1.2). Alternatively, a coil of the appropriate square or circular cross-section and of the same dimensions as the sample can be constructed. The surface should be reasonably filled with a single layer of wire (e.g. > 10 turns on a 2.5 cm cylinder) and ideally the windings should be in grooves cut with their planes perpendicular to the cylinder or cube axis, connected by a longitudinal groove, to ensure an axial field. However, there will be little error introduced if a continuous winding of fine wire is used. The equivalent moment P_c of the coils with a given current is given by

$$P_c = NAi \tag{9.22}$$

where A is the area of the coil.

In using any form of coil it is important to avoid stray fields from the leads by ensuring that they are closely twisted together and approach the coil on the side remote from the magnet system.

An alternative but less satisfactory method of calibration is by means of a small magnet, the magnetic moment and moment of inertia of which are accurately known. The magnetic moment P can be then determined by suspending the magnet, with its axis horizontal, on a torsionless fibre and timing its oscillations in the horizontal geomagnetic field, or better still, at the centre of a coil, the field of which can be accurately calculated from the coil geometry, number of turns and the current through it. The coil is placed with its axis horizontal and parallel to the Earth's horizontal field B_H and the period of oscillation of the magnet determined with the coil current in one direction and then reversed. If the coil field is $B(< B_H)$ and the period of oscillation T_1 and T_2 in the fields $B_H + B$ and $B_H - B$ respectively, then the relationship

$$T = 2\pi\sqrt{\left(\frac{I}{PB}\right)} \tag{9.23}$$

leads to

$$B_H = B\left(\frac{T_2^2 - T_1^2}{T_2^2 + T_1^2}\right) \tag{9.24}$$

and P can then be determined by using B_H and Equation 9.20 or by using T_1 or T_2 and $B_H + B$ or $B_H - B$, the field being derived from Equation 9.24. It is convenient for the calculation of I if the magnet is of rectangular or circular cross-section, and it should be magnetized along its long axis, otherwise T may be too short for easy counting of oscillations.

For day-to-day monitoring of magnetometer sensitivity it is convenient to apply a field gradient by using a fixed coil near the magnetometer carrying a known current, or by positioning a small magnet in a known and repeatable orientation near the magnetometer. If the deflection produced is noted when the absolute calibration is carried out and again at a subsequent time, the change in sensitivity (expressed as the field at the magnetometer/unit output) will be in the inverse ratio of the initial and subsequent deflection. It is not necessary to know the field produced by the coil or magnet.

(*b*) *Measurement of NRM*

There are three basic methods of NRM measurement with magnetostatic instruments, designated here three-component, direct and off-centre.

(*i*) *Three-component*

This method can be used in any of the sample positions shown in Fig. 9.9 and also with a parastatic system, and consists of determining the orthogonal components X, Y and Z of the sample magnetic moment P_t. It is convenient to have one of the orthogonal axes parallel to the active magnet axis for the zero position of the sample: then rotation about a vertical axis (positions (a), (b), (c), Fig. 9.9) or horizontal axis (position (d)) gives deflections of the system proportional to $P \sin\theta$, where θ is the angular setting: deflections at 0°, 180° and 90°, 270° are proportional to the two orthogonal magnetization components of P, the moment in the plane of rotation. By adjusting the sample orientation so that deflections proportional to the two other pairs of components can be obtained in the same way, a total of four readings proportional to each of the components X, Y and Z is obtained. The multiple readings of the components give an average value of each, eliminate induced magnetization in any external field and reduce the effect of inhomogeneity of NRM.

It can easily be seen that if d_x, d_y, d_z are the average deflections corresponding to X, Y and Z, then

$$D = \tan^{-1}\left(\frac{d_y}{d_x}\right) \tag{9.25}$$

$$I = \tan^{-1}\frac{d_z}{(d_x^2 + d_y^2)^{1/2}} \tag{9.26}$$

and

$$P_t = s(d_x^2 + d_y^2 + d_z^2)^{1/2} \tag{9.27}$$

where s is the sensitivity expressed in the appropriate form and P_t is the dipole moment of the sample. Equation 9.25 only gives an angle between $+90°$ and $-90°$, and the quadrant in which D lies and the sign of I can be determined from the polarity of the magnet system, the sense of the sample rotation and signs of the individual deflections.

A variation of this method is continuous rotation of the sample with each of the axes of X, Y and Z in turn parallel to the rotation axis. This is an alternative way of reducing the effects of inhomogeneity and, with semi-automatic setting of the rotation axes, can be a more rapid method (Collinson, 1970).

For the static measuring method only a simple rotatable sample holder is required, suitably graduated at 90° intervals and provided with a means of altering the magnetometer–sample distance. The method is generally more suitable for magnetometers at which the operator sits, since with high-sensitivity instruments it is desirable to alter the rotation axis by remote control. Although such a device could be built it would be somewhat complicated and there may be problems in making it sufficiently non-magnetic for the measurement of weak rocks.

(ii) Direct measurements

This method is suitable for feedback and short period ($\leqslant 2$ s) direct-reading magnetometers using sample position (a). The sample is rotated to appropriate azimuths about horizontal and vertical axes until the horizontal component or total vector of NRM is parallel or perpendicular to the lower magnet axis, as indicated by a null or maximum deflection. Since the rate of change of deflection with angular setting is greater at a zero reading than at d_{max}, greater accuracy will be achieved by recording the null position. By using graduated scales on the rotation axes and combining readings taken to eliminate induced magnetization and minimize inhomogeneity, D and I can be computed directly from the angular settings of the sample. An application of this method is described by Roy *et al.* (1972).

(iii) Off-centre

This measuring method was devised by Blackett (1952) and is illustrated (for a dipole sample) in Fig. 9.16. As in the three-component method, with the dipole on and rotated about the vertical axis of the system (Fig. 9.16(a)) the deflections will describe a sine-wave of amplitude proportional to P_{H}, the horizontal component of the sample magnetic moment. The field due to P_{v}, the vertical component, exerts no torque on the system in this position. Thus

$$sd_{\mathrm{H}} = B_{\mathrm{H}} = \frac{\mu_0 P_{\mathrm{H}}}{4\pi z^3}\sin\theta \tag{9.28}$$

where d_{H} and B_{H} are the deflection and deflecting field due to P_{H}, and z is the distance from the dipole to the lower magnet. The effect of the upper magnet is at present neglected; s is the sensitivity expressed as tesla/metre deflection, the field acting on the lower magnet only.

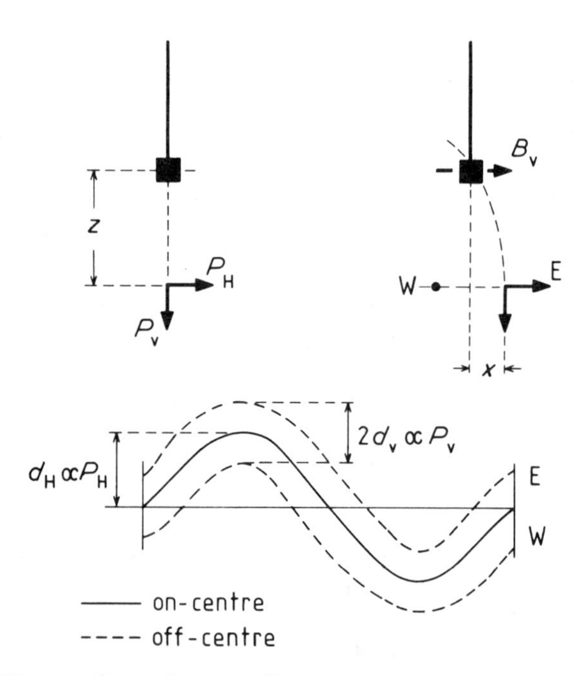

Fig. 9.16 Diagram illustrating the off-centre method of measurement. W and E indicate that the horizontal transverse axis of the sample is perpendicular to the magnetic moment of the lower magnet of the magnet system.

If the dipole is now traversed a small horizontal distance $x \ll z$ perpendicular to the axis of P, the moment of the lower magnet, there will be a horizontal component B_v of the field due to P_v at the magnet, the magnitude of which will be independent of θ (Fig. 9.16(b)). The sine wave described by the deflections due to P_H will be essentially unchanged except that it will be displaced above or below the zero axis due to the constant contribution of B_v: the sign of the displacement will depend on the direction of x and the polarity of the active magnet. The amount of displacement is related to P_v by

$$sd_v = B_v = \frac{3\mu_0 P_v x}{4\pi z^4} \tag{9.29}$$

where d_v is the deflection due to B_v (Fig. 9.16(c)).

Thus, rearranging and dividing Equation 9.29 by Equation 9.28

$$\frac{P_v}{P_H} = \tan I = \frac{B_v z}{B_H 3x} = \frac{d_v z}{d_H 3x} \tag{9.30}$$

The 'on-centre' and 'off-centre' sine waves are normally defined by taking readings at 0°, 90°, 180°, 270° with $x = 0$ and $x = x$; d_v is computed by subtraction of the deflections obtained at the same value of θ, and $d_H = (d_x{}^2 + d_y{}^2)^{1/2}$, where d_x and d_y are the deflections obtained at $\theta = 0°$, 180° and 90°, 270°, and from which D can be obtained, as for the three-component method.

For measuring cylindrical samples rather than dipoles, Blackett (1952) tabulates functions F_H and F_v, dependent on z and the cylinder dimensions, for modifying Equations 9.28, 9.29 and 9.30. The form of these functions and the modified equations are given in Appendix 7, together with the method of allowing for the upper magnet and other measurement details.

The off-centre method has certain advantages. The sample remains in one position in the holder (cylinder axis vertical) and is turned to different azimuths, transversed horizontally (to two possible positions, ~3–4 mm and ~6–8 mm, on each side of the magnet system according to whether z is small or larger) and raised up to an appropriate and adjustable distance below the lower magnet. These manipulations are comparatively easy to carry out remotely in sensitive instruments. The method is also suitable for disc-shaped samples using appropriate values of F_H and F_V or an approximation derived by Irving (Section 9.1), who showed that the disc magnetization can be represented by an equivalent dipole placed a distance $z_0 = (z^2 + r^2)^{1/2}$ below the magnet where r is the disc radius.

(c) *General comments*

For measuring weak samples the usefulness of an astatic or parastatic magnetometer is ultimately governed by random zero drift. The effect on measurement accuracy can be reduced at the expense of increased measuring time by taking zero recordings before and after each deflection and computing an average zero: this can only be done if there is a facility for withdrawing the sample from and re-introducing it to the magnet system. A less satisfactory alternative is to take zero readings before and after a series of deflections and assume a linear variation of the zero to compute a zero for each deflection.

When near the limit of measurement it is necessary to have the sample near to the magnet system to obtain significant deflections. How close this can be depends on the size, shape and magnetic homogeneity of the sample and ultimately on mechanical considerations, i.e. the size of the magnet system enclosure. A rule of thumb is that $z_{min} \approx d$, where d is a typical dimension of the sample and z_{min} is the distance of the centre of the sample to the centre of the active magnet. At these distances ($\sim$2.5 cm) there may be fields of the order of 100 μT at the sample due to the magnet system, assuming $P \approx 10^{-2}$ A m^2 (positions (a) and (b), Fig. 9.9, and parastatic systems). However, the induced magnetization will not cause a deflection of the system since it is parallel to the axis of the inducing magnet, unless there is inhomogeneity or anisotropy of susceptibility in the sample. An advantage of position (d), Fig. 9.9, is that there is zero field due to the magnet system in the central plane of the sample, and only a small field in the volume occupied by the sample. If inhomogeneity or anisotropy of susceptibility are present their effects, together with those of NRM, can be minimized by taking readings with the sample in appropriate orientations relative to the magnet system. The dominant effect is to introduce a $\sin 2\theta$ term when the sample is rotated through 360° (three-component and off-centre methods) in addition to the $\sin \theta$ term arising from the NRM. By averaging readings at azimuths differing by 180° the $\sin 2\theta$ terms are removed. Another simple precaution to minimize the effects of inhomogeneity of NRM, for instance in position (a), Fig. 9.9, is to take readings with the sample upright and inverted.

(*d*) *Applications*

Although astatic and parastatic magnetometers are now rarely used for routine palaeomagnetic measurement they remain useful and versatile instruments for a variety of measurements.

The remanent magnetism of irregularly shaped samples and regular samples of non-standard size can be measured, as well as mechanically weak samples, powders and deposited material, e.g. from depositional remanence experiments. If there is a coil system around the instrument, low field susceptibility can be measured by comparing deflections obtained in zero ambient field and in an applied horizontal field perpendicular to the axis of the sensor magnet (positions (a) and (b), Fig. 9.9), or parallel to the magnet system stem (position (d)). With an insulated sample holder of appropriate design low-temperature experiments (down to liquid nitrogen temperature) can be carried out, for instance testing for magnetite or haematite by their transitions at −140°C and −20°C respectively (Section 13.2).

For most applications the off-centre measuring method is the most useful since a complete measurement of NRM can be made with minimum disturbance of the sample, without inverting or turning on its side a fragile sample or one contained in a low-temperature holder. A quick determination of the NRM of large strongly magnetic samples can be carried out by using position (d), Fig. 9.9, with the sample at an appropriate distance to give on-scale deflections, and applying the appropriate form of Equation 9.21, precautions being taken to eliminate the effect of induced magnetization in the Earth's field. A rapid, approximate method of defining the polarity and axis of magnetization is to orient the sample until the deflection is a maximum, when the sample magnetization will be vertical and the polarity can be deduced from the sense of the deflection and the polarity of the magnet system.

Inhomogeneity of NRM through variation of NRM direction in a sample can be detected with an astatic magnetometer. Roy and Lapointe (1978) demonstrated such inhomogeneity in 2.5 cm sandstone samples by measuring them with first the top surface near the lower magnet and then the bottom surface. In each position the NRM of the sample region near the magnet was measured, and further information was obtained by slicing the samples into successively smaller volumes. Irving *et al.* (1966) describe calculations to determine the azimuth and displacement of the equivalent dipole in inhomogeneous disc-shaped samples.

More details of the use of astatic and parastatic magnetometers for the measurement of various magnetic properties are given in the appropriate chapters.

9.2.5 Use of fluxgate sensors

In recent years the improved sensitivity (~ 0.1 nT) of fluxgate detectors has led to their use as a rock-magnetometer sensor in both static and dynamic instruments. Using two sensors in an astatic configuration (i.e. 'back to back'), the ambient field can be effectively backed off and the rock or other sample introduced in any of the positions suitable for the astatic magnet system. Alternatively, a single sensor may be used in a zero-field environment; the precise backing-off of any residual field, if necessary, is a facility usually provided with commercial fluxgate magnetometers.

Present technology does not allow sensitivities comparable to astatic magnet systems to be achieved, but a direct-reading astatic fluxgate system can provide a rapid and convenient method of measuring the NRM of large and irregularly shaped samples. For a usable sensor sensitivity of 0.1 nT (B_{min}), a rule of thumb gives $J_{min} \approx 10^{-5}$ A m^2 kg^{-1}, based on $(B_x{}^2 + B_y{}^2 + B_z{}^2)^{1/2} = 5B_{min}$, and $z_{min}/d = 1.5$, where d is a typical sample dimension and z_{min} is measured to the centre of the sample. $B(x, y, z)$ are the field components at the sensing fluxgate due to the components of the sample dipole moment. J_{min} is independent of sample size down to a practical minimum which depends chiefly on the size of the fluxgate sensor.

Some references to the static use of fluxgates for NRM measurement are Helbig (1965), von Meitzner (1965) and Sharma (1968).

9.3 Spinner magnetometers

9.3.1 General design features

The principle of these instruments is the generation of an alternating voltage by the continuous rotation of a magnetized sample within or near a coil or fluxgate system. For a given sensor configuration the amplitude of the output voltage is proportional to the component of magnetic moment perpendicular to the rotation axis ($P_\perp$), and the phase of the voltage is utilized to relate the direction of the measured component to a reference direction in the sample. The total vector is determined by spinning the sample about a second orthogonal axis, although in practice the sample is rotated successively about three axes to obtain average values of the NRM components and reduce the effect of inhomogeneity.

For instruments using a coil sensor the output voltage V_s is of the form

$$V_s = f(P_\perp, \omega, n, r_1, r_2, l) \tag{9.31}$$

where ω is the rotation frequency and n, r_1, r_2 and l are the number of turns on the coil and its internal and external radius and length respectively. Rotation frequencies were formerly in the range 100–400 Hz to improve the sensitivity and to provide a reasonable frequency for the electronic detection circuits, and some effort was also expended in designing pick-up coils which maximized the signal/Johnson noise ratio. Present magnetometers, including the commercial instruments, favour lower frequencies (5–80 Hz) and rely on modern signal discrimination circuits to reduce the effects of noise.

If a fluxgate sensor is used the output signal is independent of ω and

$$V_s = f(P_\perp, s, z) \tag{9.32}$$

where $P_\perp$ is the moment perpendicular to the rotation axis (which is itself perpendicular to the fluxgate axis) and the sample is a distance z from the fluxgate sensor of sensitivity s.

The direction in the sample of the rotating moment in the plane perpendicular to the rotation axis is measured relative to a fixed direction in the plane by means of a reference signal obtained from the rotation system, and therefore, if required, of identical frequency. In early magnetometers a sinusoidal reference voltage was obtained either from coils surrounding a small magnet attached to the shaft of the rotation system, or photoelectrically by reflecting a light beam off a black and white sinusoidal pattern painted on the shaft or sample holder. The phase difference between the reference and

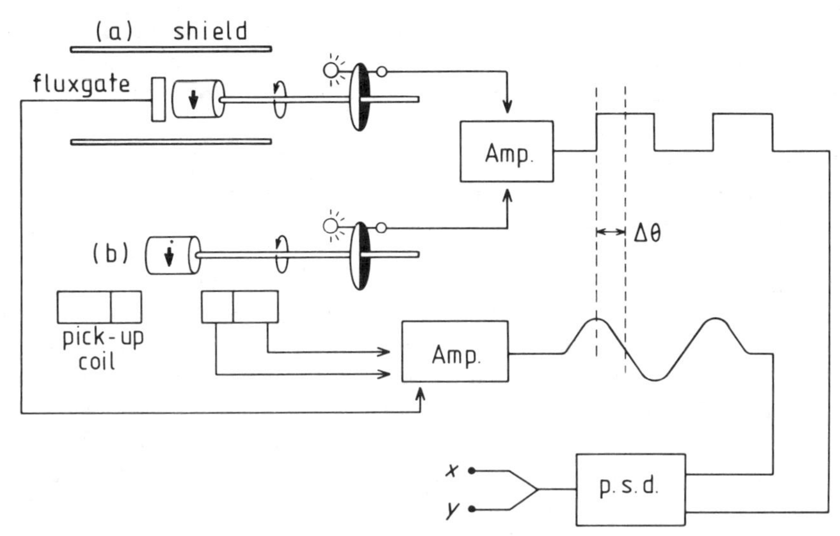

Fig. 9.17 Block diagram of spinner magnetometer showing alternative detector systems: (a) fluxgate detector within magnetic shield; (b) compensating pick-up coil system.

signal voltages is proportional to the angle between the direction of the measured magnetization and some fixed direction across the sample holder, oriented such that when $P_\perp$ and fiducial direction are parallel the phase difference is zero. Signal amplitude and signal phase for different sample rotation axes can then be combined to determine the magnetic moment vector in terms of D and I. In later magnetometers the phase information is used to resolve the signal voltage into orthogonal components proportional to components of magnetization parallel and perpendicular to the reference direction on the holder. The total moment and its direction are then determined from the average values of X, Y, Z. Figure 9.17 shows the general layout of spinner magnetometers.

During the development of spinner magnetometers efforts were made to increase the minimum measurable sample magnetization by increasing the rotation speed and careful design of the pick-up coil system. Since the use of electric motors for rotation limited the operating frequency to ~ 100 Hz and were also sometimes a source of electromagnetic noise, several laboratory magnetometers were built using an air turbine of the Beams type (Beams and Pickels, 1935) to rotate the sample. The turbine rotor is in the form of a cylinder with a conical base, on the surface of which are cut a ring of small flutes. The stator is a re-entrant, hollow cone containing a ring of small angled holes which coincide with the rotor flutes when it is placed in the stator. The axes of the rotor and stator are vertical and when air is forced through the stator holes the rotor lifts slightly and rotates at high speed through the action of the angled air jets on the rotor flutes. A stator excess air pressure of ~1 bar or less is sufficient to achieve rotational speeds of up to 1000 rev s^{-1} in a rotor taking a ~2 cm sample. However, noise and run-up times are excessive at these speeds and for the usual constructional materials (Perspex, Lucite) both the rotor and some rock samples are liable to break up under rotational forces, and practical experience dictated operating frequencies in the range 250–500 Hz.

The air turbine offered simplicity and a greater flexibility in pick-up coil design compared with a motor drive although it did, of course, require a compressed air supply. Since the air line could enter the stator at any angle a solenoidal pick-up coil could be used (an optimum configuration) which, because of the need for a bevel-geared right-angle drive, was not convenient with shaft drive. Disadvantages included generation of noise signals of electrostatic origin, and uneven running and vibration if the rock samples were not accurately shaped or homogeneous in density. With the improvement of signal/noise ratios, through new developments in electronics and the greater convenience and rapidity of measurement with motor-driven systems, the air turbine fell out of favour.

It is intuitively clear that the maximum output voltage from the pick-up coil will be obtained when the coil fits as closely as possible around the rotating sample and the windings extend a substantial distance in all directions. An approximate calculation also shows that a large sample is desirable. The flux linkage ϕ due to a dipole P at the centre of a coil of N turns of radius r, given by

$$\phi = \frac{\mu_0 PN}{2r} \tag{9.33}$$

Thus ϕ and therefore $\mathrm{d}\phi/\mathrm{d}t = V$ is proportional to r^{-1}, but assuming the sample completely fills the coil, $P \propto r^3$ and therefore $\phi \propto r^2$ to a first approximation for a small winding volume. However, as already mentioned, there is now a strong tendency to standardize on 2.5 cm cylinders of rock as the measurement unit in palaeomagnetism, and the advantage to be gained from larger samples would only be utilized in a magnetometer for special measurements. It should also be noted that although the sensitivity is increased by increasing r, the signal/noise ratio may not be improved: for instance, if the dominant source of noise is strong alternating fields inducing a voltage in the pick-up coil, then the noise signal would also increase as r^2 and the signal/noise ratio would remain constant.

Although the sensitivity of a magnetometer (i.e. pick-up coil output per unit rotating magnetic moment) is maximized by using a large coil of many turns, the Johnson (resistance) noise voltage at the coil terminals becomes significant and comparable to the signal voltage from weakly magnetized samples. Johnson noise is only one source of noise in spinner magnetometers, but since ideally all other noise sources can be very much reduced or eliminated, the philosophy behind coil design when spinners were being developed in the early 1960s was to maximize signal/Johnson noise ratio.

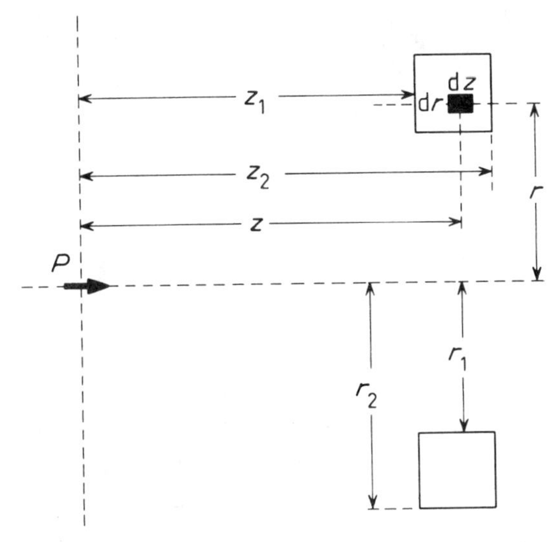

Fig. 9.18 Pick-up coil geometry, with magnetic moment P rotating about a vertical axis on the axis of the coil, and an element of the winding area dr dz centred on z and r.

The flux ϕ through a circular loop of radius r due to a magnetic dipole moment P situated on and directed along the axis of the loop and distance z from it, is given by

$$\phi = \frac{\mu_0 P r^2}{2(r^2 + z^2)^{3/2}} \tag{9.34}$$

Consider a coil of rectangular winding area with dimensions as in Fig. 9.18. If the wire has cross-sectional area A and the packing factor is σ (the proportion of the winding area occupied by copper, typically 0.6–0.7), the number of turns in the small area dr dz is

$$\mathrm{d}N = \frac{\mathrm{d}r\,\mathrm{d}z\,\sigma}{A} \tag{9.35}$$

and the flux linked with the area of the coil is

$$\mathrm{d}\phi = \frac{\mu_0 P r^2 \sigma\,\mathrm{d}r\,\mathrm{d}z}{2A(r^2 + z^2)^{3/2}} \tag{9.36}$$

and for the whole coil

$$\phi = \frac{\mu_0 P \sigma}{2A} \int_{r_1}^{r_2} \int_{l_1}^{l_2} r^2 (r^2 + z^2)^{-3/2}\,\mathrm{d}r\,\mathrm{d}z \tag{9.37}$$

$$= \frac{\mu_0 P \sigma}{2A} \mathrm{f}(r_1, r_2, l_1, l_2) \tag{9.38}$$

If the dipole is rotating at n rev s^{-1} about an axis perpendicular to z, the r.m.s. voltage induced in the coil is

$$V_s = \frac{\mu_0 \pi^2 P \sigma n}{4\sqrt{2A}} \mathrm{f}(r_1, r_2, l_1, l_2) \tag{9.39}$$

The resistance R_c of the coil is

$$R_c = \frac{\pi^2 \rho \sigma}{A^2}(l_2 - l_1)(r_2{}^2 - r_1{}^2) \tag{9.40}$$

where ρ is the resistivity of the wire. The r.m.s. random noise voltage V_n due to R_c is

$$V_n = 1.27 \times 10^{-10}(R_c \Delta n)^{1/2} \tag{9.41}$$

where Δn is the effective bandwidth of the detecting circuit. Combining Equations 9.39, 9.40 and 9.41 the signal/noise ratio is

$$\frac{V_s}{V_n} = EP\sigma^{1/2} n \rho^{-1/2}(\Delta n)^{-1/2}(l_2 - l_1)^{-1/2}(r_2{}^2 - r_1{}^2)^{-1/2}\mathrm{f}(r_1, r_2, l_1, l_2) \tag{9.42}$$

where

$$\mathrm{f}(r_1, r_2, l_1, l_2) = l_2 \log\frac{r_2 + (r_2{}^2 + l_2{}^2)^{1/2}}{r_1 + (r_1{}^2 + l_2{}^2)^{1/2}} - l_1 \log\frac{r_2 + (r_2{}^2 + l_1{}^2)^{1/2}}{r_1 + (r_1{}^2 + l_1{}^2)^{1/2}} \tag{9.43}$$

and E is a constant. Thus, the signal/noise ratio is independent of the number of turns except in so far as an improved packing factor is obtained with thicker wire.

For a solenoidal coil with the dipole (rock sample) rotating at the centre, $l_2 = -l_1 = l$, and Equation 9.39 gives

$$\frac{V_s}{V_n} = \frac{AP\sigma^{1/2} n}{\rho^{1/2}(\Delta n)^{1/2}} \mathrm{F}_1(r_1, r_2, l) \tag{9.44}$$

where

$$\mathrm{F}_1(r_1, r_2, l) = \frac{l^{1/2}}{(r_2{}^2 - r_1{}^2)^{1/2}} \log\frac{r_2 + (r_2{}^2 + l^2)^{1/2}}{r_1 + (r_1{}^2 + l^2)^{1/2}} \tag{9.45}$$

Figure 9.19 shows plots of $\mathrm{F}_1(r_1, r_2, l)$ against r_2/r_1 for different values of l/r. The optimum ratios are near $r_2/r_1 = 2$–4 and $l/r_1 = 1$–2. r_1 is taken as 4.0 cm in this example.

The other configuration of interest is two disc-shaped coaxial coils with the sample rotating between them and on their common axis. In this system, $r_1 \to 0$, and the signal/noise ratio is

$$\frac{V_s}{V_n} = \frac{AP\sigma^{1/2} n}{\rho^{1/2}(\Delta n)^{1/2}} \mathrm{F}_2(r_2, l_1, l_2) \tag{9.46}$$

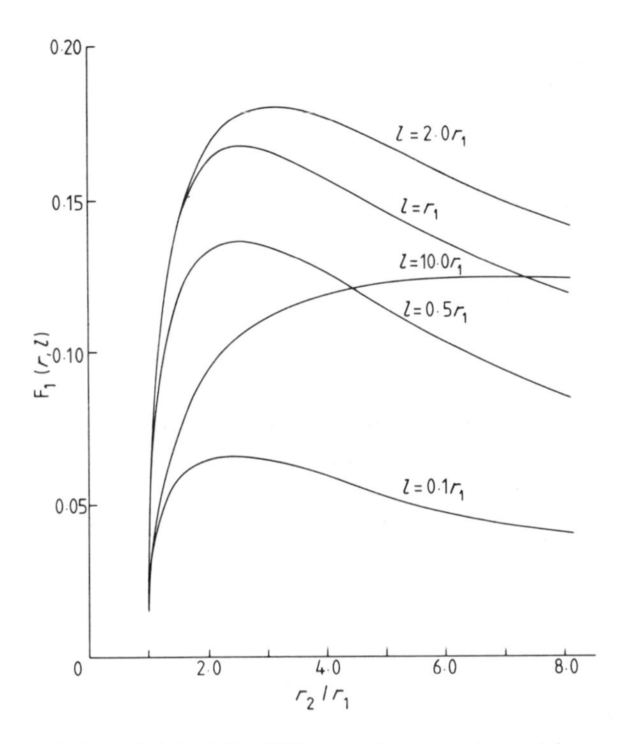

Fig. 9.19 Variation of $F_1(r, l)$ for different values or r_2/r_1 and l.

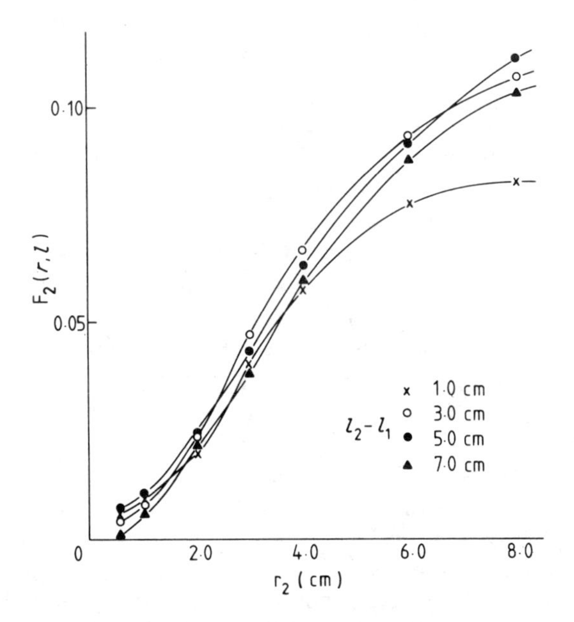

Fig. 9.20 Variation of $F_2(r, l)$ for different values of r_2 and $(l_2 - l_1)$.

where

$$F_2(r_2, l_1, l_2) = (l_2 - l_1)^{-1/2} r_2^{-1/2} \left[l_2 \log \frac{r_2 + (r_2^2 + l_2^2)}{l_2} - l_1 \log \frac{r_2 + (r_2 + l_1^2)^{1/2}}{l_2} \right] \quad (9.47)$$

The variation of $F_2(r_2, l_1, l_2)$ with r_2 and $(l_2 - l_1)$ is shown in Fig. 9.20. For practical coils, the solenoidal configuration provides a marginally better signal/noise ratio of the two designs.

The number of turns on the coils is governed by the requirement of a detectable signal from the weakest rocks to be measured and the nature of the first amplifying stage, specifically its input impedance. A certain amount of impedance matching can be achieved with a transformer between coil and amplifier. Too many turns should be avoided because of the then significant self-capacity of the coil: this limits the efficiency with which a compensating coil can be used to reduce noise signals produced by stray fields acting on the pick-up coil (see below). The number of turns used in coils described in the literature is typically 300–3000. In a solenoidal coil constructed by the author, $r_1 = 3.9$ cm, $r_2 = 5.4$ cm, $l = 4.9$ cm, and the winding was ~ 2000 turns of 33 SWG wire, with $R_c \sim 20\ \Omega$. The r.m.s. voltage output at $n = 100$ rev s^{-1} was ~ 8.0 V (A m^2)$^{-1}$, i.e. 10.8×10^{-9} V for a 2.0 cm cylindrical rock sample of specific magnetization 1×10^{-7} A m^2 kg^{-1}. At a bandwidth of 1 Hz, the noise voltage is $\sim 5.7 \times 10^{-10}$ V, giving a signal/noise ratio of ~ 20 for this rather weakly magnetized sample.

Like astatic and parastatic magnetometers in which the 'ultimate' noise level is rotational Brownian motion of the suspended system, the fundamental Johnson noise in spinner instruments is also usually exceeded by other noise signals, possible sources of which are stray magnetic fields, electrostatic fields, electric charges on the sample holder, pick-up coil vibration and electronic noise in the detector circuits.

Since stray alternating magnetic fields are likely to arise from mains electrical circuits with a frequency of 50 or 60 Hz, the rotational frequency in early spinners was chosen to avoid the mains frequency and its harmonics and detector circuits were often tuned to the operating frequency. Substantial reduction of stray field noise can also be achieved by using a secondary coil of the same area-turns as the pick-up coil and connected in series opposition with it. The coil is placed in such a position so that it is much more loosely coupled with the sample than the pick-up coil and therefore the output from the combined coil is only slightly reduced compared with that of the pick-up coil alone. A common configuration is where the compensating coil is placed coaxially and concentrically outside the pick-up coil. The increased area of the compensating coil implies fewer turns than on the pick-up coil and therefore a smaller Johnson noise contribution. With exact equality of area-turns and a uniform ambient alternating field the output voltage of the combined coil due to the field is ideally zero. In practice, small adjustments are required to accommodate non-uniformity of the field, changes in field gradient and the correction of small phase differences between the voltages induced in each coil due to self-capacity, and noise reduction factors of 100–1000 are possible. The arrangement is closely analogous to the two opposed magnets of an astatic magnet system. Coil vibration in the ambient magnetic field, caused by the sample drive system, is a possible source of coherent noise. It can be minimized by careful attention to pick-up coil and drive mounting or by nulling the ambient magnetic field at the coil system.

Hall (1963) investigates the conditions for optimum signal/noise ratio for a double coil, i.e. a pick-up coil with compensating coil wound directly on it and of the same length l. It can be shown that for zero net output in a uniform alternating field the outer radii of the pick-up and compensating coils are related by

$$c = (2a^3 - 1)^{1/3} \tag{9.48}$$

where a and c are the ratios of the outer radii of the two coils to the inner radius of the pick-up coil. Incorporating this condition into the expression for the signal/Johnson noise ratio of the double coil of the type shown in Fig. 9.19, Hall derives optimum coil dimension ratios that are similar to those derived from the curves in Fig. 9.19 and he also analyses the relation between coil resistance and minimum detectable signal and the effect of sample position within the coil. Grenet (1966) shows that there is some advantage in signal/noise ratio to be gained from a coil in which the length l increases as the radius increases.

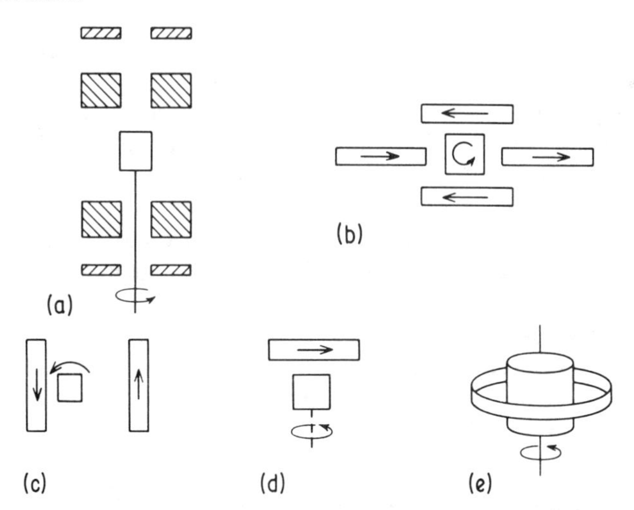

Fig. 9.21 Pick-up coil and fluxgate configurations for spinner magnetometers: (a) PAR pick-up and compensation coils; (b) four-fluxgate system of Sharma (1968); (c) Foster (1966) fluxgate system; (d) Schonstedt single fluxgate; (e) Digico ring fluxgate.

Where fluxgate sensors are used it is necessary to reduce the ambient field to a low value and to minimize variations in both direct and alternating ambient fields. In the first instrument to use fluxgate sensors dynamically (Foster, 1966) a simple 'astatic' arrangement of two sensors was used, and in a later version a four-sensor arrangement of increased sensitivity (Fig. 9.21). The Schonstedt, Digico and Adkin instruments (see below) rely on a multilayer magnetic shield to attenuate ambient direct and alternating fields over the single fluxgate sensors used. In addition to the compensating coils in the PAR magnetometer, an aluminium eddy current shield surrounds thc pick-up coils for additional attenuation of stray fields (Section 5.2.1). It is usually desirable to reduce the ambient field over the sample to a low level in all spinners since anisotropy of susceptibility in a sample can cause spurious signals. For instance, a field parallel to the sample rotation axis can induce a magnetization in the plane of measurement, a feature which is utilized in the measurement of anisotropy with spinner magnetometers (Section 2.8).

In recent spinner magnetometers the coherent electromagnetic noise problem has receded somewhat with the advent of low operating frequencies ($\sim$10 Hz), and random noise is more efficiently dealt with by phase-sensitive detection. Another common technique still employed is to increase the time constant of the detector circuit when the signal/noise ratio is low, which reduces random noise of frequency less than the time constant T as approximately $T^{1/2}$. Values of T up to 1000 s have been used. Since the time of measurement of an NRM component must be at least as long as the time constant used the measurement of very weakly magnetized samples can be time consuming.

The accumulation of a non-uniform distribution of electric charges on the sample and sample holder is also a possible source of noise, particularly in dry air conditions. Methods used to reduce or eliminate the effect include spraying with an 'antistatic' conducting film and the use of a small radioactive source (polonium 210) on the inside of the pick-up coil to ionize the air in the vicinity of the sample and allow the charges to leak away (Philips and Kuckes, 1967). A medical humidifier emitting a spray of fine water droplets was adapted by Graham (1968) to provide a conducting film on and humid environment around the rotating sample.

Shielding against electric fields, if necessary, is provided by enclosing the coil system in an earthed electrically conducting sheath of copper or aluminium foil. It may be advantageous to cut narrow gaps in the sheath parallel to the rotation axis to keep any current loops small.

9.3.2 Development of spinner magnetometers

Since there are now a number of commercially-made spinner magnetometers available and laboratory-built instruments are less common it is not proposed to describe in detail the design and construction of examples of the latter. In this section the development of spinner magnetometers is described, with brief descriptions of any special design features or improvements in particular instruments.

Johnson and McNish (1938) described the first spinner magnetometer designed specifically for small, weakly magnetized rock samples, although Chevallier (1925) appears to have been the first to use a very simple rotating sample instrument for measuring the NRM of lava samples from Mount Etna. Johnson designed a pick-up coil, of length small compared with the internal and external radii, for maximum signal/Johnson noise ratio, connected directly to the grid of the first value in the amplifier. The coil had many turns of fine wire to make the noise contribution from the valve small compared with that of the coil. Phase discrimination was by a calibrated synchronous commutator with variable brushes, the latter being rotated about the sample drive shaft until maximum output was recorded on a milliameter. At an operating frequency of 10 Hz a specific magnetization of 5×10^{-8} A m^2 kg^{-1} was detectable in a 5 cm^3 sample with $T = 50$ s. In a later, portable instrument Johnson *et al.* (1949) used a reference voltage generated in a coil placed near a magnet rotating with the drive shaft to drive one side of an electronic wattmeter, the signal voltage driving the other. By altering the reference voltage phase by rotating the reference pick-up coil around the shaft the voltmeter output could be brought to zero, when the two signals were 90° out of phase. The wattmeter is really an early form of phase-sensitive detector and acts as a narrow band detector which enables signals to be measured which are the same order of magnitude as random noise in the circuit. The difficulties of using narrow band filters to reduce noise is illustrated by the amplifier circuit used. The dependence of signal phase and intensity on frequency was such that a 2% change in rotation frequency resulted in a phase change of 10° in the output signal (and thus a 10° error in the measured direction of the component of NRM being recorded) and 5% in signal intensity.

Bruckshaw and Robertson (1948) used a similar reference voltage generator with phase discrimination by an a.c. potentiometer circuit in which the reference and signal voltages produced zero output, as shown by a vibration galvanometer, only if the two signals were in antiphase. The advantage of these null methods of phase measurement is that they depend only on phase relationships between the signals and not on signal strength, and therefore not on constancy of amplifier gain characteristics.

The 1950s saw the introduction of higher rotation speeds using Beams-type air turbines, e.g. Graham (1955) 282 Hz and Gough (1956) 185 Hz, and the generation of the reference signal by a photoelectric method, the simplest form of which is a reflecting sine-wave pattern of one wavelength painted round the circumference of the sample holder. Electronic phase-sensitive detection was also introduced at this time. A 'mechanical' phase-sensitive detector (p.s.d.) was employed by Griffiths (1955), a cam-operated switch on the drive shaft earthing the detector coil amplifier output for alternate half-cycles. The portion of the cycle cut off depends on the switch position and is varied by rotating the switch relative to the shaft. With all p.s.d.s the usual procedure was to adjust the output to zero by adjusting the reference signal phase, to determine the angular orientation of the NRM component being measured, and then shift the reference signal phase by 90° when the output of the detector, now acting as a half-wave rectifier, was proportional to the NRM intensity.

By the early 1960s spinner magnetometers were capable of measuring specific magnetizations down to 1.0×10^{-7} A m^2 kg^{-1} in ~10 cm^3 samples, a complete determination of NRM taking 10–20 minutes at the limit of measurement and less for stronger samples. The measurement limit was usually set by stray-field noise or, particularly with air-turbine drives, electrostatic noise.

With improvements in electronic techniques of low-noise, low-frequency amplification and signal extraction rotation speeds became slower during the next decade, which also saw the appearance of fluxgate sensors and commercially-made magnetometers. Princeton Applied Research, followed by the Schonstedt Instrument Company (both in the USA) produced motor-driven instruments (105 Hz or 15 Hz and 5 Hz respectively) with somewhat improved performance, i.e. measuring time for a given weak magnetization. Both these instruments resolve the p.s.d. output into the orthogonal components of the sample magnetization vector perpendicular to the spin axis. The PAR instrument, based on a design by Philips and Kuckes (1967), used a conventional compensated pick-up coil system (Fig. 9.21) and eddy-current shield and Schonstedt employ a fluxgate sensor, a method pioneered by Foster (1966). Foster used an astatic arrangement of fluxgate probes (Fig. 9.21) to eliminate the d.c. signal due to the Earth's field, whereas Schonstedt use a single probe and magnetic shield. Another commercial instrument is the UGF-JR2, manufactured by the Institute of Applied Geophysics, Prague. The rotation speed is 85 Hz and the two flat pick-up coils (uncompensated) are surrounded by a triple-shell Permalloy shield. A time constant of up to 100 s is used and the instrument has the distinction of being the only magnetometer (at the time of writing) to be calibrated in SI units.

The instruments of Dianov-Klokov and Anasov (1967) and Bruce (1967) are of some technical interest. In the former, the pick-up coils are cooled in liquid nitrogen to reduce Johnson noise (V_n). Since V_n is proportional to the square root of the absolute temperature and the coil resistance also decreases with temperature, a reduction in V_n by a factor of about 5 can be achieved at 77 K, assuming R decreases by a factor of about 7. Bruce (1967) describes an ingenious shielded pick-up coil system which achieves good flux linkage with the sample (rotated by air turbine), with most of the sample holder and turbine outside the coil: this feature decreases electrostatic noise and the effects of holder contamination.

A recent investigation of pick-up coil design is by van Oorschot (1976), who treats the coil and pre-amplifier as a unit and optimizes the signal/noise ratio at the output of the latter. The effect of increasing the wire cross-section as the coil radius increases is also investigated. As expected this significantly improves the signal/noise ratio of a compensated pick-up coil but has less effect on a single coil and is probably not worth the added constructional complication in either case. The optimum coil dimensions derived for a split solenoidal coil are similar to these derived from Fig. 9.16: the wire gauge is chosen to provide the correct resistance for matching to the pre-amplifier. No compensating winding is used and stray field signals are reduced by eddy current shielding, electronic filtering and phase-sensitive detection. The reference signal for the latter is obtained from an oscillator rather than the conventional optical or magnetic methods (van Oorschot and Ridler, 1976). The drive motor speed is locked on to the reference frequency (85 Hz) by a servo-mechanism. With a 3 s time constant the minimum measurable magnetization in a 2.5 cm sample compares favourably with current commercial spinners.

In the Digico magnetometer, developed in 1970 by L. Molyneux at the University of Newcastle upon Tyne, an entirely new concept was introduced whereby the signal from the sensor is processed by an on-line computer to produce phase and intensity information (Molyneux, 1971).

A circular fluxgate sensor surrounds the rotating sample ($\sim$ 7 Hz). By means of a disc with 128 radial slots, rotating on the same shaft, and an analogue-to-digital converter each cycle of the alternating output from the sensor is defined by 128 twelve-bit numbers proportional to the amplitude of the waveform at approximately 3° intervals: each number is stored in the computer. For each complete revolution of the sample the waveform amplitudes corresponding to each of the 128 intervals are added to the appropriate store and thus a number proportional to the mean amplitude for each interval is obtained. The mean waveform is then Fourier analysed and the phase of the first harmonic is computed and printed out together with the X and Y, Y and Z or X and Z components of magnetization, according to the spin axis. Random noise is reduced by increasing the total spin time, measured in terms of 2^n spins, where $n = 4$–11. This range corresponds to a one-axis measuring time of about 2 s to 5 minutes. After four or six spins declination, inclination and mean volume magnetization and its components $\bar{X}$, $\bar{Y}$ and $\bar{Z}$ are pointed out in units of 10^{-6} G (10^{-3} A m^{-1}) based on a 2.5 cm diameter cylindrical sample of pre-programmed height. The computer can also be programmed for subsequent calculation of bedding corrections, Fisher's statistics and pole positions. The options available for other types of measurements are described in the relevant chapters.

The Digico magnetometer takes much of the tedium out of routine palaeomagnetic measurements, and the instant availability of NRM intensity and direction and its components substantially reduces overall measurement time and enables anomalous data and accidental measurement errors to be spotted without delay. Comparison of X, Y, Z values obtained from different spins also enables an estimate to be made of inhomogeneity of NRM. The good signal/noise ratio available makes feasible the measurement of more weakly magnetized material than was hitherto practicable.

The Schonstedt and UGF magnetometer are also now available with computer processing. A feature of the former instrument is the continuous monitoring of the signal/noise ratio during sample measurement. The sample spin continues until the ratio reaches a value pre-set by the operator, when the spin is stopped and the signal processed. This system has the advantage of producing results with an approximately uniform and predictable level of error. Other facilities available are the estimation of inhomogeneity of NRM by comparing directions obtained from the different sample faces presented to the fluxgate sensor, and the correction of sample NRM data for weak magnetic contamination of the sample holder. The UGR-JR4 instrument has integration time of 1, 10 and 100 s, with facilities for on-line and off-line computer processing.

Kono, Hamano, Nishitani and Tosha (1981) have designed an ingenious spinner in which the sample is simultaneously rotated about two orthogonal axes near a fluxgate sensor. The output signal, which is processed by an on-line computer, contains three frequencies, ω_1 and $\omega_1 \pm \omega_2$, assuming a dipole sample, where ω_1, ω_2 are the angular velocities of the rotation of the sample. Analysis by computer of the amplitudes and phases of the three frequency components enables the total vector to be determined. In its present form, the noise level is equivalent to about 1.0×10^{-5} A m^2 kg^{-1} in a standard 2.5 cm sample after processing signals obtained from 256 sample rotations. The magnetometer appears to be useful for the rapid measurement of igneous rocks and has potential as well for rapid a.f. demagnetization by surrounding the sample with a suitable coil.

9.3.3 Calibration and measurement

The absolute calibration of sensitivity of spinner magnetometers is not straightforward and commercial instruments are usually provided with 'standard' magnetic moments for sensitivity and angular calibration. A coil of the same size and shape as a rock sample carrying direct current can be used, but since the coil is continuously rotating during calibration, it is usually necessary to have either a slip ring system with its attendant complications or a self-powered coil containing its own small battery. The difficulty with the latter method is the magnetic moment of the battery and resistor(s) and stray fields from the leads. These unwanted moments were eliminated by Robertson (1978) by carrying out two calibration runs, the coil connections being reversed between runs. The coil current is determined from the voltage drop across the resistor. With a 100-turn coil of 2.4 cm diameter powered by a zinc-carbon cell giving 10–1000 μA, moments in the range 10^{-7}–10^{-5} A m^2 are obtained with suitable resistors. With zero current the residual moment of the coil components is $\sim 10^{-7}$ A m^2. It is also necessary to check, as Robertson did, that the relative permeability of the components is sufficiently near unity so that induced moments can be neglected.

With slow-speed spinners an absolute calibration may be carried out with an externally powered coil by allowing the leads to twist. In the Digico instrument a calibration run using 2^5 spins results in that number of twists in the leads, which can easily be accommodated in $\sim$50 cm of fine wire.

In some magnetometers it may be possible to use a stationary coil carrying alternating current. However, since the detector circuits usually require the reference signal for their operation it is necessary to be able to detach the sample holder and place the coil in the correct orientation in the sample position, allowing the drive shaft still to rotate. Alternatively, the test coil can be supplied by an oscillator, the output of which is also used to actuate the reference circuits (de Sa and Molyneux, 1963). Reasonably close reproduction of the operating frequency is necessary, particularly in magnetometers with a coil sensor and/or narrow band filters in the detector circuits. The coil method can be used in van Oorschot and Ridler's (1976) magnetometer, in which the reference voltage is generated independently of rotation.

Noltimier (1964) proposed the use of a closed wire loop spinning in a direct applied field B. Neglecting inductance effects the effective magnetic moment generated is

$$P = \frac{\pi^2 r^4 \omega B}{2R} \sin 2\omega t = P_0 \sin 2\omega t \tag{9.49}$$

where r and R are the loop radius and resistance and ω the circular frequency. Provided $r < \sim 1$ cm and $R > 1\,\Omega$ and the spin frequency is of the order of 100 Hz or more, inductance effects are negligible. If $r = 1$ cm, $f = \omega/2\pi = 100$ Hz and $R = 1\,\Omega, P_0 = 3 \times 10^{-8}$ A m^2 for 1 mT applied field. This calibration technique has some disadvantages. A single loop moment is not a good representation of a rock sample magnetization (a multi-turn coil would enhance the effects of self-inductance, increasing the impedance and reducing the coil current for a given B and ω), and the sin 2ωt factor in Equation 9.49 implies that the magnetometer must run at half speed to obtain a signal at the operating frequency. The method is not suitable for slow-speed instruments nor where a fluxgate sensor is used, because of the presence of the applied field.

The above techniques are time-consuming and for day-to-day monitoring of sensitivity the small 'standard' magnet supplied with most instruments is satisfactory. It is only necessary to ensure that it is magnetically stable, and a small piece of permanent magnet material or strip of recording tape magnetized along its length may also be used. The latter is also useful for calibration of the reference direction of the magnetometer sensor, since its magnetic axis is accurately known. It is important to note that such 'standard' magnetic moments cannot, in general, be used to standardize or calibrate different magnetometers because of the different coupling between magnet and sensor in different instruments. For instance, the calibrating moment provided with the Digico instrument is not the moment which can be calculated from the magnetometer output. The latter is appropriate to a 2.5 cm cylindrical sample and is not correct for the same moment in the form of the small strip of magnetic tape in the test sample.

Consider a cubic sample which is rotated about a fixed axis parallel in turn to three mutually perpendicular axes in the sample parallel to the cube edges. There are eight different ways in which the sample can be oriented for each of the three spin axes, a total of 24 orientations in all. However, with respect to the induced signal in the sensor, six pairs of the 24 orientations are the same, since one of a pair can be produced by rotating the other through half a revolution of the rotation axis.

The usual procedure for NRM measurement is to spin each sample in six mutually orthogonal orientations such that any two of X, Y and Z are in turn perpendicular to the spin axis. Since two orthogonal components of magnetization are obtained from each spin, the orientations are chosen to provide four determinations each of X, Y and Z, usually two of each sign for each component (i.e. orientations giving 180° phase difference in the output signals). This procedure tends to average out any random noise on the signals, the effects of inhomogeneity of NRM and of any remanent magnetism in that part of the sample holder permanently attached to the rotating shaft. These effects will cause more or less scatter in the four values of each NRM component. In samples of reasonable intensity of NRM the scatter will mainly reflect and provide information on the extent of inhomogeneity of magnetization.

The six sample orientations used should ideally

(a) Provide two values each of $\pm X$, $\pm Y$ and $\pm Z$
(b) Distribute the displaced dipole caused by inhomogeneity of NRM as uniformly as possible in space
(c) Mean out possible errors arising from imperfect calibration of NRM direction of the magnetometer
(d) Measure each NRM component in each 'channel' of the magnetometer, i.e. the sample is placed in the holder with X, Y and Z each parallel and perpendicular to the reference direction

These requirements cannot usually be satisfied with a six-spin measuring procedure and (c) is usually not achieved; this should not degrade the data significantly since instrument misalignment errors should be small. (b) can be achieved sufficiently to average out the effects of 'normal' inhomogeneity encountered in rocks. It is easy to arrange for positive and negative values of X, Y and Z to be obtained, but whether (d) is possible depends on the method of holding the sample during rotation. It is possible when the sample can be placed in any of the three mutually orthogonal orientations in the holder (PAR, Schonstedt, UGF-JR4), but not if the orientations are restricted as in

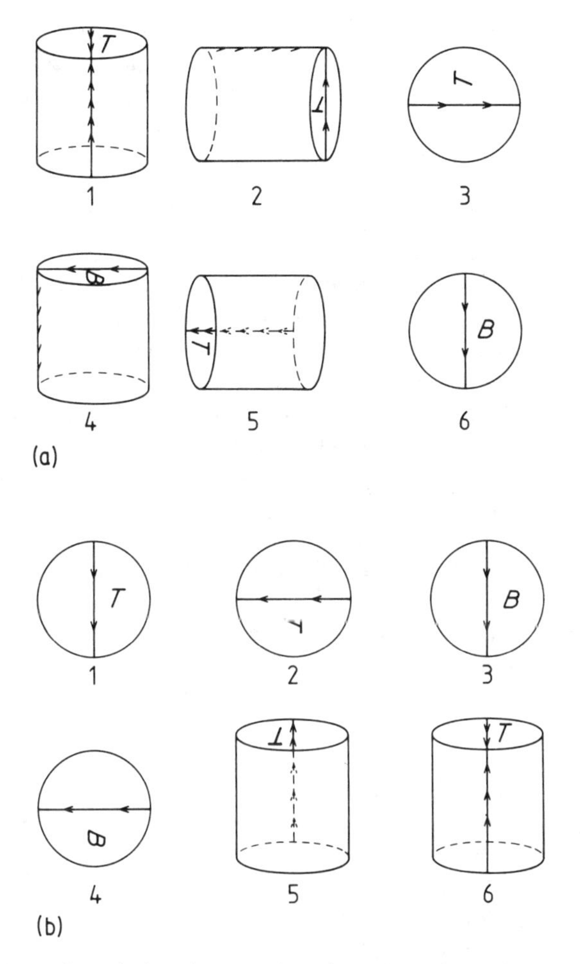

Fig. 9.22 Examples of six-spin sample orientations for spinner measurements: (a) PAR; (b) Digico.

the Digico instrument, where Z (parallel to the axis of the cylindrical sample) is only measured in one channel.

The sample orientations chosen depend to a certain extent on the sensor or sensor configuration. For instance, in the Schonstedt instrument, where the sample is rotated close to a single fluxgate about an axis perpendicular to the fluxgate axis, it is advantageous to choose orientations such that each of the six faces of the cubical sample holder are in turn presented to the fluxgate. Figure 9.22 shows examples of six-spin sample orientations.

Hummervoll (1974) points out that it may be possible to dispense with some repositioning of the sample by repeating the measurement of a sample in one orientation with the rotation of the drive shaft reversed and the phase of the reference system shifted through 90°. In principle, only three reorientations of the sample in the holder are then necessary, with a consequent saving in measurement time. This procedure effectively rotates the sample through 180° about an axis perpendicular to the rotation axis, and the second measurement provides values of the two components being measured of opposite sign to those obtained in the first measurement. This procedure may be satisfactory if inhomogeneity of NRM is not severe, but may not be if a single fluxgate sensor is employed or there is magnetic contamination in the rotating sample mounting. The author quotes results showing that the method gives good results with samples tested on a Foster (1966) fluxgate instrument.

The way in which the sample is held while rotating bears on the convenience and accuracy of measurement. Where the sample is first secured in a sample holder before being placed in another holder on the rotating shaft there is the advantage that the three-rotation axes in the rock will be accurately orthogonal, but the disadvantage that the measured NRM will be the vector sum of that of the rock and of any magnetic contamination of the separate holder. Any remanence of the shaft holder can usually be averaged out in the measuring procedure or, as in the PAR instrument, the total holder signal can be backed off prior to measurements. In the Schonstedt computer magnetometer, the holder magnetization components can be stored in the computer and the rock signals are automatically corrected for them. In the Digico and UGF-JR4 magnetometers the sample is placed, without holder, on a platform at the top of the vertical shaft and thus the only source of spurious signals is any platform remanence, which can be eliminated or very much reduced by choosing appropriate rock orientations. This advantage outweighs the disadvantage of having to set each orientation axis by eye: any errors incurred here are likely to be random over the six spins. It is, of course, also advantageous to minimize the amount of material supporting the sample from the standpoint of reducing electrostatic noise.

Convenience of measurement is a feature of magnetometers which is not often stressed. In earlier instruments, time constants were often quite large (10–50 s) even for specific magnetizations of 10^{-5}–10^{-6} A m^2 kg^{-1}, and stopping and starting rotation and sample reorientation time did not add significantly to measurement time per sample. The introduction of slow rotation speeds and lower time constants made sample manipulation time more of a consideration both from the point of view of total measurement time and of tedium of operation where, as is usual, many samples are involved. Thus, in this respect the Digico magnetometer, in which the sample rests under gravity on the rotating platform and can be rapidly reoriented, has the advantage over the UGF-JR4 magnetometer, where the magnetic shield must be removed, the pick-up coils opened, and a sample locking screw loosened prior to reorientation.

The useful sensitivity of spinner magnetometers, i.e. the weakest magnetization that can be measured in a standard sample to provide an NRM direction of acceptable accuracy, say $\pm 5°$, is an elusive quantity to define and one that is by no means constant. There is no generally accepted way of expressing this sensitivity, as perusal of the maker's information (as at 1980) soon shows. Examples are 'the instrument provides a measurement range of from 5 to less than 2.5×10^{-11} A m^2 total moment', with information on expected directional accuracy with a given signal/ratio (Schonstedt DSM-1), 'nominal sensitivity 2.4×10^{-6} A m^{-1} at 100 s time constant' (UGF-JR4), and 'the noise level is guaranteed to be less than 2.5×10^{-5} A m^{-1} for any of the orthogonal components (at a 95% probability level) when measuring over 2^7 spins' (Digico).

Of these statements, those concerning the Digico and Schonstedt instruments are the most informative, in that they provide information as to the time required to measure a weakly magnetized rock and the accuracy to be expected. For other instruments there is little information on noise level, and it is the overall noise present, both random and systematic, and its variable nature which make statements about 'sensitivity' of doubtful use. For day-to-day performance it is the factors which have already been discussed, i.e. electronic, electromagnetic and electrostatic noise and contamination of the sample holder which govern the effective sensitivity of magnetometers and the reliability of measurements of weakly magnetic material. It is the author's experience that there can be a day-to-day variation in the overall instrumental noise level of a magnetometer by a factor of up to ten. Only by tracing and, where possible, eliminating the causes of the noise can the ultimate sensitivity of the magnetometer be approached. With currently available spinner magnetometers, the practical lower limit of NRM for routine measurements is in the region of 5×10^{-8} A m^2 kg^{-1}, with a measurement time of 10–30 min per sample and an accuracy of direction of 5–10°.

At this level of magnetization, special precautions are necessary if meaningful results are to be obtained. Extrinsic contamination of the rotating platform and any sample holder, due to steel abrasion during construction, handling, atmospheric particles and adhering rock particles can usually be eliminated by chemical (e.g. 10% hydrochloric acid) and/or ultrasonic cleaning. Intrinsic magnetization due to ferromagnetic contamination of the holder material can often be removed by alternating field demagnetization and subsequent storage in field-free space may be advantageous. Electrostatic noise can usually be eliminated by treating surfaces with an antistatic spray or by the methods described earlier in this chapter.

Possible magnetic contamination of rock samples as a result of drilling and slicing, which may leave small, barely visible particles of steel adhering to or embedded in the rock surface, should always be considered. Where possible the use of phosphor-bronze cutting surfaces will eliminate such contamination. Steel contamination may be tested for by re-measuring a sample after removal of the surface layer by grinding with a non-magnetic agent such as bauxilite (Al_2O_3), or by treatment in dilute acid, although the latter procedure may dissolve surface magnetic grains or carbonate cement in sandstones.

Whether random magnetometer noise is essentially being measured or a true but weak NRM can be tested by repeat measurement of samples. This will also provide information on the accuracy to be assigned to the measured direction and intensity of any weak NRM present. It may be necessary to allow for permanent residual magnetism of the sample holder, if used, by a blank measurement without the sample. Strangway and McMahon (1973) describe the successful measurement of very weakly magnetized samples of the Green river sediments (Colorado) with NRM in the range 0.5–2.0 × 10^{-8} A m^2 kg^{-1}, using the Princeton Applied Research magnetometer.

With weakly magnetized samples, it may be worth considering whether one NRM measurement using a time constant T or two measurements at $T/2$ is preferred. The total measuring time is approximately constant in each case. If the noise on the signals is proportional to $T^{-1/2}$ the fractional error in X, Y and Z is the same for each method, but the double measurement provides information on repeatability.

9.4 The cryogenic magnetometer

This type of magnetometer, which came into use in the early 1970s, depends on various effects associated with superconductivity, namely persistent currents, the Josephson weak link and flux quantization. It is also known as a SQUID magnetometer, an acryonym for Superconducting Quantum Interference Device, derived from a phenomenon which is analogous to optical interference and which occurs in some superconducting circuits.

9.4.1 Operating principle

A detailed description of the complex physics invoved in SQUID operation is beyond the scope of this book, and the following account is intended to clarify the basic principles of the technique. For a more detailed analysis of the operation of SQUID devices the reader is referred to papers by Deaver and Goree (1967), Gallop and Petley (1976), Goree and Fuller (1976) and Petley (1980).

The principle of SQUID operation may be understood by reference to a superconducting ring of area A which has been cooled through the critical temperature in the presence of a magnetic field B_e directed along the ring axis. If B_e changes after the superconducting state is achieved a circulating current I is set up in the ring of magnitude and sense exactly that required to cancel the change in flux $\Delta\phi = \Delta(B_e A)$ linking the ring. However, if I exceeds a critical value I_c, the critical current density in the ring is exceeded and the superconducting property suppressed. The ring becomes resistive, I decreases and the flux linking the ring inceases or decreases according to the sense of

ΔB_e. The change in flux is quantized and $\Delta\phi = n\phi_0$, where n is an integer and ϕ_0 is the flux quantum. It can be shown that $\phi_0 = h/2e = 2.07 \times 10^{-15}$ Wb.

When I decreases because of loss of superconductivity, it falls below the critical value I_c and therefore the superconducting state is regained, and as long as B_e continues to change sufficiently for I to become greater than I_c the flux linkage with the ring changes, becoming constant when B_e remains constant and $I < I_c$. It can be seen that the circulating current resulting from a change ΔB_e in the external field is a measure of the field change and the sensitivity of the SQUID sensor relies on the detection of extremely small current changes.

To achieve this sensitivity the cross-section at a point in the circumference of the ring is reduced to a very small area, of the order of 1 μm across. This is the Josephson weak link or Josephson junction. A very small change in applied flux of the order ϕ_0 can now cause the critical current density in the ring to be exceeded at the junction, and the suppression of superconductivity. Suppose now that the ring is driven by a sinusoidal external field superimposed on B_e and of sufficient amplitude to cause I to exceed I_c. At each point in the cycle where I_c is exceeded and a flux quantum enters the ring the change in flux linking the ring can be detected by a voltage spike induced in a suitably placed pick-up coil. If B_e now changes, the resulting current in the ring will cause the resistive switching to occur at a different point in each cycle and alter the character of the output voltage. There is an analogy here with the operation of a fluxgate magnetometer, in which the points in a magnetizing cycle at which a high-permeability core reaches and leaves saturation is varied by a weak field which alters the core magnetization. However, there are two important differences between fluxgate and SQUID operation. The fluxgate ceases to operate if the external field saturates the core, whereas when I_c is exceeded in the SQUID flux enters the ring and the system resets itself to a lower current value. Also, a fluxgate measures absolute field values but the SQUID output is a measure of the change of field in the ring from the field present when it became superconducting.

Current SQUID magnetometers employ a drive frequency of 20–30 MHz with circulating currents of the order of 1 μA corresponding to a change of one flux quantum in the detector. For reasons associated with thermal noise the detector must be very small and posses a small self-inductance (<1 μH) and there are two common configurations. One is a quartz cylinder ($\sim$2 mm in diameter) on the curved surface of which a thin layer of superconductor is deposited. A longitudinal slot is cut in the layer except for a very narrow ($\sim$0.1 μm) bridge (Dayem bridge) at the centre, which forms the weak link. The drive coil, from which the output signal is also obtained, is wound on top of the superconducting layer. An alternative form of detector is a ring in which the weak link is formed at the point of a fine screw bearing on a flat surface.

For an estimate of SQUID sensitivity consider a cylindrical detector of diameter 2 mm. A field change of $\sim 7 \times 10^{-10}$ T along the cylinder axis corresponds to a change of one flux quantum. Since it is possible to detect a change of $10^{-3}\ \phi_0$, a field of 7×10^{-13} T can be detected by the sensor.

In principle the weak link sensor can be used directly for magnetic field measurements but it is advantageous on several counts, particularly for NRM measurements, to detect the field initially by means of a superconducting pick-up coil. This is linked by superconducting leads to a field coil tightly coupled to the SQUID detector. This system acts as a form of zero-frequency field transformer, based on the condition that the flux linkages within the circuit must remain constant. Thus, for a given field more flux is linked with the sensor because of the large pick-up coil area relative to the effective SQUID area, and substantial field amplification by a factor of 20–50 can be obtained. The use of pick-up coils is also a much more satisfactory (electromagnetically) and convenient way of detecting the field of a rock sample than placing it near the sensor.

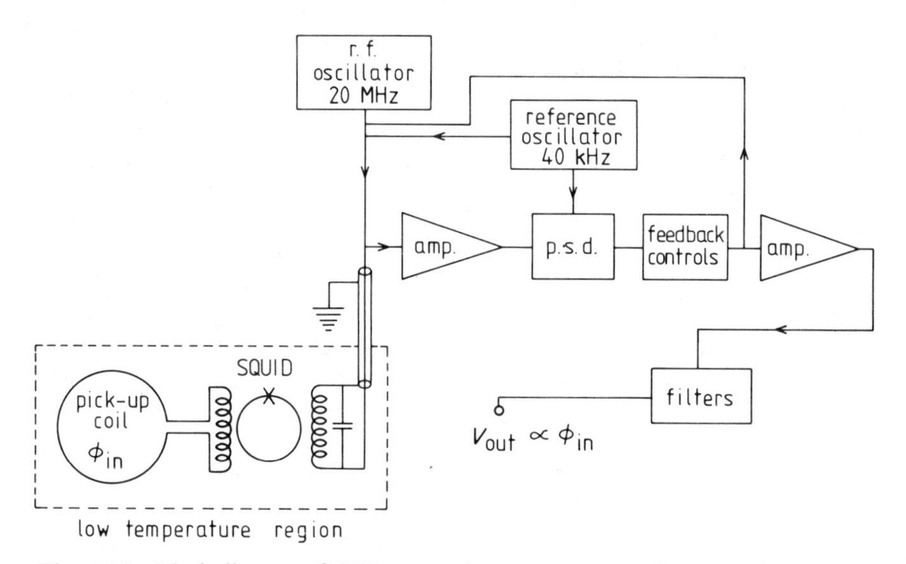

Fig. 9.23 Block diagram of CCL cryogenic magnetometer. (Courtesy of Cryogenic Consultants Ltd, London)

The output obtained from the SQUID is non-linear with the applied field and it is usual to apply feedback to the sensor via a coil, the current or voltage in which is then proportional to the applied field. A block diagram of a cryogenic magnetometer is shown in Fig. 9.23.

9.4.2 The SCT and CCL cryogenic magnetometers

At the time of writing there are two commercially built SQUID rock magnetometers on the market. The Superconducting Technology instrument

(USA) has been available for about 12 years and more recently Cryogenic Consultants Ltd (London) have entered the market. The construction of both instruments is similar and they only differ in some details: a diagram of the CCL instrument is shown in Fig. 9.24. The CCL instrument is pre-cooled with liquid nitrogen, contained in the upper tank ($\sim$20 l capacity), which cools the helium chamber by means of the radiation shield and assists in reducing helium loss during operation. Both the nitrogen and helium chambers are surrounded by vacuum spaces which can be pumped out prior to cooling. The pick-up coils and SQUID detectors are shielded from external magnetic fields by a niobium superconducting shield and also, in the CCL instrument, by a magnetic shield of Mumetal. The sample access tube is 3.5 cm in diameter and is normally at a temperature above 0°C. Liquid helium is transferred from a storage Dewar to the magnetometer chamber by reducing the pressure in the latter. About 5 l are required to cool the chamber before liquid helium begins to collect and the chamber capacity is $\sim$24 l. Helium gas is exhausted from the top of the chamber and passes to the atmosphere via a flowmeter, by means of which the boil-off rate may be monitored. In the author's magnetometer it is normally 0.2–0.5 l/min of helium gas, corresponding to about 1.2 l of liquid helium per day.

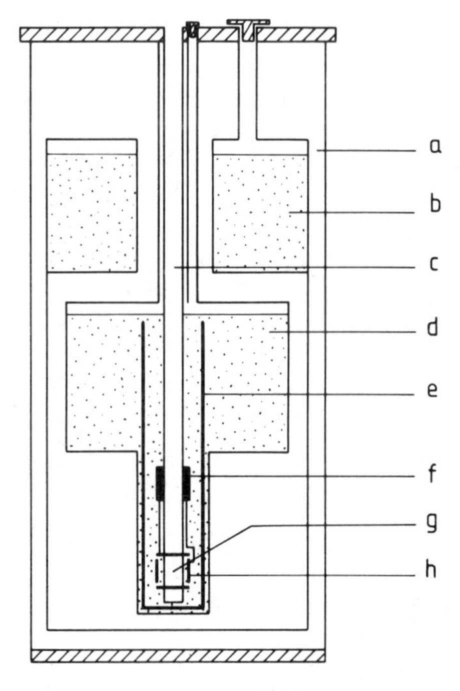

Fig. 9.24 CCL two-axis cryogenic magnetometer. a, vacuum space and super-insulation; b, liquid nitrogen; c, sample access tube; d, liquid helium; e, mumetal and superconducting shields; f, SQUID units; g, measurement position; h, pick-up coils.

The construction of the SCT magnetometer is simpler in one respect in that no liquid nitrogen is used, the helium chamber being enclosed in 'super-insulation'. There is some sacrifice in helium consumption, and the boil-off rate is around 2–3 l/day. If it is economically justified some users of cryogenic magnetometers collect the helium gas, compress it into cylinders and return it to the supplier. It may also be possible to use a continuous refrigeration system, available with both magnetometers, although these are expensive and have sometimes been found to be a source of noise in the SQUID output.

The standard SCT instrument has an access 6.3 cm in diameter, with a helium chamber capacity of 30 l. A single superconducting shield surrounds the SQUIDS and pick-up coils.

Both magnetometers can be provided with pick-up coils for measurement of NRM along one sample axis, or for simultaneous measurement along two or three mutually perpendicular axes. The common configuration is a pair of coils for each axis, one of each pair on either side of the access tube to detect the radial components of NRM and a pair (four in the CCL instrument) coaxial with the tube and symmetrically placed about the measuring position for the axial NRM component. Each coil pair is connected to its own SQUID detector, forming the field amplification system already described. The pick-up coils, leads and SQUID coil are, of course, all superconducting. The output signal from each SQUID is proportional to the flux change produced by the appropriate NRM component and is independent of the rate of change of flux, i.e. of the rate at which the sample is inserted into the pick-up coils. However, if the rate of change of flux is too fast there may be loss of lock in the feedback loop. In practice this maximum allowable 'slew rate' is not normally restrictive, and loss of lock is only likely to occur when strongly magnetized samples are rapidly inserted into the measuring space.

The sample is lowered to the measuring position and, if necessary, oriented to different angular positions there by a sample insertion system mounted on top of the magnetometer. This takes the general form of a carriage moving on vertical guides of sufficient height so that the sample holder, fixed at the lower end of a vertical stem attached to the carriage at its upper end, can be raised above the magnetometer top plate and lowered to the measuring position. Provision for rotating the sample to different (orthogonal) azimuths about the vertical axis will also usually be required in one- or two-axis instruments, and also a 'background' measuring position. The latter is a position where the sample magnetization does not couple significantly with the pick-up coils and where readings of the 'zero-level' SQUID outputs can be recorded immediately before and after measurement of the sample.

Under ideal conditions of very low external noise the noise level of the two cryogenic magnetometers is stated to be $(5\text{–}10) \times 10^{-12}$ A m^2 total moment, using an averaging time of ~ 1 s. For a standard cylindrical sample 2.5 cm in diameter and 2.2 cm high, this noise level is equivalent to a specific magnetization of $\sim (2\text{–}4) \times 10^{-10}$ A m^2 kg^{-1}. However, as in the case with all rock magnetometers, the useful limit of performance depends on the level of external sources of both random and systematic noise. Depending on the measurement procedure used, total measurement time is ~1–2 minutes per sample.

Probably the most common cause of random noise is fluctuations in the ambient laboratory magnetic field. Although the superconducting and Mumetal shields provide a high degree of protection against field variations, the exceptionally sensitive SQUID detector may still respond to residual variations. These can be further decreased by operating the magnetometer inside a single-or double-walled ferromagnetic shield fitting closely round it, or in a shielded room. Other potential sources of noise are mechanical vibration, r.f. interference, and temperature variations in the liquid helium. The latter may occur if there are pressure variations in the helium chamber, for example from wind blowing across the opening of the exhaust vent, or pressure variations in a gas collection system. A possible source of vibration is the sample touching the side of the access tube during insertion and withdrawal. It may be helpful to fix a guide to the stem above the holder to maintain the holder in a central position.

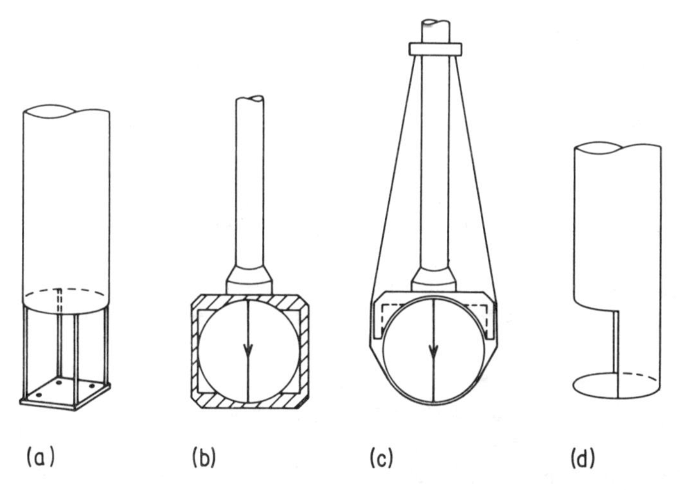

Fig. 9.25 Some examples of sample holders for the cryogenic magnetometer: (a) and (d) are mounted on and cut out of Mylar tube respectively; (b) is as narrow as possible consistent with sample location; in (c) the sample is held in place with an elastic band; (d) is only suitable where the sample is measured with its axis vertical.

If the ultimate sensitivity of the magnetometer is approached sources of systematic and random noise which are otherwise usually negligible must be eliminated. Among the latter are magnetization of the sample holder and the residual magnetic field at the measuring position. The holder should be of minimum size and mass, and among suitable materials are Perspex (lucite), wood, glass, quartz or nylon. Frequent washing or treatment in an ultrasonic bath is desirable and, if necessary, a.f. demagnetization of intrinsic ferromagnetic contamination. Some examples of holders for supporting a cylindrical sample with its axis vertical or horizontal are shown in Fig. 9.25.

Although in principle the effect of a residual field at the measuring position and resulting induced magnetization can be meaned out by measuring the samples in appropriate orientations, a fewer number of sample manipulations and saving of time is achieved if any induced magnetization is negligibly small. As an example, if a magnetization of 1×10^{-8} A m^2 kg^{-1} is required to be measured, with reasonable accuracy, the induced component should be $< 1 \times 10^{-9}$ A m^2. A weakly magnetized sediment may have a mass susceptibility of $\sim 1 \times 10^{-8}$ m^3 kg^{-1}, resulting in the above induced magnetization in a field of ~ 100 nT. This field is probably a useful maximum value to aim for the measuring position. It should not be forgotten that many rock-forming minerals have a significant paramagnetic or diamagnetic susceptibility of $\sim 10^{-8}$ m^3 kg^{-1} or more (Table 2.1) and that the holder material may also contribute in this context. However, the much smaller mass of the latter relative to the sample ensures a smaller contribution to induced magnetization from this source.

The method of achieving a low residual field in the measuring space depends on the shielding provided. The SCT magnetometer has a superconducting shield only, and thus traps the field present when cooled through the critical temperature. A suitably low field can be obtained in a Helmholtz coil system. In the currently available CCL magnetometer there is a Mumetal shield nested between two superconducting shields. This sytem has some disadvantages, in that it may not be possible to completely demagnetize the Mumetal shield. Any remanent magnetization remaining varies with temperature and results in a substantial change in the residual field during pre-cooling with liquid nitrogen, necessitating further demagnetization. The axial residual field is usually the stronger, and demagnetization can be carried out by means of a coil placed around the barrel of the instrument. The author has also had some success with magnetizing the shield in the appropriate direction to adjust the remanence during cooling, using either a solenoid inserted in the access tube at the measuring position or the external coil.

Measuring procedure with the cryogenic magnetometer depends on the number of axes along which NRM can be simultaneously measured, the degree to which the pick-up coils can accommodate inhomogeneity of NRM, and whether it is considered desirable to mean out induced magnetization. In both magnetometers the response of the pick-up coils is uniform to about 2% over the volume of a standard sample, but experience suggest that it is advantageous to use a procedure whereby at least four readings each of the X, Y and Z components are obtained, particularly when within a factor of 10 of the noise level of the instrument.

There are advantages in operating the magnetometer in a low magnetic field environment, using a Helmholtz or other coil system or a small shielded room. This eliminates the enhanced magnetic field at the mouth of the internal shielding, and also provides a low ambient field at the sample during loading and re-orientation.

9.4.3 Other cryogenic magnetometers

A SQUID magnetometer designed for measurement of small archaeological samples (typical dimension ~0.3 cm) is described by Walton (1977), a diagram of which is shown in Fig. 9.26. The measuring position is in a re-entrant cavity 1.0 cm in diameter on the underside of the helium chamber and the sample is introduced from below. The pick-up coils are placed close to the walls of the cavity and consist of a pair of axial and radial coils connected in series to form a continuous circuit. When the sample is inserted into the cavity the output signal is proportional to the sum of the vertical component of NRM and the component of the horizontal NRM perpendicular to the horizontal pick-up coils. If the sample is then rotated about the vertical axis the phase and magnitude of the horizontal component can be determined from the angular positions of the sample for minimum and maximum output. In a later modification the sample is rotated continuously at ~3 Hz and the resulting signal computer-analysed.

The magnetometer is mainly used for palaeointensity determinations by the Thellier method. A small furnace was originally placed in the cavity for in-situ heating of the sample, using alternating current at 33 kHz to avoid interference with the SQUID. A 150-turn coil wound on the outside of the cryostat supplied the (vertical) applied field and the whole instrument is operated inside a double-walled Mumetal shield, maintaining the internal field at ~1 μT. However, better results were obtained with the furnace placed just outside and below the cavity, along with vertical and horizontal applied field coils. The capacity of the liquid helium chamber is 7 l, which enables the magnetometer to be used for about two weeks.

The noise level in the magnetometer is equivalent to a total moment of $\sim 1.0 \times 10^{-4}\,\mathrm{A\,m^2}$, which in a 3 mm sample is about $4 \times 10^{-8}\,\mathrm{A\,m^2\,kg^{-1}}$.

Equipment designed by J. Shaw and J. Rogers at University College, Cardiff, Wales consist of an integrated magnetometer and magnetic cleaning system (thermal and alternating field). At the centre of the system is a vertical silica access tube, at the lower end of which is the pick-up coil and measuring position. The pick-up coil is a single coil with its axis inclined at 45° to the vertical. The sample is continuously rotated at 33.3 Hz about a vertical axis at the coil centre and the SQUID output consists of an alternating signal due to the horizontal NRM component superimposed on a steady signal due to the vertical NRM component (Fig. 9.27). The amplitude and phase of the alternating signal determines the components of the magnetization perpendicular to the spin axis.

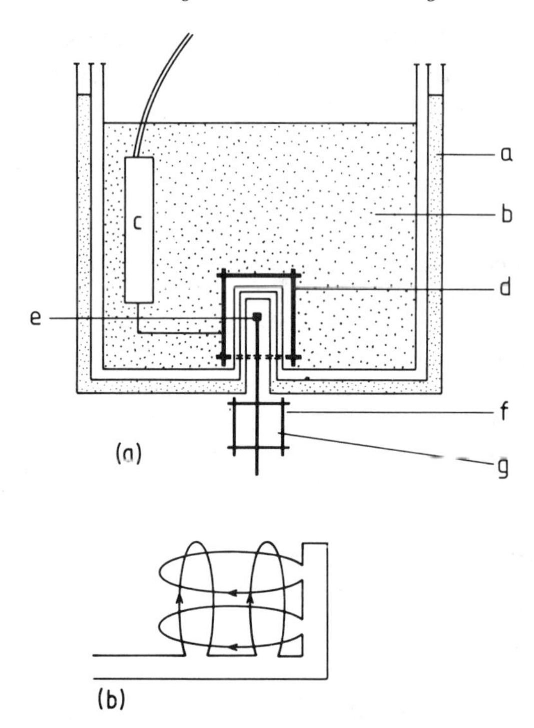

Fig. 9.26 (a) Small-sample cryogenic magnetometer (Walton, 1977). a, liquid nitrogen; b, liquid helium; c, SQUID; d, pick-up coils; e, sample; f, applied field coils; g, furnace; (b) pick-up coil system.

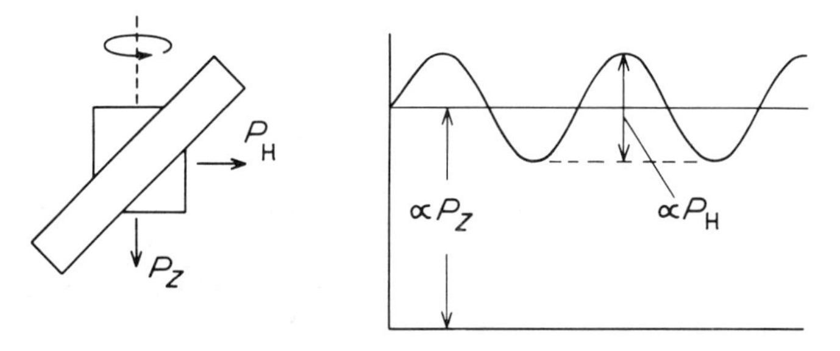

Fig. 9.27 Single 45° sense coil and output signal due to P_Z and P_H

Surrounding the upper end of the central silica tube are a demagnetizing coil and furnace and the whole system is enclosed in a Mumetal shield. Thus NRM measurement and (single-axis) alternating field or thermal demagnetization can be carried out on a sample without removal from the sample holder and with the sample experiencing only a weak magnetic field during and between measurements, thereby reducing the effects of viscous magnetization.

The system was designed to measure cylindrical samples 1.0 cm in diameter and 1.0 cm high. The noise level is equivalent to a total moment of $\sim 1.0 \times 10^{-11}$ A m^2, i.e. $\sim$3–4 $\times 10^{-9}$ A m^2 kg^{-1} specific magnetization.

9.4.4 Reliability of weak NRM measurements

With the development of the cryogenic magnetometer and the measurement of very weakly magnetic material (e.g. limestones) the NRM of which sometimes approaches the instrumental noise level, some interest has been generated in assessing the equality of the basic NRM measurement in these cases and defining acceptance criteria. Heller (1977) and Channell (1977) combine the six magnetization components (two each of X, Y, Z) they obtain from two measurements with a three-axis cryogenic magnetometer to obtain eight estimates of the NRM vector. The circular standard deviation ψ about the mean of these vectors is then computed

$$\psi = 81k^{-1/2} = 81[(N-R)(N-1)]^{1/2} \tag{9.50}$$

where $N = 8$ is the number of vectors and R their vector sum (Section 12.3). Using Watson's (1956) test of randomness the value of ψ is calculated for a 95% probability that the eight vectors are randomly distributed, for which $\psi = 57.4°$. Therefore if for a specimen $\psi > 57.4°$ that specimen is excluded from the mean for the site and the same criterion is used at each site, i.e. the site is rejected if the distribution of the N vector directions is random at the 95% probability level.

This procedure, as applied to specimen NRM determinations, can be criticized (Harrison, 1980) on the grounds that the eight values of the NRM vector are not independent, nor the twelve values used by Lowrie and Alvarez (1977) in a similar procedure, derived from six determinations of orthogonal magnetization component pairs obtained with the standard Digico measurement. Channell's measurement gives only two independent measurements of NRM and the Digico magnetometer four. However, Lowrie, Channell and Heller (1980) point out that their method, though not based on rigorous statistical principles, does provide a criterion which appears in practice to be satisfactory, and they also observe that their measurement procedure would strictly involve computing a statistically based parameter ψ from only two observations. Lowrie *et al.* also show that a closely linear relationship exists between the simple parameter, S_m/M, derived by Harrison (1980) as a measure of internal consistency of an NRM measurement, and ψ, at least up to $\psi \approx 50°$. $S_m = [(S_x^2 + S_y^2 + S_z^2)/3]^{1/2}$, where S_x^2, S_y^2, S_z^2 are the variances of the means of the magnetization components X, Y, Z, and $M = (\bar{X}^2 + \bar{Y}^2 + \bar{Z}^2)^{1/2}$ the mean magnetization.

No rejection criterion can allow for all the sources of noise on the magnetometer output signal. Apart from truly random noise there may be systematic contributions from holder magnetization, inhomogeneity of NRM and short-period viscous magnetization. Inhomogeneity can be the source of large differences among values of X, Y, Z obtained at different specimen orientations in the magnetometer, but the mean vector can be reasonably accurate and yet be rejected. On the other hand, closely consistent X, Y and Z readings may be due to an unsuspected dominant viscous magnetization which will be accepted.

The satisfactory measurement of very weakly magnetized rocks and the derivation of a meaningful site NRM direction is probably assisted by the use of some sort of rejection criterion, but it also requires constant vigilance to minimize the effects of holder contamination, VRM and sources of random noise. Repeat measurements of the same sample, storage of samples in field-free space and frequent measurement and cleaning of the sample holder (even if the measuring procedure is designed to eliminate holder remanence) all play their part in ensuring reliable results. It is also important to avoid magnetic contamination of the specimens during collection, cutting to shape and handling, and during storage through accumulation of ferromagnetic dust particles.

9.5 Other magnetometers

9.5.1 Archaeomagnetic instruments

The main requirements in a magnetometer for the measurement of NRM of different types of archaeological material are that it can accommodate large and irregularly shaped samples and that the rotation rate is slow enough for the measurement of fragile samples. To avoid irregularity of shape causing significant errors in the measured NRM it is usually necessary to sacrifice some sensitivity by increasing the sample–sensor distance, although this loss may be partially offset by the increased sample size and therefore increased total magnetic moment of the sample.

Although some measurements have been carried out using astatic or parastatic magnetometers, most workers in archaeomagnetism have favoured spinner instruments.

In his pioneer work in archaeomagnetism Thellier (1967) perfected a large sample spinner magnetometer with a novel method of measuring the NRM. A Helmholtz coil pair of radius 19.8 cm formed the pick-up coil, surrounded by compensating coils of twice the size. The sample (up to $\sim 1000\ cm^3$) is placed inside two orthogonal coil pairs fixed to the rotating platform which rotates at 5 Hz about a vertical axis. The axes of the coils are horizontal and the sample is aligned with its reference axis parallel to one of the coil axes and accurately known currents can be passed through the coils by means of slip rings on the rotation shaft. The output of the pick-up coils is amplified and (in the original instrument) passed to a vibration galvanometer. The magnetic moment in the plane of the sample perpendicular to the roation axis is determined by adjusting currents through each coil pair surrounding the sample until the pick-up coil output is zero. Each of the horizontal components of NRM is then proportional and opposite in direction to the effective moments of the corresponding coil pair, which in turn are proportional to the currents the coils carry. The absolute intensity of NRM can be determined by prior calibration. A specific magnetization of $\sim 5 \times 10^{-7}$ A m^2 kg^{-1} in a large sample was measurable with this instrument, and the large pick-up coil ensured that errors arising from irregularly shaped and inhomogeneous samples were acceptably small.

A somewhat similar null-detection method was used by Aitken, Harold, Weaver and Young (1967). The alternating voltage produced in the pickup coil by the rotating sample (5 Hz) was nulled by passing an alternating current of appropriate magnitude and phase through another coil placed near to and coaxial with the pick-up coil. The magnitude of the nulling current is proportional to the magnetization component perpendicular to the rotation axis and the phase, adjusted and measured with a Magslip generator, determines the direction of the component.

The pick-up coil is of the Rubens design (Section 5.1.3) of side 60 cm, providing a central cubic volume of side 30 cm within which equal elements of the sample magnetization induce equal output voltages. A compensating coil of side 90 cm is placed coaxially and concentrically outside the pick-up coil. For small samples a pick-up coil of side 30 cm is placed inside the 60 cm set, the latter then being used as the compensating coils. The improved flux linkage results in an increase in the signal/noise ratio by a factor of three.

An option available with the Digico magnetometer consists of a large access system capable of taking samples lying within a cylinder 5.0 cm in diameter and 5.0 cm long. The rotation frequency is 5.3 Hz, and the stated noise level, as defined for the Digico instrument in Section 9.3.3, is 2×10^{-8} A m^2 kg^{-1} in a 5.0 cm sample.

9.5.2 Long-core measurements

In studies of palaeosecular variation and geomagnetic reversals it has always been an attractive idea to measure direction and intensity of NRM continuously down a long core of rock or sediment. The potential advantages are an essentially continuous record of NRM down the core and a considerable saving in time of measurement compared with that involved in measuring separate samples taken from the core. However, there may be some sacrifice in resolution since the sensors will normally detect the field due to a longer section of core than could be sampled with a single sample. The recent interest in secular variation studies using lake bottom sediments, sampled in the form of continuous cores up to 6 m long, gave an added impetus to continuous measurement and two magnetometers have been adapted for this purpose.

Before describing techniques it is necessary to enquire which components of

NRM can be measured by different types of instrument in a long core of material. With static detecting systems (e.g. fluxgate or SQUID sensors) both axial and radial components (perpendicular to the core axis) can be measured in principle by using sensors close to the core and aligned parallel and perpendicular to the core axis. Declination information is easily obtained in this way, but data on inclination require continuous integration of the axial field down the length of the core. The sensitivity available will be limited, and it is necessary at any position in the core to take several readings at different azimuths around the core to measure the total vector and obtain a reasonable sampling of the core magnetization at that position.

With dynamic (spinner) techniques the only practicable rotation axis is the long axis of the core and only declination and horizontal intensity data can be obtained.

The output of a magnetometer during continuous core measurement can be considered as being proportional to the running average of the appropriate magnetization vector over a certain length of the core. The system response, i.e. the response of the detector to elements of magnetization at different distances from it, acts as the weighting function. The resolution of the system, the length of the core contributing the bulk of the observed signal, depends on the size of the detector and its distance from the core centre and under optimum conditions is usually of the order of the core diameter.

The Digico magnetometer has been adapted for the measurement of declination down long cores, i.e. the component of NRM perpendicular to the core axis (Molyneux and Thompson, 1973). The sensing head is essentially the same as for sample measurement and consists of a cylindrical magnetic shield, open at both ends, containing the ring fluxgate sensor. The head is mounted on a carriage which can be traversed horizontally or vertically and the core, contained in a plastic tube and also mounted with its axis horizontal or vertical, passes through the centre of the sensor head. The core rotates at about 3 rev s^{-1}, and readings of intensity and declination components are taken at chosen intervals along the core. The equipment accommodates cores up to 6.0 cm in diameter and 6 m long (horizontal) and has a stated noise level for 2^7 spins equivalent to $\sim 2 \times 10^{-8}$ A m^2 kg^{-1} in a core of maximum diameter. The effective sampling length, or 'resolution' of the sensor is approximately 6 cm in a 6 cm diameter core. Figure 9.28 shows an example of the variation of the radial NRM along a core as measured with the Digico instrument and by measurement of extracted samples. Heye and Meyer (1972) describe the use of a similar magnetometer for measuring cores on board ship.

The second instrument that has been adapted for long-core measurements is the SCT cryogenic magnetometer. Instead of a vertical measuring chamber closed at one end the magnetometer has horizontal access and it continues through the instrument, allowing the core to be passed through either in a series of steps for spot readings or traversed at a slow constant rate for a continuous record. The latter procedure is suitable for a three-axis magneto-

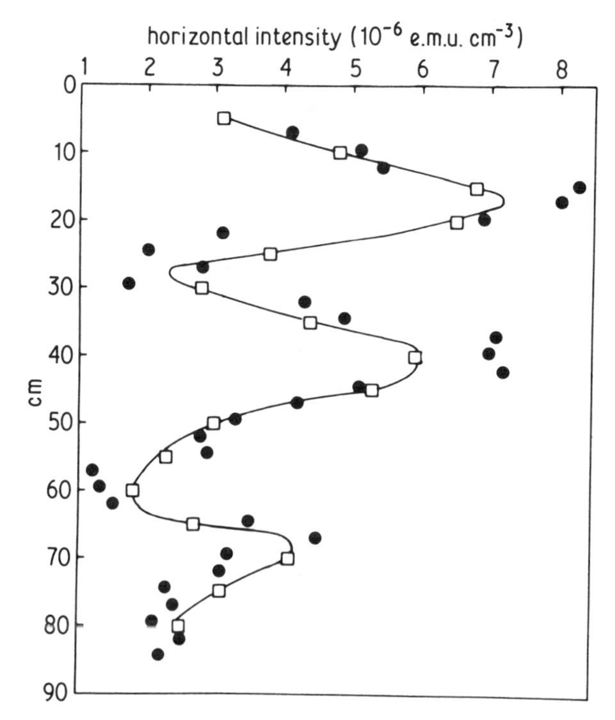

Fig. 9.28 Comparison of measured horizontal NRM intensity changes in a 1 m sediment core. Circles show measurements on separate samples, squares are whole core measurements. (Reproduced with permission from Molyneux *et al.*, 1972)

meter and the former for a two-axis instrument (axial and one radial sensor) with axial rotation of the core through 90° at each measuring position to determine declination.

The record of NRM obtained from a long-core measurement is a smoothed version of the record which would be obtained if the core was cut into many thin slices and each slice measured individually. Dodson, Fuller and Pilant (1974) describe a 'deconvolution' procedure for deriving spot readings along the core from the smoothed data. The instrument output can be regarded as a convolution of the fields due to a series of slices with the wave-form representing the instrumental response to one of the slices. By deconvolution of the output with the small slice response an approximation to point readings of NRM along the core can be obtained. Thus under ideal conditions it is possible to combine the rapidity and convenience of long-core measurements with the high resolution obtainable from the separate measurement of closely spaced, small samples.

Although deconvolution can, in principle, be applied to the complete NRM vector, i.e. declination, inclination and intensity down a core, in practice it would be a cumbersome procedure and Dodson *et al.* illustrate the technique using only the axial magnetization component. After slicing a 1 m core (of the Tatoosh intrusion) into 2 cm lengths the samples were measured on a spinner magnetometer and the core then 'reconstituted' by placing the samples in their correct relative position and orientation, securing them with adhesive tape and sealing rigidly into a quartz tube. The response of the magnetometer to the axial magnetization of a slice as it is traversed through the detector was determined with one of the 2 cm samples. The results obtained with the technique are shown in Fig. 9.29, and it can be seen that the deconvolved data is a good replica of that obtained by separate sample measurements. The procedure as described was not a full deconvolution, i.e. an infinitely close set of readings was not obtained but rather a smoothed representation of the test

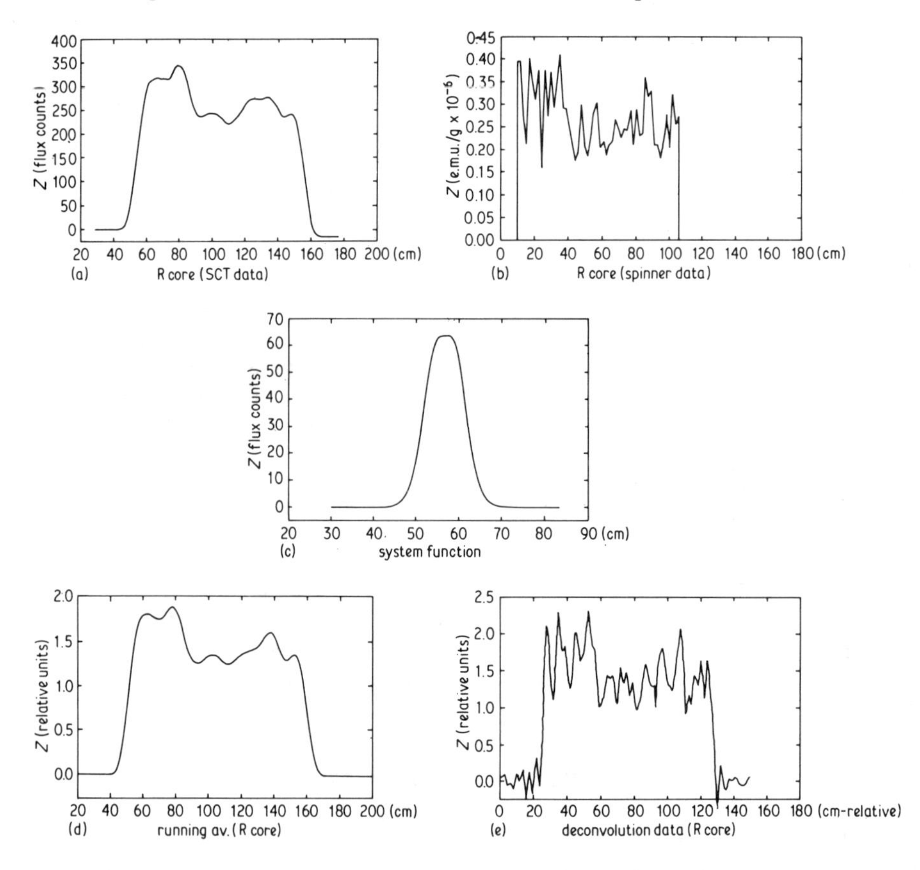

Fig. 9.29 Deconvolution technique for long cores: (a) axial NRM variation down the core measured by magnetometer; (b) individual 2-cm specimen measurements; (c) instrument response for 2-cm specimens; (d) synthetic long-core result, obtained by convolving the specimen response with the individual specimen measurements; (e) deconvolved data, which is a good approximation to the result shown in (b). (From Dodson *et al.*, 1974)

samples. The resolution chosen was three to five times better than in the instrumental response curve obtained from the core. Dodson *et al.* (1977) describes results of applying the technique to sediment cores from Lake Michigan, USA.

Barton (1978) and Barton *et al.* (1980) have considered the measurement of inclination and intensity of NRM in cores with the normal SCT cryogenic magnetometer, in which the measuring chamber is closed 7.5 cm below the sample measuring position. Because of the different shape of the response curves of the axial (vertical field) and radial (horizontal field) SQUID detectors as a sample is traversed through the optimum measuring position (Section 9.4), the measured inclination is steepened or flattened according to whether a sample is more or less than 4 cm from the measuring position. The error to be expected in a core which spans the normal measuring region was estimated as follows. A 2.2 cm cylindrical rock sample with an accurately known inclination of $-53.1°$ was measured at 1.0 cm intervals over a length centred on the normal measuring position of the magnetometer and the error in the measured inclination determined for each measuring position. From the response curves of the sensors the contribution to the inclination error of the core slice at each measuring position was estimated by assuming a linear dependence of the error on the normalized response for each position. The inclination error to be expected from various lengths of core, each of which has some part spanning the sense region, was then calculated from the sum of the separate error contributions over the core length divided by the sum of the normalized responses. The result for all such cores was an expected inclination flattening of $\sim 4°$. The effective sampling length of a core is ~ 5 cm on either side of the normal measuring position. A rough calculation suggests that the error would reach a maximum of $\sim 5°$ for an inclination of 45° and decrease towards zero for inclinations approaching 0° or 90°.

It is also necessary to apply a correction to the measured magnetization of the core to obtain the true (volume) magnetization (Barton, 1978). The principle of the method is shown in Fig. 9.30. If a slice of the core has its ends at a distance x_1 and x_2 from the centre of the sense region and S is the normalized response of a dipole at a distance x from the centre, then a small length $\mathrm{d}x$ of the slice gives a signal $S\,\mathrm{d}x$. The signal from the whole core is given by integration of $S\,\mathrm{d}x$ over its length, i.e. the shaded area A under the normalized response curve in Fig. 9.30, whereas if the response was uniform over its length the signal would be $1.0(x_2 - x_1)$. Therefore we have measured magnetization/true magnetization $= A/(x_2 - x_1)$. The response can be derived from the cumulative response curve of the magnetometer. It is usually sufficient to use a mean response curve derived from the axial and transverse curves.

The above procedure assumes uniformly magnetized cores and this also applies to the derivation of the inclination error.

Because of the different characteristics of the axial and radial detector coils in cryogenic magnetometers there will usually be a difference in the resolution

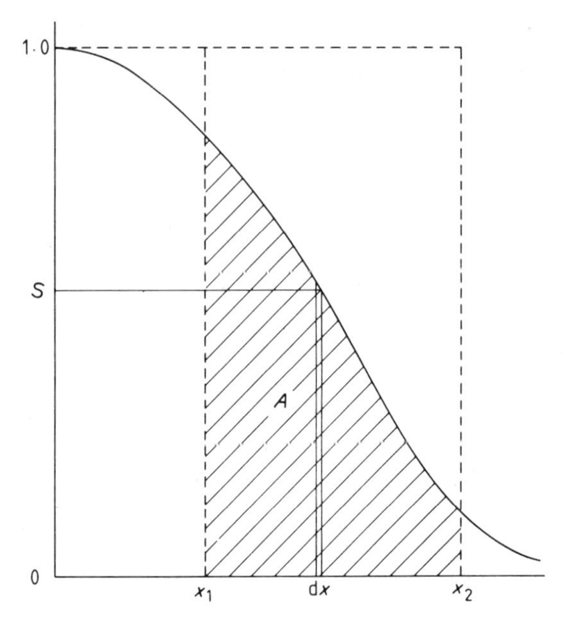

Fig. 9.30 Principle of correction of measured intensity of magnetization in long-core measurement (Barton, 1978).

obtainable of the axial and radial components of magnetization. Dodson *et al.* (1977) derive resolutions of 3 cm and 6 cm for the axial and radial magnetizations respectively with the SCT magnetometer, using their deconvolution procedure.

9.5.3 Ballistic magnetometers

The basic principle of these magnetometers is the measurement of charge transfer produced by the change in magnetic flux linked with a coil system when a magnetized sample is inserted into, withdrawn from or rotated through 180° within the system. These instruments were used in early palaeomagnetic investigations (Chevallier, 1925; Grenet, 1933, Thellier, 1967; Nagata, 1961) and, using a ballistic galvanometer connected directly to the sense coil, were capable of measuring the NRM of many igneous rocks. The difficulty of signal amplification restricted their development for measuring more weakly magnetized material and they fell out of favour with the development of other magnetometers.

Consider a coil of N turns of average radius r, at the centre of which is a dipole moment P directed parallel to the coil axis. The flux ϕ linked with the coil is

$$\phi = \frac{\mu_0 PN}{2r} \tag{9.51}$$

and this is also the change of flux through the coil when P is removed to a point far from the coil. If a galvanometer is connected to the coil a quantity of charge q passes through the galvanometer when P is removed, given by

$$q = \frac{\phi}{R+G} \tag{9.52}$$

where R and G are the coil and galvanometer resistance respectively. (It is assumed here that the axial width and radial depth of the coil are small.) Therefore

$$q = \frac{\mu_0 PN}{2r(R+G)} \tag{9.53}$$

or

$$P = \frac{2rq(R+G)}{\mu_0 N} \tag{9.54}$$

Although maximum charge transfer occurs when $R = G$, this results in heavily over-damped motion of the galvanometer coil and it is usual to arrange for $(R+G)$ to be equal to the critical damping resistance R_c of the galvanometer. Putting some representative values in Equation 9.54, $r = 3\,\text{cm}$, $R_c = 1000\,\Omega$, $N = 10\,000$ and $Q = 10^{-9}\,\text{C}$ per scale division, then $P \approx 5 \times 10^{-6}\,\text{A m}^2$ per scale division deflection of the galvanometer. For a standard 2.5 cm sample this is equivalent to a specific magnetization of $\sim 2 \times 10^{-4}\,\text{A m}^2\,\text{kg}^{-1}$. Inspection of Equation 9.54 shows that sensitivity can be improved by using larger samples in a coil of larger radius, and an additional factor of two could be gained by rotating the sample through 180° within the coil rather than removing it. The limiting sensitivity of direct-reading ballistic magnetometers is probably about $10^{-5}\,\text{A m}^2\,\text{kg}^{-1}$ using a galvanometer of period 20–40 s.

The main design feature of the sense coil is to maximize the number of turns while maintaining the coil resistance at a value which contributes the appropriate amount to the total (critical damping) resistance of the galvanometer circuit. Noise arising from variations in the ambient (uniform) field can be reduced by using two identical coils mounted co-axially and connected in series opposition: if the sample is moved from inside one to inside the other the signal is effectively doubled (the flux change is $+\phi$ to $-\phi$) compared with that obtained by removing the sample from, or moving it into, a single coil. In a system developed by E. Thellier and described by Daly (1967) each coil has an additional compensating coil wound coaxially outside it, connected in series opposition: this four-coil system effectively reduces noise caused by variations in the non-uniform ambient field (e.g. $\mathrm{d}B/\mathrm{d}x$).

Ballistic magnetometers using directly connected galvanometers are described by Thellier (1967) and Nagata (1961). Magnetometers in which the pick-up coil is moved relative to a stationary sample are described by Nagata and Kinoshita (1965) (for measuring the NRM of rocks under pressure) and Vlasov *et al.* (1964) for measuring the magnetization of laboratory-deposited sediments.

A semi-automatic recording device attached to a ballistic magnetometer is described by Daly (1967). The deflection of the sensing galvanometer is recorded by a light beam falling on a split-photocell, the output of which supplies a recording light-spot follower. The deflection is also recorded in the form of a number of pulses (1 pulse = 0.4 mm) obtained from a photo-diode which receives light through a toothed wheel attached to the spot-follower drive motor. A single reading consists of a 'double deflection', the sample being transferred from one coil into another and then back to the first when the galvanometer has returned to zero. To reduce the effects of noise up to 20 double deflections are automatically obtained and the pulses stored and averaged.

The sensing coil system is of the two-coil type, compensated for changes in the uniform field and its first derivative as previously described. Good sensitivity is achieved by using large samples (5 cm cubes) and 16 000 turns on the primary coils. For this sample size (~300 g), a specific magnetization of $\sim 2 \times 10^{-8}$ A m^2 kg^{-1} is the lower limit of measurement.

The instrument is also provided with coils for the measurement of isotropic and anisotropic susceptibility (Section 2.8).

A versatile ballistic magnetometer designed mainly for a variety of rock-magnetic measurements is described by West and Dunlop (1971). The two identical sensing coils, connected in series opposition, are of internal diameter 5.6 cm (accommodating a tubular furnace, water jacket and applied field coils) and 6.0 cm long with a 16 cm centre-to-centre separation. Each coil has ~3000 turns of 30 SWG (28 AWG) wire (0.32 mm).

The detection circuit employs a feedback galvanometer amplifier. The signal from the sense coil is passed via a step attenuator to one of the coils of a galvanometer which has two electrically independent coils on the same suspension. The galvanometer deflection is monitored by means of a 5 m optical lever received by a split photocell system, the output of which is fed back negatively to the second coil of the galvanometer. The period of the galvanometer is thereby substantially reduced and electromagnetic damping becomes negligible. Thus the usual requirement that the critical damping resistance controls the total galvanometer circuit resistance does not now apply, and the optimum power transfer condition of equal resistance of source and detector can be achieved. The amplified output of the photocell detector is recorded by a 10 s ballistic galvanometer with a 3.5 m light path to the reading scale.

The standard sample size used in the magnetometer is a cylinder 10.0 cm long and 1.45 cm in diameter. For such a sample the minimum detectable magnetization is about 7×10^{-6} A m^2 kg^{-1}, although this sensitivity cannot be utilized in a city-centre laboratory because of imperfect compensation of varying ambient fields by the sensing coils.

As the authors point out, the amplifier output could be read with an electronic integrator rather than a ballistic galvanometer. With modern

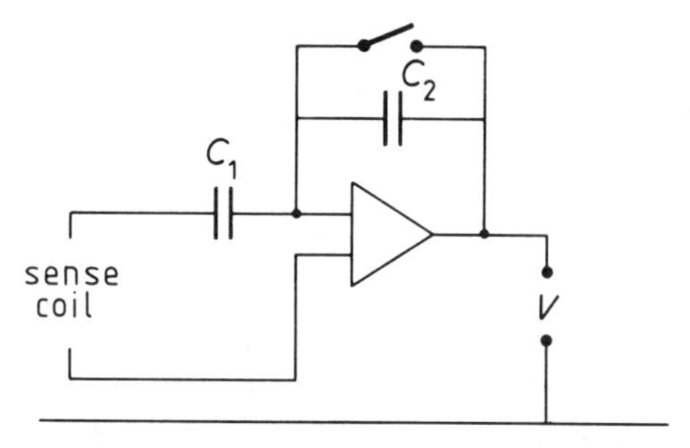

Fig. 9.31 Basic circuit of electronic integrator for ballistic magnetometer.

electronic techniques it would now be possible to dispense with both galvanometers and use an operational amplifier circuit as detector. Figure 9.31 shows the basic circuit. The effect of the amplifier, with a gain of 10^6–10^7, is to allow the charge transfer Δq resulting from the flux change through the coil to rapidly charge up the capacity to a voltage given by $V = \Delta q/C_2$. For satisfactory working, C_2 should not be less than $\sim 0.1\,\mu$F, and V, which will decay only slowly, can be read on a digital voltmeter without further amplification except where the input signal is exceptionally weak: for example with $\Delta q \approx 10^{-9}$ and $C_2 = 0.1\,\mu$F, $V = 10$ mV.

9.5.4 Experimental magnetometers

During the period when palaeomagnetic investigations were undergoing a rapid expansion several alternative techniques for the measurement of NRM were explored. For various reasons the instruments were not satisfactory for routine measurements, but an account of some of them is included here for the sake of completeness and to illustrate some of the ideas that seemed worth investigating.

An early attempt to increase the sensitivity of the astatic magnetometer was that of Kumagai and Kawai (1953), who applied the principle of resonance to the instrument. The sample was placed level with the lower magnet of the system (position (b), Fig. 9.9) and rotated about a vertical axis with the same period as that of the free torsional oscillations of the magnet system. The periodic torque exerted on the lower magnet by the field due to the horizontal component of the sample magnetic moment results in large resonant oscillations of the magnet system, dependent on the degree of damping of the system and the magnitude of the horizontal moment. With no damping and identical drive and magnet system periods the phase of the magnet system oscillations leads that of the rotating horizontal NRM component by 90°, thus enabling the declination of the latter to be determined. In practice some damping is necessary to avoid instability and this modifies the phase relationship. Using a 5 cm cubic sample, and NRM of $\sim 5 \times 10^{-7}$ A m^2 kg^{-1} was stated to be measurable. The measurement time is not stated, but is probably long because of the time required for the resonant oscillations to build up.

The resonance technique was examined further by Farrell (1967) and Graham (1967b). The resonant deflection of the driven magnet system is Q times its deflection for the static (zero frequency) case, where $Q = I\omega/\lambda$. I is the moment of inertia of the magnet system about its axis, ω its natural frequency ($= 2\pi/T$, where T is the natural period) and λ is the damping coefficient in the equation of motion. If the drive and magnet system frequencies are not identical the phase lag between the rotating NRM vector and the magnet system response differs from 90° by an amount δ dependent on the ratio of the frequencies and Q: if Q is low, δ is small for a given frequency ratio and increases with Q. Farrell (1967) used an air-damped system with $Q = 220$ at $T = 1$ s, and the author's condition for $\delta < 1°$ is that the frequencies must be matched to better than 1 part in 10^4.

The sample rotation frequency is controlled by a crystal-controlled oscillator, stable to 1 part in 10^5/°C. The natural frequency of the magnet system is adjusted initially by varying the length of the suspension fibre and then trimmed to the exact value by means of a small Helmholtz coil pair around the upper magnet. The main course of variations in the frequency is temperature changes (affecting the torsional constant of the suspension) and stray fields acting on the magnet system.

The oscillations of the magnet system are recorded with a reflected light beam, chopped at 330 Hz and falling on a split photocell bridge. The time between pulses corresponding to successive zeros in the output signals from the bridge is half the period of the magnet system and of the sample rotation shaft. Measurement of the time between a zero position and a reference pulse from the drive shaft establishes the phase of the output signal which is a measure of the angle between the component of NRM, perpendicular to the rotation axis and the reference mark on the shaft.

The sensitivity of the magnetometer as constructed was limited by vibration of the magnet system due to its dynamic unbalance. This noise level was equivalent to a magnetization of 5×10^{-7} A m^2 kg^{-1} in a 5 cm^3 sample, placed 2.5 cm from the sensing magnet. Overall measurement time for a rock sample was $\sim$25 min.

The resonance instrument described by Graham (1967b) operated at a frequency of 4 Hz with a Q of about 150. The phase of the sinusoidal output of a split photocell detector is compared with the phase of a reference signal generated photoelectrically on the drive shaft: this phase relationship defines the direction of the NRM vector perpendicular to the sample rotation axis. The d.c. output of a phase-sensitive detector is amplified and used to drive a servo motor which controls the angular position of the reference generator photocell such that the p.s.d. output is zero. The phase of the measured NRM component can then be read on a previously calibrated scale. The amplitude of the signal is obtained from a second phase detector gated at 90° from the first p.s.d.

Although the instrument performed satisfactorily for strong magnetized samples it was of little use when the signal was comparable to those produced by random noise. During the build-up of a resonant deflection a single pulse of noise could unlock the resonance, causing considerable extension of sample measurement time because of the further time required to build up a new resonant amplitude.

Graham's instrument is of some historical interest as it was the first magnetometer to be adapted for the continuous measurement of declination of NRM along the length of a sample in the form of a long core up to 3 m long. The core, rotating at 4 Hz, was supported horizontally below the magnet system and automatically traversed beneath it. Declination and signal amplitude could be continuously recorded on a chart recorder.

de Sa *et al.* (1974) attempted to use a partial resonant technique applied to an astatic magnetometer with feedback. The feedback reduces the time constant of torsional oscillations of the magnet system to about 10 ms, and the sample is rotated at about 10 Hz near the low magnet. The alternating signal obtained by means of lamp, mirror and split photocell is processed by phase-sensitive detection.

There are two basic difficulties with the resonance principle. If large resonant amplitudes are to be obtained this implies high Q-values. However, the higher the value of Q the more sensitive is the phase difference between the magnetic system and forcing oscillations (from which the direction of the NRM vector is determined) to changes in the forcing frequency and the natural frequency of the magnet system. The second difficulty is associated with the time required to establish the resonant oscillations. The complete solution of the equation for forced oscillations consists of two terms, one of which describes the steady-state oscillations of the system and the other a transient oscillation corresponding to the free period of the system. The latter term has an exponential decay factor in it, of the form $\exp(-\lambda t)$, where λ is the damping coefficient, and this factor determines how rapidly the transient terms decay to zero. The smaller the damping the longer is this time and thus also the time constant of the instrument for a sample measurement.

Graham (unpublished work) has also described a novel spinner magnetometer in which the pick-up coil rotated about a stationary sample. Its purpose was the measurement of NRM of fragile samples and of wet sediments and the changes in the magnetization of the latter as they dry out.

The sample rests on a small horizontal platform which can be rotated to different azimuths about a vertical axis. The platform is supported by an arm inclined at 45° and the pick-up coil is mounted on a drive shaft parallel to this arm. If the sample magnetization is vertical the pick-up coil output is constant and if the magnetization is at 45° the output varies sinusoidally from zero to some maximum value as the sample is rotated through 360°. If the NRM is perpendicular to the sample rotation axis there are two maxima and two minima in the coil output for 360° rotation. The direction and intensity of magnetization are derived from the outputs of two phase-sensitive detectors controlled by two reference signal generators on the drive shaft, 90° out of phase, each being fed by the signal from the coil.

The pick-up coil, rotating at 30 Hz, fitted closely round the sample with a compensating coil wound inside it, and a preamplifier inside the drive shaft amplified the signal before passing it to the primary of an air-cored transformer the secondary of which was inside it and stationary: this arrangement avoids the complications and noise of slip rings. Graham was unable to complete and test the instrument but it seems unlikely that good sensitivity would be obtainable. There are constraints on the design and size of the rotating pick-up coil and severe problems in compensating the synchronous noise signal arising from the ambient field. A simple calculation shows that, for the pick-up coil rotating in the geomagnetic field a compensation factor on $\sim 10^6$ would be required to measure satisfactorily (sample signal/'noise' ≈ 10) a rock sample of specific magnetization $\sim 10^{-5}$ A m^2 kg^{-1}. This could only be achieved by a combination of the compensating coil and spinning the coil in a very low ambient field, as indeed Graham proposed.

The possibility of using a 'pulse' magnetometer, as distinct from the spinner instrument in which there is a continuous alternating output, was first investigated by Dianov-Klokov (1960). Whereas in ballistic instruments the sample is withdrawn from or moved into a coil, resulting in a net charge transfer through the detecting circuit, in the pulsed system the sample effectively moves through the pick-up coil, resulting in equal voltage pulses of opposite sign with no net charge transfer.

Dianov-Klokov proposed an instrument in which the sample is supported at the end of a horizontal arm rotating about a vertical axis. Once per revolution the sample passes through a pick-up coil consisting of two coaxial coils, one above and one below the sample, with their axes vertical and coincident with the vertical axis of the sample. The author compares the basic signal/Johnson noise ratio of such a device with that of a conventional spinner as follows. If the arm is of length r and is rotating with angular velocity ω, then the speed of the sample is $v = r\omega$. If a typical dimension of the sample in the direction of motion is $2l$ ($2l \ll r$) and it is assumed that the flux change producing the voltage pulse takes place over a sample travel distance of $6l$, the time interval in which this flux change takes place is $6l/v = 6l/r\omega$. The corresponding time interval for one cycle in a spinner magnetometer working at the same rotational frequency is $2\pi/\omega$, and therefore the rate of change of flux, and the peak output voltage, is greater in the pulsed system by a factor of approximately $(2\pi/\omega)$ $(6l/r\omega) \approx r/l$. If the detecting circuit only operates during the period of the pulse, the effective averaging time T_1 for thermal fluctuations is $T_2(l/r)$ where T_2 is the total measuring time. Therefore the noise bandwidth, which is approximately $1/T_1$ in the pulse instrument, is increased by a factor l/r compared with Δf in the spinner, where $\Delta f \approx 1/T_2$. Since Johnson noise is proportional to $(\Delta f)^{1/2}$, the signal/noise ratio of the pulsed magnetometer shows an increase by a factor of $(r/l)^{1/2}$ over the spinner, assuming similar pick-up coils in each case. Dianov-Klokov does not report the construction of an instrument using the pulse technique, and electronic and mechanical problems associated with amplifying the signal and rotating the sample at reasonable speed and with a useful ratio of l/r probably inhibited its development. Also, in most magnetometers there are more important sources of noise than Johnson noise. The only report of a pulsed rock magnetometer in the literature is that of Cialdea (1966). The sample is supported on a bifilar suspension and executes pendulum motion above and close to two pick-up coils connected in series opposition. The plane of the pendulum is perpendicular to the horizontal axis joining the centres of the coils, whose axes are vertical. The component of NRM in the sample parallel to the horizontal axis joining the coils induces transient voltages in the same sense in each coil as the sample approaches them, the sign of the voltage changing as the sample leaves them. In the measuring method described the sample is adjusted until the output voltage is zero, when the NRM component lies in the plane of the pendulum motion. The sensitivity achieved is only referred to indirectly, and appears to be about 10^{-3} A m^2 kg^{-1} in a 30 cm^3 sample. The rate of change of flux through the coils and the coupling between coils and samples are inferior to those of the Dianov-Klokov design.

An interesting aspect of the pulse principle is the possibility of multiple sample measurement. Samples mounted on the rim of a circular rotating table would pass successively through a pick-up coil. With modern signal-processing techniques it should not be difficult to store the signals obtained from each sample for each of the chosen sample orientations and then combine the data to give direction and intensity of NRM.

An astatic coil magnetometer has been described by Dürschner (1954) in which the magnets of an astatic system are replaced by co-planar coils suspended by three conducting wires, one carrying the weight of the system and two thin slack ones. The area-turns of each coil are approximately equal in magnitude to the lower coil by means of a variable shunt resistance across the former. The sensing (lower) coil is a Helmholtz pair 10 cm in diameter into which the sample is moved on a sliding carriage. The field due to the horizontal component of the sample magnetization perpendicular to the coil axis extends a torque on the coil through its effective moment and the deflection is read by mirror, lamp and scale. The use of a Helmholtz coil for the sensing coil ensures optimum response to magnetically inhomogeneous samples, since over the region within the coil in which the field is sensibly constant all elements of the sample volume contribute equally to the torque on the system. This advantage is gained at the expense of sensitivity because of the decreased coupling between coil and sample with that obtained with a smaller coil. Up to the limit where the coil field imparts a measurable IRM to the sample sensitivity can be increased by increasing the coil field.

The main disadvantage of this type of magnetometer is the long time constant, arising from the large moment of inertia of the two coils about the suspension axis. The sensitivity of coil and conventional astatic magnetometers may be approximately compared as follows. If the coil field is B and the sample moment perpendicular to it is P_r, the resultant torque on the coil is $B P_r$. The torque exerted on a magnet system by P_r (placed below the lower magnet) is $\mu_0 P_r P_m/4\pi z^3$, where z is the distance from P_r to the lower magnet of magnetic moment P_m: z is assumed small compared with the magnet separation. For these torques to be equal, $B = \mu_0 P_m/4\pi z^3$. Typical values are $P_m = 10^{-2}$ A m^2 and $z = 4.0$ cm, giving $B \approx 15$ μT. This field is probably an order of magnitude lower than that which would impart a measurable IRM to most rocks, so the torque exerted by P_r on the coil system could be made larger (by increasing the coil field) than the torque exerted on a typical conventional magnetometer. However, this would only provide greater sensitivity if the coil system could be suspended on a suspension of the same torsional constant as the magnet system, which because of the greater mass of the coils would not usually be possible.

9.6 Small-scale magnetometers

In this section some instruments are described for measuring the remanent magnetization of very small samples, e.g. small rock chips or single mineral grains. Another potentially useful application of such instruments is small-scale 'magnetic surveying' over the surface of a rock sample to locate the carriers of remanent magnetization and determine their in-situ directions of magnetization.

As an example of the magnetic moments involved, consider a 100 μm grain of haematite with typical NRM intensity of 0.01 A m^2 kg^{-1}; the magnetic moment is $\sim 3 \times 10^{-11}$ A m^2 and its saturated IRM would be $\sim 3 \times 10^{-10}$ A m^2. For magnetite the moments would be increased by a factor of about 100. A 1 mm^3 chip of typical red siltstone might have a remanent and saturated magnetic moment of about 1.0×10^{-11} and 1.0×10^{-8} A m^2 respectively. With the exception of the cryogenic instrument (see Section 9.4) rock magnetometers cannot satisfactorily measure the weaker of the above moments since about 1.0×10^{-10} A m^2 is their reasonable lower limit: there may also be difficulty in mounting very small samples for measurement in conventional magnetometers.

The main requirements for a small-scale magnetometer are a high-sensitivity detector, the size of which, if of the fluxgate or magnetostatic type, should at least be comparable to and preferably smaller than the samples to be measured.

Fales, Breckenbridge and Debnam (1974) describe a-miniature plated wire detector (Oshima, Watanabe and Fukui, 1971) using a sensor of dimensions 0.76 mm × 0.64 mm with a sensitivity of ~2 μT. Although this device lacks sensitivity it has successfully detected large magnetic particles in the surface of a rock sample at a distance of 0.15 mm above the surface. Assuming a dipole source the strongest signal received corresponded to a particle moment of ~2×10^{-9} A m^2 about ten times above noise level. A small Hall probe ~0.05 mm across designed for solid state application was developed by Carey and Isaac (1966). It could detect a field of ~1 μT and has possible applications in the present context.

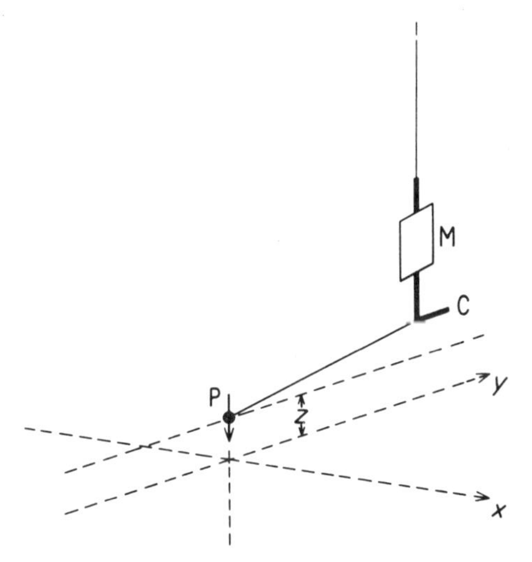

Fig. 9.32 Magnet system of small-scale magnetometer. P, spherical magnet; M, mirror; C, counterweight. The traverse axis of the sample is along x.

Figures 9.32 and 9.33 are diagrams of a magnetostatic instrument and sensor developed by Collinson and de Sa (1971) and modified by Watson (1981). The vertically magnetized spherical Magnadur magnet detector (0.66 mm in diameter, $P = 2.7 \times 10^{-5}$ A m^2) experiences a translation force and hence exerts a torque on the suspension when acted on by a horizontal field possessing a vertical gradient, for example the field due to a magnetized particle below the magnet.

If a dipole of horizontal and vertical components P_{H} and P_{V} is traversed along the y-axis at a vertical distance z below the magnet, the deflecting gradients are

$$G_{\mathrm{H}} = \left(\frac{\mathrm{d}B_y}{\mathrm{d}z}\right)_{P_{\mathrm{H}}} = \frac{3\mu_0 P_{\mathrm{H}} z}{4\pi} \frac{(z^2 - 4y^2)}{(z^2 - y^2)^{1/2}} \tag{9.55}$$

$$G_{\mathrm{V}} = \left(\frac{\mathrm{d}B_y}{\mathrm{d}z}\right)_{P_{\mathrm{V}}} = \frac{3\mu_0 P_{\mathrm{V}} y}{4\pi} \frac{(y^2 - 4z^2)}{(z^2 - y^2)^{1/2}} \tag{9.56}$$

giving the curves shown in Fig. 9.34. The corresponding torques on the suspension are $PG_H l$ and $PG_v l$, where l is the length of the arm and P the moment of the magnet.

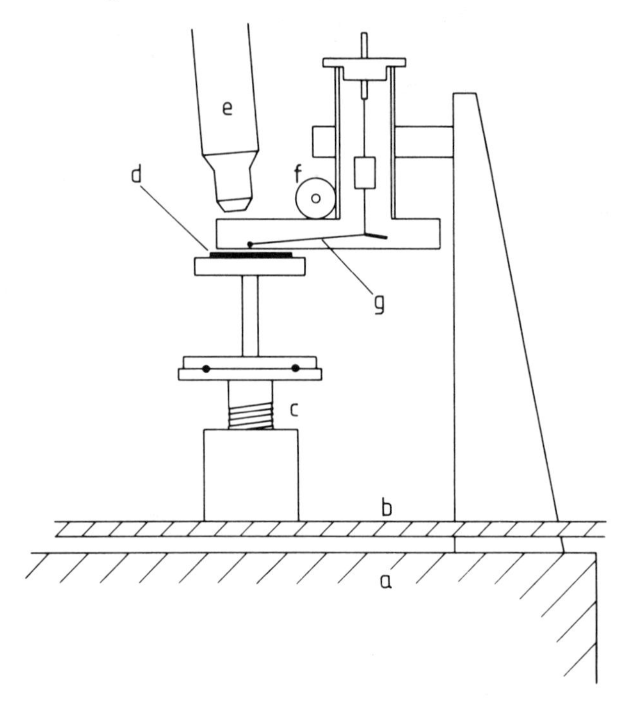

Fig. 9.33 Diagram of small-scale magnetometer. a, anti-vibration mounted concrete block supporting magnet system; b, independent mounting for sample manipulation device (c), and microscope (e); d, sample; f, feedback coil; g, magnet system.

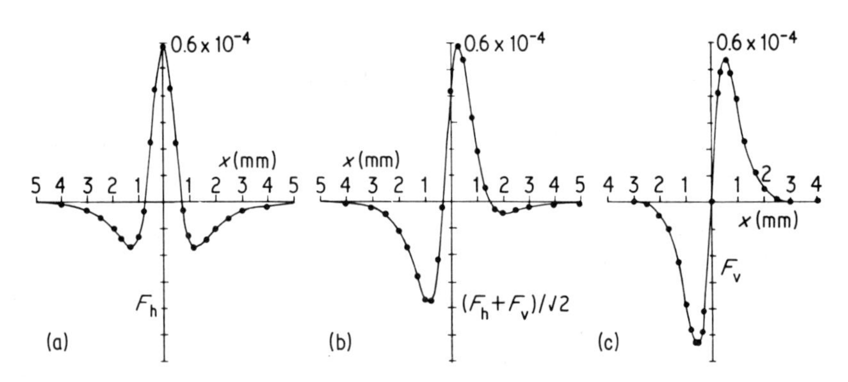

Fig. 9.34 Theoretical response curves of small-scale magnetometer: (a) horizontal dipole traversed along x; (b) the same dipole inclined at 45°; (c) the same dipole with its axis vertical. In each case $z = 1.5$ mm. (From Collinson and de Sa, 1971, copyright of the Institute of Physics)

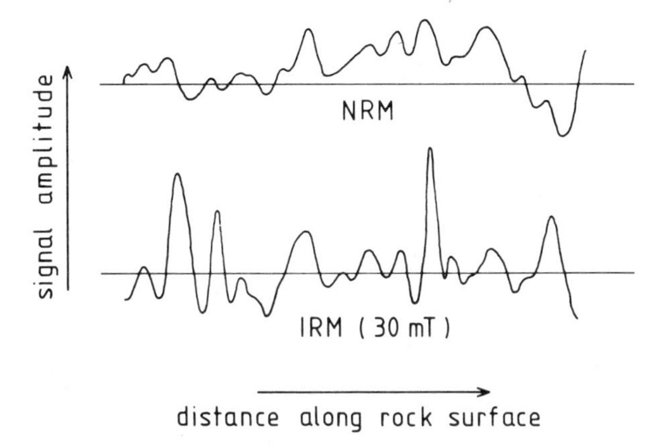

Fig. 9.35 Example of signal obtained by traversing a plane rock surface below the magnetometer probe. Upper trace, NRM; lower trace, after IRM imparted in 30 mT. The traverse length is 3.5 cm.

The system is self-astatic, i.e. will not respond to a time-varying uniform magnetic field. Photocell detection with feedback via a coil placed behind the probe is used, and particle moments down to 3×10^{-12} A m^2 can be measured with $z = \sim 1.0$ mm. Rock surfaces can be traversed underneath the magnet and simultaneously observed with a low-power microscope through a window with mineral particles visible in the rock surface. An example of curves obtained by traversing a rock surface underneath the probe is shown in Fig. 9.35.

The feature that limited the usefulness of the instrument is that it responds to both remanent and induced magnetization, the dominant inducing field for $z = 1.0$–5.0 mm being that of the detector magnet (~ 5 mT at $z = 1.0$ mm). It is not easy to distinguish the remanent and induced signals, particularly when using the 'survey' mode. In a modified sensor arrangement the magnet was horizontally magnetized which enables the distinction to be made more easily in single-particle measurements. For an application of the instrument to determining the origin of the NRM of a sandstone formation see Collinson (1972).

The response of the magnetometer to susceptibility severely limits its usefulness for some applications and in a modified version of the instrument the effect of susceptibility is much reduced. It can be seen from Equations 9.55 and 9.56 that, for a fixed z-value the sensitivity to remanence varies as P, the moment of the sensor magnet, whereas the sensitivity to susceptibility varies as P^2. Thus reducing P by a factor n will reduce the remanence and susceptibility signals by factors of n and n^2 respectively, i.e. a relative reduction by a factor n. Therefore, if a sensor of smaller magnetic moment could be used and the overall sensitivity of the system increased in the same ratio without introducing more noise, the original sensitivity and signal/noise ratio could be recovered with reduced response of the sensor to susceptibility.

In the modified instrument (Watson, 1981) the sensor magnet has a diameter of 0.35 mm and a magnetic moment of 2.8×10^{-6} A m^2, approximately ten times weaker than the original. It would have been possible to demagnetize the original sensor by the required amount, but it was considered worthwhile to make a new, smaller one because of the improved resolution available when the magnetometer is used in the survey mode. Small spherical magnets can be made by agitating by compressed air small (~ 1 mm) approximately equi-dimensional chips of Magnadur or similar magnet material in a cylindrical brass chamber, the walls of which are lined with emery paper. The air enters through a tube in the cylinder wall and leaves via a fine wire-mesh window in the upper end of the chamber.

By paying careful attention to dynamic balancing of the suspended system and supporting the instrument on a vibration-free mounting the noise level of the new magnetometer is much reduced, enabling increased gain to be used in the feedback system to restore the original sensitivity. Test with a standard sample of known susceptibility (a grain of ferrous sulphate) indicated that the susceptibility response was reduced by the expected factor of approximately ten.

Under optimum conditions the noise level of the magnetometer is equivalent to a particle moment of 1.0×10^{-12} A m^2 when horizontal and placed at $z = 1.0$ mm below the centre of the sensor magnet: the time constant is ~ 2 s. Because of the very rapid increase in the susceptibility response (as z^{-7}) as z decreases, values of z much less than 1.0 mm are not practicable.

The magnetometer is calibrated by means of a horizontal line current along the y-axis a known distance below the probe magnet or by the gradient due to a coil or magnetic dipole below the probe. The small vertical distance between probe and particle centres during measurement can be determined by focusing the microscope successively on the top of the probe and the top of the sample. A dial gauge attached to the microscope mounting is used to record this distance, which is then corrected for the diameters of the probe and particle and the optical thickness of the glass cover slip below the probe.

Small astatic magnetometers have been used for measuring NRM over small areas or volumes of rock, by placing the detecting magnet close to the rock samples. Edwards (1965) detected areas of normal and reversed NRM in vertical sections of an iron gritstone using a small astatic system. A vertical section was cut through a hand sample and the cut surface placed horizontally under the lower magnet of the astatic pair and $\sim$2mm from it. Since the gritstone when in place is magnetized nearly vertically the polarity of the NRM at points in the section can be determined by noting the deflection obtained when the section is rotated through 180° about the vertical axis of the magnet system, the initial position being such that the magnetic axis of the lower magnet is parallel to the original horizontal in the section.

Kawai, Nakajima, Yasukawa, Hirooka and Kobayashi (1973) used a small astatic magnetometer for investigating changes of NRM polarity down a sediment core. The magnets are 0.5 cm long and 0.05 cm in diameter mounted 0.5 cm apart on a vertical stem. The core is vertical and coaxial with the magnet system, with its upper, horizontal surface close to the lower magnet. Declination is measured by rotating the core to different azimuths, and readings down the core are obtained by successively removing 2.0 mm slices of the core and raising it by the same amount each time. An advantage of using small systems in these applications of the astatic magnetometer is that the field at the sample due to the sensing magnet is small and insufficient to magnetize the rock in its vicinity.

In principle it would be feasible to construct a very small spinner magnetometer for NRM measurements of particles, although there would be some difficulty in supporting the particle for total NRM vector measurement by spinning about two or three mutually perpendicular axes. An option available (at the time of writing) with the SCT cryogenic magnetometer is a model with a small diameter (~ 0.3 cm) access space, and correspondingly smaller and more closely coupled pick-up coils. The quoted sensitivity is equivalent to a minimum detectable moment of 1.0×10^{-13} A m^2.

A small-scale detector designed for mapping stray magnetic fields and domain boundaries in ferromagnetic surfaces is described by Kaczer and Gemperle (1956). A fine Permalloy reed about 2.0 cm long and 0.05 mm thick is mounted vertically with its tip just clear of the surface to be measured. The upper end of the reed is secured to a piezoelectric crystal which vibrates the reed in a vertical plane at ~ 150 Hz. The magnetization of the reed varies according to the vertical component of the magnetic field at the surface immediately below the reed tip, and this variation is detected by a pick-up coil around the reed.

9.7 Measurement of NRM in the field

In some investigations it may be desirable to measure, or at least estimate, NRM direction and intensity in the field, either at the outcrop or after drilling and cutting (or roughly shaping) standard specimens. This procedure may be useful, for instance, in locating reversal horizons and regions of normal and reversed polarity.

Measurements at the outcrop are essentially confined to the determination of magnetic polarity, and at many igneous outcrops an ordinary compass can be used to indicate the field direction immediately adjacent to the outcrop and thus the polarity of the magnetization. Doell and Cox (1962, 1967c) describe the use of two compasses mounted in gimbals at each end of a 50 cm-long rod to detect gradients in the magnetic field close to the rock face. Significant gradients (10°/m) may indicate a lightning strike and anomalous magnetization of the rock.

Doell and Cox also describe a portable fluxgate magnetometer with a cylindrical sensor 10 cm long and 1 cm in diameter. This is set up with its axis perpendicular to the local geomagnetic field and large, roughly-shaped oriented samples are placed at a suitable distance along the axis of the sensor and oriented so that maximum positive and negative readings are obtained. The axis and polarity of the rock magnetization is thus defined. Induced magnetization in the geomagnetic field is not recorded with this configuration of sample and sensor.

Sherwood and Watt (1968) use a commercial portable low-field fluxgate magnetometer (1 nT sensitivity), with the roughly equidimensional 2.0 cm specimen obtained from a drilled core placed in a holder below and close to the horizontal sensor. The geomagnetic field is backed off by means of a small solenoid surrounding the sensor, and recordings are taken with the samples at different orientations relative to the sensor. The lower limit of measurement is $\sim 3 \times 10^{-4}\,\mathrm{A\,m^2\,kg^{-1}}$ and the main factor affecting accuracy of the NRM direction in stronger samples is inhomogeneity of magnetization.

A commercially available portable fluxgate instrument is marketed by the Schonstedt Instrument Company in which the sample is successively rotated about three orthogonal axes near a sensor inside a small magnetic shield until maximum signal is obtained. The noise level is equivalent to $\sim 10^{-7}\,\mathrm{A\,m^2\,kg^{-1}}$ in a $\sim$5.0 cm sample.

A variety of 'portable' spinner magnetometers have been described, although most are essentially battery-powered laboratory instruments which can be used in the back of a truck or on board ship (Graham, 1955; Helsley, 1967b). A recently developed truly portable magnetometer is currently available from LETI (Laboratoire d'Electronique et de Technologie de l'Informatique), Grenoble, France. It is a slow-speed spinner instrument with a shielded thin-film detector. A standard 2.5 cm sample is rotated close to the detector and orthogonal components of the NRM in the sample plane perpendicular to the rotation axis are read out. The noise level of the magnetometer appears to be equivalent to $\sim 1.0 \times 10^{-8}\,\mathrm{A\,m^2\,kg^{-1}}$ in a standard sample. The power supply is a 12 V, 10 A h battery and the instrument weight is 11 kg. Details of the magnetometer and of the technique of thin film magnetometry are given by Chiron, Laj and Pochachard (1981) and Chiron and Delapierre (1979).

A portable Digico magnetometer is currently being developed, consisting of two modules. One is the measuring head and rotation system, provided with a simple read-out of the two orthogonal components of NRM being measured. A computer processor is also available which will run off a car battery or mains electricity for the normal Digico processing and information presentation (on a paper print-out). Sensitivity is comparable to the laboratory Digico magnetometer.

Chapter Ten

Field and laboratory stability tests

One of the most common problems in the interpretation of palaeomagnetic data is the presence in rocks of secondary remanent magnetizations acquired after the time of formation, which if not detected and removed prevent the direction of the primary magnetization, and thus the ancient field direction at the site, from being established. The possible origins of secondary NRM are viscous remanent magnetization (VRM), partial thermoremanent magnetization (PTRM), for example during burial and heating of sediments, isothermal remanence (IRM) through lightning strikes and chemical remanence (CRM) through post-depositional alteration and diagenesis. With the exception of IRM these secondary components are aligned along the ambient geomagnetic field at the time of acquisition, with widely different directions possible through field reversals, polar wandering and plate movements.

The generally lesser coercivity of secondary NRM relative to the primary component allows in many cases for its removal by demagnetization ('magnetic cleaning') techniques, of which alternating field and thermal clearing are the most important. There are also field tests which can be applied in appropriate geological environments which, together with observational data, can provide information on the presence or otherwise of secondary NRM.

Before these are considered, it may be useful to clarify the meaning of 'stability' of NRM. In the present context a stable rock is one in which there has been little change in its primary NRM since the time of acquisition whereas an unstable rock has acquired a secondary magnetization at some stage.

10.1 Field tests

10.1.1 Observational data

In many cases VRM in the recent geomagnetic field is the origin of secondary NRM, and the existence of consistent palaeomagnetic directions remote from the present field direction in a group of samples, perhaps of different rock types, from a geological unit argues alone for some stability against VRM. A distribution of site mean directions towards the present (dipole) field direction will suggest acquisition of VRM at some sites, while a similar trend among directions within a site will indicate varying acquisition of VRM among different samples (Figure 10.1).

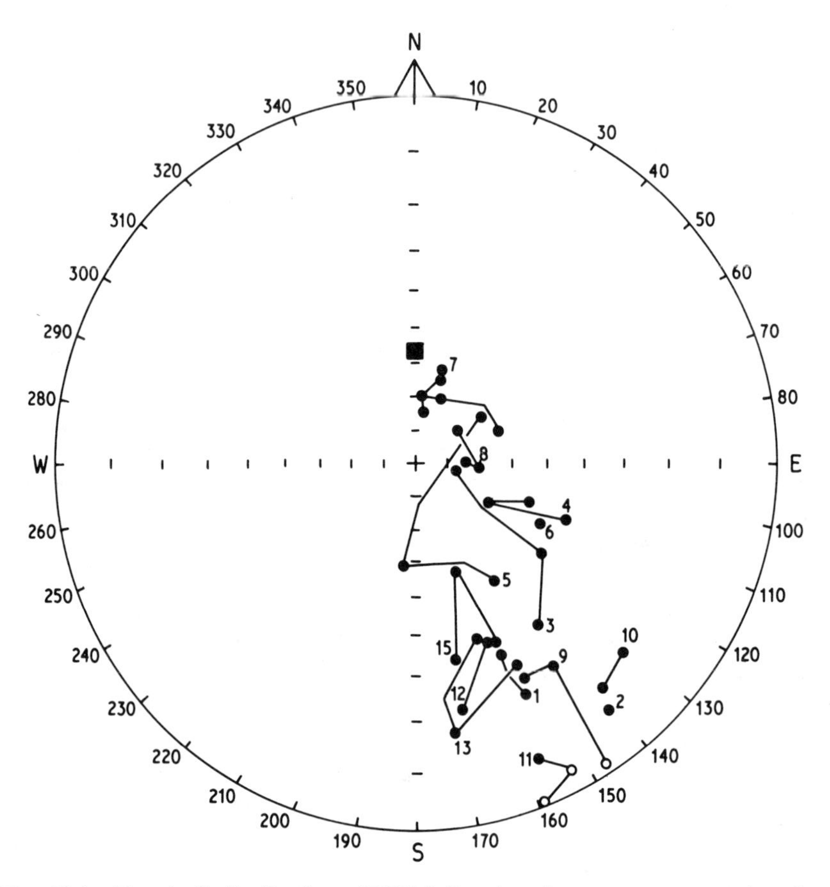

Fig. 10.1 'Streaked' distribution of NRM directions between present and ancient field directions due to VRM acquisition (Red siltstone, Chinle formation, Utah, USA)

If reversals occur in the primary NRM of a rock unit, acquisition of a uniform secondary magnetization may be indicated by the mean normal and reversed directions being less than 180° apart, and in principle the primary axis of magnetization may be approximately determined (Fig. 10.2). A good example of stability, showing only a small secondary component, is provided by Triassic redbeds of the Chugwater formation from Wyoming, USA (Fig. 10.3).

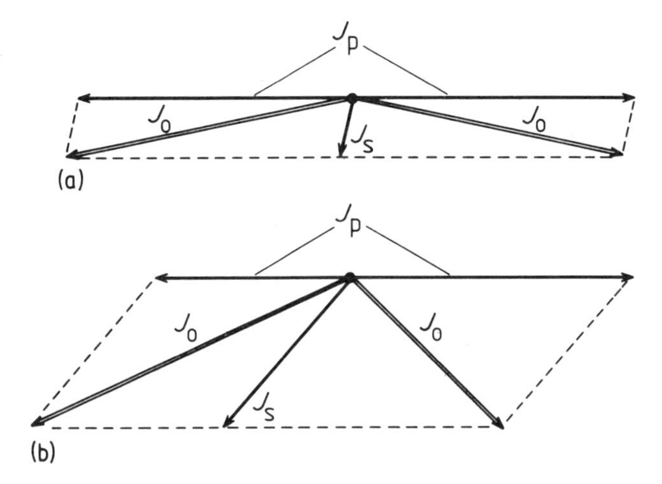

Fig. 10.2 (a) Effect of addition of small secondary magnetization (J_s) to equal, exactly opposed primary components (J_p). In principle the observed NRMs (J_0) can be used to define the primary directions; (b) if J_s is comparable to or greater than J_p, and the normal and reversed magnetizations are unequal, only an approximate direction for J_p can be derived from J_0.

10.1.2 Fold test

If a plane rock layer possesses a uniform primary magnetization and is subsequently folded, then if no secondary magnetization is acquired after folding the NRM directions of samples collected over the extent of the fold will remain uniform when referred to the local bedding but be scattered in a plane perpendicular to the fold axis when they are referred to the horizontal plane (Fig. 10.4). If a secondary magnetization is acquired after folding the combined NRM directions will be scattered relative to the local bedding, and in the extreme case of a dominant secondary NRM they will be uniform relative to the horizontal plane. Thus, the stability test consists of 'unfolding' the beds and comparing the precision of the directions so obtained (now referred to the bedding plane) with those referred to the horizontal. This test, originally proposed by Graham (1949), can also be applied to samples from an area in which the bedding varies locally. An example of the fold test is shown in Fig. 10.5 and others are described by Irving and Runcorn (1957), Gough and Van Niekerk (1959) and Purucker, Elston and Shoemaker (1980). McElhinny (1964) describes a relevant statistical test which indicates whether the difference in precision before and after unfolding is significant (Section 12.4.4). On a much smaller scale, the uniformity or otherwise of NRM directions through ripple marks and other types of disturbance in sediments can indicate the degree of stability of NRM in the rock.

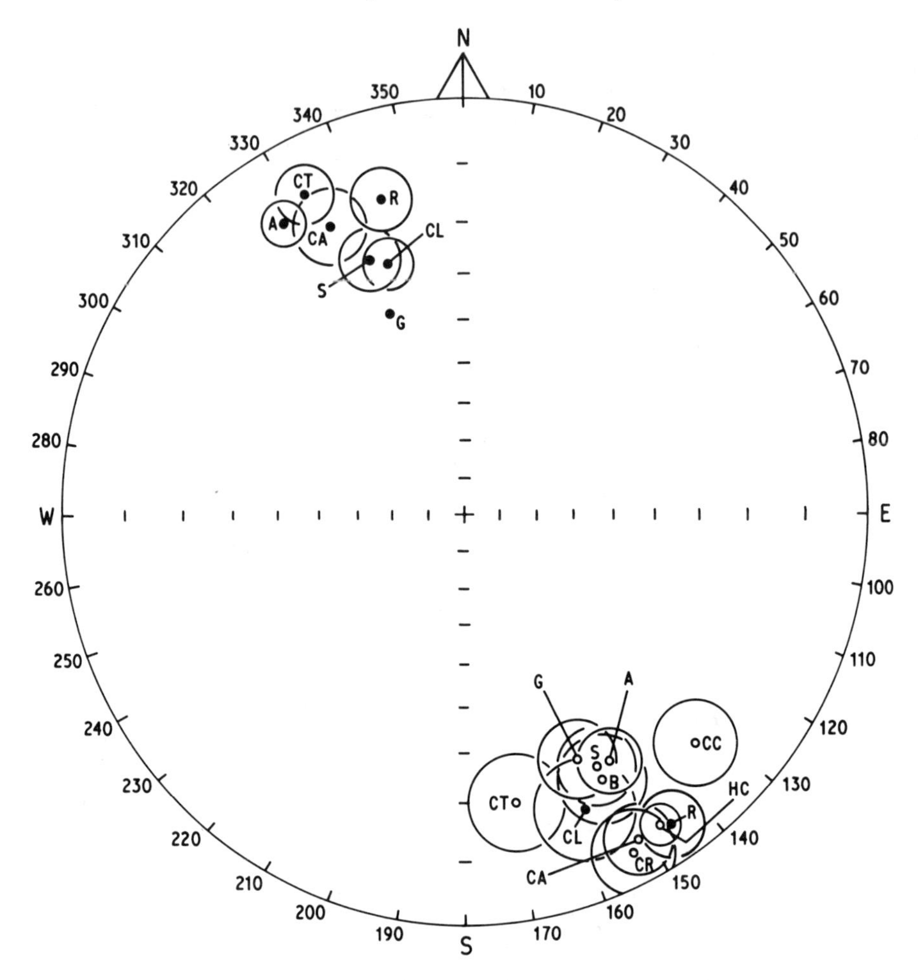

Fig. 10.3 Normal and reversed site mean directions of NRM in the Chugwater formation (Triassic), Wyoming, USA.

10.1.3 Conglomerate test

This test may be applied if pebbles or fragments from a rock layer whose NRM is being investigated can be identified in an associated conglomerate bed (Graham, 1949). If the pebbles, and therefore presumably the parent rock, have not subsequently acquired a secondary NRM, the NRM of the pebbles in the conglomerate should be randomly directed.

An example of the test in the Torridonian sandstone of Scotland is given by Irving (1957) (Fig. 10.6) and Stewart and Irving (1974), and Starkey and Palmer (1971) have examined the expected sensitivity of the conglomerate test

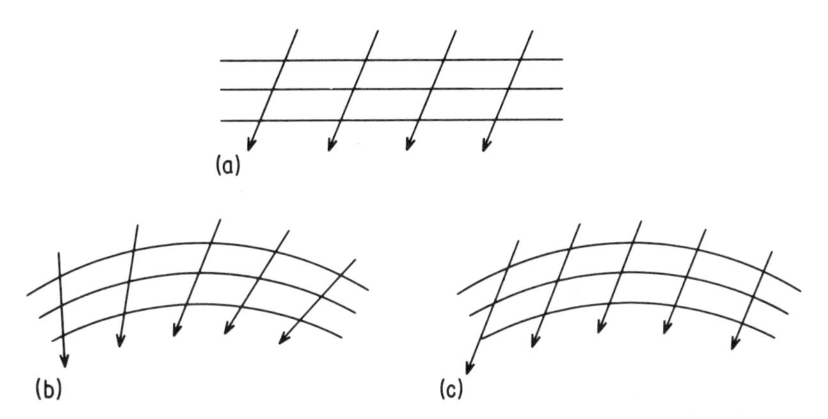

Fig. 10.4 The fold test: (a) uniformly magnetized, flat-lying beds; (b) if the NRM is unaltered during and after folding, the NRM directions in samples from different parts of the fold are scattered when referred to the local horizontal but uniform when referred to the local bedding; (c) remagnetization after folding results in uniform NRM relative to the horizontal.

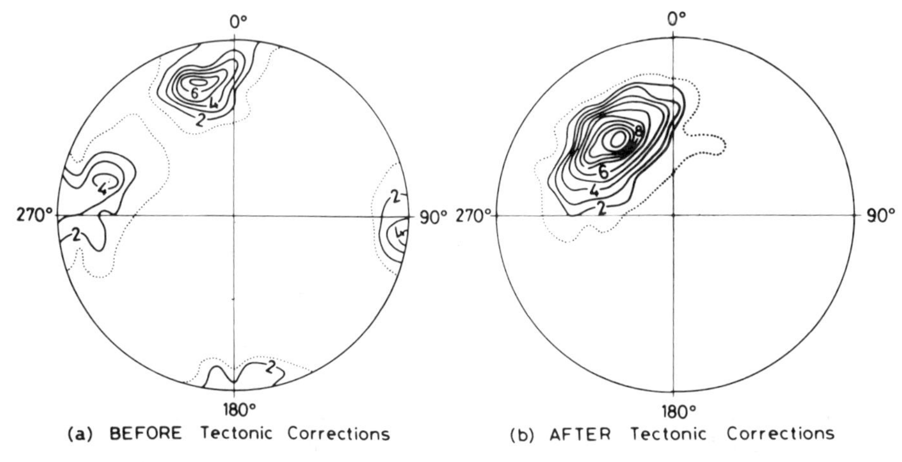

Fig. 10.5 Example of the fold test in the Scaglia Rossi limestone, Italy: (a) before tectonic corrections; (b) after tectonic corrections. In (a) NRM directions from the Gubbio section, where the bedding dips to the north-east, lie near 90° and 270°, while those from the Moria section (south-west bedding) lie near 0° and 180°. The display of NRM directions by density plot is described in Section 12.1. (From Lowrie and Alvarez, 1977)

to the strength of the secondary NRM in the pebbles. They took samples of NRM directions drawn from a random population, summed them vectorially and calculated the length R of the resultant vector. Each random direction was then biased by adding to them a common vector of varying magnitude. This is

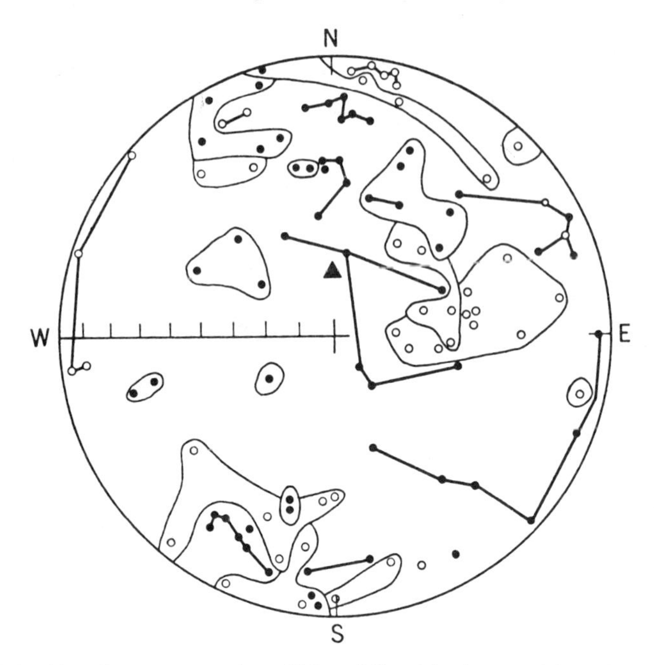

Fig. 10.6 Conglomerate test in pebbles of Torridonian sandstone in conglomerate bed of New Red Sandstone age, indicating stability of the pebble NRM. Plotted points are from specimens from the same pebble, enclosed when New Red Sandstone conglomerate age is known and linked where age is less well defined. (From Irving and Runcorn, 1957)

equivalent to superimposing a uniform secondary magnetization on the (presumed) randomly directed primary NRM of each conglomerate pebble. Comparison of R from samples of unbiased and biased directions then enables estimates to be made of the relative magnitude of primary and secondary NRM which is required for R to become greater than R_0, the length of the resultant of a given number of directions which indicates that they are random at a given level of probability (Section 12.3.3). The authors' results suggest that in a sample of 25 pebbles a secondary/primary magnitude ratio as high as 0.5 may be undetected (at a 95% probability level), and that more than 100 samples are required for a statistically significant result if the ratio is 0.25. On the basis of this study it appears that the conglomerate test should be used with caution and that a test of randomness may give a misleading result.

It is possible that a grouping of NRM directions among the pebbles could result from the process of conglomerate formation and such a result should be investigated further, e.g. by magnetic cleaning of parent rock and pebbles. The same type of test can be applied to slump or contorted beds, deformed before consolidation, but interpretation of the results is not always straightforward (Irving and Runcorn, 1957).

10.1.4 Baked contacts

When an igneous rock intrudes a formation already possessing an NRM, that NRM will in general be altered in the contact zone through the acquisition of a total TRM in the immediately adjacent zone and PTRM in the country rock as far away from the contact as significant heating occurs ($\geqslant \sim 100°C$). The simplest case is igneous intrusion into a much older sediment of high magnetic stability: if adjacent to the intrusion, after cooling, the country rock is remagnetized and when ultimately collected and measured has the same NRM direction as the intrusion, it argues for the stable magnetization of both. Further away from the intrusion there will be a transition in NRM directions to the unchanged country rock direction, as the superimposed PTRM becomes progressively weaker. There are several other possible situations, which have been well summarized by Irving (1964) and McElhinny (1973). A useful compilation of baked contact studies is that of Wilson (1962), and Everitt and Clegg (1962) report on a detailed study of the magnetic effects of an intrusion.

10.1.5 Magnetic profiles

In some geological situations, e.g. over a dyke, it is sometimes possible to compute the magnetic anomalies expected from the susceptibility and NRM of the dyke rock, on the assumption that the direction of the NRM is appropriate to the age and geographical position of the dyke and it has no secondary NRM; non-agreement between the observed and computed magnetic anomalies would suggest the acquisition of significant secondary magnetization since emplacement. Studies of this nature are described by Green (1960) and Strangway (1965).

10.2 Laboratory stability tests

Various laboratory tests have been proposed for testing for the presence in rocks of magnetic particles possessing high and low coercivity, thus indicating the rock's capability of maintaining a stable remanence.

The Königsberger ratios Q are defined as the ratio in a rock of a magnetization to the induced magnetization in the Earth's field. For the present purposes, Q can be expressed as either

$$Q_{\mathrm{T}} = M_{\mathrm{T}}/kH \qquad (10.1)$$

or

$$Q_{\mathrm{r}} = M_{\mathrm{r}}/kH \qquad (10.2)$$

where M_{T} and M_{r} are, respectively, the thermoremanent and natural remanent (volume) magnetization of the particles carrying the observed magnetization, k is their initial volume susceptibility and H the Earth's field.

For multidomain grains the theory of Dickson *et al.* (1966) gives

$$M_T = 1.57H \tag{10.3}$$

For roughly equidimensional grains of titanomagnetites, $k \approx 3.14$: thus, combining Equations 10.1 and 10.3 gives $Q_T \approx 0.5$ for multidomain particles in igneous rocks. For single-domain grains contributing a stable remanence, $Q_T \approx 1.0$, and values of Q_r approaching 0.5 suggest the presence of a large proportion of multidomain, low-stability grains. However, an assumption in this argument is that $M_T = M_r$, since only M_r (and therefore Q_r) can be measured. M_r may have been modified, for example by acquisition of VRM or viscous decay of NRM. Irving (1964) has listed some rock units with stable and unstable magnetizations showing that the former are associated with a ratio of NRM intensity/$k > 1$ and the latter < 1. In some cases, the cause for this is the presence of low-coercivity magnetite which both acquires a secondary NRM and substantially increases k.

The coercivity of remanence, i.e. the field which, when applied in a direction opposed to the NRM and then removed, reduces that NRM to zero has been used as a test of stability, mainly by Russian workers (Petrova, 1961): the coercivity of saturated remanence has also been used. If the remanence is carried by a single and the only ferromagnetic mineral in the rock and the mineral has uniform magnetic stability, then high or low coercivity would indicate stability or instability of the NRM. However, in the more common situation of ferromagnetic particles possessing a range of coercivities, low coercivity of remanence would only indicate the presence of a high proportion of low coercivity material. High coercivity of saturated remanence only demonstrates the presence of high-coercivity grains, without indicating whether they are carrying the primary NRM. A test based on the rate of viscous decay of saturated remanence has been proposed by Stacey and Banerjee (1974).

A test involving PTRM is due to Stacey (1963). The PTRM acquired in a period of t_1 at absolute temperature T_1 can, on certain assumptions, be related to that acquired in a shorter time t_2 at a higher temperature T_2 by the relation

$$\frac{T_2}{T_1} = \frac{\ln ct_1}{\ln ct_2} \tag{10.4}$$

This relation is derived from Equation 11.25, in which c is a frequency factor usually taken to be $10^{10}\ \mathrm{s}^{-1}$. Thus, in principle, acquisition of PTRM at a low temperature over a geological time interval, say 10^8 years, can be tested by heating a rock to a higher temperature for a laboratory time interval of a few minutes or hours. Substituting $t_1 = 10^8$ years and $t_2 = 1000$ s in Equation 10.4 gives $T_2/T_1 \approx 2$. (The ratio is rather insensitive to t_1: for $t = 10^7, 10^8, 10^9$ years, $T_2/T_1 = 1.88, 1.96, 2.04$.)

Therefore if the rock is heated in the Earth's field for 1000 s to twice the absolute temperature that it may have experienced since it was primarily magnetized, any PTRM acquired will be a measure of the secondary NRM expected to be present due to geological reheating. For $T_1 = 20°C$ and $100°C$, $T_2 = 302°C$ and $457°C$. Survival of all or part of the NRM and non-acquisition of PTRM or its acquisition and subsequent successful removal by thermal demagnetization will demonstrate the stability of the primary NRM and the feasibility of removal of the secondary.

Equation 10.4 is strictly only applicable to single-domain material and Stacey (1963) modifies it for the effect of temperature on energy barriers opposing wall motions in multidomain particles according to

$$\frac{T_2}{T_1}\frac{J_{s1}}{J_{s2}} = \frac{\ln ct_1}{\ln ct_2} \tag{10.5}$$

where J_{s1} and J_{s2} are the spontaneous magnetizations at T_1 and T_2.

Apart from some inadequacy in the derivation of Equations 10.4 and 10.5 (see for instance Pullaiah, Irving, Buchan and Dunlop, 1975), the test is clearly not a precise one for determining the extent of PTRM acquisition over geological time intervals, if only because the thermal history of a formation is not sufficiently well known. However, like the coercivity test, it can indicate the presence of potential carriers of secondary magnetization and provide qualitative information on their likely importance.

Chapter Eleven

Magnetic cleaning techniques

The foregoing field and laboratory tests are at best evidence for secondary NRM acquisition since some geological event or for the presence of low-coercivity material, and the more positive demagnetization (or 'cleaning') techniques which detect and remove secondary magnetizations are now a routine part of palaeomagnetic measurements.

The underlying principle of the routine magnetic cleaning techniques is based on the generally lower stability of secondary magnetizations relative to those acquired by the primary processes of chemical remanence and thermo-remanence. This allows the preferential removal of secondary NRM by the application of sufficient energy to overcome the magnetostatic energy of alignment within particles carrying secondary remanence, leaving them magnetically randomly oriented. In practice the intensity of the primary component is often decreased also, but since it is the direction of the primary NRM which is usually of interest for palaeomagnetic interpretation the intensity decrease is only important if it significantly affects the accuracy with which the surviving NRM can be measured.

The two routine cleaning techniques currently employed are alternating magnetic field and thermal methods, in which magnetic and thermal energy respectively are used to randomize the moments of particles carrying secondary NRM. Chemical cleaning, described in Section 11.3, is a more recent development of rather restricted application involving the physical removal of mineral(s) carrying unstable NRM.

11.1 Alternating field cleaning

11.1.1 Introduction

This technique is essentially the same as that used in ferromagnetic studies, in which demagnetization is achieved by cycling the material through magnetic hysteresis loops of decreasing amplitude to randomize domain moment directions. In demagnetizing rocks the situation is somewhat different since there may be single- and multidomain particles carrying remanence, and whereas the latter can be demagnetized the former cannot but can only have their magnetic moments changed in direction. Even a multidomain grain may possess a finite moment after cycling in a decreasing alternating field because in a 'non-perfect' multidomain particle (as commonly occurs in rocks) the lowest energy state does not correspond to zero net magnetic moment (Stacey, 1961). In general, demagnetization of rocks by alternating field is achieved by both multidomain particle demagnetization and randomization of particle moment directions.

The presence and character of secondary magnetization in rocks is best revealed by cycling through decreasing hysteresis loops of progressively greater initial amplitude, corresponding to applied peak fields B_{max}. Single-domain grains with an effective coercive force $B_c < B_{max}$ tend to 'follow' the decreasing alternating peak field B until $B < B_c$, when the domain moments become locked in random directions in the rock. Similarly, multidomain particles of certain coercive force will be effectively demagnetized or left with randomly directed residual moments which, like the single domains, will form a magnetically self-cancelling assembly contributing no NRM to the rock. The nature of the coercive force is different in each case, being associated with domain rotation and domain wall movement respectively.

By progressively increasing B_{max} the moments of particles or domains of successively higher coercive force are randomized or demagnetized, and remanences of differing stability can be distinguished in a rock by observing consequent changes in intensity and direction of NRM.

Although progress has been made with the theory of alternating field cleaning, applying this theory to natural samples is difficult. This is because of the dependence of effective coercive force of particles on a variety of factors, including grain size (and therefore its distribution in rocks), shape, anisotropy and the presence of intergrowths. Another factor is the modification of the field applied to a particle by the reverse field due to the induced particle magnetization. This is significant in a material such as magnetite and results in an enhanced observed coercivity of magnetite particles relative to the bulk material, whereas in haematite the effect is negligible because of its low specific magnetization.

The alternating field demagnetization of spherical magnetite particles has been examined theoretically by Stacey (1961, 1963) who showed that the critical field required to move domain walls (a measure of the grain coercive force) should vary as $d^{-3/4}$, where d is the grain diameter. Experiment shows that $B_c \propto d^{-n}$, where $0.25 < n < 1$ (Stacey and Banerjee, 1974). Stacey (1963) gives theoretical demagnetization curves for single-size grains of three different intrinsic susceptibilities (on which the internal effective demagnetizing fields depend): some results obtained by Rimbert (1956) with 0.1 μm grains are a satisfactory fit to these curves. For non-spherical particles the demagnetization characteristics depend additionally on domain orientation

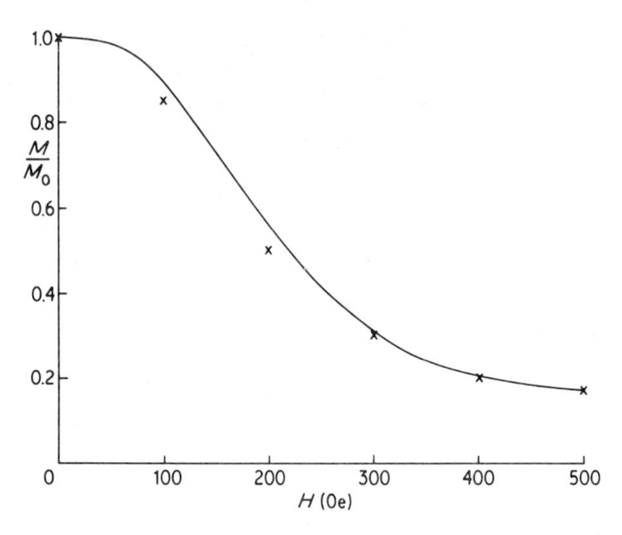

Fig. 11.1 Theoretical demagnetization curve for an assembly of grains with a log normal distribution of grain volumes (99% between 1 μm and 100 μm). Crosses are experimental points obtained from a magnetically stable basalt. (From Stacey, 1963)

relative to particle shape: if this is random an assembly of such particles behaves in a similar manner to spherical grains. To relate theory to observation it is necessary to integrate over an assumed or known grain size distribution, as described by Stott (quoted in Stacey (1963)) whose results are shown in Fig. 11.1.

In stable single-domain magnetite the coercive force is mainly controlled by anisotropy due to grain shape, and values much greater than the maximum of 30–60 mT associated with multidomain magnetite are possible. Evans and McElhinny (1969) give a theoretical maximum coercive force of ~300 mT for infinitely long needles and Strangway *et al.* (1968b) suggest that single domain, high-coercivity behaviour is possible in the elongated exsolved magnetite lamellae which commonly occur in volcanic rocks. The effect of thermal agitation in small grains, and grain interaction, in reducing coercive force is discussed by Néel (1949, 1955) and Dunlop (1965).

In haematite the NRM commonly has a high or very high coercivity arising from strong crystalline anisotropy. A secondary viscous magnetization acquired in only a few weeks or years can also require a peak alternating field in excess of 100 mT to remove it (Biquand and Prévot, 1971) while a primary chemical remanence may be stable in a peak field of 500 mT or more.

In practice it has been found comparatively easy to construct demagnetizing equipment capable of satisfactorily cleaning rock samples in peak fields of up to ~100 mT, but with higher fields the instrumental difficulties increase and also the likelihood of imparting a spurious (anhysteretic) magnetization to the rock. These are among the reasons why a.f. cleaning is favoured for the magnetite- and titanomagnetite-bearing igneous rocks while the sediments, in which haematite is the major carrier of remanence, are usually cleaned more satisfactorily by the thermal technique.

11.1.2 Practical considerations

In the design of a.f. cleaning equipment the aim is not only the successful removal of the unwanted secondary NRM but also the prevention of acquisition by the rock of spurious magnetizations arising from the cleaning procedure. The latter can be an isothermal remanence induced by a random transient 'spike' on the a.f. waveform, or an anhysteretic magnetization (ARM) if there is a direct magnetic field acting on the sample in the presence of the decreasing alternating field. Possible sources of a direct field are the residual ambient (Earth's) field and even harmonics in the a.f. waveform.

The requirements for satisfactory cleaning of igneous rocks are therefore:

(a) A coil and power supply to provide a maximum peak alternating field of ~100 mT of reasonable homogeneity over the rock sample.
(b) A method of stepless and transient-free reduction to zero of the coil current.
(c) An a.f. waveform in which the second and higher even harmonic content is as low as possible.
(d) A zero or very low direct field of external origin at the sample.

Requirements (a) and (b) are considered later and we first consider (c) and (d) and some other practical aspects of a.f. cleaning.

(a) Direct fields

A rough calculation shows that (c) is likely to be the most demanding requirement with regard to minimizing direct fields at the sample. The important feature required of the demagnetizing field waveform is that positive and negative peak values should be of equal amplitude (neglecting the decrease from cycle to cycle due to the progressive decrease of the field). Odd harmonics of the sinusoidal fundamental component, regardless of their phase relation to the fundamental, distort the waveform but do not cause inequality of the positive and negative peak amplitudes (Fig. 11.2(a)), whereas even harmonics (in particular the second) unless of a restricted range of phase difference relative to the fundamental do cause amplitude inequality. In the worst case (Fig. 11.2(b)) a second harmonic content of amplitude a fraction p of the fundamental amplitude B_{pk} is equivalent to a direct field $h = pB_{pk}$ superimposed on the alternating field. If $p = 0.001$ and $B_{pk} = 50$ mT, $h = 50\ \mu$T or approximately the value of the uncancelled Earth's field at mid-latitudes. In practice, however, waveform asymmetry and its effects may not be as serious as the above calculation suggests. The phase difference between the second harmonic and fundamental may be more favourable than that shown

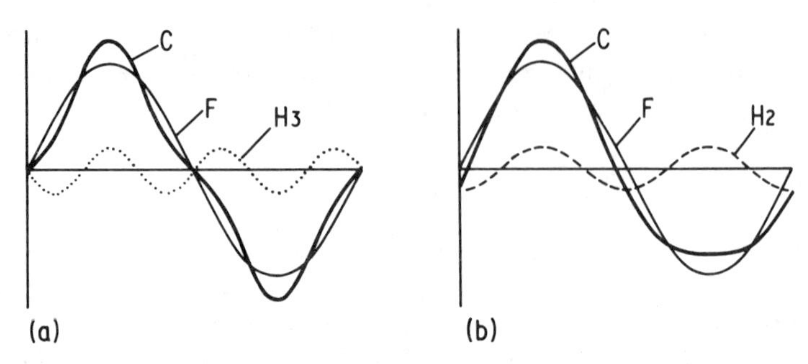

Fig. 11.2 Effect of (a) odd and (b) even harmonics on a sinusoidal waveform. F, fundamental; H_2, second harmonic, H_3, third harmonic; C, combined waveform.

in Fig. 11.2(b), the coil circuit can be tuned to the fundamental frequency and thus attenuates higher harmonics, h is proportional to B_{pk} and decreases with it and is therefore less effective in producing an ARM than a constant direct field h and the sample can be rotated or 'tumbled' during cleaning.

Hailwood and Molyneux (1974) have investigated the above factors in terms of the 'asymmetry' of the mains supply. They define asymmetry by

$$\frac{B_1 - B_2}{(B_1 + B_2)/2} \times 100\,\% = \frac{200(B_1 - B_2)}{B_1 + B_2}\,\% \tag{11.1}$$

where B_1 and B_2 are the larger and smaller of the distorted waveform amplitudes. Note that asymmetry as defined above does not provide information directly on second harmonic content. A waveform with 1% second harmonic content of phase difference $\pi/4$ relative to the fundamental (i.e. the maximum amplitudes of the second harmonic displaced by $\pi/4$ from those of the fundamental) has zero asymmetry: if the phase difference is such that the maxima and minima of the fundamental are in phase or antiphase with those of the second harmonic (Fig. 11.2(b)) there is the maximum asymmetry of 2%. Thus waveform asymmetry is more informative than second harmonic content when estimating direct field magnitudes.

The above authors find that asymmetry in the mains supply (at the Nuffield Palaeomagnetic Laboratory of the University of Newcastle upon Tyne) equivalent to 0.1%–1.0% second harmonic content of the 'worst' phase is not uncommon. They also demonstrated that by tuning the demagnetizing coil to the mains frequency (50 Hz) the equivalent 100 Hz content of the waveform was reduced by a factor of ~50: attenuation factors of 25–50 (between mains input and coil) should be obtainable in most laboratory demagnetizers. Details of the asymmetry detector used by the authors are given in Appendix 8.

It can be seen that even when the above factors are taken into account the residual direct field due to the alternating field may still be typically 100–1000 nT at peak alternating fields of ~50 mT. Thus, little is gained by cancelling the Earth's field to much better than ~ ± 50 nT. If the geomagnetic field is nulled over the sample by means of a Helmholtz or other coil system, as is usual in laboratory-built demagnetizers, no more current control is needed than is provided by a constant voltage or constant current power supply stable to one part in 10^3. At the above low level of direct field only a small minority of rocks are likely to acquire a significant ARM in any case.

The commercial demagnetizers presently available (in which the alternating coil current is generated independently of the mains) claim 100 dB attenuation of second harmonic (Schonstedt) and 70 dB (Highmoor) in the coil relative to the fundamental, i.e. 0.001 % and 0.03 % respectively.

Where a Helmholtz or other coil system is used to null the ambient field an alternating voltage will be induced by the demagnetizing field in the windings of any coils which are coaxial with the demagnetizing coil. Depending on the dimensions of the coil system, the number of turns on it and the demagnetizing field frequency, the peak induced voltage will be typically 5–50 volts with a 50 mT field in the demagnetizing coil. This will contribute a small (10–100 μT) additional alternating field over the sample in antiphase with the coil field which will not be important unless there is any part of the nulling coil circuit in which a partial rectifying action can occur, producing a direct field component in the waveform. The most likely source of such action is the power supply but it is unlikely to occur in conventional constant current or constant voltage units but could occur in a supply based on reversible chemical cells. A suitable choke in and/or a capacity across the coil system might then be necessary to eliminate the effect. Alternatively, if the demagnetizing coil can be oriented with its axis perpendicular to the meridian no field nulling coils will be needed along that axis.

(*b*) *Sample rotation*

There is some confusion in the literature regarding the purpose of sample rotation or 'tumbling' during a.f. cleaning. For some authors the primary object is to allow the demagnetizing field to act along as many directions in the rock as possible, and for others it is to randomize and thus greatly reduce the magnitude of any ARM induced in the sample by a residual direct field.

It might be expected that a sample could be effectively cleaned in a particular peak field by a single cleaning run with the sample NRM in a random direction relative to the alternating field. However, because of the dependence of the effect on single and multidomain grains of the direction of an applied field relative to the magnetic axis (SD) or domain directions (MD), the most efficient orientation of the field for demagnetization purposes is parallel to the NRM direction. The effect is not averaged out in an assembly of grains, essentially because their net magnetization, i.e. the rock NRM, implies a degree of alignment among the domains. Rimbert (1959) showed that in an andesite sample a 50 % greater field was typically required to produce the same relative reduction in NRM intensity if the field direction was perpendicular rather than parallel to the NRM, and Brynjolfsson (1957) demonstrated a similar effect in

Icelandic basalts. The effect has also been investigated in an early, little-known paper by Schmidlin (1937) and more recently by McFadden (1981).

Since it is not usually convenient to orient the sample with its NRM parallel to the field, and since the NRM is commonly the vector sum of two or more components of different directions and changes significantly during cleaning, the practice evolved of demagnetization along three mutually perpendicular axes in the sample at each value of the peak field: thus no direction in the rock is more than $\tan^{-1}\sqrt{2} \approx 55°$ from the field direction or sees less than 0.57 B_{max}.

Tumbling clearly allows the alternating field to act along many more directions in the sample and therefore promotes more efficient cleaning as well as reducing the magnitude of any ARM acquired: there is also a saving in time compared with the three-axis procedure.

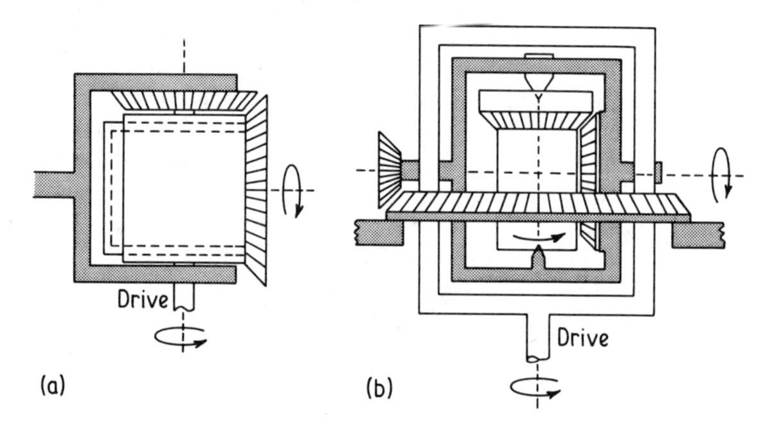

Fig. 11.3 Examples of tumbler designs: (a) two-axis; (b) three-axis.

Tumbling devices consist of a system by which the sample is rotated simultaneously about two or three mutually perpendicular axes (Fig. 11.3). Their design involves requirements which are to a certain extent mutually exclusive. The demagnetizing field should see as many axes as possible in the rock and these axes should be as symmetrically distributed as possible, and because the field is decreasing, the time constant of the system, i.e. the time required for a complete tumbling cycle, should be short compared with the field decay time. However, it is generally true that the more axes in the rock that at some time in the cycle come within a given (small) angle of the field direction the greater the time for one cycle to be completed, and some compromise is necessary.

McElhinny (1966) has analysed two-axis tumbler performance in terms of the ratio of the rotation speeds about vertical (ω_v) and horizontal (ω_h) axes, the smallest angle θ that some directions in the sample make with the field direction during a tumbling cycle and the time for a cycle to be completed. As an example, the author takes rotation ratios ($\omega_v : \omega_h$) of 15:16 and 1:2 and calculates $\theta = 4°$ and 40° respectively: since the effective field is $B_0 \cos\theta$ (B_0 is the axial field) this means that all directions in the rock see $0.998\, B_0$ and $0.77\, B_0$ respectively during a cycle. However, the 15:16 ratio requires 15 revolutions (of the vertical axis) to complete a cycle whereas the 1:2 ratio completes a cycle in one revolution. At a typical rotation speed of 150 revs min^{-1} the cycle times are 6 s and 0.4 s. Taking a field decay time of 2 minutes from 50 mT the field will decrease by 2.5 mT and 0.17 mT during one cycle, and the former value is probably unacceptably large. McElhinny further shows that with a ratio of 11:16 all directions in the rock approach to within 25° of the field direction in one revolution and therefore see 90% of B_0 and in a complete cycle (11 revolutions) the values are 5° and 99.6%. Since there is unlikely to be a significant decrease in the efficiency of demagnetization if all directions in the sample see ~80% or more of the field and the field decay is < 1 mT per tumbling cycle, the choice of ratio based on the above analysis is not too critical and ratios of ω_v to ω_h between 1:1 and 1:2 are commonly used.

Hutchings (1966, 1967) approached tumbler design with the aim of reducing the acquisition of ARM by randomizing its component in the rock such that its vector sum approaches, and is ideally, zero. This is equivalent to making the average direct field h zero, as seen by the sample, and the author introduces the relation $ht = t_0$ where t is the time after tumbling starts and t_0 is a quantity termed the hyperbolic decay time: t_0 is a function of ω_v, ω_h and vector direction in the sample, and is a measure of the rapidity with which the tumbler may be expected to average out the ARM. On the basis of his analysis, which assumes that the direct field is of constant magnitude, the author gives an optimum ratio range of between ~3:5 and ~11:20, and if $\omega_v > \omega_h$ between ~5:3 and ~3:1. These ratios are to some extent dependent on qualitative assessment of the degree of randomness of rock directions seen by the demagnetizing field: these can be computed and displayed on a stereographic projection. A tumbler with $\omega_v > \omega_h$ may have an advantage in that it can be made more compact for a given sample size thus increasing the efficiency of the demagnetizing coil by minimizing its internal radius.

It is important with two-axis tumblers that the axis of the demagnetizing field does not coincide with (or is near) the fixed rotation axis of the tumbler: if this is the case a component of the NRM in the sample is always perpendicular (or near perpendicular) to the field. The usual, and best, configuration is where the field axis is in the plane traced out by the inner rotation axis.

Hutchings (1967) has also examined three-axis tumblers and concludes that although improved isotropy of h may be obtained with them, t_0 is usually larger. For routine a.f. cleaning, the improvement to be gained by them may not offset the disadvantages of their greater size and complexity. Doell and Cox (1967b) describe a three-axis tumbler with rotation speeds relative to the vertical axis (1.00) of 1.21 and 1.61, the last axis holding the sample, and Roy,

Reynolds and Sanders (1973) use ratios 1.00 (parallel to the coil axis), 0.87 and 0.81.

A tumbling device which achieves random orientation of the sample relative to the field is an interesting idea which Morris (1970) has attempted to realize in practice. The sample is enclosed in a Perspex sphere supported on a rotating air jet and can be made to randomly alter its spin axis by varying the speed of rotation. However, perfect randomness, which also involves rapid changes of spin axis, seems unlikely to be achieved and it also seems probable that slight surface and geometrical irregularities would cause certain directions of the spin vector to be favoured.

Absolute rotation speeds of tumblers are usually in the range 2–6 rev s^{-1}, and the occurrence of a rotational frequency about any of the tumbler axes which is a sub-harmonic of the field frequency should be avoided.

(c) *Demagnetizing field: frequency and decay rate*

In most laboratory-built a.f. cleaning equipment it has hitherto been convenient to use the electricity mains as the power supply and the demagnetizing field therefore has a frequency of 50–60 Hz. When solid state oscillators and power amplifiers became available the choice of frequency was extended. The optimum frequency can be considered from the points of view of efficient cleaning and constraints set by the demagnetizing coil and its associated circuitry.

From the former standpoint the chief requirement is that the decrease in peak amplitude of the field per half cycle is small ($< 100\ \mu T$). This partly depends, of course, on the field decay rate and with a 50 Hz current 100 μT/half cycle corresponds to a decay rate of 10 mT s^{-1} much faster than is normally used, and therefore this requirement is easily met.

There is no reason to expect any marked effect on the cleaning process if higher frequencies are used. The electrical conductivity of terrestrial magnetic minerals and rocks is not sufficiently high for the skin effect to be significant, even in the kHz range. The skin effect arises when an alternating magnetic field enters an electrically conducting medium and is caused by the generation of eddy currents, the effect of which is an exponential decrease with depth of the magnitude of the field in the medium. The skin depth, δ, which varies as the inverse square root of the frequency, permeability and conductivity is the depth at which the field is attenuated by a factor e^{-1}, or 37%. For (pure) magnetite, δ is of the order of 1 cm at 1 kHz and for haematite it is several orders of magnitude larger, and thus the skin effect is not observed at these frequencies in terrestrial rocks or even in pure minerals. However, for iron or nickel–iron, $\delta \approx 1$ mm at 50 Hz and iron meteorites cannot be cleaned by normal a.f. techniques and there may also be a significant effect in those meteorites in which there are iron particles in the above size range. The iron in lunar samples is too fine-grained for the skin effect to be significant at the frequencies normally used for a.f. cleaning.

From the point of view of coil design the choice of frequency affects the Q of the coil and interacts with its inductance and tuning and with losses in the circuit elements. At resonance, the peak voltage V_{pk} developed across the coil and tuning capacity (neglecting the ohmic voltage drop) is given by

$$V_{pk} = 2\pi \sqrt{2} f L i = \frac{i \sqrt{2}}{2\pi f C} \tag{11.2}$$

where f is the frequency of the r.m.s. current i in the coil, R and L are the coil resistance and inductance and C is the tuning capacity. Thus, with $L = 0.5$ H, $f = 50$ Hz and a maximum current of 10 A, $V_{pk} = 2220$ V. Although this voltage should not cause problems with proper coil design and choice of capacitors this coil would not be suitable for $f = 500$ Hz, when an excessive voltage would be generated. The increase in Q with frequency is beneficial in the improved rejection of higher harmonics of the coil current (Section 11.1.3).

Losses can occur in the coil circuit from the self-capacity of the coil and in the tuning capacity. The effect of the former can be represented by a small capacity in parallel with the coil and a resistance in series with it, and the tuning capacity can be considered as having a small resistance across it. The effect of these losses is to decrease the Q of the circuit. In modern capacities with low-loss dielectrics the effect is usually negligible. The apparent increase in the coil resistance is of the form (Terman, 1943)

$$R_{\omega} = R(1 - \omega^2 L C_p)^{-2} \tag{11.3}$$

where C_p is the self-capacity of the coil, L its inductance and R, R_{ω} the coil d.c. resistance and its apparent resistance for a current of angular frequency ω. The increase in R, which will affect the maximum field available per unit power input, may be significant: Widdowson (1974) quotes an apparent increase of 25% in R in a typical coil between 50 Hz (when the effect was negligible) and 500 Hz. Losses can also be contributed by the leads, usually a few metres long, between coil and power supply.

For power supplies other than the mains it is usually somewhat easier (e.g. for frequency stability) to employ oscillators and circuitry working at ~100 Hz or higher, and existing equipment operates in the range 250–400 Hz.

The rise and decay times of the demagnetizing field are not critical. It is usual (and time saving) to allow the field to rise fairly rapidly (~10–30 s, according to maximum field value) and decay somewhat more slowly (~60–120 s): the decay may be linear with time, or as some investigators prefer, it may vary such that the decay rate is slow near the maximum field and increases as the field decays.

(*d*) *Demagnetizing coil design*

The chief design criteria of the coil are that it should accommodate the sample and tumbling mechanism if used and provide the required peak field from the available power supply with reasonable field homogeneity over the sample. As noted earlier there may also be restraints set by the coil inductance and it may be desirable, if possible, to minimize the self-capacity.

If a maximum peak field of ~ 100 mT or more is to be obtained with a reasonable current the only practical coil shape is a multi-layered solenoid. This can take the conventional form, or follow a design based on that of Daniels (1950) in which there is a central gap in the coil parallel to its plane. The latter design improves field homogeneity in the central region for a coil of given total length. It may also be convenient in allowing access for the drive shaft of a tumbler and in the use of a separate power source for each section of the coil.

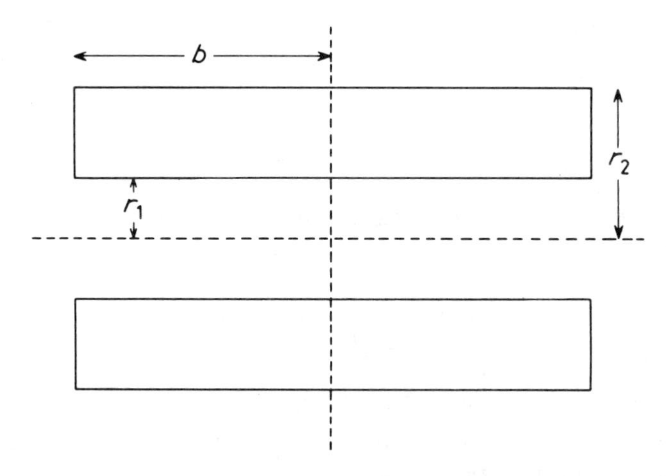

Fig. 11.4 Dimensions used for design of demagnetizing coil.

For a cylindrical coil (without gap) with rectangular winding cross-section and uniform current density the field at the centre is given by

$$B_0 = \frac{\mu_0 Ni}{2c} \log \left[\frac{r_2 + (b^2 + r_2{}^2)^{1/2}}{r_1 + (b^2 + r_1{}^2)^{1/2}} \right] \qquad (11.4)$$

where N is the number of turns, i is the current in amperes and r_1, r_2, b are as shown in Fig. 11.4 and $c = r_2 - r_1$.

The resistance of the coil is

$$R = \frac{2\pi N^2 (r_1 + r_2)\rho}{4bc\sigma} \qquad (11.5)$$

where ρ is the resistivity of the wire in Ω m and σ is the packing factor, i.e. the proportion of the winding cross-section occupied by conductor.

Following Cockroft (1928), by substituting for N from Equation 11.5 in Equation 11.4 we get

$$B_0 = \mu_0 \left(\frac{R\sigma}{r_1 \rho} \right)^{1/2} Gi \qquad (11.6)$$

where

$$G = \left[\frac{\beta}{2\pi(\alpha^2 - 1)}\right]^{1/2} \log\left[\frac{\alpha + (\alpha^2 + \beta^2)^{1/2}}{1 + (1 + \beta^2)^{1/2}}\right] \tag{11.7}$$

and $\alpha = r_2/r_1$, and $\beta = b/r_1$. Thus G is a dimensionless factor which varies only with the shape of the coil, and it has a maximum value of 0.142 when $\alpha \approx 3.0$ and $\beta \approx 2.0$. G does not change rapidly with α and β, i.e. $G > 0.127$ for $\sim 2 < \alpha < \sim 5$ and $\sim 1 < \beta < 4$.

Since the power dissipated in the coil is $W = i^2 R$ W, Equation 11.6 is another form of the well-known Cockroft relation

$$B = \mu_0 G \left[\frac{W\sigma}{r_1 \rho}\right]^{1/2} \tag{11.8}$$

It is clearly important to keep r_1 as small as possible if the power available is limited.

The resistance of the coil in terms of α, β and r_1 is, from Equation 11.5

$$R = \frac{\pi N^2 (1 + \alpha)\rho}{2\beta r_1 (\alpha - 1)\sigma} \tag{11.9}$$

In designing a coil the fixed quantities are usually the internal radius r_1, set by the dimensions of the tumbler or sample holder, and the maximum field required. (Note that in alternating field demagnetization it is the peak field that is important, i.e. $\sqrt{2}$ times the field value that appears in the above equations if the r.m.s. current is used.) Equation 11.8 then gives the approximate power required, assuming a near-optimum value of G. The packing factor is typically in the range 0.6–0.8 and ρ for copper at 20°C is $1.67 \times 10^{-8}\ \Omega$ m.

The choice of α and β is governed by the field homogeneity required, the need to maintain G near its maximum value, and to a lesser extent by the desirability of minimizing the weight of copper required. Homogeneity will improve as β increases (for fixed r) but calculations show that within the acceptable range of α and β and with typical values for r_1 the field will be uniform to better than 1% over a 2.5 cm sample, an acceptable homogeneity for routine a.f. demagnetization. (See also Gray (1921) for homogeneity calculations.)

The maximum voltage of the power supply determines the current required for the maximum field, and therefore the resistance of the coil. Then the number of turns required and the wire gauge is calculated from Equations 11.4 and 11.8. It is generally more satisfactory to use a reasonably high maximum current (~ 10 A) and thus minimize N, since the self-inductance increases as N^2 and there is a small decrease in σ as the wire gauge decreases. Since for a fixed winding volume R also increases as N^2, Q is independent of N (see Section 11.1.3).

There is no simple formula for calculating the self-inductance of a solenoidal multi-layer coil. Terman (1943) and Grover (1947) give expressions which

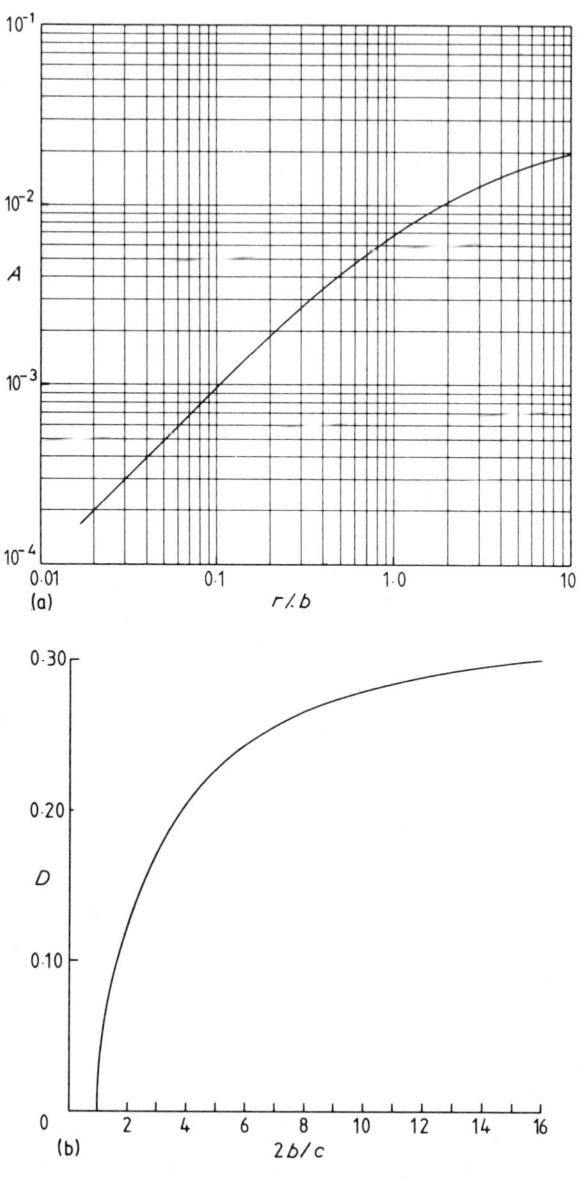

Fig. 11.5 Graphs for determining factors A and D in Equation 11.10. r is the mean radius and $c = r_2 - r_1$, the coil depth.

involve functions of coil length and diameter which are read off tables. Terman's expression can be put in the form

$$L = N^2 r \left[A - \frac{0.0063c}{b} (0.69 + D) \right] \mu H \tag{11.10}$$

where r is the mean radius of the coil (in cm), c is the depth of the winding and A and D are factors depending on the ratios r/b and $2b/c$ respectively. Values of A and D can be read off from the graphs given in Fig. 11.5. The above expression should then be accurate to within a few per cent. Terman, quoting Wheeler (1928), also gives an empirical expression for the inductance of a 'ring-shaped' coil, i.e. one in which $2b \approx c$. The expression is

$$L = \frac{0.31\, r^2 N^2}{6r + 18b + 10c}\, \mu\text{H} \tag{11.11}$$

Equation 11.11 is stated to be accurate to about 1% when the terms in the denominator are approximately equal, i.e. when $2b \approx c \approx r/2$ and in fact it also appears to be reasonably accurate for solenoid coils. The rate of temperature rise in the coil at maximum current can be approximately calculated for the worst case of no heat loss to the surroundings and continuous running, from a knowledge of the power dissipated and the mass and specific heat of the windings. If necessary the coil can be constructed with one or more longitudinal annular channels through which air can be blown, at the cost of a small loss in coil efficiency. Copper wire can also be obtained with heat-resistant insulation which again may mean a small sacrifice in performance through a decrease of the packing factor.

A reason for allowing only a moderate temperature rise is the consequent increase in coil resistance and therefore decrease in the maximum field available. The temperature coefficient of resistance of copper is $3.9 \times 10^{-3}\,°\text{C}^{-1}$, implying an increase in coil resistance of $\sim 10\%$ for a temperature rise 25°C.

As a guide to coil construction some approximate values are given for parameters derived from the above equations for a coil in which $B_{max} = 150$ mT (peak) and $r_1 = 4.0$ cm. With a packing factor of 0.70 the power required is about 450 W, say 10 A maximum current at 45 V, giving a permitted coil resistance of 4.5 Ω. Then with $\alpha = \beta = 1.5$, $G = 0.12$ and about 1000 turns of 17 SWG enamel-covered wire are required (4.4 kg). The coil inductance and resistance are 0.050 H and 3.3 Ω respectively and the temperature rise at 10 A is of the order of 10°C/min. At 50 Hz the Q factor of the coil, the significance of which is described in the next section, is 4.8.

As noted earlier, the alternative coil design consists of two coaxial solenoidal coils with a small gap between them (Fig. 11.6). Among the potential advantages of this configuration are access for a tumbler drive shaft, improved field homogeneity and cooling and the use of a separate power supply for each coil. The design, which follows that of Daniels (1950), is basically the establishment of the Helmoltz condition $(\mathrm{d}^2 H/\mathrm{d}z^2)_0 = 0$ for a coil pair whose winding depth and width are comparable to their internal radius. The condition is

$$4\left[\frac{\beta^{-4}}{(1+\beta^{-2})^{3/2}} - \frac{\alpha^3 \beta^{-4}}{(1+\alpha^2\beta^{-2})^{3/2}}\right] + 3\lambda(1-\alpha^{-2}) = 0 \tag{11.12}$$

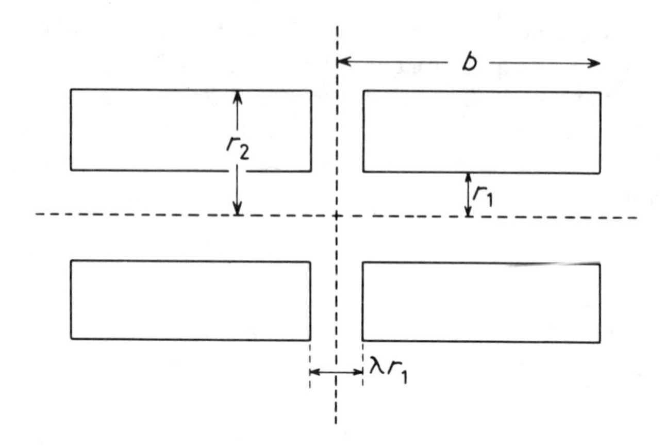

Fig. 11.6 Dimensions used for split coil design.

where λr_1 is the gap width and α and β are as before. In a coil providing 100 mT (peak) at 7.2 A (Widdowson, 1974; de Sa and Widdowson, 1975) in which $\alpha = 2.1$, $\beta = 1.6$, $r_1 = 3.5$ cm and the gap width is 1.3 cm ($\lambda = 0.37$) the effective G-value is 0.121, compared with 0.137 for the same coil dimensions with no gap. The field homogeneity is better than 99.5% over a 2.5 cm sample volume. A larger gapped coil for use with a three-axis tumbler is in use in the author's laboratory. The dimensions (each coil) are $r_1 = 8.0$ cm, $r_2 = 11.1$ cm, $2b = 12.0$ cm and gap width 1.8 cm. Each coil has 1420 turns of 17 SWG wire. The two coils, run in parallel, have inductance and resistance of 0.19 H and 4.4 Ω respectively, with a Q of 13.5. The coils will carry 12 A for short periods, giving a peak field of 180 mT. It can be seen from the coil parameters derived for a coil with no gap that the power requirement and heat dissipation for moderate field cleaning ($B_{max} < \sim 150$ mT) is not very demanding and the decrease in the G-value associated with a gapped coil is quite tolerable.

Although peak demagnetizing fields of ~ 100 mT are adequate for the routine cleaning of most igneous rocks in which magnetite or titanomagnetite carries the NRM, some investigators have examined the a.f. cleaning of the more highly coercive sedimentary rocks, using maximum peak field of ~ 500 mT (see, for example, Klootwijk and Van der Berg (1975)). The design of suitable coils still follows the procedure outlined above but with these high fields the power requirement rises (as B^2 for fixed r_1) and with it the problem of Joule heat dissipation.

Roy *et al.* (1973) use the approximate formula for the field at the centre of a solenoid coil of mean radius r

$$B = \left(\frac{\mu_0 N_i}{2b}\right) \cos \delta \tag{11.13}$$

where $\cos\delta = b/(b^2+r^2)^{1/2}$ from which the number of turns per unit length of the coil $(N/2b)$ needed to produce the required maximum field from the maximum available current can be determined. The authors show that for a given input voltage it is useful to design the coil to give as high a value as possible of B/R, where R is the coil resistance. For fixed r_1 and b, B increases with the wire diameter, which is then chosen to permit the required maximum current from the supply voltage and the required turns/cm. In their calculations the authors use the value of the resistivity of copper at 65°C, to allow the design maximum field to be reached at an elevated working temperature. The coil dimensions are $r_1 = 5.7$ cm, $r_2 = 10.9$ cm, $b = 9$ cm and it contains 2460 turns of 15 SWG wire, giving 19 mT (peak)/A. The maximum field is 380 mT (peak) from 3.95 kW input power: the Q of the coil is 20 at the working frequency of 60 Hz.

Hailwood (1972) designed a split coil for high fields with forced air cooling through longitudinal channels in the windings. The coil shape follows Daniels' (1950) design for good field uniformity ($r_2 = b = 3r_1$, gap width $= 0.18\,r_1$) and $r_1 = 2.0$ cm was chosen to accommodate a 2.5 cm cylinder without tumbling. Cockroft's expression (Equation 11.8) gives the approximate power required to produce the chosen peak field of 500 mT, and thus the maximum current and permissible maximum coil resistance are determined for a given supply voltage. The required wire gauge can then be calculated so that the winding space is filled (1930 turns of 15 SWG wire) and the total resistance (3.8 Ω) is as near as possible to that required. With an inductance of 0.064 H at 50 Hz, $Q = 5.3$, an undesirably low value which is a feature of coils with a small internal radius.

The coil used by the Utrecht group (J. D. A. Zijderveld, private communication) has $r_1 = 9.0$ cm, $r_2 = 16.0$ cm, $b = 24.0$ cm, with 9950 turns of wire of rectangular cross-section 2.2 mm × 1.6 mm: the coil constant is 24.8 mT (peak)/A (maximum field ~350 mT) and $Q = 62$ (50 Hz).

Design features of various types of coils for a.f. demagnetization are discussed by Kono (1982).

11.1.3 a.f. demagnetizing equipment

Alternating field demagnetizing equipment consists basically of the coil (and sample holder or tumbler), a capacity bank for tuning the coil, power supply and current reduction mechanism and coil system or magnetic shield for providing a low ambient field over the sample.

In some of the early equipment, usually powered from the electricity mains at a frequency of 50–60 Hz, smooth, transient-free reduction of the coil current was achieved by the use of a liquid electrolyte column with an internal electrode movable along the column length. The column was used either as a variable resistance or, more commonly, as a variable voltage divider, the voltage across its ends being adjusted by a conventional variable transformer

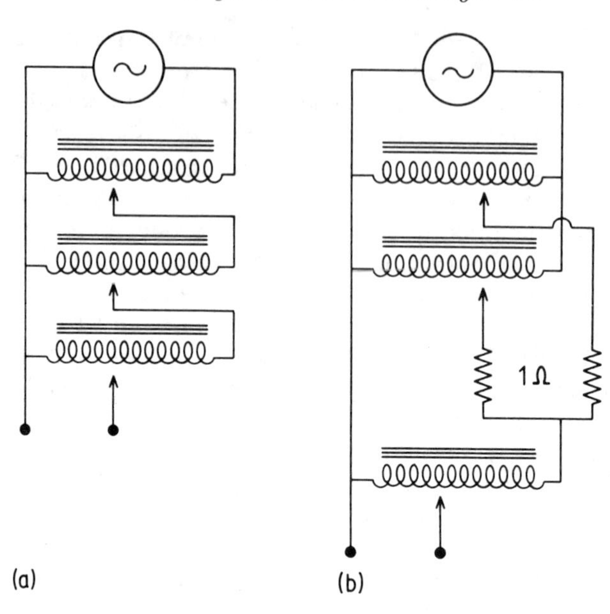

Fig. 11.7 Two circuits for current control using variable transformers.

supplied from the mains. Because of Joule heating and electrolytic action (e.g. between copper electrodes and copper sulphate electrolyte) this method was not entirely satisfactory and brushless variable transformers of, for example, the Inductrol type, in which variable flux linkage between two coils is obtained by rotating one relative to the other were favoured. Some workers (e.g. As, 1967b) have claimed that satisfactory demagnetization can be achieved using a variable transformer with brush take-off for current reduction, and this may well be possible with certain rocks using equipment in which there is efficient rejection of second harmonics in the waveform. The probability of direct current pulses (arising in the transformer) will depend on the brush configuration. Figure 11.7 shows two ways in which coupled variable transformers have been used for current control,

The development of solid state power amplifiers and current reduction devices now enables a.f. equipment to be run independently of the mains. The working frequency is usually somewhat higher, e.g. 400 Hz and 275 Hz respectively in the commercially-made Schonstedt and Highmoor demagnetizers: the latter instrument is based on the design of de Sa and Widdowson (1975), in which the current increase and decrease is controlled by an LED and photo-resistor.

Tuning the coil to resonance at the working frequency serves two purposes, to minimize the impedance of the coil (ideally to its d.c. resistance only) and so obtain the maximum current for a given applied voltage and to partially filter out second and higher harmonics of the current waveform.

The resonant frequency, f_0 Hz, of a series resonant L–C–R circuit is given by

$$2\pi f_0 = \omega_0 = (LC)^{-1/2} \tag{11.14}$$

Provided therefore that the current source can be regarded as of zero impedance as is usually the case, the required tuning capacity can be easily estimated from Equation 11.14. The exact value is best determined by adding or subtracting trimming capacities until the maximum current is obtained with a fixed input voltage. The maximum peak voltage appearing across the coil and capacity may be ~1000 V or more (Section 11.1.2(c)) and capacities should be chosen with an appropriate low-loss dielectric and high working voltage.

The performance of the tuned coil as a filter is conveniently analysed in terms of the quality factor Q of the coil, where $Q = \omega L/R$. If the working frequency is ω, then for an applied voltage V with a fractional second harmonic content of p and no tuning

$$i_\omega = V(R^2 + \omega^2 L^2)^{-1/2} \tag{11.15}$$

and

$$i_{2\omega} = V(R^2 + 4\omega^2 L^2)^{-1/2} \tag{11.16}$$

where i_ω and $i_{2\omega}$ are the fundamental and second harmonic currents. Typical Q-values for demagnetizing coils are 5–20 and therefore $R^2 < \omega^2 L^2$ and $Q^2 \gg 1$ and we can write

$$i_\omega = V/RQ \tag{11.17}$$

and

$$i_{2\omega} = V/2RQ \tag{11.18}$$

and

$$i_{2\omega}/i_\omega = 0.5\,p \tag{11.19}$$

Thus, the untuned coil attenuates the second harmonic content by a factor of 2. With a perfectly tuned coil we have

$$i = V/R \tag{11.20}$$

$$\begin{aligned} i_{2\omega} &= pV\left[R^2 + \left(2\omega L - \frac{1}{2\omega C}\right)^2\right]^{-1/2} \\ &= pV\left[R^2 + \left(2RQ - \frac{RQ}{2}\right)^2\right]^{-1/2} \\ &= pV\left[\left(1 + \frac{9Q^2}{4}\right)R^2\right]^{-1/2} \\ &\approx 2pV/3QR \end{aligned} \tag{11.21}$$

With perfect tuning therefore, the second harmonic is reduced by a factor of approximately $3Q/2$, i.e. between about 8 and 30 for the above range of Q of

5–20. The attenuation achieved in practical circuits may fall short of prediction because of stray capacity, tuning capacity losses, etc. It can easily be shown that Q is also the factor by which the voltage across the coil and capacity exceeds the supply voltage.

For a coil of fixed length and winding cross-section (and assuming a constant packing factor), $L \propto N^2$ and $R \propto N^2$ and therefore Q is independent of N, the number of turns. It can be shown that Q increases approximately as the square of the mean coil diameter, for constant α and β.

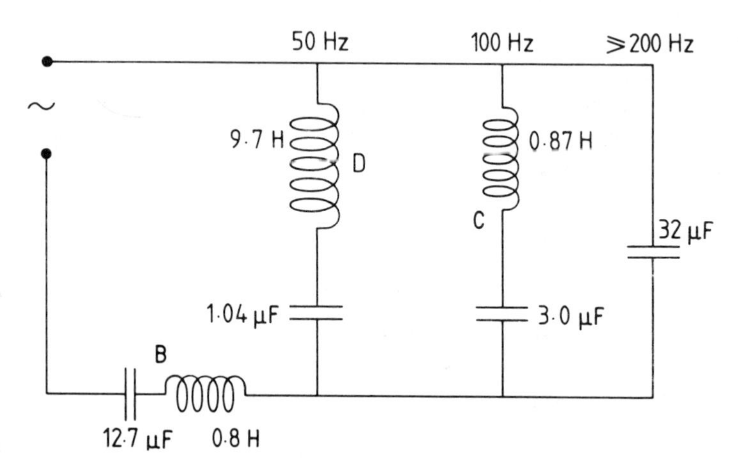

Fig. 11.8 Demagnetizing coil circuit as used in the Utrecht laboratory. D is the demagnetizing coil.

In some demagnetizing equipment (e.g. Irving, Stott and Ward, 1961; Larochelle and Black, 1965) an *L–C* circuit tuned to twice the operating frequency is connected across the tuned demagnetizing coil. The aim is to further reduce the second harmonic current in the coil by providing a short-circuit path for it outside the coil. Assuming the current source is of very low or zero impedance such an additional circuit has little or no effect. However, if an impedance is placed in series with the current source the additional tuned circuit becomes effective. This system has been adopted by the Utrecht group (J. D. A. Zijderveld, private communication) who use an *L–C* circuit tuned to the fundamental frequency (50 Hz) as the series impedance (Fig. 11.8). For the 50 Hz current there is a sacrifice of about 10% decrease in the maximum current through the demagnetizing coil, because of the ohmic resistance of circuit B (Fig. 11.8) but the impedance of B at 100 Hz ($Z \approx 380\Omega$) causes the additional tuned circuit C to attenuate the 100 Hz current in the demagnetizing coil by a factor of approximately 75. This is, of course, in addition to the (theoretical) attenuation factor of ~ 100 achieved by the tuned coil alone. This efficient rejection of second harmonic current is a major factor in the successful a.f. cleaning of rocks in high fields reported by the Utrecht group.

A wide variety of a.f. cleaning equipment is described in the literature. Thellier (1966), one of the pioneers of the technique, describes a system suitable for large samples in which the field is increased and decreased by moving the coil towards the sample until the latter is at the coil centre, and then moving the coil slowly away. As (1967b) used three variable transformers for current control (Fig. 11.7(a)) and originated the use of additional tuned circuits for even harmonic rejection. Other early equipment is described by Creer (1959) and Irving *et al.* (1961). Inductrol-powered equipment is described by Larochelle and Black (1965) and Doell and Cox (1967b), and these authors also tested the performance of the demagnetizers for their ability to minimize ARM in various test samples. Larochelle and Black demonstrated the importance of having a very low direct field parallel with the alternating field axis, and Doell and Cox noted some anomalous magnetization which persisted even when there was good zero direct field over the sample. This is now believed to be an early observation of rotational remanent magnetization (Section 11.1.4(b)).

Snape (1967) designed equipment with a working frequency of 500 Hz, using a 5 kW alternator to supply to coil current. The current was varied by altering the direct current in the alternator windings. The split, forced-air cooled coil was designed for high currents ($\sim$100 A) and contained 240 turns with $r_1 = 4$ cm. The coil produced 1.7 mT (peak)/A and the low inductance of 0.003 H ($Q = 13.5$) ensured that the peak voltage across the coil and tuning capacity did not exceed $\sim$1100 V. A possible advantage in using an alternator is a smaller second harmonic content in the current waveform of a more constant phase difference relative to the fundamental; the latter feature was used by Snape (1971) in his procedure for removing ARM from cleaned samples (Section 11.1.4(b)).

Other demagnetizers are described by McElhinny (1966), Hailwood (1972) and Roy *et al.* (1973). Three commercially built demagnetizers are available at the time of writing, the Digico, Highmoor and Schonstedt instruments. The latter is a 400 Hz demagnetizer with a maximum field of 100 mT, and it is available with a two-axis reversing tumbler to minimize rotational magnetization (Section 11.1.4(b)). The ambient field is nulled to $< 50\gamma$ over the sample by means of a five-shell cylindrical magnetic shield and there are six electronically controlled decay rates of 0.1–4 mT s^{-1}. The Highmoor demagnetizer is based on the design of de Sa and Widdowson (1975). It has a two-axis tumbler and a maximum field of 100 mT in a forced air-cooled split coil, each half of which is supplied by a separate power amplifier. Variable rise and decay rates of the field are available, and also facilities for holding the field at any point in the cycle and for terminating the cycle at any point. The Digico instrument is supplied by a two-Variac system (Fig. 11.7(b)) with a choke in series with the demagnetizing coil which has the effect of improving the Q of the circuit. The sample is rotated about two axes and the coil (maximum field 100 mT) is placed in a three-layer shield in which the residual field is $\sim$100 nT.

The rise and decay times are fixed. In the three demagnetizers the demagnetization cycle is automatically controlled.

Hummervoll and Totland (1979) have constructed an automatic demagnetizing and NRM measuring device. After demagnetizing in a split-coil at the required peak field, the coil opens and the sample is raised and secured in a holder beneath an astatic magnetometer with feedback. After automatic recording of values of $\pm X$, $\pm Y$, $\pm Z$, the sample is lowered back into the coil for the next demagnetizing step. The maximum field available is ~ 70 mT and magnetizations in the range 3×10^{-2} to 2×10^{-4} A m^2 kg^{-1} can be measured.

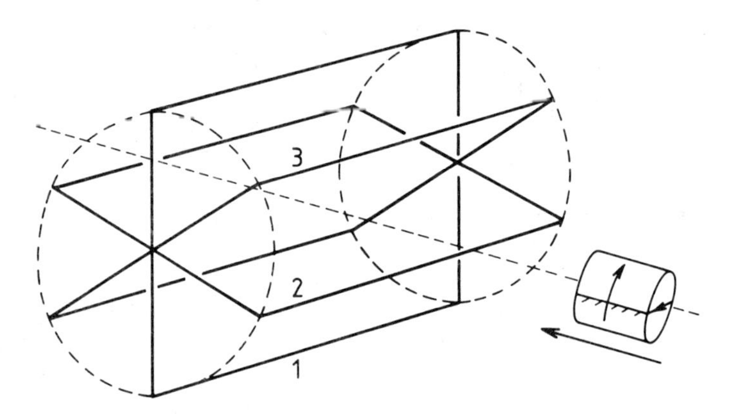

Fig. 11.9 Demagnetizer of Noël and Molyneux (1975). The numbers indicate the three coils.

A recent, new approach to a.f. demagnetizing is that of Noël and Molyneux (1975), illustrated in Fig. 11.9. A rotating magnetic field (frequency f_1) is generated at the centre of a three-coil system by feeding each coil with the alternating current of relative phase difference $2\pi/3$. If a sample is passed through the coil, in the manner shown, all directions in the z–x plane will experience an alternating field which has a maximum value at the coil centre and decreases to near zero at the coil edge. Thus, there will be a demagnetizing action for all directions in the z–x plane and if the sample is continuously rotated about the z-axis at an appropriate frequency f_2 ($f_2 < f_1$, and not an integral multiple of it), essentially all directions in the sample will see the demagnetizing field. In the equipment described, $f_1 = 50$ Hz, $f_2 = 1$ Hz and the sample moves through the coil at 0.5 cm s^{-1}. The design has advantages of speed and the capability of magnetically cleaning long cores of material. At the time of writing the chief limitation to its use is a low maximum field of 12 mT, but in principle this can be increased and E. Hailwood (private communication) has constructed similar equipment in which the peak field is 60 mT. The sample is placed at the coil centre and is rotated from 0° to 360° and then from 360° to 0° about the horizontal axis (Fig. 11.9) about once per second during decay of the field to prevent the acquisition of RRM (Section 11.1.4(b)). A two-layer magnetic shield reduces the ambient field to $\sim 200\,\gamma$.

This idea has been extended by Matsuda, Hyodo, Inokuchi, Isezaki and Yaskawa (1981). They use three orthogonal coil pairs to provide a 'tumbling' magnetic field which demagnetizes a stationary sample, thus inverting the conventional technique. The coil pairs, which enclose a central volume in which eight 2.5 cubic samples can be demagnetized, carry currents which produce magnetic fields of the form

$$\begin{aligned} B_x &= B_0 \sin \omega t \cos \omega^1 t \\ B_y &= B_0 \sin \omega^1 t \sin \omega t \\ B_z &= B_0 \cos \omega t \end{aligned} \qquad (11.22)$$

The central field at any instant is $(B_x^2 + B_y^2 + B_z^2)^{1/2} = B_0$. By choosing appropriate values of ω and ω^1 different field tumbling regimes can be obtained. Matsuda *et al.* use $f = \omega/2\pi = 60$ Hz and $f' = \omega^1/2\pi = 2$ Hz. Like Noël and Molyneux's demagnetizer the maximum field available, at 19.2 mT, is low. If this can be increased the demagnetizer appears to have advantages for cleaning fragile material which cannot be tumbled and in producing negligible ARM in samples if the external field is nulled sufficiently.

11.1.4 Some aspects of a.f. demagnetization

(*a*) *Other a.f. techniques*

Before discussing some general aspects of a.f. cleaning, brief mention is made of other possible a.f. techniques which have been investigated but have not yet been successfully utilized.

A simple technique in principle is to rotate the sample in a decreasing direct field, the axis of rotation being perpendicular to the field. All directions in the plane perpendicular to the rotation axis will see a decreasing alternating field and three-axis treatment should effectively clean the sample. In practice the problem is to align the rotation axis perpendicular to the field with sufficient accuracy such that the (direct) field component parallel to the axis of rotation is small enough to avoid the production of ARM in the sample. For example even in a modest peak field of 20 mT misalignment of the rotation axis by only 0.1° will result in a peak axial field in the rock of $\sim 35\,\mu$T: this is above the lower limit for ARM to be a problem in typical rocks, even allowing for its lower intensity when the alternating and direct fields are mutually perpendicular. In principle it appears that a two-stage procedure for each cleaning field would be satisfactory for this method. The sample is first spun about one axis and the cleaned component of NRM perpendicular to this axis then measured. This procedure is then repeated for a spin axis perpendicular to the first. Any ARM acquired during the first spin would be removed during the second spin and the ARM left in the sample after the second spin would be removed in the higher field of the next cleaning step. In practice there might be problems in obtaining a reliable direction of cleaned NRM if the ARM intensity was comparable to or greater than the residual NRM intensity.

The technique of pulsed demagnetization has been discussed but no working equipment has yet been constructed for routine use. The basis of the technique is the discharge of a capacity through a suitably designed coil which, with the capacity, forms an oscillatory circuit: the circuit parameters are chosen so that the oscillations are damped and produce a smoothly decreasing alternating current in the coil. The potential advantages include rapidity of operation, simplicity and cheapness.

In a high Q, series L–C–R oscillatory circuit excited by capacity discharge the instantaneous current i through the coil (inductance) is given by

$$i = i_0 \exp(-tR/2L) \sin \omega_0 t \tag{11.23}$$

where $i_0 = V(C/L)^{1/2}$ is the maximum current, V is the charging voltage of the capacity C and $\omega_0 = (LC)^{-1/2}$ is the natural frequency of the circuit of inductance L and resistance R. The important features of the circuit are the maximum current and number of turns on the coil (and therefore demagnetizing field) and the fractional decrease of the field per cycle.

Using constant V and C and a coil of fixed internal radius and winding cross-section containing N turns, $\omega_0 \propto L^{-1/2} \propto N^{-1}$, and the maximum field $B_0 \propto i_0 N$: therefore B_0 is independent of N (neglecting packing factor variations and other effects when N is small). The fractional decrease of the current per cycle $(i_n - i_{n+1})/i_n$ is derived from the logarithmic decrement

$$\log\left(\frac{i_n}{i_{n+1}}\right) = \pi R \left(\frac{C}{L}\right)^{1/2} = \frac{\pi}{Q} \tag{11.24}$$

where $Q = \omega_0 L/R$. If the field decrease per cycle is not to be greater than 1%, then

$$Q = > \frac{\pi}{\log 1.01} \approx 315$$

This is a high Q-value and may be difficult to achieve in a practical circuit. Since $Q = \omega_0 L/R$ and for a coil of given geometry $\omega_0 \propto L^{-1/2} \propto N^{-1}$, $L \propto N^2$ and $R \propto N^2$, then $Q \propto N^{-1}$ and a coil of few turns is required, implying a large maximum current ($\sim$100 A) if a reasonable maximum field is to be obtained.

As an example we consider a coil of internal radius 2.0 cm (which will accommodate a 2.5 cm sample for three-axis treatment) and 10.0 cm long, wound with four layers of 10 SWG (3.3 mm diameter) wire, giving $N = 120$. For this coil R and L are approximately $3.7 \times 10^{-2}\ \Omega$ and 2.6×10^{-4} H respectively, and with $C = 1\ \mu$F, $\omega_0 \approx 6.2 \times 10^4$ ($f_0 \approx 9900$ Hz) and $Q \approx 430$. If $V = 1$ kV, $i_0 \approx 70\ A$ and $B_0 \approx 110$ mT (peak), the field decay envelope is exponential and the field will decay to $1/e$ of its initial value in a time $t = 2L/R \approx 10^{-2}$ s.

It can be shown that if C is increased by a factor α and the coil radius and winding cross-section are kept constant, then the Q value can be maintained if L is reduced to L/α (by reducing N to $N/\alpha^{1/2}$). For the same charging voltage V the maximum current i_0 increases by α and the peak field obtainable by $\alpha^{1/2}$.

Although the above example suggests that a coil with the required parameters can in theory be constructed, it is the whole oscillatory circuit that must be of high Q, and this demands particular qualities in the capacity, switching device and connecting leads. Another factor which may be important is the apparent increase in R for high-frequency currents (Equation 11.3).

Another potentially useful technique, but one that has not been extensively investigated, is simultaneous demagnetization and measurement of residual NRM, i.e. the analogue of the measurement of residual NRM at elevated temperatures in continuous thermal demagnetization (Section 11.2). Shashnakov (1970) describes the use of an astatic magnetometer (time constant $\approx$ 15 s), on either side of the lower magnet of which there are identical tuned coils: one of these contains the sample and provides the demagnetizing field and the other is to compensate for the effect of the first coil on the magnet system. The magnetometer is already insensitive to the 50 Hz field because of its much greater time constant. The measuring method is not described but presumably involves measurement of three mutually perpendicular components of NRM after each increase of the field. The technique would appear to have some advantage of speed and the elimination of ARM (cf. the elimination of spurious PTRM in continuous thermal cleaning), but may lack sensitivity. A similar technique has been examined at Bergen University by R. Hummervoll (private communication), who used fluxgate detectors and a rotating sample.

11.1.4 (*b*) *a.f. cleaning–general*

In an ideal demagnetizing sequence the NRM intensity of a sample decreases (or it can increase) and its direction of NRM changes systematically as the cleaning field increases until a 'plateau' occurs in the intensity–field plot and the NRM direction remains constant. This indicates the presence of a residual, highly stable primary remanence. In a group of samples from a site the directions of NRM condense towards a well-defined mean. As already noted it is the constancy of NRM direction which is the important feature: continuing decay of intensity may well occur at high fields, but this is not important provided the residual intensity is sufficient for satisfactory NRM measurement.

Although in many cases the above sequence may approximate to what is observed in practice, the removal of secondary NRM may be more or less obscured through the simultaneous acquisition of ARM. For a given direct field the intensity of ARM increases as the field increases whereas the NRM is usually decreasing, and thus the ARM/NRM ratio steadily increases with the alternating field, and therefore also the likelihood of anomalous results from the cleaning process. These will be apparent as increasing scatter of the directions of NRM at higher fields since ARM is expected to be randomly directed relative to the sample axes, particularly if tumbling is employed and the direct field is of second harmonic origin. Repeat demagnetization of a sample at the same field value should indicate the acquisition of ARM by greater scatter of NRM intensity and direction than that expected from measurement errors.

Although three-axis demagnetization (sample not tumbled) is now somewhat uncommon, some workers favoured the technique after the existence of rotational remanence was established (see below). The usual procedure is to apply the field successively along three orthogonal axes in the sample and then measure the NRM: alternatively, the NRM along each axis is measured after the field is applied along that axis, and the three components then combined to give the total vector. The latter technique, although slower, is probably preferred.

Snape (1971) reports the acquisition of ARM by red sandstones during cleaning in three-axis (non-tumbling) equipment. He proposed a method for eliminating the ARM acquired in a certain field by partial removal and introduction of an equal and opposite ARM. The sample is turned through 180° about an axis perpendicular to the coil axis and demagnetized again in a field which produces an ARM of half the initial intensity: the required field is derived from a previously determined curve of ARM intensity against alternating field for the rock being cleaned. The author showed that, with constant direct field, ARM intensity was approximately proportional to alternating field in the sandstones and he therefore suggested the use of half the initial field for the reversed sample for routine cleaning. Apart from its inconvenience as a routine procedure the technique presupposes a constant direct field, which in this case arose from waveform asymmetry. This implies a second harmonic content of constant phase and amplitude which might be a valid assumption in the author's equipment, which was supplied by a 500 Hz alternator, but might not be in a mains-supplied unit.

Hailwood and Molyneux (1974) compared the intensity of ARM acquired in a given alternating field with a constant direct field of external origin with that acquired in the same direct field produced by appropriate asymmetry in the waveform at the same alternating field level. The authors show that the former intensity is expected to be twice the latter (based on Preisach diagram analysis), but in test on various rocks the factor was variable and somewhat greater than two in fields up to 50 mT. The other feature that distinguishes an ARM required in an external direct field from one of asymmetry origin is the more rapid increase of the ARM with peak alternating field in the case, provided saturation ARM is not approached. The form that the ARM intensity–peak field curve takes will depend on the distribution of grain coercives in the rock: if this is such that ARM is approximately proportional to peak alternating field when h is of external origin, then for h of asymmetry origin ARM $\propto \sim B^2$, since $h \propto B$.

A new type of anomalous behaviour during a.f. cleaning, which does not appear to be explicable by ARM has been encountered in some lunar samples. It is characterised by non-reproducible NRM resulting from repeated demagnetization in a given alternating field, and has been extensively investigated by Hoffman and Banerjee (1975) in a lunar basalt (15535) (Fig. 11.10(a)). The NRM appears to be roughly confined to a plane in the sample, and the authors propose a mechanism involving multidomain grains which cannot be demagnetized and the NRM of which is confined to a plane but is only loosely pinned within that plane. This preferred plane could arise from anisotropy in the rock fabric, perhaps associated with shock due to lunar meteoritic impacts (Brecher, 1976). A similar and perhaps related effect has been observed in other lunar samples in which a roughly reproducible NRM intensity and direction results from repeated demagnetization at the same peak field value but considerable scatter is observed in the mean NRM after cleaning in successively higher fields (Fig. 11.10(b)).

A hitherto unsuspected anomalous magnetization associated with a.f. cleaning was first reported by Wilson and Lomax (1972), although a similar

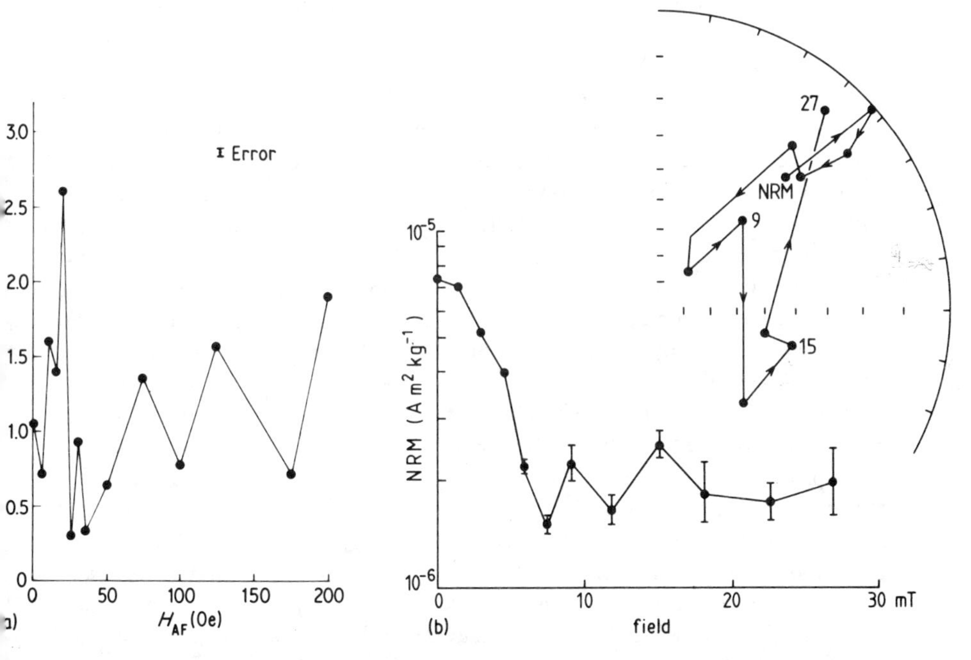

Fig. 11.10 Examples of anomalous demagnetization behaviour: (a) lunar basalt 15535 (from Hoffman and Banerjee, 1975); (b) lunar breccia 66055 (from Collinson, Stephenson and Runcorn, 1973).

magnetization noticed by Doell and Cox (1967b) in tests of their demagnetization equipment was probably of the same type. Known as 'rotational remanent magnetization' (RRM), it can be acquired along the axis about which a sample is rotating in a decreasing alternating field. In two- or three-axis tumblers RRM appears along the innermost axis. The experiments of Wilson and Lomax on basalts and synthetic magnetite showed that RRM intensity in these samples increased with increasing peak alternating field, generally showed a maximum at rotation speeds between about 0.1 and 1.0 rev s^{-1} (and was detectable even at 0.02 rev s^{-1}) and varied with the a.f. frequency and with angle θ between the rotation and alternating field vectors, with a maximum at $\theta = 90°$. In a 60 mT peak field typical RRM intensities, which in their experiments were always antiparallel to the rotation vector, were 10^{-5}–10^{-4} A m^2 kg^{-1}. The experiments were conducted in such a way as to eliminate any ARM acquired.

These results were generally confirmed by Brock and Iles (1974), who used a two-axis tumbler, but these authors found that in one basalt sample the RRM was parallel to the rotation vector: using a single rotation axis (as did Wilson and Lomax) they then found that this basalt acquired an RRM parallel to the rotation axis in low alternating fields but was antiparallel when $B_{pk} > 30$ mT. Further investigations are reported by Stephenson (1976) and Edwards (1980). Stephenson studied the generation of axial fields and moments in good conductors and in soft iron, obtaining negative results in both cases. He also showed that in two basalt samples the RRM fell to zero when the rotational and field frequency were equal and became rapidly large and negative (i.e. parallel to the rotation axis) at higher rotation speeds, and also that only relative rotation of field and sample is required, i.e. RRM can be produced by rotating the field around a stationary sample. The latter result apparently rules out association of RRM with some fundamental magnetic phenomenon such as the gyromagnetic effect, which appears to be too small by several orders of magnitude to explain observed RRM intensities. However, extensive theoretical and experimental investigations by Stephenson (1980a,b) suggest that, under certain assumptions, the gyromagnetic effect may be significant and is capable of explaining most RRM characteristics. During this work an elegant technique for measuring the torque exerted on a rotating rock in a magnetic field was developed (Stephenson, 1980c).

At present RRM is still being investigated but there is no agreement among different investigators regarding its importance in routine cleaning. Whether it appears seems to depend on both rock and demagnetizer. Some laboratories have modified the tumbling procedure in their a.f. equipment, mainly by reversing the sense of rotation every few seconds which should eliminate significant RRM acquisition if the field decay is not too rapid. Wilson and Lomax (1972) suggest a two-axis tumbler with 1:2 rotation ratio (vertical: horizontal) and rotation reversal every two seconds (Fig. 11.11). They also emphasize the point that the axis of the alternating field should be in the plane defined by the horizontal rotation axis. Alternatively, three-axis cleaning

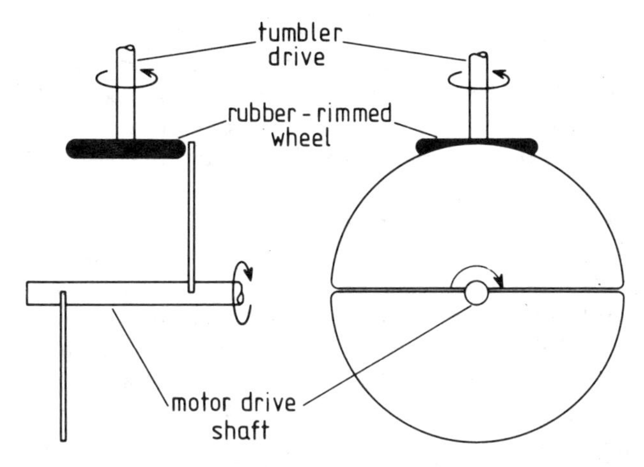

Fig. 11.11 Device for periodically reversing the rotation of a tumbler. The two semi-circular plates are fixed to the shaft as shown, each turning the rubber-rimmed wheel in the opposite sense.

without tumbling can be reverted to, a perfectly satisfactory if somewhat lengthier procedure, provided that direct fields of external or second harmonic origin are small enough to prevent ARM acquisition. Hillhouse (1977) proposed the following five-step procedure where tumbling is used:

1. Demagnetize at required peak field B.
2. Measure the NRM, J_1.
3. Demagnetize again in B but with the tumbler rotation reversed.
4. Remeasure NRM, J_2.
5. Average the vector components of J_1 and J_2 and calculate the NRM of the sample, now free of RRM.

This should be satisfactory provided significant ARM is not acquired in steps (1) and (3).

11.2 Thermal demagnetization

11.2.1 Introduction

Magnetic grains possess a characteristic time constant known as the 'relaxation time' (τ), which is a measure of the time required for the grain to achieve equilibrium with its surroundings. Values of τ range from a small fraction of a second in para- and diamagnetic materials, in which thermal equilibrium is rapidly regained after removal of an applied field, to periods in excess of 10^{10} years for grains possessing a highly stable TRM or CRM.

The expression for τ is of the form

$$\tau = \frac{1}{C} \exp\left(\frac{vB_c J_s}{2kT}\right) \tag{11.25}$$

where C is a frequency factor ($\sim 10^{10}$ s^{-1}), v is the volume of the grain of coercivity B_c and spontaneous magnetization, J_s, k is Boltzmann's constant and T is the absolute temperature. Thus τ is strongly dependent on T, and is related directly to the coercivity. Unstable components of NRM in rocks are associated with relaxation times in the approximate range of 100 s (leading to rapid acquisition of viscous remanence) to $\sim 10^8$ years, at normal temperature.

If the temperature of a rock is raised until the relaxation time of some fraction of the particles carrying NRM has been reduced to a few minutes or less, the NRM of these particles will be 'unblocked' and their contribution to the NRM of the rock lost. If the rock is then cooled in an ambient magnetic field the particles will acquire a PTRM in the field, whereas cooling in zero field will result in random orientation of particle or domain moments at room temperature and therefore no contribution to the NRM of the rock.

The above process is the principle behind both the acquisition of an important type of secondary magnetization, namely PTRM, and its removal by thermal cleaning. In the latter the rock sample is raised to successively higher temperatures and its remaining NRM measured either at the elevated temperatures or after cooling in zero field. The lower blocking temperature components are progressively removed in this way, leaving the primary component of high stability and with a blocking temperature near the Curie point relatively unaltered. Irving and Opdyke (1965) use the terms 'thermally distributed' to describe the blocking temperature spectrum when it covers a range of temperatures up to the Curie point and 'thermally discrete' when the spectrum is restricted essentially to within ~ 100°C below the Curie point (Fig. 11.12).

Thermal cleaning has proved generally successful when applied to sediments and in particular to redbeds, in which PTRM acquired during burial and VRM acquired in the recent geomagnetic field are common sources of secondary NRM. Reasons for the successful thermal cleaning of haematite-carried NRM in redbeds have been discussed by Dunlop (1970).

It should be noted that the observed blocking temperature for particular particles in a rock may not be that at which they acquired their PTRM, because the 'unblocking' temperature increases with the time which the rock spent at the elevated temperature. An example from the theoretical work of Chamalaun (1964) and Briden (1965) shows that PTRM acquired by burial at 150°C for 10^4 years would require 450°C for removal in a typical laboratory heating time of 5 minutes. Pullaiah *et al.* (1975) have recently discussed this topic further, and in the same context Bol'shakov and Faustov (1976) have studied the removal of VRM from some Permian sediments.

11.2.2 Zero field cleaning

This is the simplest cleaning technique involving thermal energy but one which is only very rarely useful. It can be used for the removal of a viscous

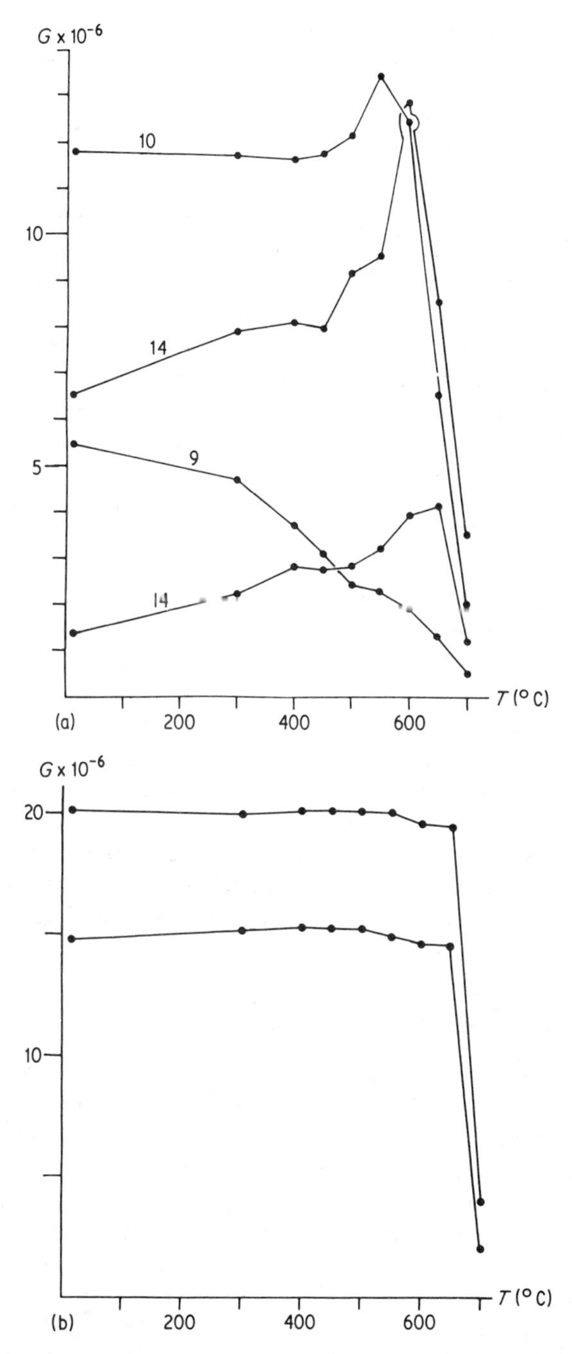

Fig. 11.12 Thermal demagnetization of redbeds of the Bloomsburg formation, USA: (a) thermally distributed, and (b) thermally discrete blocking temperature spectra. (From Irving and Opdyke, 1965)

magnetization of short time constant. The rocks are placed in zero magnetic field at room temperature, which allows relaxation and randomization of those domains and domain walls which underwent thermally activated and directionally biased changes to contribute the original VRM. Viscous components acquired in the Earth's field in periods of minutes to weeks can often be removed by this method but it cannot, of course, remove a VRM of time constant longer than the zero field storage time. However, the technique may reveal the presence of such components by their partial decay affecting the observed direction and intensity of NRM in a rock. Alternatively, the acquisition of VRM during storage of a rock in the Earth's or other weak field for periods of up to a few years will demonstrate the presence of viscous secondary magnetization at the time of collection, and a negative result of this test may also be significant (Griffiths, 1955). If significant VRM is acquired by a rock in this test, it may be possible to estimate the magnitude of the VRM acquired since the last field reversal, if it is assumed that the VRM intensity grows logarithmically with time and that the field strength remains essentially constant (Section 6.4). There are obvious dangers in extrapolating results obtained in months or years to hundreds of thousands of years, but it may at least allow an estimate of the upper limit of the VRM content of the observed NRM in a rock. Runcorn *et al.* (1971) use this principle in analysing the possible origins of the NRM of some *Apollo 12* lunar rocks.

The stability of VRM, acquired in a given period, against alternating field demagnetization varies greatly in different types of rock and VRM acquired in a short time interval (days or weeks) can show high stability. Biquand and Prévot (1971) quote examples of VRM in some basalt samples in which a 30 day VRM (acquired in the Earth's field) was not completely erased in a 42 mT demagnetizing field, and a red limestone sample in which 70% of a 30 day VRM remained after 98 mT cleaning. Zero field cleaning is clearly not satisfactory in such rocks. Dunlop and Stirling (1977) have also demonstrated the acquisition of hard VRM in sized samples of synthetic haematite (0.1–5 μ).

11.2.3 Stepwise thermal demagnetization

This is the most satisfactory and widely used technique for thermal cleaning during routine palaeomagnetic measurements. After the initial NRM has been measured, rock samples are heated to 80–100°C, cooled to room temperature in zero field and their NRM measured. This process is repeated using successively high temperatures (50–100°C steps) until there is evidence that secondary NRM components have been removed (see Chapter 12). In some instances this may involve heating to within a few degrees of the Curie point of the mineral carrying the NRM.

The main requirement for satisfactory thermal cleaning is to cool the rocks in 'zero' ambient field, to avoid the acquisition of a PTRM of sufficient intensity to contribute significantly to the remaining NRM. What constitutes 'zero' field depends on several factors: 100 nT may give entirely satisfactory results for a strongly magnetized sediment heated to 150°C whereas in a rock with very weak NRM cooled from 500°C, magnetite, perhaps formed in the rock through chemical change promoted by the heating, could acquire a dominating PTRM in 10 nT. For demagnetization of sediments it is now usual to aim for a maximum residual field of 1 nT. The design and control of suitable coil systems for cancelling the ambient field is considered in Chapter 5.

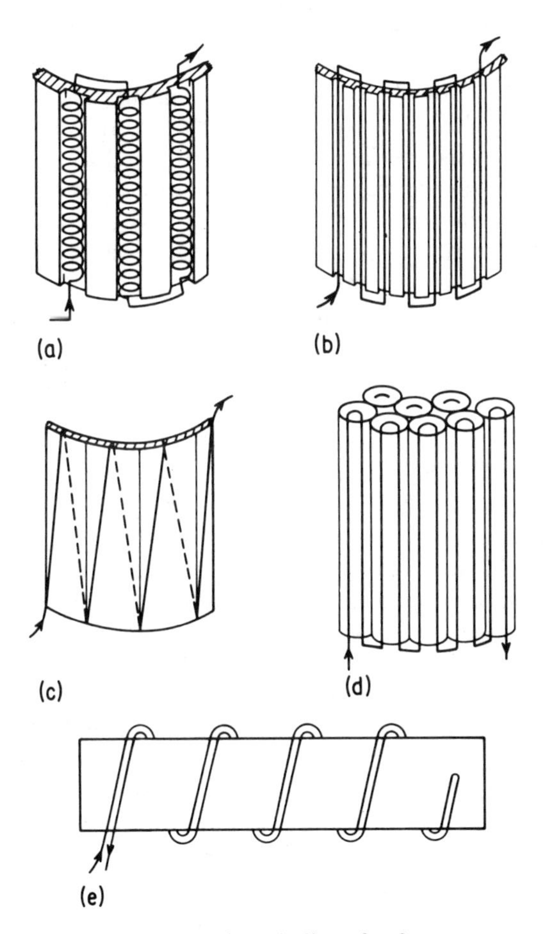

Fig. 11.13 Examples of non-inductive windings for furnaces.

The furnace should be of sufficient size to hold a reasonable number of samples (10–50) and to give a uniform temperature over them ($< \pm 10$°C at 600°C) and the winding should be non-inductive (Fig. 11.13) to reduce stray magnetic fields and be capable of heating samples to 700°C in a reasonable time (~1 hour). This temperature covers the Curie point of the common NRM-carrying minerals, titanomagnetites (≤ 575°C) and haematite (685°C). For investigations of lunar samples, a maximum temperature of 800°C is required to exceed the Curie point of iron (780°C). The most common design for an air furnace is a vertical cylinder open at the lower end which can be lowered on vertical guides to cover the samples, which are placed on shelves on the furnace base (Fig. 11.14). The windings are wound on the curved surface of a silica tube forming the inside wall of the furnace, or it is possible to buy helically-wired curved segments similar to electric fire radiants which when placed together form a cylindrical enclosure. The latter construction is used in equipment in the author's laboratory and the furnace, holding about fifty 2.5 cm samples, consumes ~3 kW at maximum power and reaches 700°C in about $1\frac{1}{2}$ hours. Although large furnaces are useful for dealing with many samples at once, their thermal inertia is considerable and may cause problems with temperature control. The best material for the shelves and shelf spacers is probably silica (~5 mm thick), although this is now very expensive: alumina may also be used but is mechanically rather weak. The author has found 2 mm titanium plate a satisfactory and much cheaper substitute for silica, if precautions are taken to check for significant ferromagnetic contamination. If present it can often be removed by thermal demagnetization. All materials used in furnace construction should, of course, be non-ferrous.

A suitable thermocouple material is Pt/13% Pt–Rh, with an output of ~6.7 mV at 700°C. The thermocouple tip may be covered with a standard rock sample to the centre of which an axial hole has been drilled: it is then likely that the rock samples will be uniformly heated throughout to the indicated temperature, by ensuring that the thermal inertia of rock and thermocouple are similar. There are several commercial temperature controllers available for automatically raising and holding the furnace temperature to a preselected value. The controller senses the furnace temperature via a thermocouple placed close to the furnace winding.

The design of furnace windings is not critical. The available maximum voltage is usually that of the electrical mains and the maximum current is dictated by that of the temperature controller being used, or of a variable transformer, and thus the resistance of the furnace winding is determined. Between 20°C and 700°C the resistance of the winding will increase (by 5–10%) If the resistance determined from the available voltage and current is the cold resistance the current (and heating rate) decreases at high temperatures, whereas if it is the hot resistance the cold current will be slightly above that calculated.

The wire diameter depends on the size of the furnace and the closeness of the turns. The minimum separation of the turns is set by the requirement of sufficient insulation between them, and wide separation is detrimental to temperature uniformity. A separation of three to six diameters is acceptable for

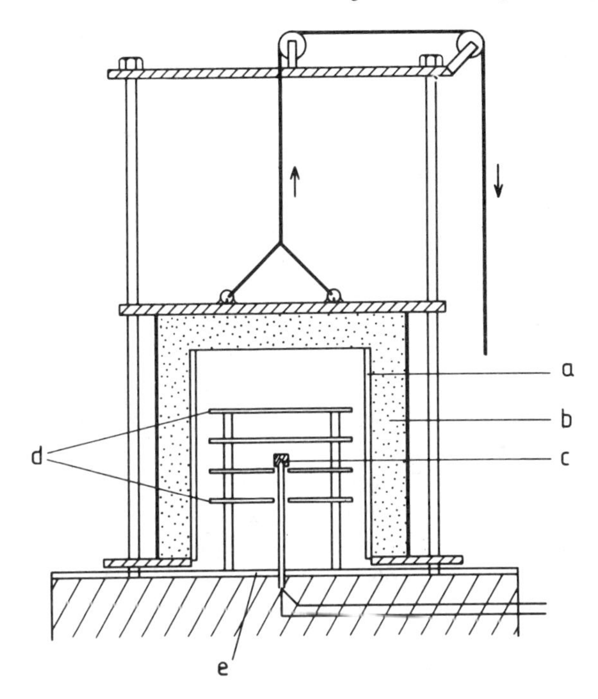

Fig. 11.14 Design of a large furnace for thermal cleaning. a, winding; b, insulation; c, thermocouple and covering rock sample; d, quartz shelves; e, heat-resisting base.

most purposes. In general, the required wire diameter increases with the size of the furnace in order to maintain a given resistance of the increased length of wire.

Thermal cleaning equipment has been described in the literature by several workers. Barbetti (1972) emphasizes uniformity of temperature within a vertical tubular furnace holding 10 samples. The winding density is increased near the ends so that the greater heat loss there is offset by greater heat production and a copper sheath is placed round the samples to promote temperature uniformity. Heating time to 500°C is 30 minutes. Information is given on the winding design and the stray (alternating) field due to the winding, which at the ends of the winding amounts to 5 μT A^{-1}. Provided any stray field does not exceed the coercive force of any grains in the rocks it is unlikely to affect the NRM even if it has a small d.c. field associated with it, since during the important period when the rocks are cooling the current can be turned off and the furnace removed from the samples. Heller, Scriba and Weber (1971) describe a furnace of 8000 cm^3 heated by a circulating gas from a remote furnace. Rapid heating is sacrificed for a good magnetic environment, the heating times to 500°C and 700°C being 2 hours and 6 hours, respectively.

Other thermal cleaning equipment has been described by Roy, Reynolds and Sanders (1972). Their furnace accommodates 120 2.5 cm × 2.2 cm samples in a magnetic field < 6 nT, with 45 of the samples in a field < 1 nT. The axis of the cylindrical furnace is vertical, with the windings in parallel vertical grooves cut into the outer surface of an alumina cylinder forming the furnace wall (Fig. 11.13(b)). This is an alternative method of non-inductive winding initially used by Thellier (1938). The windings are enclosed by an outer alumina cylinder which in turn is surrounded by curved fire bricks for good insulation and promotion of temperature uniformity. The furnace is lowered on vertical guides to cover four alumina shelves on which the samples are placed. The furnace is wound with 114 m of Nichrome wire 1.63 mm in diameter, and takes 15–16 A at 117 V (alternating), the lower current passing at 700°C when the winding resistance increases to ~7.8 Ω from ~7.3 Ω at room temperature.

A semi-portable thermal demagnetizer, incorporating a magnetic shield for ambient field reduction, is marketed by the Schonstedt Instrument Company. The horizontal, cylindrical shield encloses coaxial heating and cooling chambers, placed end to end. When one batch of 10 samples has been heated to the required temperature, it is pushed through to the cooling chamber, and another batch of samples is placed in the heating chamber. Heating time is 25–800°C in 30 minutes and fan-aided cooling from 800 to 40°C in 10 minutes. Power consumption is 1.2 kW. With the shield axis E–W, the heating chamber residual field is ~100 nT and the cooling chamber field normally ~5 nT, with the capability of reduction to ~1 nT by shield demagnetization with a built-in coil which can also be used to provide small, controlled ambient fields. An advantage of this equipment is the rapid heating and cooling, which reduces the likelihood of mineral alteration.

The degradation of magnetic phases during heating of lunar samples led Hale, Fuller and Bailey (1978) to investigate the use of microwave heating for thermal cleaning and Thellier-type palaeointensity determinations in these rocks.

Microwave heating occurs through two processes in rocks, dielectric losses in polarizable material in the oscillating electric field and Joule heating by eddy currents generated in highly conducting particles. Thus the matrix minerals are heated through the former process and the iron and nickel–iron particles by the latter. Using expressions developed by Puschner (1966) and Smyth (1950), Hale *et al.* (1978) show that most heating will occur in the iron grains. This suggests that less thermal alteration of these grains may occur than in normal heating, because of the lower temperatures and thermal activity in the surrounding matrix. However, further analysis shows that the time required for heat transfer from micron-sized grains to the matrix and the attainment of thermal equilibrium is less than a second, and therefore the achievement of 'isolated' heating of iron grains does not appear to be possible.

A diagram of the equipment used is shown in Fig. 11.15. The microwave source was taken from a commercial oven, feeding into an 'S' band waveguide with 650 W input power at 2.5×10^9 Hz.

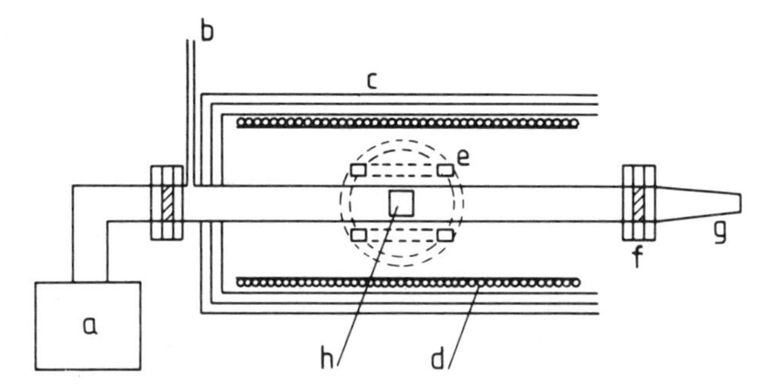

Fig. 11.15 Diagram of microwave furnace. The three-layer Mumetal shield (c) reduces the field over the sample (h) to ~ 100 nT, and this is reduced further by the coil system (e). The solenoid (d) provides an applied axial field for PTRM experiments. Other components are the source magnetron (a), vacuum or controlled atmosphere inlet (b), pressure window (f) and load (g). (Redrawn from Hale *et al.*, 1978)

Some test samples consisting of various sizes of iron grains dispersed in aluminium oxide or calcium fluoride were used to investigate thermal demagnetization of saturated IRM. Demagnetization was achieved after 14 minutes exposure. Some changes of magnetic characteristics in the samples were observed after a few minutes of microwave heating, but the magnitude of the changes was less than that produced by the conventional heating required to demagnetize the samples to the same extent. Other investigations showed that the temperature rise for a given exposure time depended on iron content and grain size, the most efficient heating occurring in 100–150 μm grains.

Applied to synthetic and natural lunar samples, microwave heating gave somewhat unpredictable results. A Thellier palaeointensity determination on a synthetic sample gave a reasonable result. For this technique the assumption is made that equal exposure times correspond to equal temperature rise, and therefore NRM lost and PTRM gained is compared after equal exposure times rather than the equal temperature intervals of the conventional method.

A lunar sample, a recrystallized breccia (77135), was also used for a Thellier determination with some success and aspects of microwave heating and TRM acquisition and demagnetization were investigated on other lunar samples. The overall result was that in some samples microwave heating did indeed produce less alteration than conventional heating and that TRM could be imparted and demagnetized. In other samples microwave heating was inefficient or unsuccessful and offered no improvement over normal methods of heating.

Microwave heating is not a practical proposition for terrestrial rocks because the electrical conductivity of titanomagnetites and haematite is too low for significant eddy current heating and dielectric loss heating is very slow.

An aspect of microwave heating of rocks which has not received much attention is whether local magnetic fields arising from the eddy currents can significantly alter the remanent magnetism of the particles. Potentially the most important effect would appear to be some sort of a.f. demagnetization at the instant when the microwaves are turned off and the currents decay to zero. A test for this would be measurement of the sample magnetization during and immediately after switching off the microwaves, but this is currently not a practical proposition.

It is also not entirely clear what part, if any, is played by the skin effect. The skin depth (Section 5.2.1) in iron at 2.5×10^9 Hz is approximately $(3.3 \times 10^{-4})\mu_r^{-1/2}$ cm where μ_r is the relative permeability of iron at this frequency, which is not known with any accuracy. Thus, the skin depth is probably of the order of 1 μ or less, suggesting that eddy currents are confined to a surface layer in particles of diameter $> \sim 5\ \mu$m. This appears to be inconsistent with Hale *et al.*'s observation that the most efficient heating occurred in a $\sim$100–150 μm grain size sample, and the least efficient in sub-micron particles. If there is a contribution to the heating from magnetic hysteresis loss this would also be reduced through failure of the magnetic field to penetrate the larger particles.

In spite of these uncertainties and other possible ones (the 'switching' time of domains and domain walls is comparable with or larger than the microwave period), Hale *et al.* have demonstrated that microwave is a potentially useful technique for extraterrestrial materials.

If during cooling of rocks the 'zero' field is insufficiently controlled, a significant PTRM may be acquired either by the mineral carrying the NRM or by a new mineral formed by chemical change during the heating. This is sometimes a problem with sediments, in which magnetite can be formed during heating to 500° C and above through the breakdown of clay minerals. A PTRM in even a very small amount of magnetite may dominate the NRM of a weakly magnetized sediment. However, the strong magnetic properties of magnetite allow its formation to be easily detected, and it is common practice to test for heat-promoted chemical or physical changes in rock samples during thermal cleaning by testing one or more of their magnetic properties after each heating. A common test is of initial susceptibility, which does not affect the NRM and is quickly carried out (Chapter 2): one part in 10^4 by weight of magnetite formed in a rock will contribute an initial specific susceptibility of $\sim 6 \times 10^{-8}$ m^3 kg^{-1}, well above that of many sediments. Another non-destructive test is of low-field IRM (using the minimum field necessary to give a measurable IRM) which may be advantageous in that it only monitors changes in materials that can carry remanence. However, before the next heating, it is necessary to remove the IRM by a.f. demagnetization which itself may introduce a spurious magnetization into the rock.

Alternatively, these tests can be carried out on separate specimens from the same rock samples in which the decay of NRM is not measured and therefore the type of monitoring is not restricted. This assumes homogeneity of magnetic and mineralogical properties, but does allow other monitoring measurements, e.g. of saturation remanence and induced magnetization to be made.

If the acquisition of PTRM is suspected, a useful precaution is to randomly orient the samples in the furnace relative to their fiducial lines and upper and lower surfaces. Assuming the stray field is constant during cooling, the PTRM will be approximately randomly directed in the samples although its intensity in each may vary, thus increasing the scatter of NRM directions without systematically affecting the mean. If the relative orientations of some of the samples is noted, vector subtraction of NRM after a heating may indicate whether a PTRM due to an external field of constant direction has been acquired.

If rocks possessing strong and weak magnetizations are being cleaned and the latter are susceptible to PTRM acquisition, it is worth checking that fields due to the former are not causing it. For two adjacent 2.5 cm samples, a horizontal specific magnetization J A m^2 kg^{-1} in one can produce a maximum field of approximately $(3 \times 10^{-4} J)$ at the centre of the other, for example, 30 nT if $J = 10^{-4}$ A m^2 kg^{-1}. Weakly magnetized rocks are not necessarily more prone to acquiring significant PTRM, because low intensity often implies low content of NRM-carrying mineral and therefore correspondingly low PTRM. However, if a mineral such as magnetite is formed the PTRM contributed by it is likely to be much greater than that contributed by the original NRM-carrying mineral, e.g. haematite in red sediments, and this PTRM will be relatively stronger in weakly magnetized rocks.

Another feature that is encountered in thermal cleaning is the occurrence of significant VRM of low relaxation time in rocks heated to 500°C and upward. In sediments this again appears to be due to the production of magnetite of appropriate grain size, and it is important to keep rocks with this property in zero field between cleaning and measurement.

11.2.4. Continuous demagnetization

In this technique a single sample is heated in a furnace and its NRM measured continuously, or at intervals, with a magnetometer adjacent to or combined with the furnace. The potential advantages of this technique are the elimination of PTRM acquisition, since the sample is not cooled before NRM measurement, less importance of chemical changes provided that any magnetic material formed is above its Curie point (or blocking temperature), and rapidity of a thermal cleaning run, albeit of only one sample at a time. Among the difficulties of the method are lack of sensitivity because of the increased separation of sample and detector caused by the furnace surrounding the former, and manipulating the sample in the furnace for a complete determination of the NRM vector. The NRM of a rock measured at a particular temperature during continuous thermal cleaning will generally be less than that measured after cooling from the same temperature because of the decay of

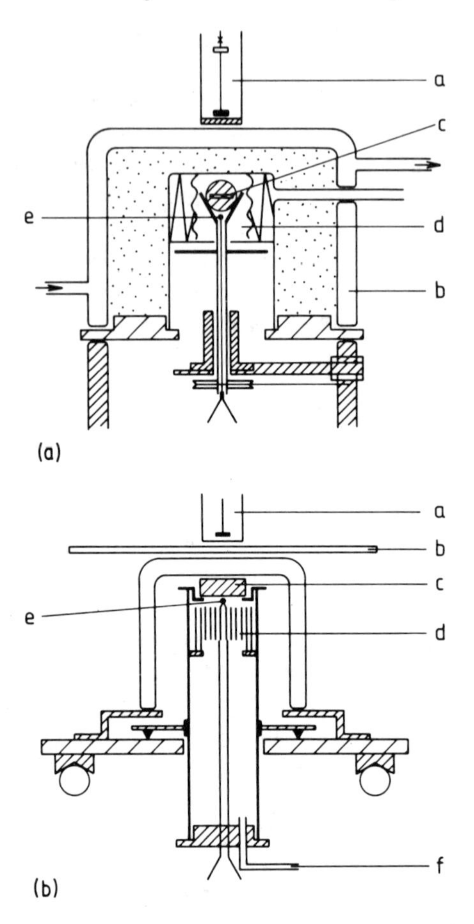

Fig. 11.16 Furnaces of Wilson (1962) (a), and Creer (1967) (b). a, Magnetometer; b, water jacket; c, sample; d, furnace winding; e, thermocouple; f, nitrogen entry.

spontaneous magnetization with temperature, and curves of thermal decay of NRM will differ from those obtained with stepwise cleaning.

The design of continuous demagnetization equipment is mainly directed towards maintaining good flux linkage between sample and detector and avoidance of thermal and magnetic disturbance of the magnetometer by the furnace. Some early equipment, using a furnace and astatic magnetometer, is described by Wilson (1962), Radhakrishnamurty and Sahasrabudhe (1965), Creer (1967) and Chamalaun and Porath (1968). The furnaces of Wilson and Radhakrishnamurty are similar in design and consist basically of a sample holder, supported on a vertical shaft, surrounded by a vertical cylindrical furnace, insulation and an outer Pyrex or quartz water jacket (Fig. 11.16 (a)). Non-inductive furnace windings are achieved by the winding shown in

Fig. 11.13(a) (Radhakrishnamurty) and (d) (Wilson) and typical power consumption at 700°C is 100 W. Pure platinum wire is used for the furnace to avoid magnetic contamination, and suitable non-magnetic thermal insulation media are magnesium oxide, quartz wool or plaster of Paris (hydrated calcium sulphate). The cylindrical sample lies in the holder with its axis horizontal and its centre below and on the axis of the astatic pair, and the holder can be turned through 360° about a vertical axis, enabling the Z component (along the sample axis) and X or Y components of NRM to be measured. The third component is measured by turning the sample through 90° about a horizontal axis with a brass or quartz 'screwdriver' inserted through a hole in the furnace wall and engaging with a slot cut in the appropriate face of the sample. Deutsch (personal communication) has used this type of furnace with the sample only 4.5 cm from the lower magnet of the magnetometer, when an intensity of $\sim 5 \times 10^{-7}$ A m^2 kg^{-1} gives a deflection of 1 mm.

Creer (1967) and Chamalaun and Porath (1968) avoided manipulation of the sample independently of the furnace by rotating the furnace about a vertical axis and traversing it horizontally for off-centre measurement with an astatic magnetometer (Section 9.2). Creer (Fig. 11.16(b)) supported the sample (disc-shaped) above a heating element consisting of a vertical cylindrical grouping of fine quartz double-bore tubes through which platinum wire is threaded to form a non-inductive winding (Fig. 11.13(e)). The furnace projects up into a cylindrical quartz vacuum Dewar and can be rotated about a vertical axis inside it. Nitrogen or air is passed up through the furnace, over the sample and out of the base of the Dewar in order to improve uniformity of sample temperature. The whole furnace is supported on horizontal rails for traversing off-centre and for withdrawal from the magnetometer to monitor zero drift.

Chamalaun and Porath use a conventional cylindrical furnace, with its axis horizontal, surrounded by a silvered cylindrical quartz vacuum envelope. The sampler holder is supported at the end of an axial quartz tube, down which thermocouple leads pass, and the tube can be rotated about its axis so that measurements can be made with the sample in the 'upright' and 'inverted' positions. As with the Creer instrument, the furnace can be rotated about a vertical axis to different azimuths and traverse horizontally on a rail system, and nitrogen or other gas can be introduced into the heating space. About 1 mm deflection with a direct reading magnetometer was achieved with a magnetization of $\sim 10^{-5}$ A m^2 kg^{-1} in a 7 cm^3 sample, using a 6 s time constant.

The movement of the whole furnace during measurement necessitates the use of highly non-magnetic construction materials if spurious signals are to be avoided. Careful attention to non-inductive furnace windings, the use of platinum wire, and quartz or high-purity brass or aluminium for construction are among the requirements.

Petherbridge *et al.* (1972) use a cylindrical furnace (inner diameter 5.0 cm, wall thickness 0.8 cm) with its axis horizontal. Three separate windings along the cylinder enable the axial temperature gradient to be reduced by having a smaller current in the central winding than in the outer pair. An improvement in the vertical temperature gradient (increasing upward) was achieved by placing the furnace tube eccentrically in an outer horizontal Vitreosil (silica) tube so that the axis of the latter is below that of the former. The volume between the cylinders is filled with kieselguhr (diatomaceous earth) the smaller thickness of which at the top allows more heat loss and a decrease in the temperature at the top of the furnace. This arrangement also allows the heated sample to be closer to the lower magnet of the astatic magnetometer.

The sample holder is mounted on a vertical silica rod passing through the furnace wall and free to rotate about its axis. Three-component measurement of NRM is then possible by rotating the sample to different azimuths about the vertical axis, and also about the horizontal axis by means of a silica 'screwdriver' inserted through the end of the furnace and engaging with a slot cut in the surface of the sample. The sample temperature is read by means of a thermocouple embedded in a silica cylinder of the same size and shape as the samples (2.5 cm cylinders), placed at the side of the rock sample. A water jacket thermally isolates the furnace from the magnetometer. The minimum distance between sample and lower magnet is 8.5 cm when the minimum measurable specific magnetization is $\sim 2 \times 10^{-6}$ A m^2 kg^{-1}. A possible source of noise of thermal origin in equipment using astatic magnetometers is magnetic fields associated with Thomson currents in the damping plate of the astatic system, if this method of damping is used. Such currents will flow if small transient temperature gradients from the furnace develop across the plate, and the close coupling between plate and magnet can lead to significant magnetic torques acting on the magnet system. The most efficient heat shield between furnace and magnetometer is a thin (2–3 mm) water jacket carrying a flow of a few cm s^{-1} of water. It is also advisable to avoid substantial areas of metal in furnace construction again to avoid magnetic fields of Thomson current origin.

A spinner magnetometer, combined with a furnace, was constructed by Stacey (1959). Figure 11.17 shows the general layout. The sample holder and upper part of the shaft are made of nimonic and Inconel respectively (high nickel content alloys), allowing temperatures up to 800°C to be used. The Inconel shaft is a thin-walled tube, to minimize thermal conduction, and the holder accommodates 2.5×3.5 cm cylindrical samples: the rotation frequency is 78 Hz. The reference signal is obtained from a small magnet attached to the lower end of the shaft generating a signal in a surrounding coil system. Any spurious signal in the sample pick-up coil due to the reference magnet is backed off by a small trimming magnet, adjustable in height and azimuth, higher up the shaft. The sensitivity was limited by permanent magnetism of the sample holder, equivalent to about 10^{-4} A m^2 kg^{-1} in a standard sample and presumably variable with temperature. This could be removed by vectorial subtraction, with an error of $\sim 2 \times 10^{-5}$ A m^2 kg^{-1}, giving reasonable

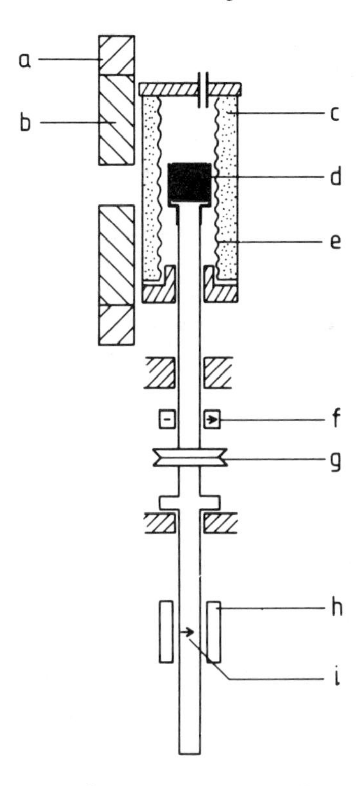

Fig. 11.17 High-temperature spinner magnetometer (Stacey, 1959). a, b, Pick-up and compensating coils; c, insulation; d, sample; e, furnace winding; f, trimming magnet; g, drive pulley; h, i, pick-up coil and reference magnet.

sensitivity for most igneous rocks for which the equipment was designed. Conventional signal processing allowed the two orthogonal components of NRM perpendicular to the rotation axis to be measured during heating. Burakov (1977) obtained improved sensitivity in his spinner system by using a ring fluxgate detector placed outside and on the surface of a vertical cylindrical water-cooled furnace within which the sample rotates at 11 Hz in the plane of the fluxgate. By means of two reference signals 90° out of phase the two orthogonal components of NRM perpendicular to the (vertical) rotation axis can be measured. Cubic samples 1.0 cm on a side are measured, and the noise level is stated to be equivalent to a specific magnetization of 1.0×10^{-7} A m^2 kg^{-1}.

Krause (1963) describes a vibrating sample magnetometer (Section 3.2) combined with a furnace. An ingenious arrangement of eight pick-up coils surrounding the vertical tubular furnace enables the three orthogonal components of NRM in the sample to be determined at any required temperature. Cubic samples 2.5 cm on a side were measured and a magnetization of about 2×10^{-5} A m^2 kg^{-1} could be theoretically detected, but in practice the limiting sensitivity was about 2×10^{-4} A m^2 kg^{-1} due to noise of electrical and magnetic origin.

The Digico magnetometer (Section 9.3) can be supplied with a measuring head for high-temperature measurement of NRM. The usual rotor assembly is inverted, with the sample held at the lower end of the shaft, around which there is the vertical tubular furnace, water jacket, ring fluxgate and a surrounding Mumetal shield. The thermocouple is placed just below the sample. The cylindrical sample has its axis horizontal and can be rotated about a horizontal axis through 90° by means of a loop of silica string passing around it and up the centre of the rotor shaft to a control mechanism mounted at the top. The greater separation of sample and fluxgate detector causes a loss in sensitivity compared with the room temperature head by a factor of ~ 20, i.e. a noise level of about 5×10^{-7} A m^2 kg^{-1} in a standard 2.5 cm sample, averaging over 2^7 spins.

A laboratory-modified Digico magnetometer for high-temperature measurement is described by Heiniger and Heller (1976). It is essentially similar to the maker's system. The ring fluxgate at 8.0 cm in diameter and encloses the water-cooled furnace, and the stated loss of sensitivity over the room-temperature instrument is only a factor of four. The sample is placed in a horizontal quartz semi-cylinder which can be rotated through $\pm 90°$ about a horizontal axis at the same time as the sample is rotating about the vertical axis. Thus, the three orthogonal magnetization components can be measured with one continuous spin.

In principle the cryogenic magnetometer is ideally suited for continuous thermal studies, but in practice there are some difficulties in heating a sample at the measuring position. As well as the space for an insulated furnace being very restricted, it is not possible to wind an electric furnace with a residual magnetic field low enough to give no spurious signals from the SQUID detectors. An early alternative method which was investigated was a flame furnace, in which a small gas flame was directed on to the sample, contained in a copper case, while in the measuring position in a three-axis magnetometer (Day, Dunn and Fuller, 1977). A tubular water jacket surrounds the sample. In a more recent version (M. D. Fuller, personal communication), a laser beam is directed on to the sample contained in a silicon holder, which promotes temperature uniformity but has low electrical conductivity. The latter property inhibits the production of magnetic fields from Thomson currents flowing in the holder material.

In another approach, which has had some success, radiation from a high-wattage lamp is focused by a concave mirror and directed down the access tube on to the silicon-cased sample at the measuring position (P. J. Wasilewski, personal communication). The sample and holder are contained in a small quartz open-ended Dewar to reduce heat loss and the whole is supported at the lower end of a quartz tube, the inside surface of which is gold plated to reflect the maximum amount of radiation on to the sample.

Among problems associated with the continuous method are thermal inertia in the sample, causing differences between thermocouple and sample temperature, and temperature gradients in the sample. The latter can be reduced by suitable design and/or by enclosing the sample in a metal container; electrolytic copper can be used in an inert atmosphere or vacuum but platinum or titanium are preferable in air, in which copper rapidly oxidizes.

11.2.5 General comments

A problem similar to that discussed in Section 14.4 that may also be encountered in thermal cleaning is the partial or complete destruction of the NRM-carrying mineral, either by chemical change or by change of phase or physical state. Examples of the former are the oxidation of titanomagnetites, or of magnetite to haematite, and in lunar rocks iron to magnetite and haematite: change of phase is probably uncommon (maghemite, γ-Fe_2O_3 reverting to haematite at 300°–400°C is a possibility), but alteration of the physical state of iron grains is believed to be important in some lunar rocks (Banerjee *et al.*, 1974).

Mineral alteration will almost certainly be reflected in changes of magnetic properties of the rock. Heating in air will promote oxidation, which is generally associated with a decrease in initial susceptibility, isothermal remanence or induced magnetization. Other tests on the rock, e.g. Curie temperature or low-temperature transition may reveal and identify the new mineral. There may also be indications of mineral alteration in the changes of direction and intensity of NRM that occur during heating. If the primary and secondary magnetizations are carried by the same mineral, a stable direction is unlikely to be reached before alteration is complete. If the magnetizations are carried by different minerals, there are several possible results depending on whether either or both minerals are changing and which mineral is carrying which magnetization.

Non-repeatability of intensity and direction of NRM after a second heating to the same temperature may indicate that more than thermal unblocking is occurring, provided that the samples are held at the temperature for only a few minutes.

Thermal cleaning of sediments is best carried out in air but other investigations may require a different atmosphere or a vacuum. Although the furnaces described above can be modified for a controlled atmosphere a different design is sometimes desirable, particularly if a vacuum is required as well. Figure 11.18 is a diagram of a system for providing a controlled atmosphere or vacuum for thermal cleaning and PTRM experiments. The furnace accommodates three 2.5 cm samples for heating up to a maximum temperature of 800°C in a vacuum of 5×10^{-5} torr. A set of manually-controlled 1 m Helmholtz coils cancels or provides the ambient field. The furnace is moved away from the samples along the quartz tube for cooling the

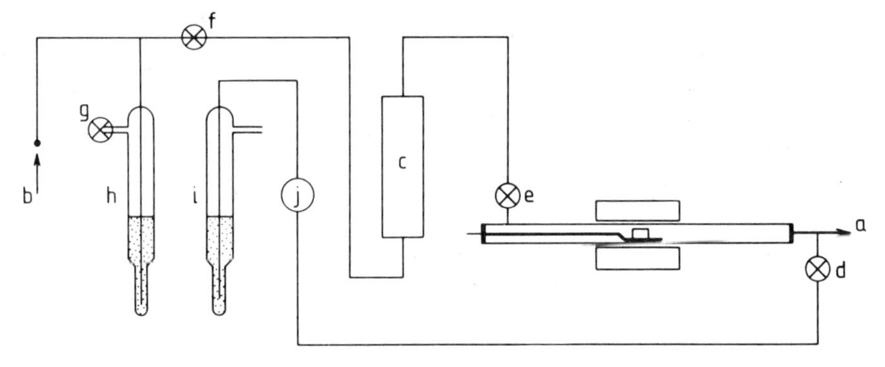

Fig. 11.18 System for heating samples in an oxygen-free argon atmosphere. a, Vacuum pump; b, argon inlet; c, titanium column; d, e, f, g, valves; h, i, silicone fluid outlet chambers; j, gas flow meter. The system is evacuated back to valve f (valve d closed) and then argon admitted, using chamber h to ensure that air is not admitted. When atmosphere pressure is reached, valve d is opened and the argon pressure increased until the required flow is indicated on the meter. The titanium powder column is heated to 600°C to remove residual oxygen from the argon.

samples, and cooling can be accelerated if required by allowing water to fall on the quartz tube (the very low expansion coefficient allows this) or, in the later stages, the backing pump can be used to draw air over the samples. In the author's laboratory the system has been adapted for heating lunar samples in an oxygen-free argon atmosphere. The inlet tube for the argon includes a section containing titanium sponge which, heated to 700°C by a furnace, removes residual oxygen from the argon.

An alternative procedure for inhibiting mineral alteration on heating, for example the partial oxidation of titanomagnetites, is to confine the sample in a sealed, evacuated capsule. This is not always satisfactory because of occlusion of gases at high temperatures and evacuation while the sample is at a moderate temperature (100–150°C) may be a partial answer to this. Taylor (1979) describes such a method, and in addition he includes some titanium in the capsule, separated from the sample, as an oxygen 'getter'. Some success is reported with this method in preventing oxidation of lunar samples during palaeointensity determinations (Suguira *et al.*, 1979).

Larson *et al.* (1975) describe a system whereby the partial pressure of oxygen (the 'fugacity', $f(O_2)$) can be varied at will over a wide range of very low and known values. This is achieved by first evacuating to $\sim 10^{-5}$ torr and then admitting controlled amounts of CO_2/N_2 and H_2/N_2 gas mixtures. Oxygen fugacities in the range 10^{-3}–10^{-28} atmospheres can be obtained, enabling many magnetic minerals to be maintained within their stability fields during heating. Although the authors describe the use of the method in a magnetic balance for determining J_i–T curves and Curie points in small, powdered samples it may also be suitable for small (~ 1 g) rock samples.

11.3 Other cleaning techniques

11.3.1 Chemical demagnetization

This technique is of somewhat limited application and relies on selective destruction by chemical reagents of minerals carrying different magnetizations in a rock. If the most resistant mineral is also carrying the primary NRM, chemical 'cleaning' can be achieved and the technique is also capable of detecting different NRM components in rocks and identifying the minerals that carry them (Section 13.3).

Carmichael (1961) used acid solution to separate large and small ilmeno–haematite lamellae occurring in the Allard Lake ilmenites and showed that the NRM was carried in the latter, and similar experiments were carried out by Kawai (1963) and Domen (1967) (Section 13.3). This suggested a possible cleaning technique if the primary and secondary NRM in a rock are carried by different minerals, or by different phases of the same mineral. The latter situation has proved to be most profitable, in particular the occurrence of black polycrystalline and red pigments form of haematite in red sediments. Using the technique described in Section 13.3, Collinson (1965b) selectively dissolved out the pigment with hydrochloric acid in samples of the Triassic Supai formation (western USA), when their NRM direction, initially some 80° from the primary NRM direction established from other rocks of the same age, migrated to within $\sim 20°$ of this primary direction. In these rocks the secondary magnetization is apparently carried by the pigment form of haematite, a situation which has subsequently been found in other sediments (Collinson, 1974).

The author's technique in this field suffered from limited application because of the requirement of thin (~ 2 mm) discs (and therefore of low magnetic moment) of porous rock for good penetration of the acid. Burek (1969) used a thicker sample and forced 3N hydrochloric acid through it under pressure in a special chamber. For fine-grained samples a hole was drilled through the sample and acid forced through from the inside. In a later system an oil pressure cell was used in which good penetration of fine-grained samples was achieved with a pressure of 50 bars (Burek, 1971) and a similar cell is described by Smith (1979).

An intensive study and detailed interpretation of chemical cleaning was carried out by Park (1970) and Roy and Park (1974) in their studies of the Hopewell Group sediments of New Brunswick (Canada). Their samples were in the form of standard (2.5 cm) cylinders with an axial hole 1 cm in diameter. Using long solution times (up to 4500 h) in 8–10N hydrochloric acid they were able to isolate three components of magnetization, only one of which was distinguishable by thermal cleaning (Fig. 11.19). A feature of their technique is the long solution times: since pigment and specularite are chemically identical it might be expected that they would both ultimately be dissolved by a reagent that attacked either one of them. However, Roy and Park report a variable rate of attack by hydrochloric acid on specularite, and clearly some specularite in the Hopewell sediments is extremely resistant.

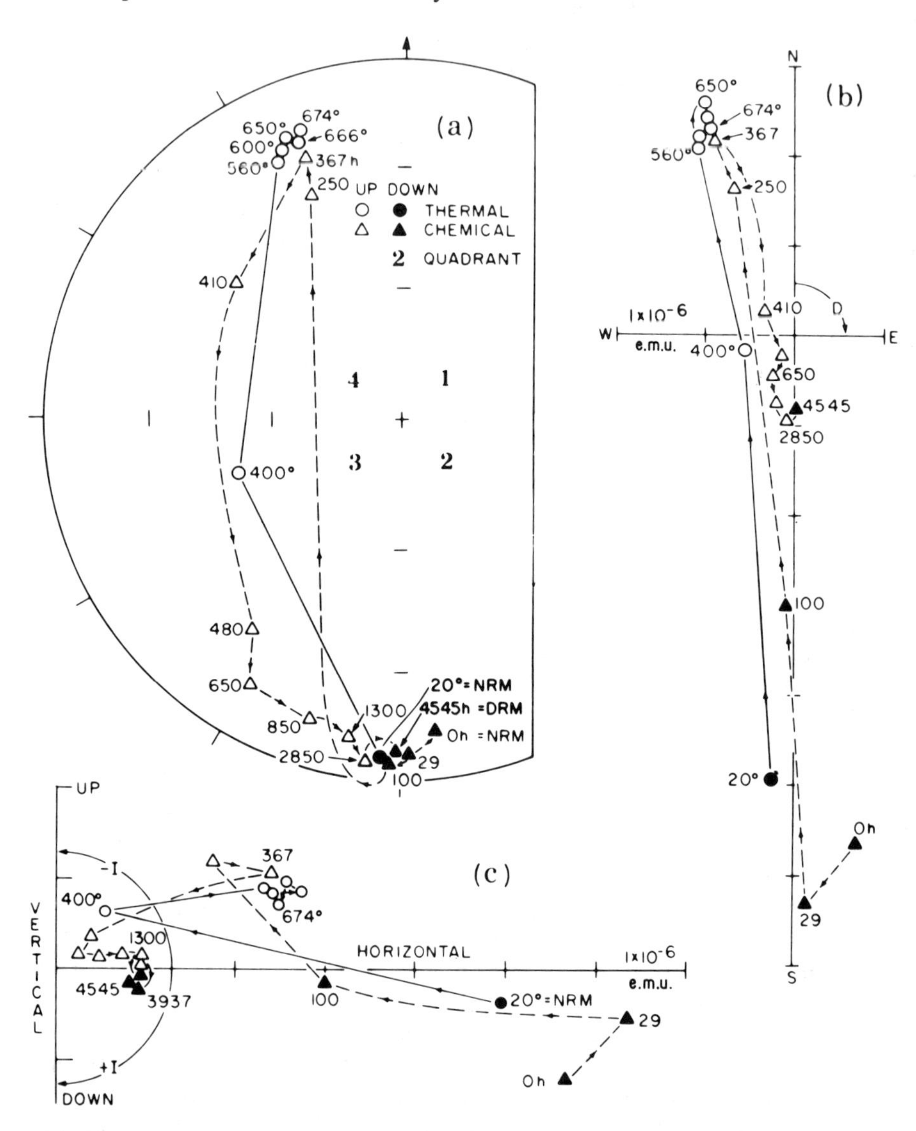

Fig. 11.19 Thermal (circles) and chemical demagnetization (triangles) of two samples from the same core. The numbers against the points are °C or hours. Whereas thermal cleaning reveals only the northerly-directed NRM component, prolonged chemical leaching reveals this and a second southerly-directed NRM. (b) and (c) show the vector end-points projected on the horizontal plane and the vertical plane respectively. (From Roy and Park, 1974)

Henry (1979) has investigated various aspects of chemical cleaning in order to achieve the most efficient technique. The samples were cut from a slab of Pre-Cambrian Freda sandstone (from a medium-grained red arkose containing specularite, red haematite cement and titanomagnetite).

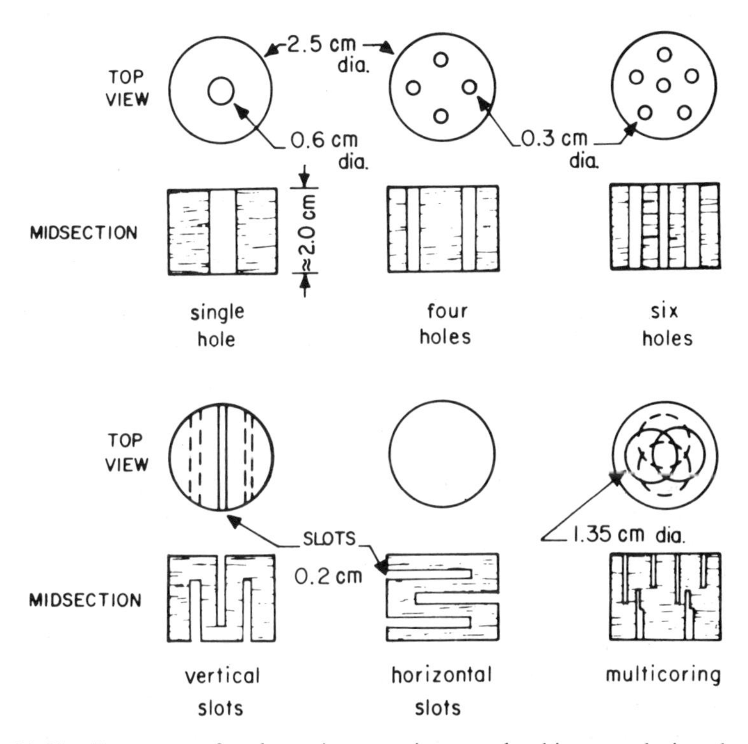

Fig. 11.20 Treatment of rock specimens to improve leaching rate during chemical demagnetization. (From Henry, 1979)

The samples, 2.5 cm in diameter and 2.0 cm high, were further cut in various ways to increase the surface area for acid attack. The cuts investigated included one axial hole 0.6 cm in diameter, four and five 0.3 cm holes, vertical and horizontal slots 0.2 cm wide, and interlocking circular slots cut inward from each end of the sample with a 1.4 cm diameter drill (Fig. 11.20). The last method preserves the most sample volume while providing the greatest increase in surface area and proved to be the most satisfactory, as measured by the time required to reach a stable end point, i.e. when all forms of haematite have been removed. In general, any axially symmetric distribution of holes perpendicular to bedding planes provides the most uniform leaching, while allowing penetration of the acid along the bedding planes. This latter feature may be advantageous in that solution takes place along paths followed by the iron-bearing solutions from which secondary haematite was deposited.

Henry also investigated the effect of acid strength and temperature on the leaching rate and on the separation of NRM components. As might be anticipated, the leaching rate increases markedly with acid normality and temperature. However, rapid attack on both forms of haematite is not desirable from the standpoint of revealing components of NRM carried by different mineral phases, since this depends on a different solution rate of each phase. The optimum acid normality and temperature will vary between rocks since it depends on porosity and permeability, grain size, pigment/specularite ratio, etc. In the Freda sandstone the optimum values were found to be 6N–8N and 80–90°C, with an associated leaching time to the stable end point of 100–200 h.

Evacuating the samples before leaching did not significantly improve the rate of attack, and in view of the complications involved in admitting hot hydrochloric acid into an evacuated chamber (where it boils even at room temperature) this technique does not appear to be worth pursuing.

It is probably a worthwhile precaution to carry out chemical cleaning in a low magnetic field environment. This eliminates the acquisition of VRM during the process and a possible spurious magnetization noted by Reeve (1975). If his leached sandstone samples were insufficiently rinsed before measurement, a magnetic iron phase was redeposited from the residual solution in the rock pores on drying and either a CRM or VRM acquired in the ambient field. Reeve therefore recommends forcing water through samples, if possible, to remove all iron solutions before dying and also storage of samples in field-free space while drying out.

Morris and Carmichael (1978) increased the surface area for acid attack in their samples of west African sediments by drilling four opposing pairs of annular cuts into each cylinder, the axes of the cuts making approximately equal angles with each other. They show an interesting comparison of thermal and chemical demagnetization.

There are no reports in the literature of acids other than hydrochloric being used for chemical cleaning, probably because in many cases the rate of attack by HCl has proved to be suitable. Another reagent which might be of use is oxalic acid, the use of which is described by Leith (1950) in connection with the removal of pigment and other soluble material from rock powders. He recommends heating (90–100°C) in 5% oxalic acid in the presence of metallic aluminium.

11.3.2 Low-temperature demagnetization

This technique, in which a rock sample is taken through one or more cycles between room temperature and liquid nitrogen temperature (−196°C) in zero magnetic field, is not in general use for routine palaeomagnetic measurements. It is mainly confined to those rocks in which the NRM is carried by the titanomagnetite series of minerals: these minerals undergo a transition in their magnetic properties at $-140°C$ (for magnetite) and at lower temperatures with increasing titanium content (Ozima, Ozima and Nagata, 1964; Syono and Ishikawa, 1963). The removal of a low field (1.2 mT) IRM in a basalt sample was demonstrated by Ozima *et al.* (1964) by cooling in liquid nitrogen ($-196°C$) and warming to room temperature in zero field. IRM in 10 mT was not completely removed. Investigations into the low-temperature demagnetization of the IRM of magnetite and the NRM of some magnetite-bearing diorite samples are described by Merrill (1970). In the latter samples there was progressive demagnetization with each low-temperature treatment, although the first cycle was most effective. Multiple treatments could be equated with a.f. cleaning in a peak field of about 15 mT. Some other relevant work in this field is that of Creer and Like (1967) and Brecher and Arrhenius (1974).

Creer and Like showed that some Tertiary basalts containing titanomagnetite with $x = 0.77$ and 0.64 (in $x\mathrm{Fe_2TiO_4} \cdot (1-x)\mathrm{Fe_3O_4}$) showed complete reversibility of their NRM variations during cooling to $-196°C$ and warming to room temperature. Low-titanium ($x < 0.1$) dolerite samples from the Whin Sill were partially demagnetized, the NRM after one cooling cycle being in the range 0.6–0.9 of the initial NRM.

Brecher and Arrhenius (1974) applied the technique to some samples of carbonaceous chondrite meteorites in order to attempt the recovery of a stable primary remanence from these thermally unstable samples. Merrill (1970) showed that single-domain magnetite is largely unaffected by thermal cycling, and since a soft, secondary NRM is likely to be carried by multidomain magnetite the residual NRM after several low-temperature cycles is expected to be largely primary in origin. The normalized loss of NRM after cycling is thus a measure of the multidomain magnetite content of the samples, although in some cases the situation is complicated by the presence of nickel–iron particles carrying remanence. Most of the chondrite samples showed a stable component of remanence amounting to 50% or more of the initial NRM intensity: in a few samples the NRM decreased to 10–20% of its initial value after one cooling cycle.

At the transition temperature magnetocrystalline anisotropy passes through a minimum value and the effect on the NRM appears to depend on how all the grain anisotropies add to determine the grain coercivity at that temperature (Merrill, 1970). It is thus possible for the NRM of those grains which show stability to 10 mT at 20°C can be released at $-140°C$, and conversely the coercivity of a low-stability NRM at room temperature could be increased at the low temperature. In contrast to a.f. demagnetization low-temperature cleaning is dependent on the source of coercivity rather than just the magnitude.

Since the transition temperature in titanomagnetites is lowered by increasing titanium content and is $\sim -196°C$ (liquid nitrogen) for $x = 0.1$–0.2, low-temperature demagnetization is expected to be more apparent in low-titanium rocks.

Low-temperature demagnetization has also been reported in rocks in which the NRM is carried by metallic iron. Jahn (1973) examined the effect in samples of dispersed multidomain iron grains carrying different types of remanence. Saturated IRM showed a decrease after one cooling cycle (to $-196°C$) but ARM and TRM were unaffected with no change in magnetization during cycling. Experiments on the NRM of two lunar igneous rocks (*Apollo 12*) have been reported by Runcorn *et al.* (1971) who found that the magnetization decreased irreversibly by a factor of ~ 2 after several cooling cycles. No information about NRM direction was obtained.

Where the NRM of iron grains is affected by low-temperature cycling the magnetization decays more or less steadily as the temperature decreases, and in the absence of a structural transition there is no sudden change in NRM. The decrease in magnetization appears to be due to a change in anisotropy constant with temperature, resulting in a change in domain structure of multidomain grains. As with the titanomagnetites, low-temperature demagnetization in iron-bearing samples is therefore expected to preferentially occur in multidomain grains.

It is important for detecting low-temperature demagnetization effects that the cooling and warming takes place in magnetic field-free space. It has been demonstrated that a remanence can be acquired by titanomagnetites when warmed through the transition temperature in a weak ambient field (Ozima *et al.*, 1963), which may obscure changes in NRM due to demagnetization. Such a remanence can be easily detected if the cycling is carried out in an ambient field by cycling samples placed with their NRM parallel with and opposed to the field. Assuming equivalent demagnetization in each sample, the observed decrease in NRM will be less in the former relative orientation than in the latter.

The type of container described in Section 4.3 on low-temperature measurement of NRM with an astatic magnetometer is convenient for cooling standard rock specimens and synthetic samples. Since some samples may break up under thermal stresses if placed directly in liquid nitrogen, a useful precaution is to place the sample in a semi-insulating container before cooling, to lessen the thermal shock. Ample time must be allowed for the sample to achieve thermal equilibrium.

11.3.3 Pressure demagnetization

This is another process of little practical use in routine palaeomagnetic studies but of some theoretical and indirect historical interest. The latter association arises from the early observations of Graham, Buddington and Balsley (1957) that uniaxial pressure applied to a rock sample could reversibly alter the direction of its NRM. In some samples there was an irreversible change and naturally occurring pressure, for instance due to burial, was seen as a potential agent for masking the true primary NRM direction. The general effect of pressure on rocks in the absence of ambient field is demagnetization of the NRM and in this section some of the observations and techniques are described. Other aspects of the magnetic effects of stress in rocks and the equipment used to measure pressure effects are described in Section 6.3.

Kume (1962) reported that IRM (45 mT) in synthetic maghemite ($\gamma - Fe_2O_3$) and magnetite and the NRM of an andesite sample were partially demagnetized by application and removal of hydrostatic pressure up to 10 kbar. Low coercivity components of magnetization were removed, the demagnetization achieved at 10 kbar being equivalent to a.f. demagnetization in approximately 15 mT. IRM acquired in 10 mT by magnetite and haematite crystals was shown by Shimada, Kume and Koizumi (1968) to be almost completely removed by treatment at 10 kbar, while a 70 mT IRM was much more stable. A TRM imparted to a haematite crystal and a sample of magnetite powder was almost unaffected by a pressure of 5 kbar. In all the tests the pressure (hydrostatic) was applied for 5 minutes. These results broadly confirm earlier experiments by Girdler (1963) on haematite TRM and on the NRM of some basalt, gabbro and dolerite samples after application of hydrostatic pressures up to 10 kbar for 1 hour.

The above experiments are the mechanical analogue of stepwise thermal demagnetization and the mechanical equivalent of progressive thermal demagnetization, i.e. measurement of NRM during the application of pressure, shows that reversible as well as irreversible decreases of NRM occur in igneous rocks (Ohnaka and Kinoshita, 1968). Significant changes were observed during uniaxial pressures of less than 1 kbar, the magnitude of the changes being greatest when the pressure axis was parallel to the NRM.

It is preferable for experiments on pressure demagnetization to be conducted in zero magnetic field, since the application of pressure to a rock in an ambient field can result in the acquisition of a pressure or 'piezoremanent' magnetization (PRM). Although by carrying out the tests in a suitable manner the effect of any PRM may be eliminated from the observations there are sufficient uncertainties about the characteristics of PRM and its behaviour in any particular rock to make it desirable to exclude PRM from the experiments from the outset.

Chapter Twelve

Presentation and treatment of data

12.1 Presentation of NRM directions

The basic data of palaeomagnetism are a group of directions and intensities of demagnetization measured in rock samples from the same collection site. After magnetic cleaning a well-defined mean primary NRM direction ideally remains which, with other site means from the same rock unit, can be interpreted through an overall formation mean direction in terms of polar wandering and/or plate movements.

The initial reference system for cylindrical specimens for the declination (D) and inclination (I) of the NRM is usually a direction marked on the upper surface of the specimen and the plane of the upper surface of the specimen respectively, whereas for the calculation of a pole-position a reference system common to all specimens is required. This is commonly the local horizontal plane at the site and geographic North, or the regional bedding plane in sediments. The trigonometrical calculations for converting directions of NRM derived from the magnetometer to directions referred to the regional system are now commonly done by computer, although the corrections can also be done with the aid of the projections described below. The two methods are described in Appendix 5. Schmidt (1974) discusses the use of orthogonal transformations in handling palaeomagnetic data, and the same author and Zotkevich (1972) describe methods of correcting NRM directions obtained from plunging folds, i.e. folds in which the folding axis is not horizontal. Schmidt shows that if the axis is assumed horizontal the inclination after 'unfolding' will be correct but there will be an error in declination.

For the display of NRM directions and their initial assessment it is convenient to show the three-dimensional information inherent in D and I on a two-dimensional plot or projection, i.e. to transfer points distributed on the surface of a sphere on to a plane. Two commonly used projections are the Lambert equal-area and stereographic projections. Their geometrical construction is shown in Fig. 12.1 and polar and equatorial nets in Fig. 12.2. In each projection D is measured clockwise 0–360° round the circumference of

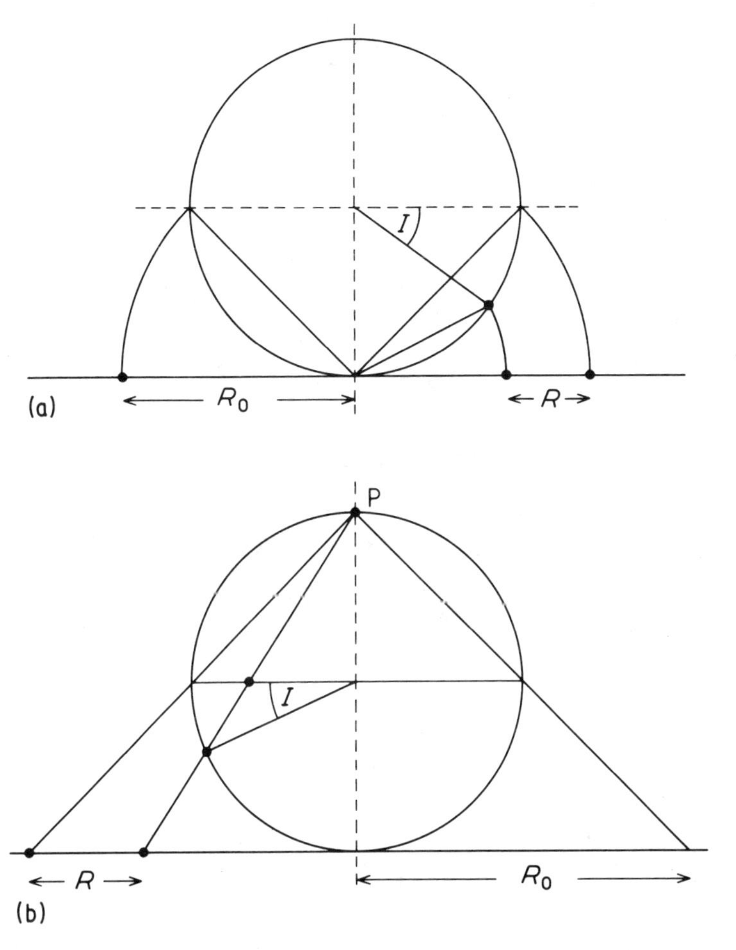

Fig. 12.1 Construction of equal area (a) and stereographic projection (b) for the point A on the sphere.

the projected circle and I is measured inward along radii, 0° at the circumference and 90° at the centre. In the stereographic projection an inclination I is represented by a point on a radius a distance R in from the circumference where

$$R = R_0\left[1 - \tan\left(\frac{\pi}{4} - \frac{I}{2}\right)\right]$$

in the stereographic projection and $R = R_0[1 - \sqrt{(1 - \sin I)}]$ in the equal-area projection, where R_0 is the radius of the projected circle. In polar projections it is usual to project the lower hemisphere only, and points on the upper hemisphere (negative inclinations) are transferred to the lower hemi-

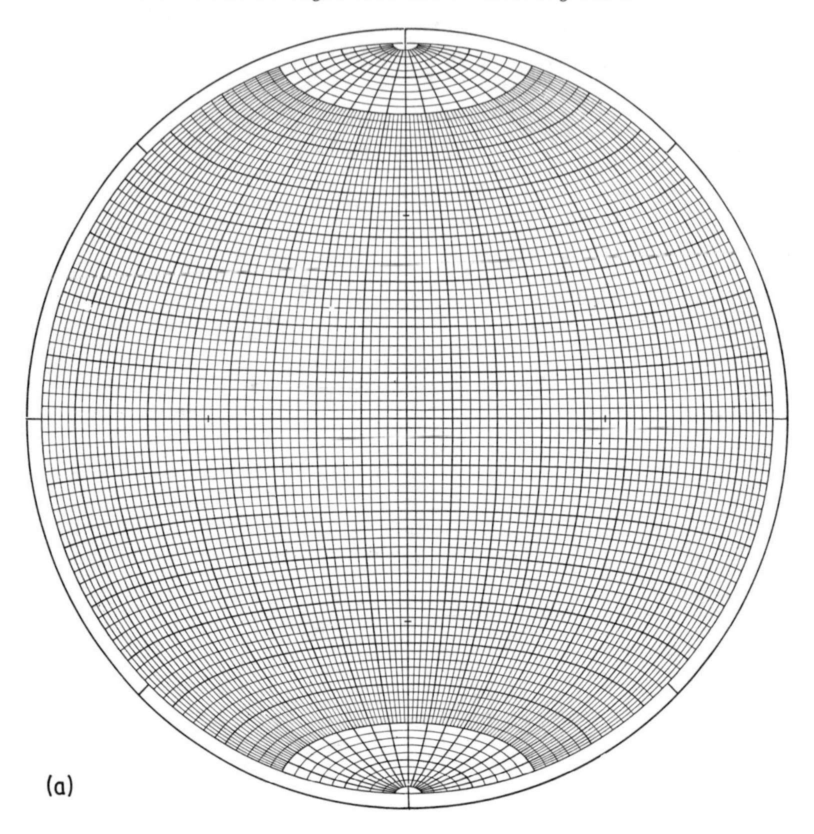

Fig. 12.2 Polar and equatorial nets: (a) Lambert equal area projection; (b) (see pages 364 and 365) stereographic.

sphere (by changing the sign of I) and a different symbol used to distinguish them from positive inclinations (Fig. 12.3). Conventionally, positive and negative inclinations are shown by full and open circles respectively. Alternatively, the pole of the projection can be chosen so that positive and negative values of I can be shown directly, e.g. by using an equatorial projection. Equatorial equal-area and stereographic projections showing great and small circles are sometimes known as Schmidt and Wulff nets respectively.

Another method of displaying a distribution of directions on a projection for visual assessment is by means of a density plot. On an equal-area projection the area is divided up into elements of the order of 1% of the total and the number of points occurring in each element recorded. Then differently shaded areas can be defined within which there is a given density of points, or contour lines can be drawn which define these areas. This type of display is most informative for points confined to either the upper or lower hemisphere, or for groups of points in either hemisphere if they well-separated on the projection.

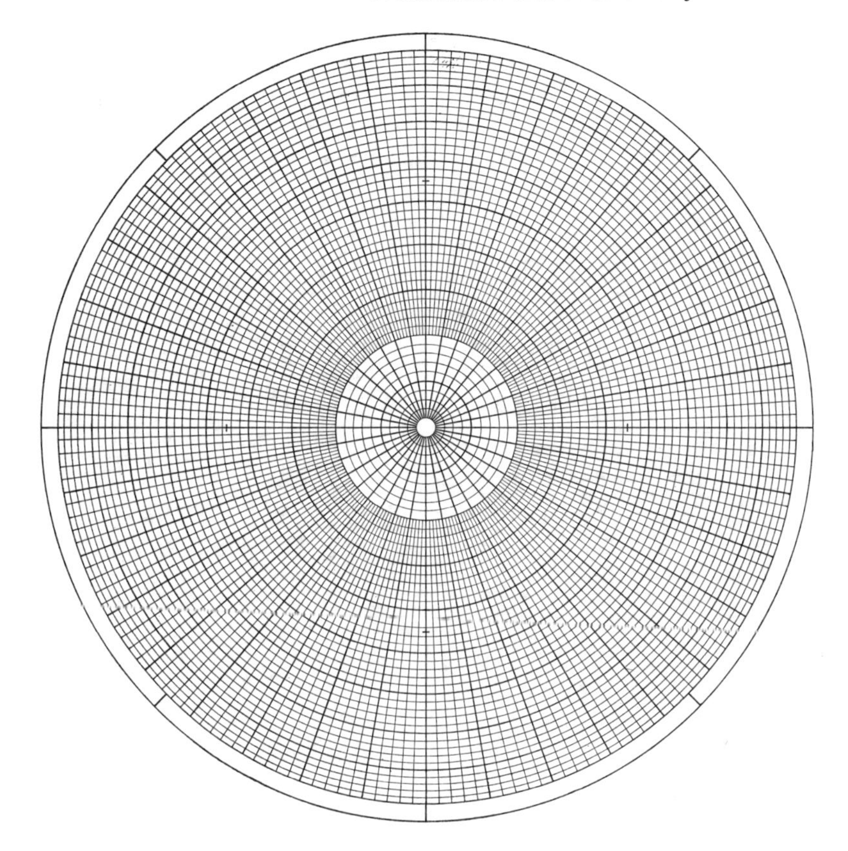

Some examples of density plots are given by van der Voo (1969), Klootwijk (1971) and Lowrie and Alvarez (1977) (see also Fig. 10.5).

On the stereographic projection circles on the sphere project as circles on the plane, a convenient feature when showing a circle of confidence around a mean NRM direction. Great circles on the sphere also project as arcs of circles on this projection. In the equal-area projection, a circle projects as an ellipse, with its minor axis along a radius and equal to the circle diameter: also, as the name implies, equal areas on the sphere project as equal areas on the plane, which may be advantageous for qualitatively assessing the scatter of a group of points. For the same reason it is useful for displaying the directions of principal axes in anisotropy investigations.

The plane of projection is chosen according to the information required and is normally either the site horizontal or the regional or local dip plane. Foss (1981) describes graphical methods, for use in conjuction with a stereographic projection, for estimating the magnitude and direction of the difference

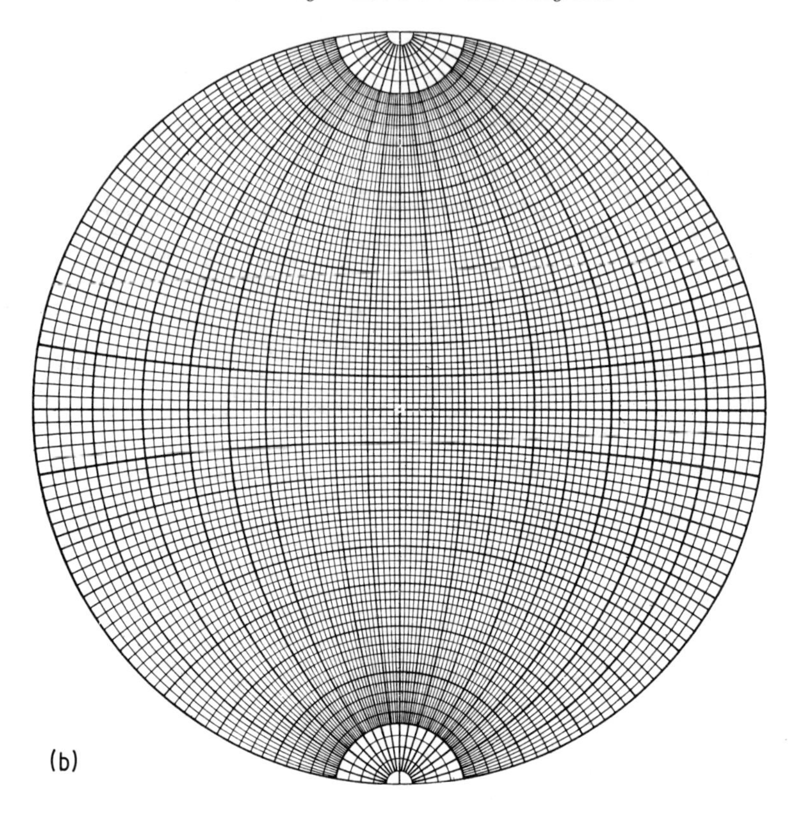

(b)

between two vectors and for resolving a vector along two or three prescribed directions.

12.2 Demagnetization data and their interpretation

The changes in NRM directions that occur during demagnetization can be conveniently displayed on one of the above projections for preliminary assessment, and the corresponding intensity changes may be plotted against peak alternating field or temperature. A linear, normalized or logarithmic intensity scale (ordinate) may be used, although a logarithmic scale is often the most informative, particularly when the intensity decays by an order of magnitude or more before a weak, stable primary NRM is revealed which was previously dominated by a much stronger secondary component. For comparison of the shapes of a series of decay curves a normalized intensity scale is convenient which can be linear or logarithmic.

The ultimate aim of magnetic cleaning is to remove all secondary magnetizations from each sample collected from a geological unit so that only

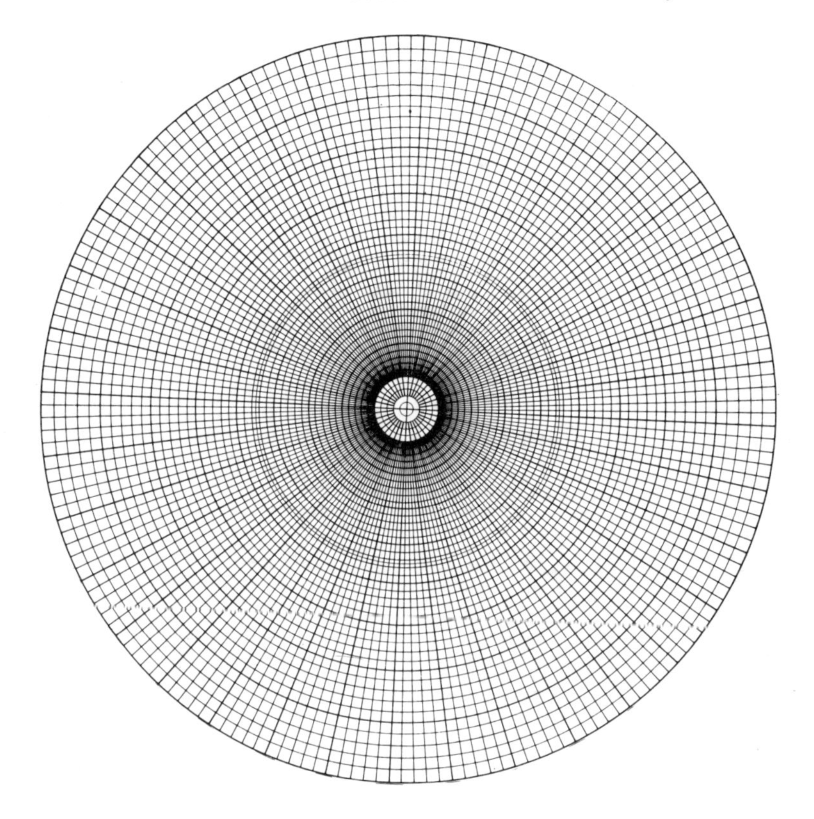

the primary NRMs are combined to provide the mean direction of NRM. The problem in achieving this object stems from the origin of the secondary NRM and it is pertinent to enquire whether satisfactory cleaning can be expected to be achieved by uniform treatment of all samples or whether different cleaning treatment for each sample is desirable.

The known origins of secondary NRM suggest that either technique may be satisfactory, depending on the type of secondary NRM present. For instance, if samples from different sites possess grains with a similar distribution of relaxation times, any VRM acquired by each sample will be removed by the same alternating field or thermal treatment. PTRM, however, acquired by a rock stratum as a consequence of burial or uplift, might be acquired at different temperatures by different samples and ideally requires that each sample be cleaned at the appropriate temperature or corresponding alternating field. The same is true of IRM due to a lightning strike: the applied field, and therefore the optimum cleaning field or temperature, will depend on the distances of the samples from the strike point.

Another factor which may contribute to the choice of optimum cleaning

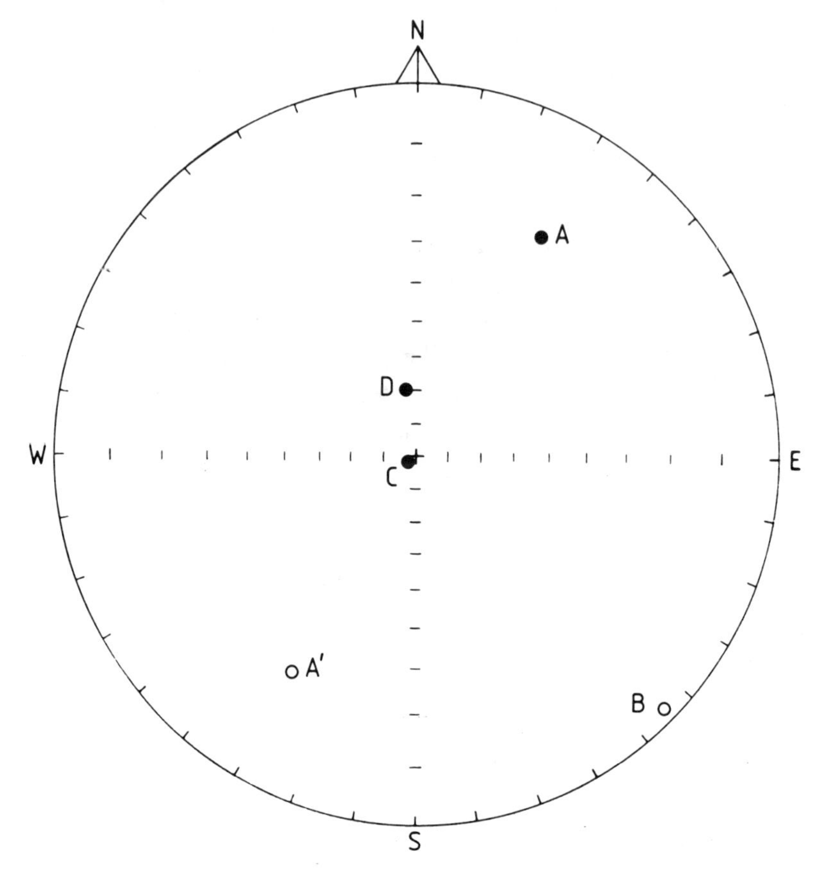

Fig. 12.3 Examples of directions shown on a stereographic projection, with the plane of projection horizontal. A, A′ are 'normal' and 'reversed' points, declination 30°, inclination +22°, and 210°, −22°; B is 134°, −2°; C is 225° +76° and D is the direction of the Earth's present field at Newcastle, 352°, +70°.

procedure is the acquisition by the rocks of a magnetization derived from the cleaning technique. In alternating field cleaning an ARM can be acquired if the waveform of the field is not sufficiently free of harmonics or a PTRM acquired in thermal or a.f. cleaning if the ambient field is not sufficiently near zero (Sections 11.1 and 11.2). The intensity of these spurious magnetizations generally increases with increasing magnitude of alternating field or temperature, and it is advantageous to clean at the minimum field or temperature consistent with satisfactory removal of secondary NRMs. Although these cleaning-induced magnetizations are usually expected to be randomly directed among different samples if the optimum experimental technique is used, the resulting increased scatter of NRM directions observed in a group of samples may obscure any improved grouping due to secondary NRM removal.

It can be seen from the foregoing that it may be difficult to establish an optimum cleaning procedure for a particular group of samples by simple visual examination of direction and intensity plots. However, this is still a widely used procedure, in which 'pilot' samples are selected for detailed thermal or alternating field cleaning, from the results of which an optimum field or temperature is qualitatively selected at which the remainder of the samples are treated.

The shortcomings of this procedure has led to the development of procedures for interpreting demagnetization data in a more rigorous way, both for extracting the primary NRM direction and for obtaining information on any secondary magnetizations present.

12.2.1. Analysis of demagnetization data

A method of data presentation due originally to Wilson (1961a) and Zijderveld (1967) combines intensity and directional changes on the same diagram. The end of the total magnetization vector is projected as points on two orthogonal planes, the horizontal plane containing the vector $(X^2+Y^2)^{1/2}$ and either the vertical and $X(X^2+Z^2)^{1/2}$ or vertical and $Y(Y^2+Z^2)^{1/2}$ plane, where X, Y, Z are magnetization components referred to the diagram axes. The $+X$, $-X$ and $+Y$, $-Y$ directions are often shown as N, S, E and W respectively on the diagrams. These planes are shown as adjacent areas with a common axis on the diagram (Fig. 12.4), and the axes are scaled in suitable intensity units. The intensity of each projected NRM component is then proportional to the distance from the origin of the corresponding points. As demagnetization proceeds the points on each plane will trace out paths according to the changes in D, I and magnetization intensity. It is usual to denote projections on the horizontal and vertical planes by solid and open symbols respectively. Some workers use a modified Zijderveld diagram in which inclination data are plotted as the end-point of the total intensity vector in the vertical plane containing the declination direction (e.g. Roy and Lapointe, 1978). The following discussion applies to results of alternating field demagnetization but the procedures and results are equally relevant to thermal or chemical demagnetization.

In the simplest case of magnetic cleaning a single secondary magnetization J_s is progressively removed without alteration of the primary NRM J_p, i.e. the coercivity spectra of J_p and J_s do not overlap. In the case where the two magnetizations are in different directions in the sample the cleaning process is indicated by a steady migration of the NRM direction (along a great circle on a projection) until the primary NRM direction is reached, with a corresponding plateau occurring on the intensity versus alternating field/temperature plot. More information is obtained from the Zijderveld diagram. The points will move along straight lines to positions corresponding to the components of the primary NRM vector: if the demagnetization is continued and the primary

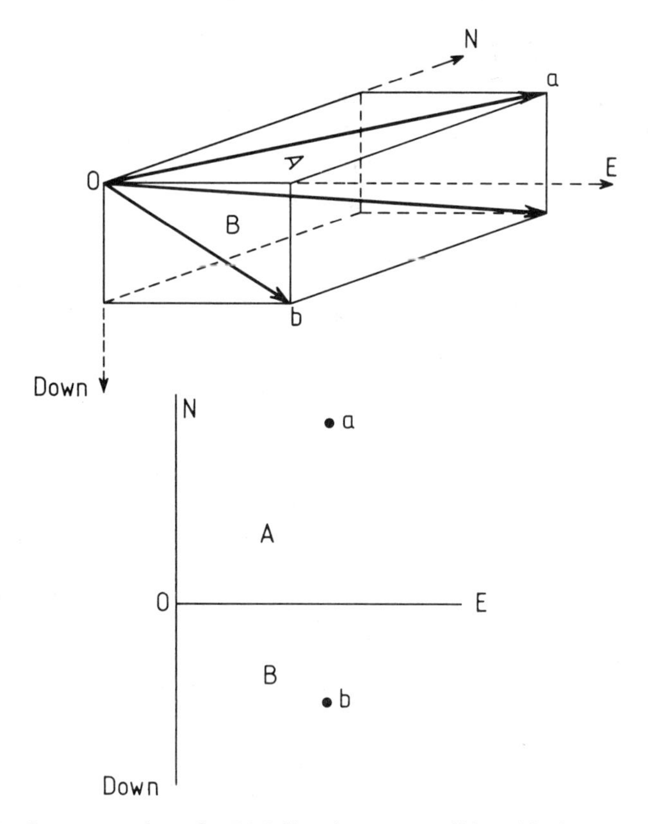

Fig. 12.4 Representation of NRM directions on a Zijderveld plot.

NRM decays the points will subsequently fall on lines directed towards the origin, the direction of the combined vector remaining constant. Thus, a Zijderveld plot in which each sector consists of two straight lines indicates the presence of two components of NRM with non-overlapping coercivity spectra (Fig. 12.5(a)). The direction and intensity of the soft NRM J_s can be determined from the angles that the straightline segments make with the axes of the diagram. Note that if the common axis (either E–W or N–S) is horizontal (as is usual), then the breakpoints of the linear segments fall on a common vertical line.

The declination D of the soft component is the angle measured eastward from N in the horizontal plane to the line in that plane corresponding to the removal of the soft component. If I_s is the angle between the common horizontal axis and the line corresponding to demagnetization of the soft NRM in the vertical plane, then the inclination I of the soft component is given by $\tan I = \tan I_s |\cos D|$ for a 'vertical and North' projection plane and $\tan I = \tan I_s |\sin D|$ for a 'vertical and East' projection. In the modified diagram in

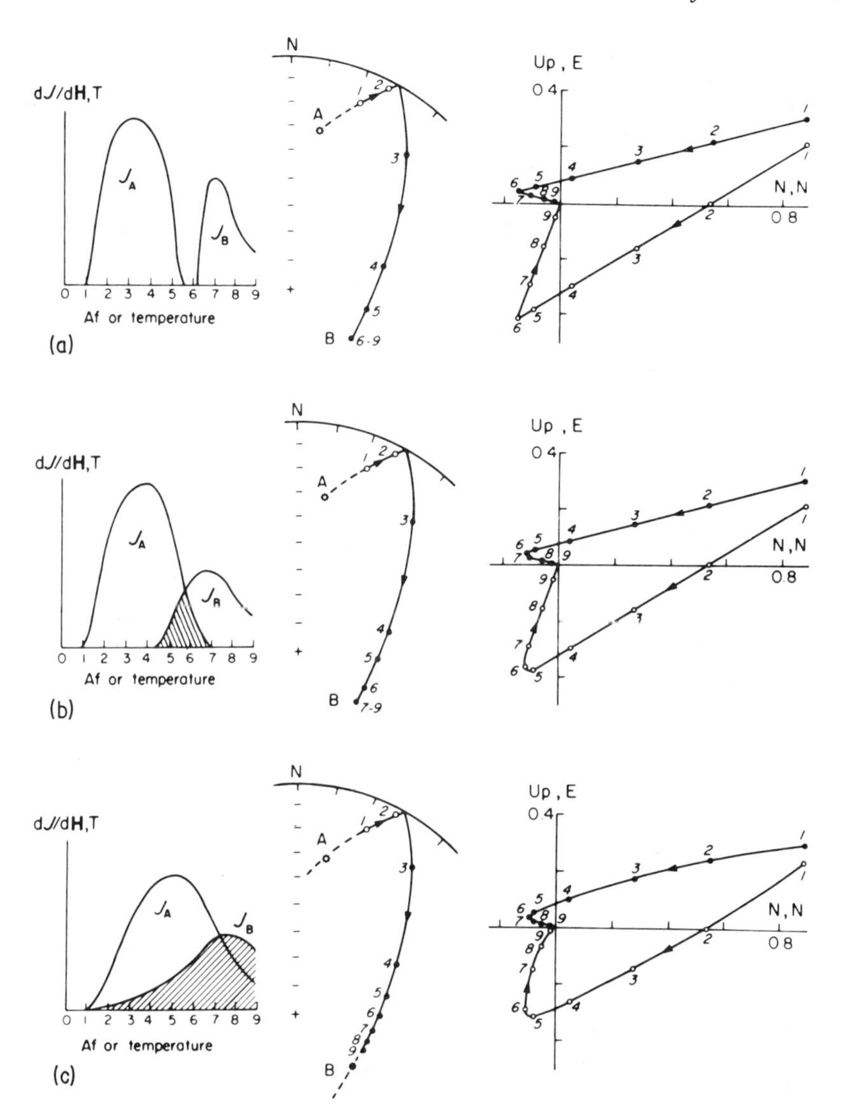

Fig. 12.5 Examples of coercivity or blocking temperature spectra and resulting NRM behaviour on cleaning displayed on stereographic projection and Zijderveld diagram: (a) non-overlapping primary and secondary spectra; (b) partial overlap; (c) total overlap. (From Dunlop, 1979)

which inclination is plotted in the vertical-declination plane, the inclination is given by the angle between the vector and the horizontal axis. The direction of the hard component of NRM is, of course, given by the stable direction reached when J_s is removed, and its intensity and rate of decay with demagnetizing field is that of the NRM when the stable direction is first reached and after subsequent demagnetizing steps. The normalized decay curve of the soft component and its coercivity spectrum can be determined from the corresponding linear segments on the Zijderveld plot. Over the total demagnetization interval represented by the end-points of the lines, the normalized intensity after each demagnetizing field is given by the ratio of the length of either of the lines between the breakpoint and the point corresponding to the demagnetizing field and the total length of the line. The absolute intensity of J_s can be obtained by transferring the end-points of the linear segments to the origin and combining the values of X, Y and Z then defined by the starting points of the lines.

The above ideal case of secondary magnetization is rarely encountered, and a more common problem is the separation of J_p and J_s when their coercivity spectra overlap to a lesser or greater extent. Partial overlap is indicated on the Zijderveld diagram by the 'rounding' of the breakpoints over the demagnetizing interval when both components are being removed (Fig. 12.5(b)). However, the direction of J_s is still well defined and the decay curves for J_s and J_p can still be determined in the way described if the reference point is taken at the point of intersection of the extended straight-line segments. This procedure is the graphical analogue of the analytical method described by Buchan and Dunlop (1976) for deriving decay curves.

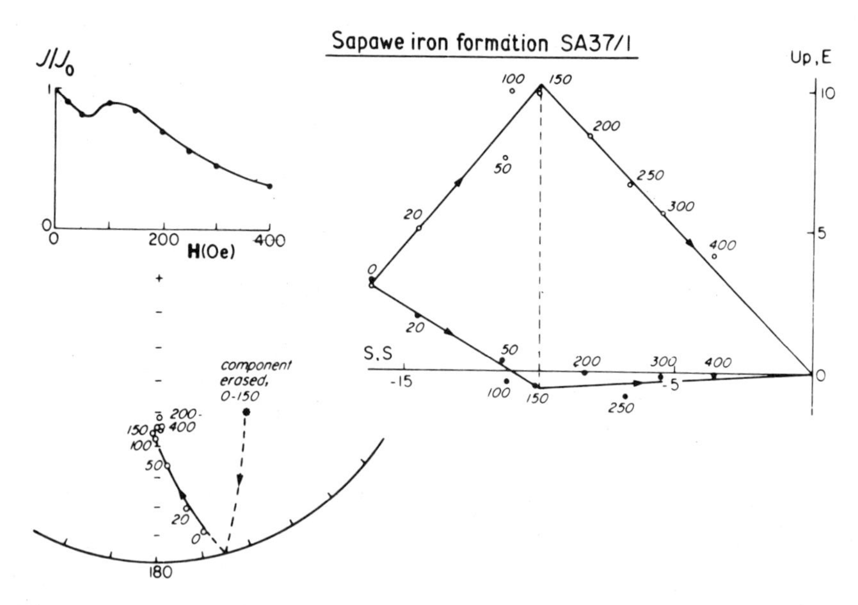

Fig. 12.6 Example of type (a) behaviour (Fig. 12.5) in an Archean iron formation, Lake Superior region, Canada. In Figs 12.6, 12.7 and 12.8 the numbers against the points are peak alternating fields in gauss (1 G = 10^{-4} T) and the Zijderveld axes are scaled in units of 10^{-6} volume c.g.s units (10^{-3} A m^{-1}). (From Dunlop, 1979)

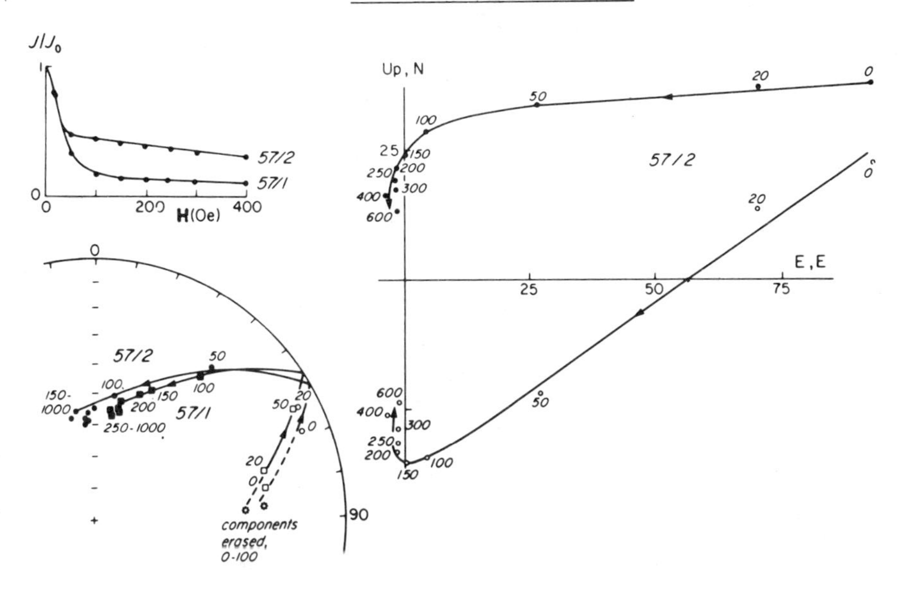

Fig. 12.7 Example of partially overlapping coercivity spectra in granite samples, Lake Superior region. (From Dunlop, 1979)

In Fig. 12.5(c) there is a much greater overlap of coercivities and J_s is only very poorly defined and J_p not at all, as no stable end-point is reached. Figures 12.6, 12.7 and 12.8 show examples of the three types of behaviour observed in rocks, taken (with Fig. 12.5) from a useful account by Dunlop (1979) describing the use of Zijderveld diagrams.

It is now common procedure in the computer processing of demagnetization data to determine the magnetic vector removed by each demagnetizing step. This provides some of the information given by the Zijderveld plot, namely whether a magnetization of constant direction is being removed. However, it should be noted here that constancy of the erased vector direction does not necessarily imply removal of a secondary NRM only. Figure 12.9 shows that if the decay of J_p and that of J_s bear a constant relation to each other (admittedly a rather unlikely situation) the removed vectors are parallel but their direction is not that of J_s.

The general problem of resolving the directions of two superimposed magnetizations is encountered in various degrees of complexity. If a stable end-point is observed, i.e. a constant NRM direction above a certain temperature or alternating field, the two directions can in principle be resolved and this is also possible if there is a range of field of temperature over which J_p is constant. In their study of the Haliburton intrusions (Ontario), Buchan and Dunlop (1976) were able to determine the direction and intensity of the

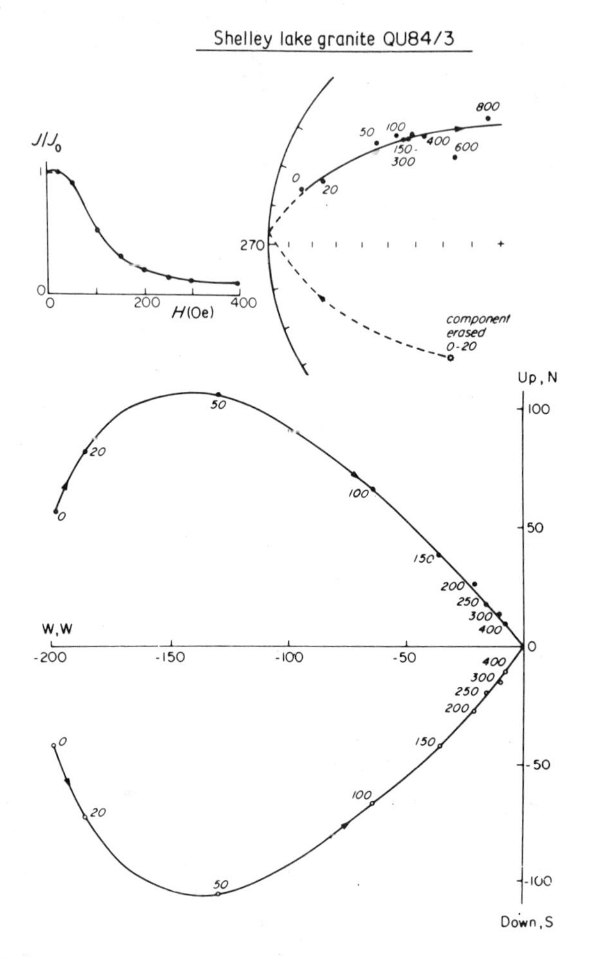

Fig. 12.8 Two (or three) NRM components with overlapping coercivity spectra in Archean granite sample. (From Dunlop, 1979)

secondary NRM in a number of samples in which the primary component showed a constant intensity (indicated by vector subtraction) in the a.f. range 15–80 mT. Figure 12.10 shows the geometry involved. The intensity of J_s in the field range can be determined after first applying the sine rule to calculate J_p from J_{15}. The usefulness of this type of procedure in establishing J_s depends on the amount of natural and/or instrumental 'noise' on the measured NRMs and on the relative magnitudes of J_p and J_s.

Although the direction of J_p is often of chief interest in palaeomagnetic studies it is often desirable to document the magnetization history of a formation as fully as possible and establish the directions of any secondary magnetizations present. Considering the simplest case of a primary NRM with

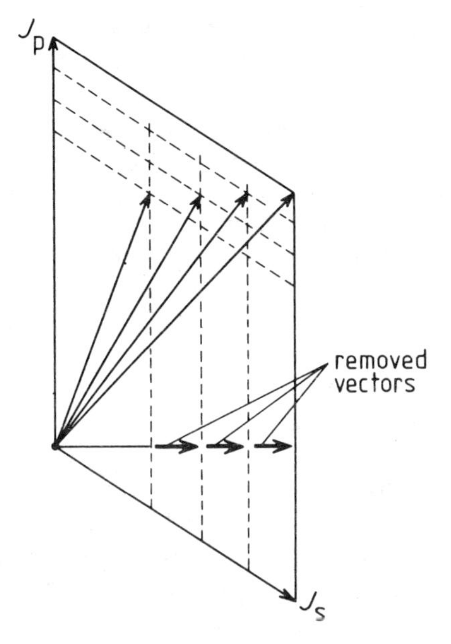

Fig. 12.9 Removal of magnetization of constant direction which is not that of secondary component.

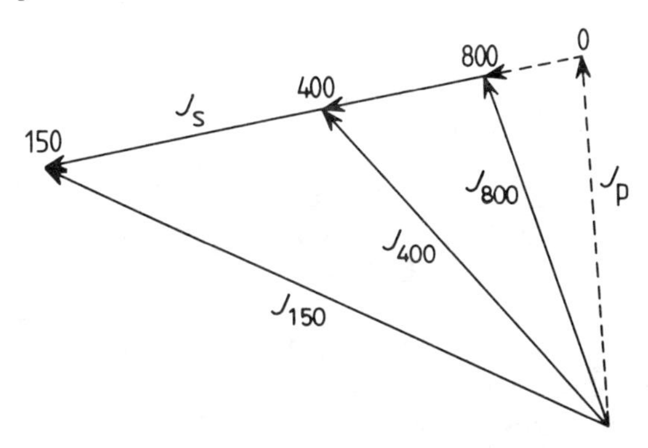

Fig. 12.10 Determination of direction and intensity of secondary NRM when primary NRM is constant over an a.f. range. (After Buchan and Dunlop, 1976)

a single secondary NRM superimposed, the above procedures fail to establish the direction of J_s in a variety of situations, e.g. if both coercivity spectra have the same lower limit but J_p has a higher upper limit, if the magnetometer noise level is reached before a stable end-point is achieved or if the coercivity spectra cover the same range.

In these cases it may be profitable to make use of 'remagnetization circles'. If J_p and J_s are being demagnetized simultaneously the resultant vectors after each cleaning step still define a great circle, the circle of remagnetization.

However, whereas the primary and secondary directions can be uniquely defined if a stable end-point is reached, with appropriate coercivity (or blocking temperature) ranges, they are not defined by the above remagnetization circle but merely lie somewhere on it.

In a uniformly magnetized rock stratum possessing a primary and secondary NRM, circles of remagnetization (referred to a common reference plane) obtained by magnetic cleaning of different samples will coincide. However, if there is dispersion of one component of magnetization relative to the other in different samples, the remagnetization circles will not coincide but will tend to converge to an intersection point. The convergence point, or its antipole, indicates the direction of one of the magnetizations. Such dispersion can arise in several ways. If the NRM of pebbles in a breccia is initially randomly oriented prior to the acquisition of a secondary magnetization, then converging circles will be obtained on demagnetization (Halls, 1978). Post-folding partial remagnetization can also provide the necessary dispersion, and secular variation of the magnetic field in which either J_p or J_s were acquired. Depending on the conditions in each case, the resulting converging remagnetization circles can be used to obtain the direction of J_p or J_s or both components, the only condition being that the coercivity or blocking temperature spectra of the two components are not identical. It may be necessary to select a suitable range of temperature or field, e.g. greater than that required to remove a VRM in the samples.

The method of using converging remagnetization circles, obtained from samples taken from sites where dispersion is of structural origin, is illustrated in Fig. 12.11, taken from Halls (1976). In this hypothetical application the primary and secondary directions can be determined because of the convergence of the remagnetization circles both before and after unfolding. Figure 12.12, also from Halls' paper, shows the method applied to some Pre-Cambrian igneous rocks from Lake Superior. The plot is obtained from a.f. cleaning of samples in which a stable end-point is apparent, and it is therefore also possible in this case to obtain normalized demagnetization curves for the primary and secondary components.

The use of converging remagnetization circles for deriving information of primary and secondary NRM in rocks was mainly developed by Russian workers before routine magnetic cleaning of samples was a common procedure (Khramov, 1958, 1967). Other references describing the technique include Creer (1962b), Jones, Robertson and McFadden (1975) and Lovlie (1978).

The scatter which is invariably present in palaeomagnetic data results in lack of precision in the definition of remagnetization circles from demagnetization data and in the point of convergence of the circles. Creer (1962b) describes a relevant problem in connection with the NRM of Devonian sandstones from Wales. Samples from five sites showed mean directions distributed approximately along a great circle which also passed near the mean Permian and

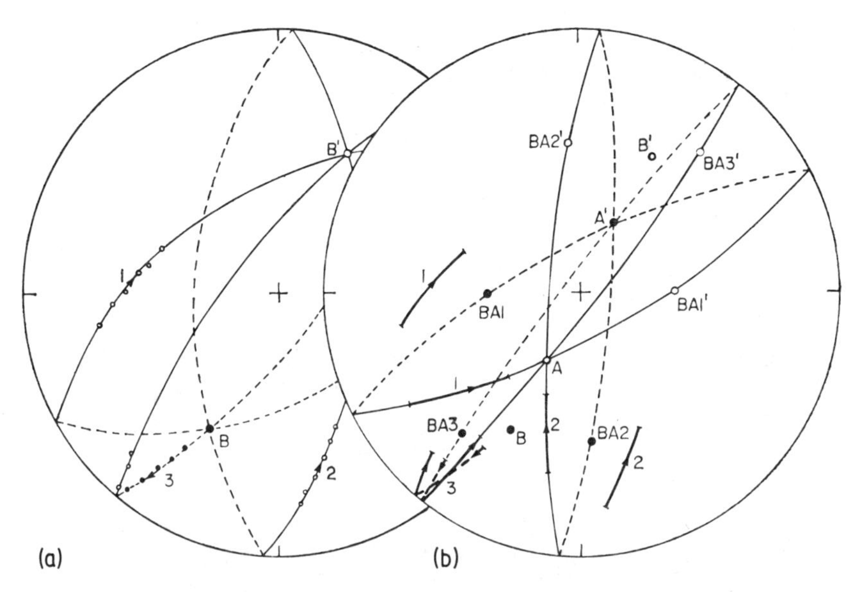

Fig. 12.11 Converging magnetization circles: Wulff net diagrams illustrate schematically how the directions and relative ages of two magnetization components can be resolved when sample cleaning yields successive magnetization vectors which define arcs of great circles, but no stable end-points. In the hypothetical example given, remagnetization circles are defined by sample data from three sites (1, 2, 3) in a structurally disturbed area. Arrows indicate sense of directional swings on cleaning; dashed/solid parts of circles are plotted on the lower/upper hemispheres; and solid/open dots are downward/upward magnetizations. In (a), B and B′ are the common intersection points of remagnetization circles before structural unfolding (SU). In (b), A and A′ are the intersection points of the same circles after SU, and BA1 and BA1′ are the positions of poles B and B′ for site 1 etc. On each of the three great circles in (b) there are two possible directions of the primary and rotated secondary components. The ambiguity is removed because the data points defining these circles must lie between the true primary and secondary directions. Hence A and B are respectively the directions of the primary and secondary components. (From Halls, 1976)

Triassic directions for the region, suggesting the acquisition of secondary NRM near these times. Using a method devised by Watson (1960) for finding the best-fitting great circle through a series of points on a sphere, Creer showed that there was a slightly greater probability of remagnetization in Triassic than in Permian times at four of the sites. Halls (1976) extends this procedure in order to calculate the best-fitting convergence point of several great circles. The latter problem is similar to the fitting of a great circle to data points, since if the circles converge at a point their normals define another great circle. Thus, in the presence of scatter the best-fitting great circle through the normals

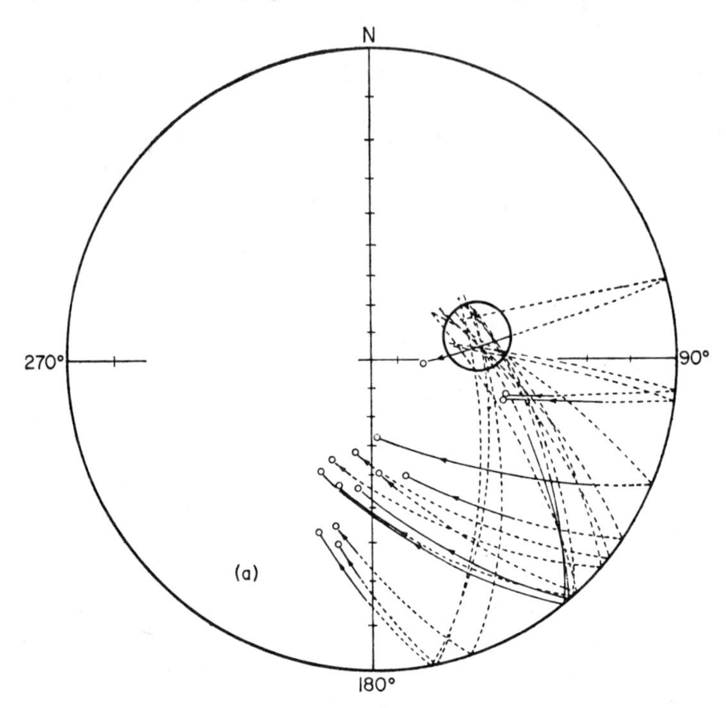

Fig. 12.12 Remagnetization circles obtained by a.f. demagnetization of Pre-Cambrian igneous rocks from the Slate Islands, Lake Superior, Canada. Solid portions of circles defined by data, dashed parts are extrapolated and arrows indicate direction of vector movement on cleaning. The final vectors are stable end-points. (From Halls, 1976)

defines the most probable convergence point. Onstott (1980) has applied statistical methods due to Bingham (1974) to this problem.

If a secondary magnetization postdates folding in a rock formation it may be possible to isolate the primary (P) and secondary (S) NRM directions by a procedure applied to the Catombal sediments, New South Wales, Australia by Williamson and Robertson (1976). They first determine an approximate primary direction by 'removing' mathematically successive proportions of the secondary NRM in each sample. A bedding correction is applied to the sample (or site) directions after each removal to refer them to the original, unfolded plane and the optimum S direction is chosen as that which gives a well-defined minimum dispersion of the unfolded directions. The S directions used are obtained from the post-folding apparent polar wandering path for the region. When the approximate P and S directions have been determined the magnitude of the S component at each site can be calculated, and a second estimate of the P direction is made by reducing the S component at the sites by amounts proportional to the magnitude of S at each site and repeating the search for minimum dispersion. If desirable, the procedure can then be repeated until there is no further improvement in the observed dispersion. The potential advantage of this procedure is the possibility of 'removing' a secondary component which is magnetically harder than the primary component, and the assignment of a time of remagnetization based on geological considerations can be tested quantitatively. The use of a computer makes the mathematical procedure quite tractable.

Stupavsky and Symons (1978) describe an analytical method of extracting primary and secondary magnetization directions and magnitudes from demagnetization data. The assumption is made that the NRM components decay exponentially with alternating field strength and therefore the resultant vector R can be expressed in the form (for two components)

$$R_B = J_p \exp(-0.7B/B_p) + J_s \exp(-0.7B/B_s) \qquad (12.1)$$

where J_p, J_s are constant vectors corresponding to the primary and secondary magnetizations and B_p, B_s are the alternating fields required to reduce J_p and J_s to half their initial values. A least-squares method is then used to find the best values of J_p and J_s, using Equation 12.1 for different values of the demagnetizing field B. B_p and B_s are found by an iteration technique after processing trial values of these fields. The analysis also includes calculation of precision parameters for the computed values of P and S, enabling models with more than two components to be tested and the optimum model chosen on the basis of its statistical significance.

The procedure was tested on groups of samples from two rock units from which best mean directions had been obtained by conventional averaging of data after demagnetization. At the 95 % confidence level the means obtained by the two methods were indistinguishable, although the analytical procedure was only carried out on a single specimen from each site. Hence a potential advantage of the procedure may be in obtaining a comparatively reliable mean direction from a limited number of samples.

Although two-component magnetizations are often encountered in palaeomagnetic studies, more complicated, multicomponent NRMs are not uncommon and are not normally easily separable. Such magnetizations arise, for instance, from more than one partial remagnetization, growth of new minerals through weathering and reversals of the geomagnetic field. Good separation of components may also be hindered by acquisition of spurious magnetization during demagnetization and decrease in the intensity of NRM to near or less than the magnetometer noise level.

Hoffman and Day (1978) describe a method in which difference vectors are displayed on a stereographic (or equal area) projection, rather than the NRM directions themselves. The principle of the method for three NRM components is illustrated in Fig. 12.13. Case (a), with no overlapping of coercivity spectra, would be amenable to analysis by, for instance, visual assessment of a Zijderveld diagram, and the vector difference display indicates the directions

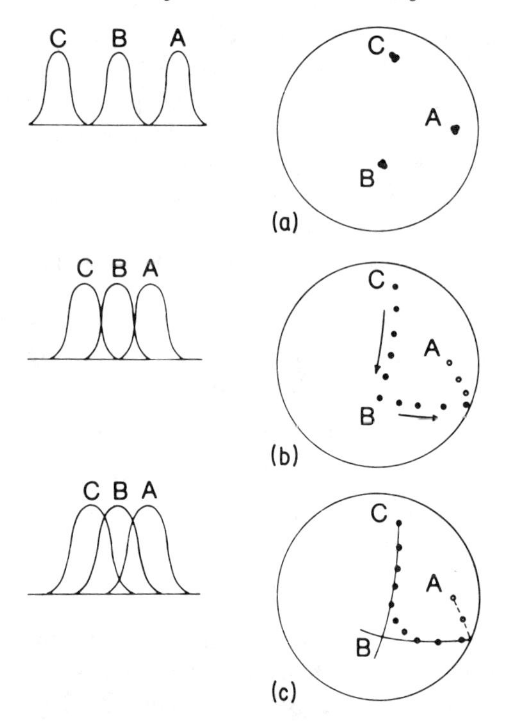

Fig. 12.13 Diagram showing coercivity or blocking temperature spectra and corresponding difference vector paths for three remanence components A, B and C with different degrees of overlap. (From Hoffman and Day, 1978)

of the three NRM components. Where A overlaps with B and B with C but not A with C (case (b)), the B direction is indicated by the discontinuity in the difference vector path, each segment of which is a great circle. The Zijderveld diagram would only show the B direction if there was a reasonable coercivity range over which it alone was removed. Case (c), in which there is overlapping of coercivities of A and C as well, results in a smoothing out of the difference vector path, from which the B direction can in principle be recovered by extrapolation of the two great circle segments. Hoffman and Day illustrate the use of the technique in extracting three components of NRM present in a Jurassic pillow basalt from California. Halls (1978) describes a procedure which combines difference and resultant magnetization vectors to determine intermediate NRM directions.

The application of 'principal component analysis' to interpretation of palaeomagnetic data has been introduced by Kirschvink (1980). The procedure is essentially that of fitting by least squares the best straight line to, for instance, segments of Zijderveld displays or the best plane to remagnetization circle data. A computerized search procedure is adopted which locates regions of the data which are linear or planar to a given precision, from which remanence directions and poles of demagnetization planes can be derived with associated angular errors.

12.2.2 Stability indices

An alternative approach to the analysis of demagnetization data is the calculation of a numerical 'stability index', derived mathematically from the magnitude of directional and/or intensity changes during magnetic cleaning of different samples. Two types of index have in fact been proposed, namely those which describe the overall palaeomagnetic stability of different rock samples and those which can be used to determine an optimum cleaning field or temperature for each sample. The latter indices are mainly of use as an aid to deciding at which level of cleaning the primary NRM of a sample is revealed.

Before considering various indices of palaeomagnetic stability that have been proposed, some other parameters that are used to indicate general stability characteristics are described.

In rocks in which a large secondary component is dominant over a weak, very stable primary NRM, the field or temperature at which the secondary remanence (and essentially the total NRM) is removed indicates the maximum coercivity of the secondary remanence. It is therefore a crude indication of the sample's overall stability: it says nothing, of course, about the characteristics of the primary NRM except to emphasize its stability and relative weakness. For purposes of comparison between samples and the better distinction between different NRM components and carriers, a 'coercivity spectrum' is more informative. In the present context this shows the proportion of the total NRM intensity carried by grains of different coercivities or blocking temperatures and can be derived from curves of a.f. or thermal demagnetization of NRM (Section 13.5).

Another quantity used to describe stability against alternating fields is the median destructive field (MDF), which is the demagnetizing field required to reduce the initial NRM intensity by half. If all NRM vectors in the rock are approximately parallel or at least in the same octant in space the MDF will provide some measure of the average stability of the NRM carriers. The presence of a secondary magnetization opposed in direction to the primary can give misleading values of MDF, giving a substantially higher or lower value than when the magnetization vectors are well grouped according to whether the secondary NRM is less or greater than the primary. However, the behaviour of the intensity and direction of NRM in these cases will warn of the relative orientation of the components. An example where MDF can be correlated with other properties of magnetic particles is given by Ade-Hall, Aumento, Ryall, Gerstein, Brooke and McKeown (1973): theory suggests that magnetic stability should decrease with increase in grain size of multidomain titanomagnetite and an approximate correlation between MDF and reciprocal grain size was observed in the titanomagnetite-carrying NRM in basalts from the mid-Atlantic ridge.

Other field values which have been used to indicate stability are 'effective' and 'critical' demagnetizing fields. These are respectively the field at which the NRM intensity is reduced to e^{-1} of its initial value and the field below which the NRM direction remains constant within 10°.

The direct field which, when applied antiparallel to the NRM of a rock reduces the NRM to zero, has sometimes been used as a measure of stability, chiefly in early investigations by some Russian workers. They sometimes gave the name 'disruptive field' to this back field. This field has some significance if the remanence is carried by all particles present of a single magnetic mineral, but is less useful if more than one mineral is carrying NRM, particularly if a substantial proportion of the particles are unmagnetized or so weakly magnetized as to contribute negligibly to the observed NRM. The back field may then only be a measure of the coercivity and intensity of IRM of the weakly magnetized material. The technique also obscures rather than reveals any primary NRM in the samples. There is also a discussion in the Russian literature of the effect of applying direct fields at varying angles to the NRM (Sholpo, Afremov and Pershin, 1972).

There is little consistency in the literature of rock magnetism in the use of the terms 'coercive force', 'coercivity' and 'coercivity of remanence'. The coercive forces indicated by the above procedures are not to be confused with the more common use of the term, associated with the stability of saturated IRM, saturated ARM or total TRM. These coercive forces and their spectrum in a sample derive from all particles capable of carrying remanence, whereas for estimating palaeomagnetic stability it is only the coercive force spectrum of particles actually carrying remanence which is of interest.

Irving *et al.* (1961) might be considered pioneers in attempting to determine optimum cleaning levels when they proposed the criterion of minimum dispersion of a group of NRM directions for deciding the best cleaning field or temperature. However, for reasons already given a distribution of secondary NRM coercivities is likely and the true mean direction and precision at a site is unlikely to be revealed by treatment at a uniform demagnetization level. Dagley and Ade-Hall (1970) attempted to allow for variable coercivities by combining the means of pairs of consecutive directions obtained from the demagnetizing curve of a core specimen with similar means obtained from other cores from the site. All possible pair mean directions were then combined and the resulting site mean corresponding to minimum dispersion was chosen as the best NRM direction for the site.

The first attempt at a quantitative description of stability of NRM in rocks and the determination of optimum cleaning fields was by Tarling and Symons (1967). For a series of cleaning steps 1,2,3,4 . . . n (alternating field or thermal) applied to a sample the quantity $F = R^{\frac{1}{2}}/\theta_{63}$ is computed for all possible combinations of NRM directions for $n > 3$ ($n = 1$ is the initial NRM). θ_{63} is the circular standard deviation of the NRM directions occurring in the range R, i.e. the corresponding alternating field or temperature interval. Thus, F is computed for steps 1,2,3, 2,3,4, 3,4,5 . . . . $(n-2)$, $(n-1)$, n; 1,2,3,4, 2,3,4,5 . . . $(n-3)$, $(n-2)$, $(n-1)$, n through to the complete range 1,2,3,4, . . . n, and the maximum value of F so obtained is designated the stability index S_{TS} for the sample. The range of alternating field or temperature which yields the stability index is that in which the stable component of NRM is best isolated. If S_{TS} is determined for selected pilot samples, then the optimum treatment may be indicated for the remainder of the samples from the site, but ideally S_{TS} should be determined for all samples and the 'best' NRM direction for each extracted accordingly for determinations of the site mean.

The circular standard deviation is chosen to describe directional scatter because of its comparative insensitivity to the number of observations n, usually small (5–10) in stepwise demagnetization. Numerically, values of S_{TS} vary between $\sim$15 or more for extremely stable remanence and $\sim$0.5 for very unstable samples.

Apart from the likelihood that successive directions during demagnetization will not be Fisherian, particularly when large changes occur, and the somewhat arbitrary square root factor in the numerator the stability index can also be criticized in that it is assumed that the hard component will give a small θ_{63} over a wider value of R than a soft component. A strong but soft VRM superimposed on a weak primary NRM may provide a stability index associated with a low field or temperature interval and indicating high stability: however, this index will not be relevant to the primary component, which may not provide an index suggesting high stability because of random measurement errors or equipment-induced spurious magnetizations.

Andel (1970) discusses S_{TS} and the significance of θ_{63} in this context and concludes that S_{TS} cannot be uniquely defined and is of limited value, mainly because of its dependence on the magnitude of demagnetization steps and of the highest field or temperature used in the cleaning procedure. He shows analytically and by laboratory tests that F tends to a maximum value at the highest values of field or temperature.

Although constancy of NRM direction during cleaning is a primary indication of palaeomagnetic stability (as distinct from the idea of stability as used in rock magnetism, i.e. the level of coercive force in a mineral or sample), the neglect of intensity information in deriving S_{TS} can be criticized. In the case of a weak primary NRM with a strong parallel or antiparallel secondary component superimposed, derivation of S_{TS} would give confusing results and intensity changes during cleaning would be more informative of the NRM components present. For this reasons, other workers have emphasized the importance of intensity data in deriving stability indices.

A simple index based on intensity changes was proposed by Russian workers (Petrova, 1961; Sholpo and Yanovskii, 1967) who used the quantity

$S_R = J_p/(J_p + J_s)$ where J_p and J_s are the intensity of primary and secondary magnetization in a sample. Thus $0 < S_R < 1$, the extreme values indicating dominant ($S \to 0$) or negligible ($S \to 1$) secondary components. In practice it is rarely possible to cleanly separate and designate intensities to primary and secondary NRM because their coercivity spectra commonly overlap.

A measure of stability against incremental demagnetization is the stability factor S_W of Wilson, Haggerty and Watkins (1968). It takes the form

$$S_W = \frac{R_{20}}{R_{20} + r} \qquad (12.2)$$

where R_{20} is the length of the vector representing the NRM remaining after demagnetization in a 20 mT field and r is the non-vector sum of the vector changes in the NRM after demagnetization in 10 mT and 20 mT. S_W lies between zero and unity.

The Wilson factor was adapted for more general use by Ade-Hall (1969) who used it in the form

$$S_A = \frac{J_n}{J_n + \sum_{1}^{n} \Delta J} \qquad (12.3)$$

where J_n is the magnetization remaining after the nth demagnetization step and $\Sigma(\Delta J)$ is the non-vector sum of all changes which have occurred in the sample magnetization during the n steps. Thus again $S_A = 1$ for an entirely stable sample and $S_A \to 0$ for large changes in intensity or direction. The cumulative nature of S_A and the non-vectorial treatment of ΔJ ensures a better representation of palaeomagnetic stability in some special cases, e.g. when an initial increase of intensity occurs during cleaning followed by a decrease (Ade-Hall, 1969), but the cumulative feature also nullifies its use as an indicator of optimum demagnetization level in palaeomagnetic studies. However, the factor was originally devised by Wilson *et al.* purely as a measure of rock-magnetic stability, to be used in correlations of this and other magnetic properties in lavas.

Intensity changes are also incorporated in an index used by Murthy (1971). The index is

$$S_M = (J_{opt}/J_0)/\theta_{63} \qquad (12.4)$$

where J_{opt} is the NRM intensity remaining after the optimum cleaning field has been applied, J_0 is the initial intensity and θ_{63} is the circular standard deviation of the directions of remanence during cleaning up to the optimum field. Like the Tarling-Symons index, S_M increases from zero as stability increases.

A palaeomagnetic stability index due to Briden (1972), derived for each demagnetization step, is based on the 'difference index' (DI) of two vectors, $\mathbf{J}_n$ and $\mathbf{J}_m$

Table 12.1 Difference index and stability index for some simple vector relationships (Briden, 1972)

Configuration	DI	S_B
Two identical vectors	0	1
One vector cf. zero	1	0
Two equal antiparallel vectors	2	−1
Two equal vectors at 60°	1	0

$$\mathrm{DI} = \frac{(\mathbf{J}_n - \mathbf{J}_m)}{J_n} = \frac{J_{n-m}}{J_n} \tag{12.5}$$

Table 12.1 shows values of the DI for some simple cases, and also of the derived stability index S_B for the demagnetization step when $\mathbf{J}_n$ changes to $\mathbf{J}_m$

$$S_B = I - \mathrm{DI} = 1 - \frac{J_{n-m}}{J_n} \tag{12.6}$$

and $-1 \leqslant S_B \leqslant +1$. To eliminate dependence of S_B on the magnitude of the alternating field or temperature interval the use of standard intervals is proposed, namely 10 mT or 50°C. It is usually reasonable to linearly interpolate if the field or temperature intervals are greater than the above. By inspection of S_B values obtained for each sample during the cleaning sequence or by plotting S_B against cleaning field or temperature, a range of field or temperature can be selected to give maximum values of S_B for each sample and the best estimate of the primary NRM direction.

A stability index of appealing simplicity is that of Symons and Stupavsky (1974) who proposed the rate of change of remanence direction with demagnetizing field as an indication of palaeoemagnetic stability. This index (S_{SS}) does not include intensity changes but can claim some advantage in being independent of the number and inequality of demagnetizing steps. For alternating field cleaning S_{SS} is determined for each sample from the change in declination D and inclination I of the NRM for each increase in field $B_i \rightarrow B_{i+1} = \Delta B$. Mathematically S_{SS} can be expressed in the form

$$S_{SS} = \left(\frac{\Delta \mathbf{r}}{\Delta B}\right)_{\Delta B \rightarrow 0} = \frac{\mathrm{d}\mathbf{r}}{\mathrm{d}B} \tag{12.7}$$

where $\mathbf{r}$ is the unit vector along the remanence direction. S_{SS} can be measured in millidegrees/mT or, as the authors propose, for c.g.s. units, millidegrees/Oe. $S_{SS} = 0$ corresponds to absolute stability and a marked minimum value of S_{SS} during a cleaning sequence is indicative of the isolation of a stable component.

Since currently available magnetometers generally provide the cartesian components X, Y, Z of an NRM direction, S_{SS} is conveniently calculated directly from them, since $x = X/R$, $y = Y/R$ and $z = Z/R$, where $R = (X^2 + Y^2 + Z^2)^{1/2}$, are components of the unit vector representing the remanence direction. Then

$$S_{SS} = 57.3 \times 10^3 \left[\left(\frac{dX}{dB}\right)^2 + \left(\frac{dY}{dB}\right)^2 + \left(\frac{dZ}{dB}\right)^2 \right]^{1/2} \text{ mdeg/mT} \quad (12.8)$$

It is assumed that each derivative is constant between adjacent steps. The derivatives can be obtained from D and I, if required, by

$$\frac{dX}{dB} = -\sin D \cos I \frac{dD}{dB} - \cos D \sin I \frac{dI}{dB} \quad (12.9)$$

$$\frac{dY}{dB} = \cos D \cos I \frac{dD}{dB} - \sin D \sin I \frac{dI}{dB} \quad (12.10)$$

$$\frac{dZ}{dB} = \cos I \frac{dI}{dB} \quad (12.11)$$

where dD/dB and dI/dB are measured in radians/mT and X, Y and Z are in the North, East and vertical down directions. The index is associated with each demagnetizing field B_i rather than a field interval, by averaging the values of S_{SS} obtained for the steps B_{i-1} to B_i and B_i to B_{i-1}. For Equations 12.9, 12.10 and 12.11, the appropriate values of D and I are D_i and I_i.

When using the stability index values of S_{SS} are plotted against the corresponding demagnetizing field and the plot is inspected for the minimum value of S_{SS}. The field, or range of fields, associated with minimum S_{SS} is then the optimum for cleaning that sample and the corresponding values of D and I used in calculating the site mean. Values of $S_{SS(min)}$ lie typically between 5 and 10 mdeg/Oe (50–100 mdeg/mT) for highly stable rocks and ~500 mdeg/Oe (5°/mT) for very low stability.

Lowrie and Alvarez (1977) use a simpler form of this index. They compute the angle $d\theta$ between two vectors on a demagnetization curve from $d\theta = \cos^{-1}(l_i l_j + m_i m_j + n_i n_j)$, where the direction cosines of the two vectors (Equation 12.18) are (l_i, m_i, n_i) and (l_j, m_j, n_j). These $d\theta/dB$ is computed, where dB is the increment in demagnetizing field which produced the vector change, and minimum values of $d\theta/dB$ indicate the most stable region. $d\theta/dB$ is also used as a data rejection criterion, samples showing values $>10°$/mT being considered unusable.

A comprehensive examination of factors relevant to palaeomagnetic stability and its quantitative description is given by Giddings and McElhinny (1976). They develop an index for describing the stability against alternating field demagnetization of magnetite-bearing igneous rocks, based on a combination of the Briden (1972) index, the observed maximum coercivity of multidomain magnetite and two experimentally determined demagnetizing fields. The latter are B_s, the field at which secondary components of NRM are essentially removed and directional changes become minimal, and a higher field B_p at which direction and intensity begin to change in a random fashion. The stable range R is then defined as

$$R = B_p - B_s \tag{12.12}$$

The proportion of the coercive force spectrum in a sample that lies in the stable range is given by

$$C_s = R/B_p \quad (0 \leqslant C_s \leqslant 1) \tag{12.13}$$

and indicates the importance of secondary magnetization in the coercivity spectrum. B_p is a measure of the hardness of the primary remanence. Assuming that the primary and secondary components of NRM in rocks in which magnetite carries the remanence are carried by single- and multidomain grains respectively, a factor C_p is then defined which relates B_p to the observed maximum coercivity of multidomain magnetite, i.e. 80 mT (Evans and McElhinny, 1969)

$$C_p = B_p/80 \quad \text{for} \quad 0 \leqslant B_p \leqslant 80 \tag{12.14}$$

and

$$C_p = 1 \quad \text{for} \quad B_p > 80$$

Combining C_s and C_p, an overall 'coercivity index' is then given by

$$\text{CI} = C_p C_s \tag{12.15}$$

where $0 \leqslant \text{CI} \leqslant 1$.

Finally, CI is combined with the mean Briden index (MBI) defined by

$$\text{MBI} = \frac{1}{n}\left[\sum_{i=1}^{n} (S_B)_i\right] \tag{12.16}$$

where S_B is the Briden index (Equation 12.6) and n is the number of S_B values in the stable range R. The overall palaeomagnetic stability index S_{GM} is given by

$$S_{GM} = \text{MBI} \times \text{CI} \tag{12.17}$$

where $0 \leqslant S_{GM} \leqslant 1$, the extreme values corresponding to complete instability ($S_{GM} = 0$) and extremely high stability ($S_{GM} = 1$). If S_B is negative, as is possible, and leads to a negative MBI a formal value of MBI $= 0$ is adopted, although the authors point out that in this situation C_s is also tending to zero.

S_{GM} cannot be used directly to define an optimum demagnetizing field or field interval. It describes, by a number between 0 and 1, the palaeomagnetic stability of a sample based on as large a proportion as possible of its coercivity spectrum. The number then classifies the sample according to suitable, subjective stability classes. An important parameter in the derivation of S_{GM} is the stable range R (which is itself a crude indication of optimum cleaning field), and the authors propose a quantitative procedure for its selection. Their method is based on the use of the Tarling-Symons index. Using the mean defined by the directions in the observed stable range region, a subset of sequential directions within 40° of this mean is selected and a new mean calculated from the subset. This procedure is repeated from new directional subsets lying successively within 35°, 30°, 25°, 20° and 15° of their mean directions, the last subset formally defining the stable range. For partially stable rocks where only two directions can be considered as closely grouped, R is only defined if the two directions are separated by less than 15°.

The index has been shown to be useful in quantitatively comparing the stability of two groups of magnetite-bearing rocks: Giddings (1976) and Giddings and Embleton (1976) show that S_{GM} for collections of samples from Pre-Cambrian dykes from different blocks in Australia have quite different distributions.

Although considerable ingenuity has been exercised by the contributors to the problem of quantifying palaeomagnetic stability and defining optimum cleaning levels, a survey of the literature shows that the various stability and optimum cleaning indices are not widely used. Among the reasons for this may be the suspicion that, because there are so many factors involved in extracting the primary NRM from samples in which there are one or more secondary components, a mathematical solution to the problem is not possible. It is clear that the angle between the relative intensity and coercivity spectra of primary and secondary components of NRM all play a part in the behaviour of the NRM vector during cleaning and may affect the chosen stability index in a confusing way. To the above factors should also be added the possible acquisition of spurious magnetizations through imperfections in the equipment. Many workers still apparently feel that either visual examination of intensity plots and stereograms of directional changes during cleaning is as satisfactory a way as any of deciding the optimum cleaning procedure for a suite of rocks, or the procedures described earlier in this chapter are employed.

Another factor which may contribute to reluctance to use stability indices is time. One of the reasons for computing an index is the assumption that uniform treatment of all samples is not justified and that each specimen should be examined individually. Thus, in a collection of 60 specimens in which 15 pilot specimens are taken through 10 demagnetizing steps and the remaining 45 treated with one demagnetization, there is a total of 195 specimen demagnetizations. Detailed treatment of all specimens involves 600 demagnetizations. This involves a substantial increase in measurement time and also in a.f. cleaning time: it may not be so important in thermal cleaning since up to ~50 specimens may be heated together in some furnaces.

12.3 Statistics

As with all experimental data, measurements of directions of NRM are subject to random (and possibly systematic) errors. However, in addition to these errors arising, for instance, from magnetometer noise, inaccuracies in sample orientation and imperfections in demagnetizing procedures there are also usually true differences in NRM directions among a group of specimens from a core or hand sample, site or geological formation. Among the causes of these differences are secular variation of the geomagnetic field, variation in the strength of residual secondary magnetization and, particularly in depositional remanence, failure of the DRM to be aligned accurately parallel to the ambient field. The overall effect of these errors and perturbations of NRM directions in a group of specimens, samples or sites is a scatter of the NRM vectors, sometimes small, sometimes extending over tens of degrees in declination and inclination.

Procedures are therefore required for calculation of the mean direction of NRM and of quantities to describe the dispersion of the NRM vectors and the error to be attached to the mean direction. The original directional data may be from a group of specimens, samples or sites in which stable endpoints have been obtained after magnetic cleaning, or the data may originate from the procedures described in the last section, e.g. analysis of Zijderveld diagrams or the intersection of remagnetization circles. Various other statistical procedures associated with interpretation of palaeomagnetic results are also described in this section.

12.3.1 Mean NRM direction, precision and circle of confidence

The initial stage in the interpretation of palaeomagnetic data is the determination of a mean direction of NRM, whether it be of a group of specimens taken from a hand sample or core, a group of samples from a collection site or a group of sites in a geological formation.

At the first statistical level it is normal practice to give unit weight to each specimen direction contributing to a mean, without regard to intensity of NRM. The only rocks in which a correlation might be expected between NRM intensity and closeness of the measured NRM direction to the true direction (i.e. that of the aligning field) are those in which the variation in intensity among samples is due to variation in number of poorly aligned magnetic particles, the origin of the NRM being depositional and the number of particles in each sample being very small ($\sim$ 5–50). The greater the number of particles, the greater the intensity and the closer the resultant direction to that of the aligning field. However, the above situation is very rare (and could be verified if suspected) and in the great majority of rocks the assignment of unit weight to each sample direction seems well founded.

The statistics associated with a distribution of points on a sphere (in the present context, the ends of a number of NRM (unit) vectors originating at the centre and grouped about a mean vector) is due to Fisher (1953) and the reader is referred to his paper for the mathematical details.

Consider the NRM directions to be represented by N unit vectors: then the best estimate of the mean direction is that of the resultant, of length R, of the N

vectors. Using appropriate orthogonal reference axes an individual direction can be expressed in terms of its direction cosines l, m, n where

$$l = \cos D \cos I$$
$$m = \sin D \cos I \tag{12.18}$$
$$n = \sin I$$

D and I are the declination and inclination of the direction referred to the same axes. The direction cosines of the resultant vector, l_r, m_r, n_r are proportional to the sums of the separate direction cosines, and are given by

$$l_r = \frac{1}{R}\sum_{i=1}^{N} l_i, \quad m_r = \frac{1}{R}\sum_{i=1}^{N} m_i, \quad n_r = \frac{1}{R}\sum_{i=1}^{N} n_i$$

and

$$R^2 = (\Sigma l_i)^2 + (\Sigma m_i)^2 + (\Sigma n_i)^2 \tag{12.19}$$

The declination and inclination of the mean are

$$\tan D_r = \frac{\Sigma m_i}{\Sigma l_i}, \quad \sin I_R = \frac{\Sigma n_i}{R}$$

The probability distribution proposed by Fisher for points on a sphere (the ends of N unit vectors) is of the form

$$P = \frac{\kappa}{4\sinh \kappa} \exp(\kappa \cos \psi) \tag{12.20}$$

where ψ is the angle between the true mean direction and one of the N vectors and κ is the precision parameter, a measure of the scatter of the vector directions. The density of directions is axially symmetrical about the mean, and the probability of finding a direction in a belt which makes an angle between ψ and $(\psi + d\psi)$ with the mean is

$$P\,d\psi = \frac{\kappa}{2\sinh \kappa} \exp(\kappa \cos \psi) \quad \sin \psi \, d\psi \tag{12.21}$$

The coefficient of the exponential term in Equation 12.20 ensures that the probability integrates to unity over the whole sphere.

The best estimate k of the precision parameter κ is given by Fisher as

$$k = (N-1)/(N-R) \tag{12.22}$$

for $k > 3$. k is usually in the range 10–1000, the higher the value the more tightly grouped the set of directions. $\kappa = 0$ (or $k \to 1$) implies a random (or near-random) group of directions. A quantity analogous to the standard deviation for a normal distribution is given approximately by $81/k^{\frac{1}{2}}$ degrees, i.e. the angle from the mean direction beyond which only 37% of the directions are expected to lie: this is sometimes known as the 'circular standard deviation'. The corresponding angle for 5% of the directions is given approximately by $140/k^{\frac{1}{2}}$ degrees.

The accuracy of the mean direction derived from N vectors with resultant R can be expressed as the semi-angle α of a cone about the observed mean within which the true mean lies with any given probability $(1-P)$: for $k > 3$

$$\cos \alpha_{(1-P)} = 1 - \frac{N-R}{R}\left[\left(\frac{1}{P}\right)^{\frac{1}{N-1}} - 1\right] \tag{12.23}$$

P is normally taken to be 0.05, i.e. there is a 95% probability that the observed mean is within $\alpha°$ of the true mean (the 'circle of confidence'). When α is small ($< 5°$), the approximate relation $\alpha_{95} = 140/\sqrt{(kN)}$ may be used. The analogous quantity to the standard error of the mean in the normal distribution, α_{63}, is given approximately by $\alpha_{63} = 81/\sqrt{(kN)}$.

12.3.2 Testing 'goodness of fit'

A pertinent question is whether a typical group of palaeomagnetic directions in fact show a Fisherian distribution. Fisher (1953) gave no physical or geophysical basis for his distribution function (Equation 12.20), although it has some basis in statistical mechanics and the same function has been examined for a two-dimensional distribution of unit vectors by Gumbel, Greenwood and Durand (1953), who called it the 'circular normal distribution'. Runcorn (1967b), arguing by analogy with the derivation of the normal distribution as applied to NRM components X, Y and Z, has suggested that it is a valid one. Wilson (1959) proposed the use of the analogue of the Gaussian standard deviation, $\delta = \cos^{-1}(R/N)$ (the angular standard deviation) and the standard deviation of the mean $\sigma = \delta/N^{\frac{1}{2}}$, claiming that with these quantities there was no supposition regarding a particular distribution. Runcorn (1960) has commented on this. For N large, such that $N \approx N-1$, θ_{63} and α_{63} are indistinguishable from δ_{63} and σ_{63}, but the latter suggest greater accuracy for R and N. As judged by eye, many groups of cleaned directions bear some similarity to a Fisherian distribution and the statistics provide a uniform way of dealing with and comparing data. However, where circles of confidence or other statistics are used to draw important conclusions from different groups of directions it may well be a wise precaution to examine the original data, and if necessary use the statistical results cautiously. Some examples of Fisherian distributions are shown in Fig. 12.14.

Watson and Irving (1957) have given expressions for testing the goodness of fit of a group of NRM directions to a Fisherian distribution. The directions should, of course, be azimuthally symmetrical about the mean: this can be

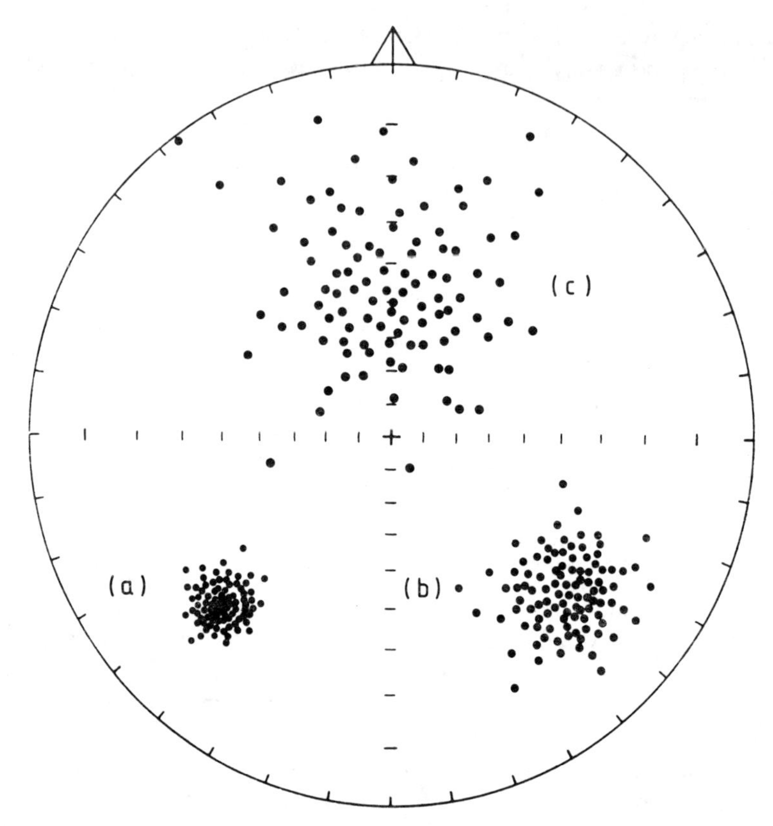

Fig. 12.14 Examples of Fisherian distributions for $N = 100$ vectors. (a) $k = 200$, (b) $k = 50$, (c) $k = 10$.

tested by calculating the statistics

$$\chi^2 = \sum \frac{(f_0 - f_e)^2}{f_e} = \sum \left(\frac{f_0^2}{f_e} \right) - N \qquad (12.24)$$

where f_0 and f_e are the observed and expected frequencies in an azimuthal interval and N is the total number of directions. χ^2 is the basis of a statistical method for testing the probability of whether a series of observations fits a distribution by measuring the divergence between the observed and expected frequency of events in a number of intervals. It is associated with the number of 'degrees of freedom' F of the system, which is derived from the number of observations and unknowns involved. In the present context, $F = I - 3$, where I is the number of azimuthal intervals chosen, and the table of χ^2 is compared with the tabled value, which corresponds to the expected χ^2 for a given probability level, and if the observed χ^2 is equal to or less than the expected value the data are taken to fit the distribution with the given probability or better.

The same test can be applied to the observed distribution of θ, the angular distance of the individual directions from the mean. The Fisherian distribution predicts the number of directions falling within an angular distance between θ_1 and θ_2 from the mean given by

$$N\{\exp[-k(1-\cos\theta_1)]-\exp[-k(1-\cos\theta_2)]\} \qquad (12.25)$$

where N is the total number of directions. The χ^2 test can then be carried out on the observed and expected distribution for suitable intervals of θ: the intervals should be chosen so that five or more directions occur in each. Watson and Irving (1957) give an example of the fit to the Fisherian distribution of a group of 70 NRM directions obtained from a single fine-grained red sandstone sample.

12.3.3 Test of randomness

If a group of directions is widely scattered over both hemispheres it may be desirable to test whether the sample is from a random population or whether there is a significant mean direction at some probability level (Watson, 1956). For true randomness, κ and R are zero, but in practice $R \neq 0$ and k, the best estimate of κ (for $N \geqslant 3$), is not (and cannot be) zero.

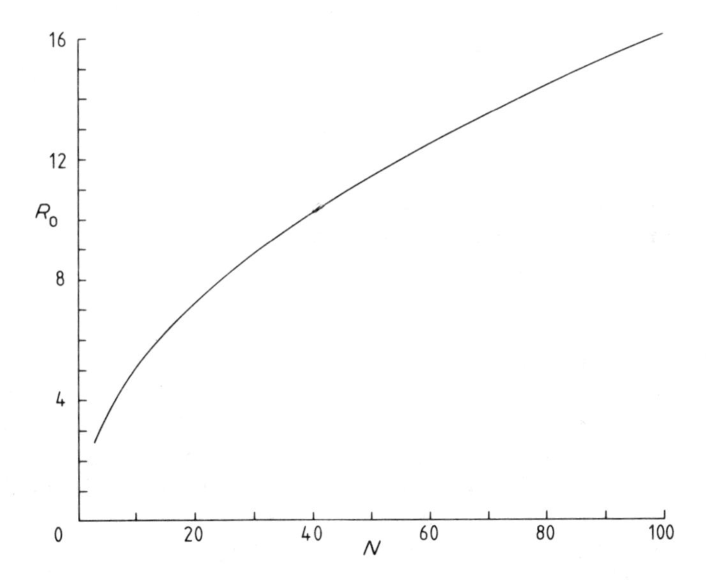

Fig. 12.15 Values of R_0 for $N = 3$–100.

For a widely scattered group of directions R will be small relative to N, and assuming a random population a value R_0 may be calculated which will be exceeded by R with any chosen probability. Figure 12.15 is a graph of R_0 for

$N = 3$–100 at the 5% probability level. As an example, in a group of 80 directions sampled from a random population, there is only a 5% probability that R will exceed 14.4, and an observed $R \leqslant 14.4$ implies that a mean direction computed from the group is not significant at this level. It should not be forgotten that the distribution of a near random group of directions will be far from Fisherian and statistics and interpretations based on such directions should be treated very conservatively.

12.3.4 Combination of observations from a rock unit

The statistical procedure described is applicable to any set of unit vectors, the distribution of which can be reasonably approximated to the Fisherian. The procedure for the first level of sampling, i.e. the derivation of a mean direction of NRM of several specimens taken from a hand sample or a field-drilled core, is straightforward. The optimum procedure for determining the mean direction, and its associated error, of a group of samples taken from a site, and then of a number of sites within a rock unit (e.g. a group of lava flows or a sedimentary formation) is less obvious and depends partly on the distribution of directions observed within and between samples and between sites.

The aim of a palaeomagnetic survey is the determination of the mean direction of primary NRM for a rock unit and its associated error. A typical sampling scheme and the sources of scatter among NRM directions observed in specimens, samples and sites have already been described in Section 8.1. The end product of the palaeomagnetic measurements is a set of directions of 'cleaned' NRM obtained from several sites within the rock unit, those from each site being obtained from several samples. The question then arises of the procedure for combining the directions in the way most likely to provide the best representative mean direction for the unit, and its uncertainty, and which makes optimum use of the available data. In many cases the estimated mean direction will be only little affected by the statistical procedure adopted, whereas the effect on the circle of confidence around it may be significant, an important consequence where means from contemporaneous rock units are being compared.

The simplest procedure for computing the error in the mean is to combine all the directions obtained from the specimens from each sample and the samples from each site and treat them as if they were from a single population, each vector being given unit weight. This procedure has little to commend it theoretically, particularly for sediments, since it implies that each observation is a spot reading of the ancient field direction and that dispersion between specimens from a sample and between samples from a site is due to the same cause, namely secular variation of the field and measurement and orientation errors. In sedimentary and igneous rocks, sampling is rarely stratigraphically or temporally random and in practice conditions which would validate the above procedure, namely the same mean direction and precision among samples and sites, are seldom encountered.

A more satisfactory procedure is to use Fisher's method to first calculate sample means, combine these for site means and then combine the several site means to obtain a mean direction and circle of confidence for the rock unit. Unit weight is given to all mean directions used. This procedure is most satisfactory if the precision and number of the observations from each sample and site are both similar, but is in any case an improvement on combining specimen directions only.

Watson and Irving (1957) describe a two-level analysis in which some allowance is made for disparity in numbers of observations and also of precision of observations within each site and of the site means about the overall mean direction for the rock unit. Their procedure is expected to be more suitable for igneous rocks, in which the within-site values of k are expected to be similar. For sediments, the formation mean direction obtained from sites at which different lithologies have been sampled is best derived from the site mean directions in the usual way, giving each mean direction unit weight.

12.3.5 Statistical significance of the fold test

If k_1, k_2 are the estimates of precision of NRM directions before and after correction for folding, then as shown by Watson (1956)

$$\frac{k_2}{k_1} = \frac{\text{variance with } 2(N-1) \text{ degrees of freedom before folding}}{\text{variance with } 2(N-1) \text{ degrees of freedom after folding}} \tag{12.26}$$

where N is the number of NRM directions. The right-hand term has the variance ratio or f-distribution. When $f = k_2/k_1 \gg 1$, then the two populations do not have the same directional dispersion, within a given confidence limit, i.e. the fold test is significant within that limit. Figure 12.16, which is based on the table given by McElhinny (1964), shows values of k_2/k_1 which must be exceeded for a given value of N if the difference in precision is to be significant at the 95% and 99% level of probability.

McFadden and Jones (1981) claim that this significance test is too stringent, because of some shortcomings in the assumptions on which it is based. They propose a test involving the mean directions of NRM obtained from each limb of the fold. After confirming that the precision of the two populations is the same, the mean directions of the group of sites from each limb are tested to see if they are significantly different at the required probability level (McFadden and Lowes, 1981).

Some other work on the application of statistics to palaeomagnetic data is that of Gidskehaug (1976), Engebretson and Beck (1978) and McFadden (1980). Methods of smoothing palaeomagnetic data (e.g. declination and inclination curves obtained from discrete sampling of lake sediment cores) and of matching curves from different cores and estimating the significance of similarities in wave form between them are given by Clark and Thompson

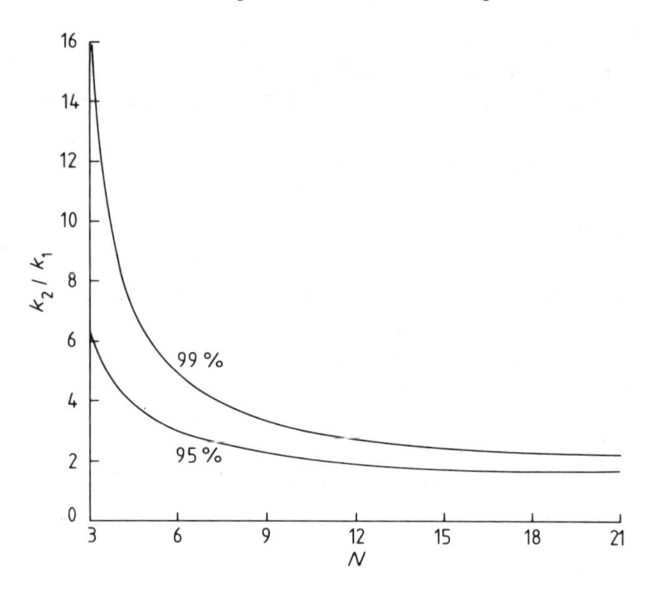

Fig. 12.16 Significance of the fold test. Values of k_2/k_1 against N for 95% and 99% probability level. (From McElhinny, 1964)

(1978, 1979) and Parker and Denham (1979). Onstott (1980) describes the application of the Bingham distribution function (Bingham, 1974) to palaeomagnetic studies.

12.3.6 Calculation of pole positions

The acquisition of palaeomagnetic data provides mean directions of NRM and associated errors for a wide variety of rock units of differing ages and from different continents. It is clearly desirable to be able to compare in a meaningful way NRM directions obtained from formations of the same age from sites widely separated on the same continent and from sites on different continents. It is also of interest to examine the significance of changes with geological time of mean NRM directions from particular continental regions where a significant proportion of the geological column can be sampled.

Since the NRM direction in a rock unit is ideally the direction of the ambient geomagnetic field at the site at the time of formation of the unit, some constant feature associated with the field at different places on the Earth's surface should provide a common reference system with which directions from widely separated areas can be compared. Historically, the observation that rocks up to a few million years old are magnetized in a direction consistent with a time-averaged geomagnetic field that is close to that due to a geocentric axial dipole suggested interpretation in terms of an ancient geomagnetic pole. Thus, the mean NRM direction at a site is used to calculate the position of the magnetic

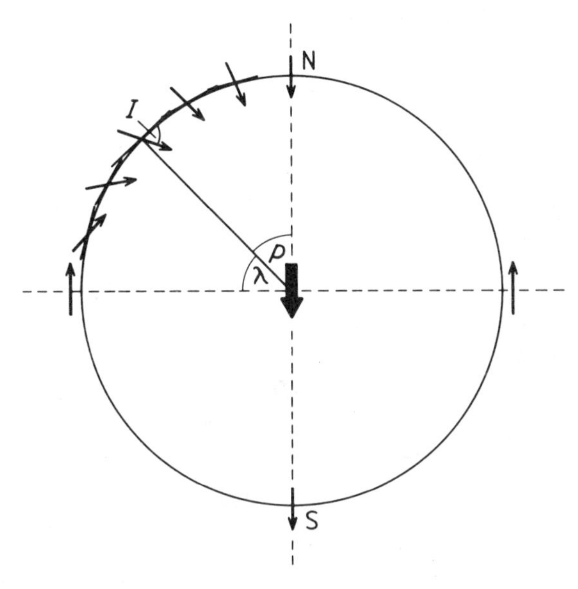

Fig. 12.17 Dipole field directions at the Earth's surface along a meridian. The relationship between the angles I, λ and p is shown.

pole (N or S) which, on the axial dipole assumption, defines the dipole axis which would produce the field (and therefore NRM) direction observed at the site.

The relevant geometry of the dipole field is shown in Fig. 12.17. If the latitude of the site is λ (relative to the equator defined by the dipole axis) the inclination I of the dipole field at the site is given by

$$\tan I = 2\tan \lambda \tag{12.27}$$

An NRM direction D, I observed at a site corresponds to a magnetic pole at an angular distance p from the site along the great circle defined by the declination D, where

$$p = [90^\circ - \tan^{-1}(\tfrac{1}{2}\tan I)] \tag{12.28}$$

Thus p is the apparent or ancient magnetic colatitude (palaeolatitude) of the site: on the further assumption of coincidence of dipole and rotation axes, p is also the ancient geographic colatitude. Application of Equation 12.28 enables a polar wandering curve to be constructed, i.e. a curve joining pole positions derived from NRM directions from rocks of different ages from the same continental region.

Comparison of pole positions derived from rocks of the same age from different continents provides crucial evidence for continental drift (or plate movements). Since a dipole field has a unique N or S pole, the non-coincidence of pole positions derived from contemporaneous rocks from sites on different continents implies changes in relative plate positions during geological time.

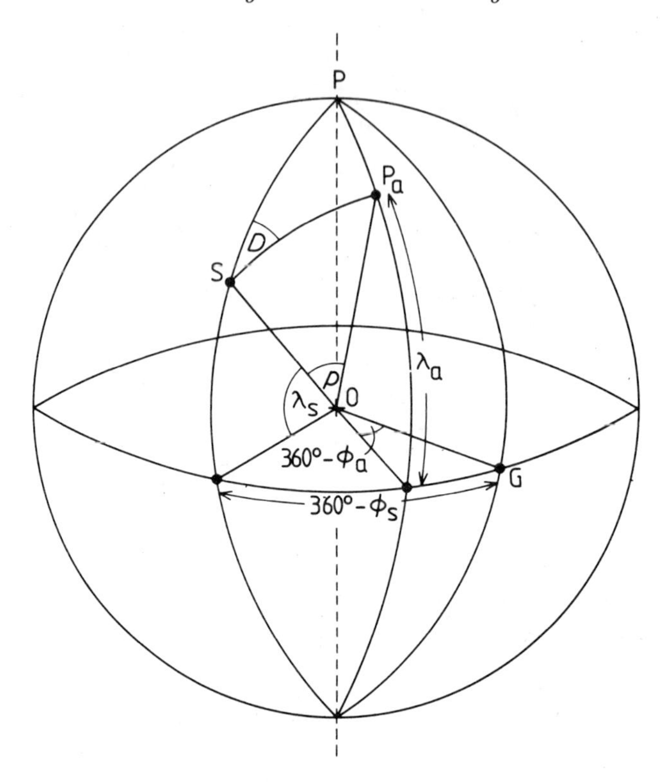

Fig. 12.18 Geometry for calculation of pole positions. P is the present geographic pole, S is the collection site and P_a is the ancient pole. PG is the Greenwich meridian.

For plotting pole positions for both polar wandering curves and for testing plate motions it is convenient to refer the ancient pole positions to present geographical reference axes, i.e. to calculate the present latitudes and longitudes of the ancient poles. The geometry is shown in Fig. 12.18. If the present latitude and longitude of the site are λ_s and ϕ_s and the mean declination and inclination of the site NRM are D and I then the latitude λ_a and longitude ϕ_a of the ancient pole are given by

$$\sin \lambda_a = \sin \lambda_s \cos p + \cos \lambda_s \sin p \cos D \qquad (12.29)$$

$$\phi_a = \phi_s + \beta \quad (\text{for } \cos p > \sin \lambda_s \sin \lambda_a)$$

or

$$\phi_a = \phi_s + (180° - \beta)\,(\text{for } \cos p < \sin \lambda_s \sin \lambda_a) \qquad (12.30)$$

where

$$\sin \beta = \sin p \sin D / \cos \lambda_a \qquad (12.31)$$

Latitude is measured between 0° and ±90°, positive in the northern hemisphere and longitude is measured eastward from Greenwich, 0° to 360°.

p is the ancient colatitude of the site, given by

$$\tan I = 2\cot p \tag{12.32}$$

and $-90° < \beta < +90°$.

Depending on the data used in their derivation, two pole positions are distinguished. A 'virtual geomagnetic pole' (VGP) is calculated from a spot reading of the palaeomagnetic field, i.e. one for which secular variation is not meaned out obtained, for instance, from a single laval flow. A pole based on a mean NRM direction obtained from several laval flows or from several specimens or samples of a sediment is expected to provide a pole free of short-period (10^3–10^5 years) perturbations, and such a pole position is termed a 'palaeomagnetic pole'. Palaeomagnetic poles are the basic data for polar wandering curves and the detection of plate motions.

Since a mean direction of magnetization has an associated error an ancient pole position derived from it is also subject to an angular error, within which there is a certain probability that the true pole lies. It is convenient to derive the polar error from the circle of confidence around the mean NRM direction, of radius α_{95}. The error dp in the ancient colatitude p is given by

$$\mathrm{d}p = \tfrac{1}{2}\alpha_{95}(1 + 3\cos^2 p) \tag{12.33}$$

and dp lies along the great circle joining the sampling and the palaeopole P. The error in the declination of the mean direction corresponds to an error dm in P along an axis perpendicular to dp, where

$$\mathrm{d}m = \alpha_{95}\sin p/\cos I \tag{12.34}$$

Except when $I = 90°$, d$p \leqslant$ dm and the polar error (dp, dm) defines an oval of confidence within which there is a 95% probability that the true pole lies. For a given α_{95}, dp and dm increase with increasing palaeolatitude of the site and thus more observations are required (to reduce α_{95}) to establish a pole with a given oval of confidence when the pole approaches the sampling site.

An alternative procedure for determining a palaeomagnetic pole and its associated error is to combine VGPs derived from site or sample means. It is assumed that the VGPs exhibit a Fisherian distribution and that the mean pole and its error can be calculated by Fisher's method. Instead of the value of declination and inclination of NRM used in determining the pole from the mean direction, the basic data used in Fisher's analysis of VGPs are their longitudes and latitudes.

It should be noted that certain assumptions are made concerning the distribution of NRM directions for each of the above procedures. In the first the usual assumption of a Fisherian distribution is implicit whereas in the second procedure a non-Fisherian, oval distribution of directions is necessary if the VGPs are to have a Fisherian distribution and a Fisherian analysis is to be valid. One of the few reports of a markedly oval distribution of NRM directions is that of Creer, Irving and Nairn (1959) is their study of the whin sill of the north of England. They show that to provide a Fisherian distribution of poles the axes of the oval distribution of NRM directions will vary with the palaeolatitude of the rock unit according to

$$\Delta D = \theta_{63}(1 + 3\cos^2 p)^{-\frac{1}{2}} \qquad (12.35)$$

and

$$\Delta I = 2\theta_{63}(1 + 3\cos^2 p)^{-1} \qquad (12.36)$$

where p is the ancient colatitude and θ_{63} is the circular standard deviation of the VGPs about the mean corresponding to ΔD and ΔI. Inspection of Equations 12.35 and 12.36 shows that for $p = 90°$, i.e. zero palaeolatitude, $\Delta D = \frac{1}{2}\Delta I$ and that the distribution becomes increasingly circular as the pole and sampling site approach coincidence.

In practice the mean pole and associated error obtained from either procedure differ only slightly. A minor advantage of averaging VGPs is the automatic correction for different site latitudes and longitudes, but this is only significant for a wide spread of sites. In the literature there is generally little evidence for oval distributions of directions nor is it clear how they would arise. One possible origin is a variable but weak secondary NRM of viscous origin causing restricted 'streaking' of points towards the local field direction.

Chapter Thirteen

Identification of magnetic minerals and carriers of NRM

13.1 Introduction

Identification of the magnetic mineral(s) in a rock and the carriers of remanent magnetization is often necessary for a better understanding of the significance of the NRM and of the rock's magnetization history. If several magnetic minerals are present, only one of which is carrying a primary NRM, experiments designed to identify the latter should clearly involve the remanent magnetism only and not other properties, e.g. induced magnetization, to which all magnetic minerals present contribute.

One problem is the widely differing magnetic properties of different minerals. A red sandstone may contain 2% by weight of haematite, which appears to be the dominant magnetic mineral, yet if magnetite is present to the extent of only 0.01% it is capable of contributing the majority of the observed NRM and yet be below the limit of detection by conventional methods such as X-ray and polished section techniques.

In many red sandstones a different problem arises, namely distinguishing which of the two forms of haematite carries the NRM. In redbeds haematite usually occurs as both the red pigment which gives the rocks their distinctive colour and as black, polycrystalline particles (specularite).

In igneous rocks the NRM is normally carried by the magnetite or titanomagnetite content, although the contribution of each grain varies because of size, shape and compositional differences. In highly altered igneous material and metamorphic rocks there may be pyrrhotite, maghemite and haematite present, one or more of which may be contributing to the NRM.

Techniques for identifying minerals or mineral phases which carry the NRM in a rock are first considered and this is followed by some aspects of the identification of minerals by magnetic methods, and methods of estimating the amount of magnetic minerals present in a rock. The standard identification techniques of the microscope, thin and polished sections, electron microprobe analysis and other techniques of the petrological and analytical laboratory are not considered here, and the reader is referred to other published work for such information. Among many useful books in which the above topics are dealt with are those of Zussman (1967), Maxwell (1968), Uytenbogaardt and Burke (1971) and Jeffrey (1978).

13.2 Thermal analysis

At and above their Curie temperatures T_c remanence-carrying minerals lose their distinctive properties of hysteresis and the ability to carry permanent magnetism and become paramagnetic regardless of their physical state. However, in appropriate circumstances the NRM can vanish below T_c, at the 'blocking' temperature T_B; this is the temperature below which the remanence is blocked in against thermal randomizing influences. In SD grains T_B depends on the grain volume and in MD grains on the temperature below which domain walls are trapped in energy minima, and also possibly on the applied field. T_B is also the temperature at which a PTRM is acquired if the grain is heated to $T > T_B$ and allowed to cool in an ambient field.

The NRM of many rocks shows a range of blocking temperatures, reflecting different properties of the remanence-carrying particles. This is indicated by the decay of NRM as the rock is heated to successively higher temperatures and the acquisition of PTRM over a range of temperatures on cooling in a field. If $T_B \approx T_c$, TRM will be lost and gained over a narrow temperature interval just below the Curie point.

If the NRM of a rock vanishes on heating to a temperature T_1 it may be concluded that any mineral present for which $T_c < T_1$ cannot be carrying that part of the NRM remaining above T_c. If the NRM vanishes at $T_B = T_c$ of a mineral known to be present then this is evidence that that mineral is contributing the NRM. However, it is necessary to be aware of possible physical and chemical changes produced by heating, e.g. oxidation of magnetite to haematite, giving a spurious value of T_B or T_c: such changes can often be detected by measuring appropriate magnetic properties (saturation magnetization, saturation IRM) before and after the heating.

At low temperatures magnetite and haematite undergo transitions in their magnetic properties which can sometimes be used as an aid to determining whether these minerals carry NRM in a rock. The transition in magnetite occurs at around $-140°C$, when the magnetocrystalline anisotropy constant K_1 goes through zero, and there is a decrease in the NRM of magnetite as it cools through this temperature. This may not always occur if the grain size is very small (in the SD range).

It has been reported that in titanomagnetites the transition temperature falls with increasing Ti^{4+} content. Nagata (1967a) states that 10 mol % of Fe_2TiO_4 is sufficient to lower the transition temperature below that of liquid nitrogen ($-196°C$), but there is no general agreement on this point (Creer and Like, 1967).

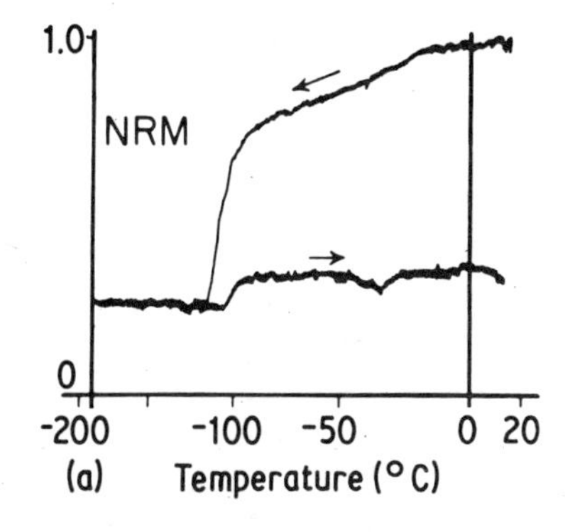

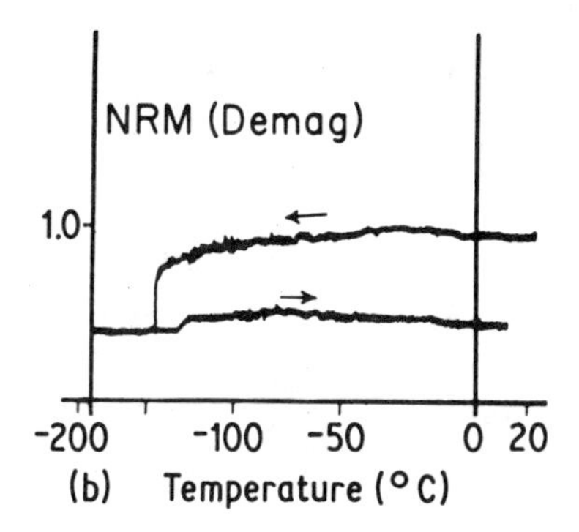

Fig. 13.1 Low-temperature behaviour of NRM of Marquette iron ore, Quebec: (a) initial; (b) after 7 mT demagnetization, indicating the proportion of the NRM carried by magnetite. (From Fuller and Kobayashi, 1967)

Most of the reported results of low-temperature investigations refer to mineral identification (as distinct from minerals carrying NRM) and it is the saturated IRM of the rock and its variation with temperature which is usually studied. This procedure provides stronger magnetizations with which to work, and some of the complications present when the much weaker NRM is studied become less important. These complications include irreversibility of cooling and warming curves and the acquisition of TRM by minerals with Curie temperatures below room temperature. Among the latter are members of the haematite–ilmenite and titanomagnetite series with high content of ilmenite and ulvöspinel respectively. An example of identification of magnetite as an NRM carrier by low-temperature observations is given by Fuller and Kobayashi (1967) in the Marquette iron ore (Fig. 13.1). They also indicate the wider potential of the technique through the study of transitions and 'memory' effects after partial demagnetization of NRM. Mauritsch and Turner (1975) show that magnetite occurring in limestones at a concentration of 10^{-4}–10^{-5} can be detected through the occurrence of the transition when samples are given a saturation IRM at -196°C and are then allowed to warm to room temperature.

The measurement of low-temperature variation of NRM is described in Section 4.3.

The Morin transition in haematite occurs at about -20°C (T_M) below which temperature the spin axis is perpendicular to the basal plane (it is parallel to it above 20°C) and the weak ferromagnetism vanishes (Morin, 1950; Nagata, 1961). However, the transition is sensitive to chemical composition, and T_M may be changed or suppressed completely, for instance by a small content of Ti^{4+} or other cation and grain size also appears to be important (Yamamoto, 1968).

Two good observations of the Morin transition in the NRM of natural materials are those of Creer *et al.* (1972) in bottom sediments from Lake

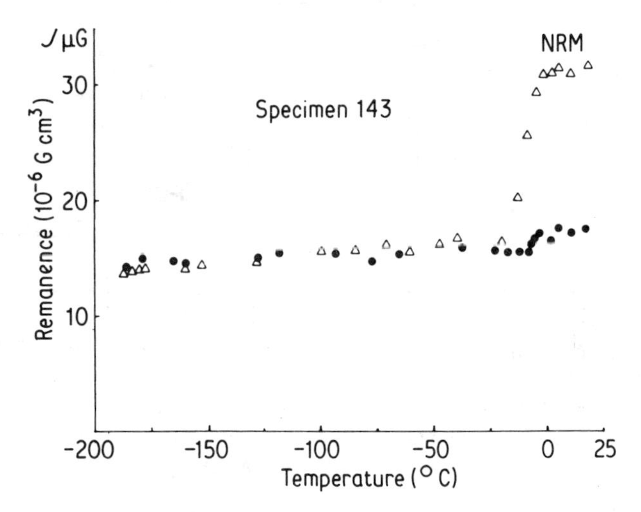

Fig. 13.2 The Morin transition in a sediment sample from Lake Windermere, showing that haematite is contributing to the NRM. (From Creer *et al.*, 1972)

Windermere, England, and Fuller and Kobayashi (1967) in samples of the Quebec iron ore. In the latter rocks there is clear evidence of haematite carrying the bulk of the NRM even though magnetite is also present. Similarly, the Windermere sediments contain both magnetite and haematite but the well-defined Morin transition (Fig. 13.2) is unequivocal evidence for haematite as the dominant NRM carrier.

13.3 Chemical methods

This technique depends on the penetration into a rock sample of a suitable reagent which differentially dissolves the magnetic minerals present: simultaneous monitoring of the NRM will then indicate the contributing mineral(s) if the action of the reagent on the different minerals is known. Since the rock must remain intact and be sufficiently porous for the reagent to penetrate in a reasonable time the technique has been more commonly applied to sedimentary rocks, although some of the earlier applications were to igneous material. Carmichael (1961) used 12N hydrochloric acid to separate different-sized ilmeno–haematite lamellae from the Allard lake ilmenites and showed that the NRM was carried by the smaller lamellae, and Kawai (1963) showed that information about the carrier of NRM in samples containing magnetite and titanomagnetite could be obtained from their different solution rates in concentrated hydrochloric acid. The rock samples were in the form of thin sections ($\sim 80\,\mu$m thick) and the rate of solution of titanomagnetite grains was found to decrease with increasing titanium content, pure magnetite being dissolved most rapidly. The author also reports that in hydrochloric acid the rate of solution of ilmeno–haematite particles also decreases with increasing titanium content. Domen (1967) used a similar technique on thin (~200 μm) basalt sections.

The author's technique with red sediments, in attempting to determine whether the NRM is carried by the pigment or specularite form of haematite, involves the use of thin (2 mm) discs of rock immersed in 10N hydrochloric acid (Collinson, 1965b). The finely divided interstitial pigment and the thin coating of it on other particles is much more rapidly dissolved than the specularite particles. The discs are periodically removed from the acid, washed and dried, and measured with an astatic magnetometer. The method met with some success (Fig. 13.3), the chief limitations being a minimum intensity of NRM ($> 10^{-5}$ A m^2 kg^{-1}) for a satisfactory signal/noise ratio for measurement and the disintegration of the rock if carbonate is a major constituent and contributory cementing agent. In the latter case initial slow leaching out of the carbonate by solution in weak acid may prevent disintegration, or the use of a supporting cage of thin glass rods supported on the upper and lower faces of a plastic ring containing the disc. The NRM measurements are most easily carried out with an astatic magnetometer.

The basic technique of acid solution has also been developed for magnetic cleaning of sandstones, and this application is described in Section 11.3. It can

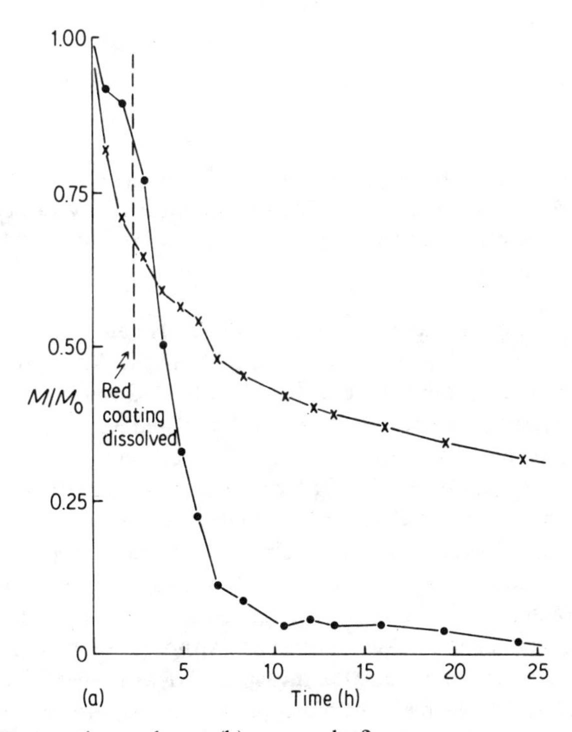

Fig. 13.3 (For caption and part (b) see overleaf)

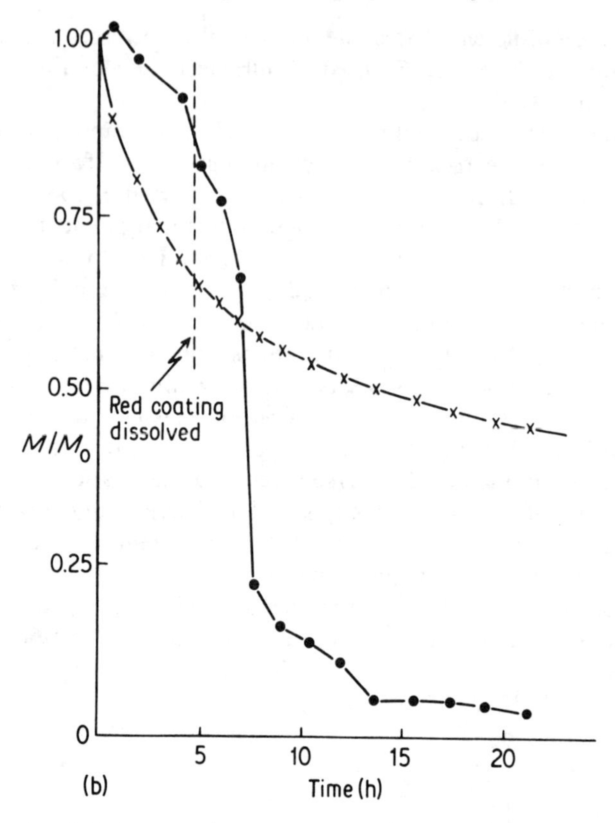

Fig. 13.3 Chemical demagnetization (full circles) of samples of the Chugwater sandstone (USA), indicating that specularite is the dominant NRM carrier. The crosses show the relative decrease in total iron content. (From Collinson, 1965b)

also be adapted for the identification of minerals carrying magnetizations other than the natural remanence, i.e. VRM, IRM, ARM and TRM, by imparting the appropriate magnetization before acid treatment.

A different type of application is that of Merrill and Kawai (1969) who used the technique to determine the presence of a self-reversal mechanism in a batholith from N. California. The NRM of the rocks is carried by magnetite and titanohaematite and it was suspected that the latter carried a self-reversed NRM. Thin (100 μm) sections of the rocks were prepared and the NRM in the plane of the sections measured with an astatic magnetometer. The sections were then treated with concentrated hydrochloric acid and periodically washed, dried and their NRM remeasured. Although there were difficulties with the measurement technique the measured NRM in one section reversed its polarity after acid treatment for one hour. Since magnetite is attacked more rapidly than titanohaematite, this was interpreted as destruction of the magnetite and its NRM, initially determining the NRM direction of the rock, and the persistence of the reversed NRM of the titanohaematite. Assuming contemporaneous magnetization of the two minerals there is strong evidence of self-reversal in the titanohaematite. The authors report that the rate of acid attack decreases with increasing titanium content in the ilmenite–haematite series.

13.4 The depositional magnetization technique

Another approach to identifying the carrier(s) of NRM is to crush the rock, separate out the magnetic minerals and then 'reconstitute' their NRM by sedimentation of each of the separates in a column of liquid in a magnetic field; each resulting sediment will possess a depositional remanent magnetization (DRM) contributed by the separated mineral. Comparison of the thermal or alternating field demagnetization characteristics of the NRM of these sediments with those of the original rock may then provide evidence regarding which mineral is carrying the NRM in the whole rock.

Hargraves and Young (1969) describe a technique they applied to the magnetic carriers in the Lambertville diabase, from New Jersey, USA. The plagioclase, pyroxene and oxide mineral fractions were separated using bromoform and Clerici's solution and the separates were allowed to settle in a 120 cm column around which a solenoid was wound, providing an axial field of 3.0 mT. A variety of liquid media were used, namely melted paraffin wax, a water soluble wax and silica gel. Silica gel has the advantage of withstanding the high temperatures necessary if thermal demagnetization is to be carried out.

Collinson (1974) used a similar technique to show that specularite carries the NRM in several red sandstone formations possessing stable NRM. Specularite (black, polycrystalline haematite particles) is separated from the crushed rock by a combination of heavy liquid and slow-speed centrifugal separation (Section 6.1) and then sedimented in acetone in a 1.0 mT horizontal field. The sediment is collected in a shallow, detachable quartz dish at the bottom of the acetone column. The residue from the separation, in which the major magnetic constituent is the pigment form of haematite, both free and as a coating on quartz and other grains, is treated in the same way. After evaporation of the acetone the sediments are sufficiently compacted for a.f. and thermal demagnetization and NRM measurement, the latter with an astatic or cryogenic magnetometer or possibly a slow-speed spinner with a suitably modified holder. Figure 13.4 shows some results obtained.

The technique also affords a method of estimating the specific magnetization of the specularite in a rock sample. In the samples examined by the author the specularite deposited in a 20 cm column of acetone in a 1 mT field was essentially completely aligned along the field (Collinson, 1974). Thus a knowledge of the DRM total moment combined with a ferric iron analysis of

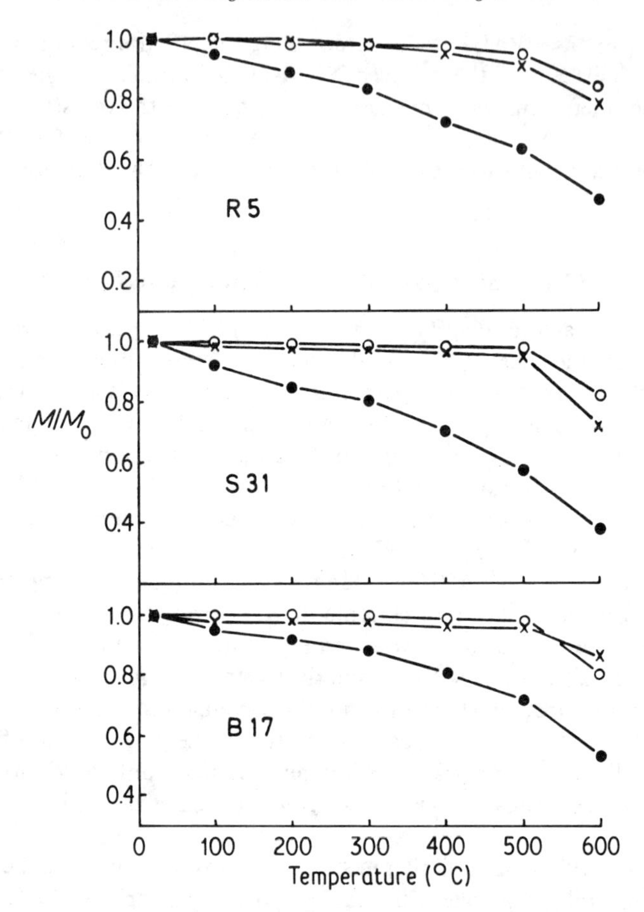

Fig. 13.4 Thermal demagnetization of NRM of red sandstone samples (open circles) and DRM of extracted 'pigment' (full circles) and specularite (crosses). There is strong evidence that specularite is the dominant NRM carrier in these rocks. (From Collinson, 1974)

the specularite separate enables the specific magnetization to be estimated, on the reasonable assumption that the major source of Fe^{3+} in the separate is specularite. A representative result was 0.30 g of separate containing 40% of α-Fe_2O_3 had a magnetic moment after deposition of 4×10^{-6} A m^2. Thus $J_r \approx 0.03$ A m^2 kg^{-1}.

A necessary precaution in the sedimentation technique is to establish that the aligning field does not impart a measurable IRM or VRM to the sediments. If it does, such magnetizations can usually be satisfactorily removed by a.f. demagnetization in a peak field equal to or slightly greater than the aligning field.

The method can also be applied to the investigation of magnetized particles in already disaggregated material, e.g. soils, muds and sands. Using this technique Runcorn *et al.* (1970) detected the presence of iron grains possessing a highly stable NRM in lunar dust returned by the *Apollo* missions. The DRM intensity, which was likely to be near saturation (i.e. particles completely aligned), was $\sim 2.0 \times 10^{-4}$ A m^2 kg^{-1}. Since the metallic iron content of the dust is $\sim 1\%$ by weight this implies a very low average specific magnetization of the iron grains of ~ 0.02 A m^2 kg^{-1}.

Collinson (1979) has also used the technique to investigate the possibility of obtaining lunar magnetic field palaeointensity data from lunar fines. The moments of the constituent iron particles are aligned as described and the resulting DRM used as the NRM for a Thellier palaeointensity determination (Section 14.4).

13.5 Additional data on magnetic minerals

In rocks possessing primary and secondary components of magnetization and more than one magnetic mineral or mineral phase, the techniques described above can often also provide evidence of which mineral or phase is carrying each type of NRM. This information is of considerable interest in connection with the pigment and specularite in red sandstones and siltstones, and Collinson (1974) shows an example of a primary NRM carried by specularite and a reversed secondary magnetization carried by the red pigment in a sample of the Taiguati formation from Bolivia (Fig. 13.5). The Dunnet Head sandstone from north-east Scotland has an NRM which is highly scattered in direction and the author was able to show by the DRM technique that the major carrier of the scattered magnetization is the pigment form of haematite (Collinson, 1980).

With the advent of more sensitive magnetometers, in particular the superconducting instrument, it is sometimes possible to hand pick magnetic particles from coarser-grained rocks and measure their magnetic properties individually. Wu, Fuller and Schmidt (1974) selected grains (minimum size 1 mm) in thin sections of a granodiorite intrusion, using a microscope, marked their orientation and then cut or drilled them out. Using the superconducting magnetometer the NRM of the particles and its stability were measured and the authors were able to show that the stable NRM of the intrusion was carried by plagioclase crystals containing magnetite inclusions. Larger, multidomain magnetite associated with biotite and hornblende crystals was found to be magnetically soft, as might be expected. A similar technique is described by Geissman (1980). Buchan (1979) investigated larger, oriented mineral separates (0.1–0.2 cm^3 in volume) from the Bark Lake diorite (Canada), using a standard spinner magnetometer, and Larson (1981) has developed a technique ('selective destructive demagnetization') in which the magnetic properties of rock thin sections of one particle thickness are measured before and after removal of different types of mineral grains. Grain removal is by physical

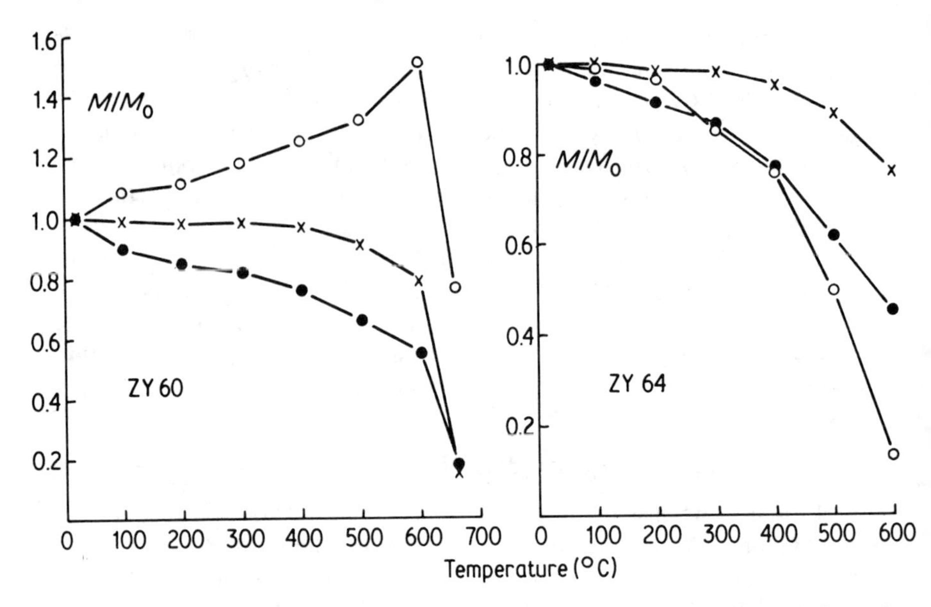

Fig. 13.5 Thermal demagnetization of NRM of rock and DRM of specularite and pigment from Taiguati formation, Bolivia. Symbols as in Fig. 13.4. In ZY60 there is evidence of a reversed NRM carried by the pigment; and also of an NRM carried by pigment in ZY64. (From Collinson, 1974)

means (drilling, cutting) or by chemical leaching, carried out under a low-power microscope.

The use of a small-scale astatic magnetometer for the measurement of individual particle properties has already been described (Section 9.6).

It is sometimes possible to estimate the quantity of a magnetic mineral in a rock from the magnitude of a magnetic property exhibited by it. The main requirements are that one magnetic mineral dominates in contributing to a particular magnetic property of the rock and that the magnitude of that property per unit mass of the mineral is known.

A simple technique of some use for igneous rocks is the value of their specific initial susceptibility. Provided the magnetite or titanomagnetite concentration is not too large and the mineral content can be regarded as a non-interacting assembly of approximately equidimensional particles of high intrinsic susceptibility, then, as described in Section 2.3, the observed specific susceptibility depends essentially on the proportion of the mineral present in the rock. This situation obtains for magnetite and also for titanomagnetite if the ulvöspinel content is not too high, i.e. $x > \sim 0.3$ in $x\mathrm{Fe_3O_4}\cdot(1-x)\mathrm{Fe_2TiO_4}$. From the relation derived in Section 2.3, $p \approx \chi/(5.8 \times 10^{-4})$ where χ is the specific initial susceptibility of a rock sample and p is the weight fraction of the magnetic mineral content. The relationship fails if there is a significant proportion of aligned, elongated or superparamagnetic particles present.

A technique for estimating the haematite content of sediments by relating their high field susceptibility to ferrous and ferric iron content is described by Collinson (1968). Unlike in most igneous rocks, in sediments there is often a substantial contribution from the paramagnetic (and sometimes diamagnetic) mineral content to the initial and high field susceptibility in addition to that contributed by haematite. The magnetic properties of haematite are somewhat variable, depending on source, method of preparation, etc., but the high field susceptibility appears to be one of the most fundamental properties and has a value close to $2.5 \times 10^{-7}\ \mathrm{m^3\,kg^{-1}}$. The expected susceptibility of minerals containing paramagnetic iron has been derived in Section 2.2, where the values of 2.07 and $2.26 \times 10^{-8}\ \mathrm{m^3\,kg^{-1}}$ were obtained for the contribution to susceptibility of each 1% by weight of ferrous and ferric iron respectively expressed as their equivalent oxides, FeO and Fe_2O_3. Thus, in conjunction with an analysis of the rock for ferrous and ferric iron content an equation can be formulated relating the observed high field susceptibility of the rock to the contributions of ferrous and ferric iron and haematite, the weight content of the latter being the only unknown. J_i–B curves are plotted for each rock sample in a maximum field $\geqslant 0.8$ T. In rocks of very low iron content ($< \sim 0.5\%\ Fe_2O_3$) a correction is made for diamagnetic minerals by estimating the quartz and carbonate content and assigning them an average susceptibility of $-6.3 \times 10^{-9}\ \mathrm{m^3\,kg^{-1}}$, which is sufficiently accurate for this purpose. The susceptibility is calculated from the straight portion of the J_i–B curve, usually from $B > 0.3$ T. Then expressing the ferrous and ferric iron content of the rock as weight per cent of FeO and Fe_2O_3 and omitting a factor of 10^{-8} from each term of the equation, the weight per cent of haematite is obtained from the equation

$$\chi = 2.07a + 0.25b + 2.28(c - b) \tag{13.1}$$

where a is the percentage content of ferrous oxide and b and c are the percentage content of ferric oxide in the form of haematite and total ferric oxide respectively.

While no great accuracy is expected from the method it gives reasonable results when applied to red sediments, in which clay minerals provide the paramagnetic content (Collinson, 1968). The results are in agreement with those provided by a method of qualitatively estimating the relative content of haematite and paramagnetic minerals in sediments, namely the shape of the J_i–T curve, the decay of induced magnetization with temperature. If haematite is the dominant contribution to J_i, then J_i decays rather slowly up to ~ 500°C followed by an increasingly rapid fall as the Neel point is reached at ~ 685°C. If paramagnetism is dominant J_i decays according to the Curie law, i.e. inversely as the absolute temperature. Figure 13.6 shows some results obtained from three samples.

A simple and informative technique – coercivity spectrum analysis – for estimating the relative content of magnetite, black polycrystalline haematite

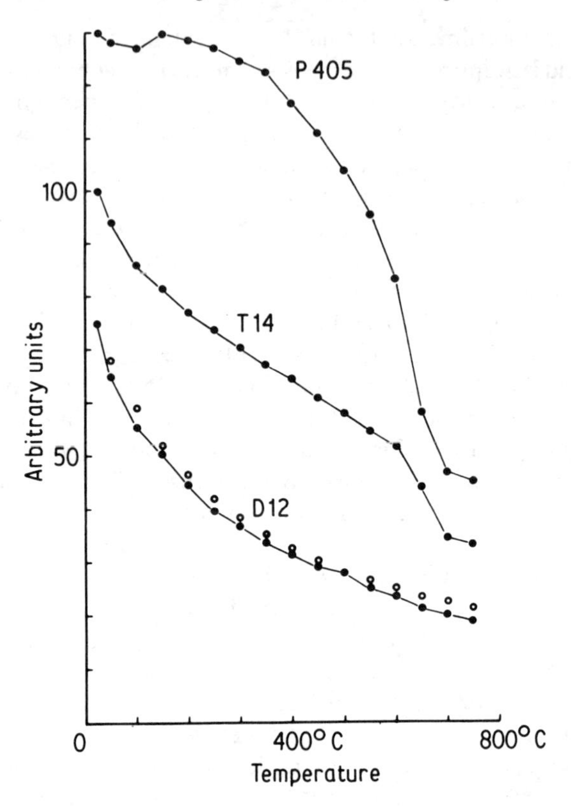

Fig. 13.6 J_i–T curves obtained from three sandstone samples with differing paramagnetic iron content. The lower curve is compared with the decay expected in a paramagnetic sample (open circles). (From Collinson, 1968)

(specularite) and red haematite (pigment) in sediments and its change on heating the rocks is described by Dunlop (1972). It is based on the differing coercivity of remanence of the three minerals, approximately in the range 0–0.1 T for magnetite, 0.1–0.3 T for specularite and 0.3–1.8 T for haematite pigment.

For igneous rocks the coercivity spectrum can be defined by a.f. demagnetization of NRM or a laboratory-produced remanence. This is unsatisfactory for haematite-bearing rocks because a.f. demagnetizing equipment rarely has the capability of achieving fields greater than 0.5 T (peak), well below coercivities observed in haematite. However, direct magnetic fields up to 1.5–2.0 T are not difficult to produce over a small volume and a coercivity distribution can be determined by measurements of isothermal remanence (IRM) of samples in increasing fields up to the saturation value, or in the highest field available if saturation is not achieved.

In Dunlop's technique, J_r–B (applied field) curves are measured up to

$B = 1.8$ T and the incremental isothermal remanence in intervals of 0.1 T is derived from the curves. A more precise coercivity spectrum would be obtained from successive values of the slope of the curve, corresponding to small increases in B, but this is laborious and probably not justified unless J_r is measured for a large number of values of B.

Figure 13.7 shows some results obtained by Dunlop from sandstones, both unheated and heated. The evidence provided on the production of magnetite in

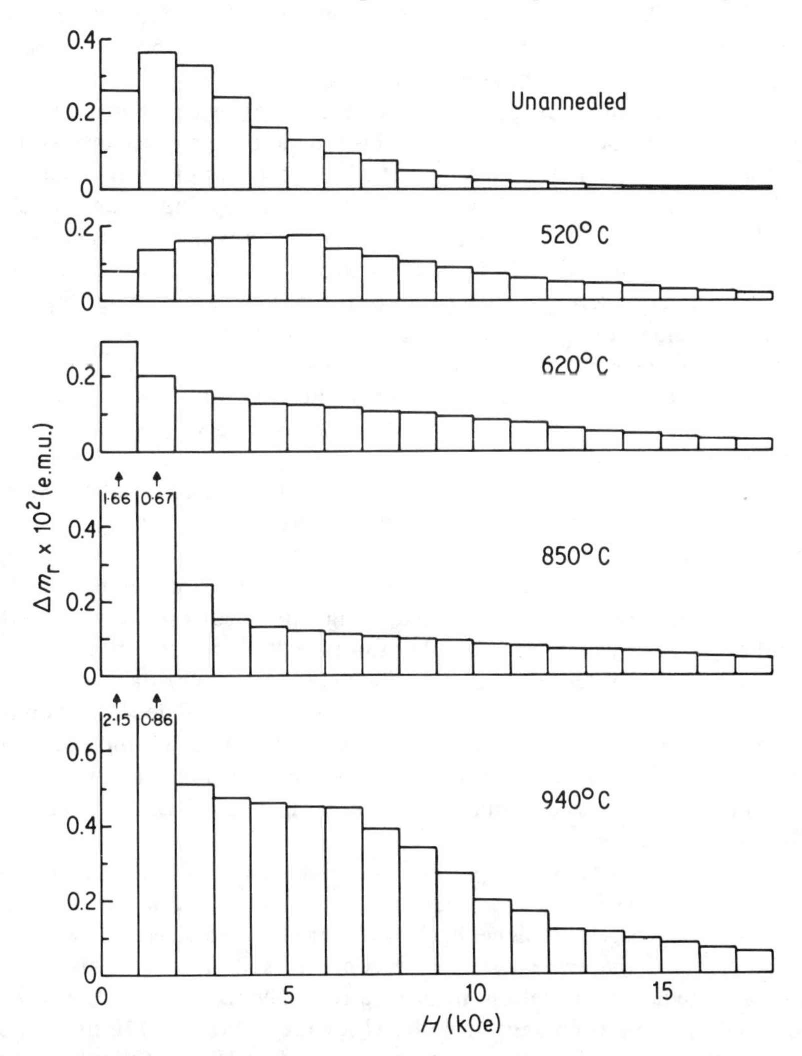

Fig. 13.7 Magnetic mineralogy of unheated and heated red sediments: change in coercivity spectrum of a sandstone on heating in air. Heating to 520°C appears to destroy about two-thirds of the original magnetite but heating to > 620°C results in substantial magnetite production. (From Dunlop, 1972)

heated sandstones is of particular interest, as it generally confirms that magnetite formation is often a result of breakdown of iron silicates (clay minerals) rather than reduction of haematite.

The author points out that some before- and after-heating measurements can be carried out on the same sample if measurements are made along mutually perpendicular directions in the sample. Thus, a room-temperature J_r–B curve and curves after heating at two high temperatures can be obtained from the sample in this way, avoiding possible confusing results if duplicate samples of the same rock have markedly different coercivity spectra. However, repeated heating of a sample may promote physical changes in magnetic minerals, e.g. in the pigment grain size in redbeds, which also alter their coercivity. Using the present technique this process has also been investigated by Dunlop, who observes a hardening of the coercivity of pigment on heating.

The coercivity spectrum is used in a different way by Sholpo and Shcekin (1977). They plot the rate of change of IRM with field during magnetization, dJ_r/dB, and plot it against B. Then different magnetic minerals tend to show different values of some characteristics of the curve, i.e. maximum slope, the maximum value of dJ_r/dB and the applied field at which the latter occurs.

Rotational hysteresis, the measurement of which is described in Chapter 3, is another property with applications to mineral identification. It is defined as the work done per unit mass due to irreversible processes as a magnetic mineral sample is rotated from 0° to 360° in a constant magnetic field. The relationship which is relevant to mineral identification is the dependence of rotational hysteresis W_r on the applied field B, and it has been shown that different magnetic minerals generate distinctive W_r–B curves. This arises because high values of W_r for a particular mineral are obtained when B is near the anisotropy field, given by the ratio of the magnetic anisotropy constant to the saturation magnetization. For all other values of B, W_r tends to be low. Thus, W_r–B curves for different magnetic minerals generally show a more or less well-marked peak value, the peak occurring at a value of B depending on the magnitude of the anisotropy field occurring in the minerals. Another feature of the curves which may be diagnostic is the D-value of a curve, the ratio of W_r in the highest value of B available to the peak value of W_r (Day *et al.*, 1970; Manson *et al.*, 1979).

Reference W_r–B curves for polycrystalline magnetite, titanomagnetite and maghemite are given by the above authors and are shown in Fig. 3.17. Titanomagnetites peak at higher fields than magnetite, which is consistent with the observations that increased titanium content results in a higher anisotropy field and reduced saturation magnetization. The D ratios (at $B = 2$ T) also increase with Ti content, from 0.037 for magnetite to 0.116 for TM 60 (60 mol % ulvöspinel). However, it can be seen from Fig. 3.17(c) that the D ratio is dependent on the highest value of B used and lacks the more fundamental significance of the field at which the value of W_r occurs. The magnitude of the W_r 'tail' in titanomagnetites at high fields ($> \sim 1.5$ T) generally reflects the presence in the sample of a small proportion of material of high anisotropy: this may be due to impurities or strain, or very fine grain size.

Day *et al.* have also demonstrated a correlation between the oxidation state of a collection of igneous samples as determined optically and their D ratios observed at $B \approx 2.0\,\mathrm{T}$, the latter increasing with increasing oxidation.

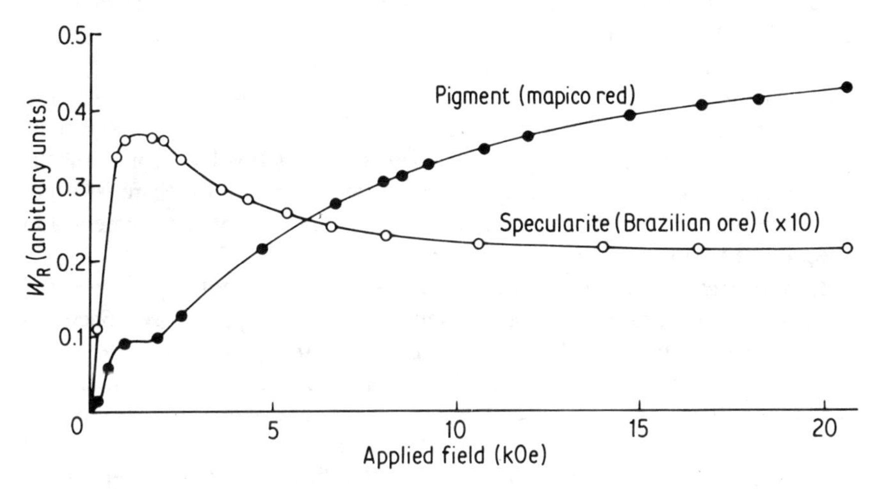

Fig. 13.8 W_R–B curves for synthetic haematite pigment (2–4 μm) and specularite ($\sim$ 40 μm). Specularite data are $\times$ 10. The data are normalized to 1 mg of sample, and 1 arbitrary unit = 0.84 ergs/rev = $0.84 \times 10^{-7}\,\mathrm{J\,rev^{-1}}$. (From Brooks and O'Reilly, 1970)

In view of the complex magnetic characteristics of haematite it is not surprising that its rotational hysteresis behaviour is less straightforward than that of the titanomagnetites. W_r–B curves have been measured in polycrystalline haematite (specularite) by Day *et al.* (1970), Brooks and O'Reilly (1970) and Cowan and O'Reilly (1972) and in single crystals in and perpendicular to the basal plane by Vlasov *et al.* (1967b). Synthetic samples of the finely divided pigment form of haematite ($\sim$1 μm) have been measured by Brooks and O'Reilly (1970) and Cowan and O'Reilly (1972), who have also measured haematite-bearing sandstones. As observed with measurements of other magnetic properties of the mineral the results obtained are often dependent on the mode of origin, grain size, etc. but an overall systematic difference is apparent in W_r–B curves obtained from pigment and specularite samples. Curves obtained by the above authors are shown in Fig. 13.8.

Brooks and O'Reilly (1970) have developed a technique for estimating the pigment and specularite content of red sandstones, based on W_r–B curves obtained from these rocks and from samples of the two mineral phases. Referring to Fig. 13.9, it can be seen that 1 mg of pigment and specularite contribute 0.042×10^{-7} and 0.032×10^{-7} J/rev at 0.15 T and 0.38×10^{-7} and 0.017 J/rev at 2.0 T. Therefore, if 100 mg of a sandstone gives x_1 and x_2 J/rev at 0.15 T and 2.0 T, then omitting a factor of 10^{-7} on either side

$$\begin{aligned} 0.042\alpha + 0.032\beta &= x_1 \\ 0.38\alpha + 0.017\beta &= x_2 \end{aligned} \tag{13.2}$$

where α and β are the percentages by weight of pigment and specularite in the rock. Thus the total α-Fe_2O_3 content and the pigment/specularite ratio can be estimated. The results quoted are in broad agreement with Fe_2O_3 analysis by chemical methods and with qualitative estimate of pigment/specularite ratios. However, since the reference W_r values are obtained from a single synthetic pigment (2–4 μm) and a single specularite sample (Brazilian iron ore), the results should probably be accepted with some caution until there is firmer evidence that the curves of Fig. 13.8 are representative of the pigment and specularite that occurs in red sandstones.

Investigations of the magnetic properties of lunar samples has generated interest in rotational hysteresis in metallic iron and troilite. W_r–B curves obtained from samples of the fines and a rock chip appear to be consistent with those for iron and troilite and with the estimated content of these minerals in the lunar material. The evidence of the curves obtained from the lunar material is that iron is the major contributor to W_r in the rock sample (14053) and iron and troilite in the fines (Runcorn *et al.*, 1971; Collinson, Runcorn, Stephenson and Manson, 1972).

13.6 The magnetic state of magnetic minerals

In this section some methods and techniques are described which provide evidence for the presence of single domain and multidomain particles and for particles in the superparamagnetic state.

In practice there is a distinction to be drawn between true single and multidomain particles, the different properties of which depend mainly on physical size, and single and multidomain behaviour, when particles of apparently multidomain size show magnetic properties consistent with those expected in single-domain grains. Thus, tests based on magnetic properties for SD and MD particles are really tests for SD and MD behaviour. Such tests are of most interest in the magnetite- or titanomagnetite-bearing igneous rocks in which the magnetic minerals commonly occur in a wide range of sizes from much less than 1 μm up to several millimetres across. A normal particle-size analysis under the microscope may fail to reveal very fine-grained SD magnetite yet these grains may be the dominant contributors to the observed magnetic properties of the rock. Therefore the SD or MD nature of the thermoremanence is not established by optical methods, and it is desirable to use the characteristics of the rock magnetization itself to deduce its origin.

Stacey (1967) suggested that the Königsberger ratio Q_T (Section 10.2), observed in an igneous rock, is indicative of whether SD or MD material carries its TRM. Using the data of Parry (1965) the author derived the grain size dependence of Q_T for dispersed magnetite particles in the range 120–1.5 μm. Theory gives $Q_T = 0.5$ for multidomain magnetite (and also titanomagnetites) which is in agreement with Stacey's values for grain diameters $> \sim 20$ μm. For diameters < 20 μm, $Q_T \gtreqless \sim 1.0$, which is generally observed in igneous rocks of use in palaeomagnetism. Stacey therefore proposed that multidomain TRM is rare and that in many igneous rocks the TRM is carried by particles smaller than ~ 20 μm, in which 'pseudo-single domain' behaviour occurs (Dickson *et al.*, 1966), and that where $Q_T < 1.0$ was observed multidomain TRM dominates. This criterion might well be valid if a single mineral of uniform magnetic state was present in a rock, a situation which is probably rare in practice. Abundant MD magnetite might govern Q_T, giving a low value, and yet very fine-grained magnetite might be the dominant carrier of TRM. The presence of minerals other than magnetite, e.g. of the ilmenite–haematite series, could also give a Q-value not representative of the magnetite present.

Some information on the presence of SD and MD grains is provided by the ratio of saturated remanent magnetization to saturated induced magnetization, J_{rs}/J_s, the ratio being higher in SD than in MD material. In an assembly of randomly oriented uniaxial SD grains, $J_{rs}/J_s = 0.5$, which is the maximum expected value of the ratio. In MD material the ratio is usually < 0.1. Pearce, Gose and Strangway (1973) demonstrate the different SD and MD iron content of the basaltic lunar samples ($J_{rs}/J_r < 0.02$) and the soils and breccias ($J_{rs}/J_r = 0.03$–0.10).

The difference in stability against a.f. demagnetization of TRM acquired in a weak and strong field by SD and MD magnetite is the basis of a test for SD and MD behaviour developed by Lowrie and Fuller (1971).

Experiments on dispersed and single-crystal magnetite, the results of which are qualitatively confirmed by theory, show that whereas the stability of TRM of SD grains increases with decreasing applied field the stability of MD grains increases with increasing field. One form of the test consists of imparting a TRM to the rock under investigation, first in a weak field (comparable to the geomagnetic field) followed by a.f. demagnetization, and then in a stronger field (1–10 mT), followed again by demagnetization. The stabilities of the TRMs are then compared and the state of the TRM carrier deduced accordingly. An alternative procedure is to compare the stability of the NRM of the rock and the saturated IRM. The reasonable assumptions involved are that the NRM is a TRM acquired in the geomagnetic field and that the saturated IRM has the same stability as the strong-field TRM. Apart from the greater simplicity of this procedure it has the added advantage of avoiding the need to heat the rock, with the associated danger of mineral alteration. In another procedure (Johnson, Kinoshita and Merrill, 1975) ARM is used instead of TRM, again avoiding the danger of mineral alteration through heating. Banerjee, King and Marvin (1981) show that ARM and initial susceptibility measured in samples taken from a long sediment core can be used, when plotted against each other down the core, to characterize the grain-size variation of the remanence-carrying magnetite.

Although less securely based than for magnetite, theory indicates that the maximum size of single-domain haematite is ≈ 0.15 cm, and so in the great majority of haematite-bearing sedimentary rocks SD structure is expected in the mineral.

The detection of magnetic minerals in the superparamagnetic state depends on the much higher magnetic moments of the superparamagnetic particles compared with the atomic magnetic moments associated with true paramagnetism.

The Langevin function, $L(a) = \coth a - 1/a$, is linear with a only when $a = pB/kT$ is small, where p is the atomic magnetic moment (Section 2.2). This condition leads to linear dependence of volume magnetization M on the applied field H and the associated paramagnetic susceptibility $\chi_v = N\mu_0 p^2/3kT$ (Equation 2.6). The effect of the large increase in p is therefore two-fold, namely for a particular value of B/T, a may no longer be small and there will be an increase in initial susceptibility. Thus, compared with normal paramagnetism saturation will be approached at much lower values of B/T and susceptibility, or the magnetization induced in a given field, will be much greater for particles exhibiting SPM.

Creer (1961) and Nagata *et al.* (1970) have demonstrated the presence of SPM in this way in rocks containing haematite and iron particles respectively. Creer studied a red mudstone from the Keuper Marl. At 9.9×10^{-8} $m^3\,kg^{-1}$, the initial susceptibility was high for the estimated haematite (and clay mineral) content and a plot of M against B/T showed marked departure from linearity at $B/T \gg 0.05$ T K^{-1}, compared with ~ 0.5 T K^{-1} for paramagnetic salts.

Measurements of the initial susceptibility of the lunar fines give values which are too high to be explained on the basis of an assembly of approximately spherical iron particles, the susceptibility of which is governed by the particles' demagnetization factor N (Equation 2.9). There is no reason to suppose that a substantial proportion of the particles are elongated (thus reducing N and increasing χ) and aligned parallel to one another, and superparamagnetism of some of the iron present seemed a more likely explanation. By plotting specific magnetization against B/T for a sample of *Apollo 11* fines, Nagata *et al.* (1970) confirmed the dominance of super-paramagnetism above 100°C: the limiting value of B/T for linearity is very low, at approximately 50 μT K^{-1}. Further evidence for the presence of super-paramagnetic particles is seen in the J_i–B curve for fines, sample 10084 (Runcorn *et al.*, 1970). Multidomain iron saturates in fields of the order of

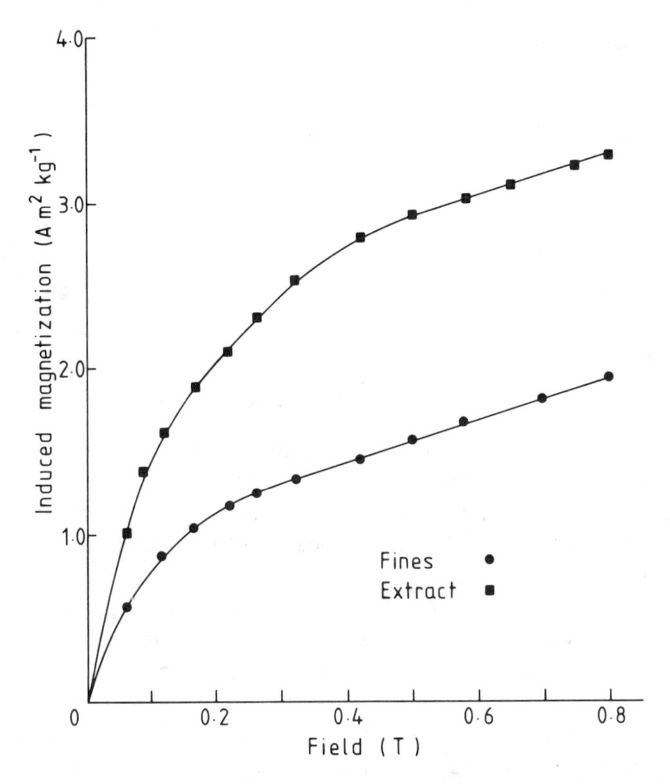

Fig. 13.9 J_i–B curves obtained from lunar fines and the extracted magnetic fraction.

100 mT, and in Fig. 13.9 it is apparent that saturation is only being approached at 0.8 T. In assessing this type of evidence it is necessary to consider whether the high field part of the curve can be due to paramagnetic minerals. If this is the case the appropriate part of the curve should be a straight line, the slope of which (i.e. the susceptibility) should be such that the paramagnetic minerals present can account for its magnitude. In Fig. 13.9 the upper parts of the curves shown are only very slightly curved but the derived susceptibility is $\sim 2 \times 10^{-6}$ m^3 kg^{-1}, much too high for the known paramagnetic mineral content of the fines.

Radhakrishnamurty and Deutsch (1974) show that variation of the shape of hysteresis loops with temperature and the variation of susceptibility with temperature can provide evidence of the domain nature and oxidation state of magnetite and titanomagnetite.

The foregoing techniques provide qualitative information on grain size, and various procedures have been investigated for quantitative determinations ('magnetic granulometry'). Among the techniques, which are based on the comparison of various magnetic properties in a sample, are those of Creer (1961), Stephenson (1971), and Dunlop (1976).

Chapter Fourteen

Intensity of NRM

14.1 Introduction

In contrast to the wealth of information provided by study of the direction of NRM in rocks, interpretation of the intensity of their magnetization in terms of the geomagnetic field intensity at the time they acquired their remanence is not straightforward, and detailed knowledge of the ancient geomagnetic field intensity and its variation is at present rather incomplete. Several techniques have been developed for relating the intensity of TRM to the strength of the field in which it was acquired and these are described in Section 14.4: in this section some other aspects of NRM intensity are discussed.

The difficulty in interpreting the observed NRM intensity arises from the many factors which control it. These are the nature, quantity, grain size and intensity and stability of magnetization of the remanence-carrying mineral(s) in the rock, and in some sediments the degree of alignment of the magnetization axes of particles contributing a depositional remanence (DRM). Among other factors this alignment depends on the strength of the geomagnetic field at the time when the DRM was acquired. The field strength will also, of course, directly govern the intensity of TRM and CRM acquired by minerals.

Thus the ambient field strength during NRM acquisition, about which one might hope to obtain information from NRM intensity, is only one of several factors controlling the observed intensity. The number of other variables involved make it difficult to derive information about any one of them.

In view of the foregoing it is not surprising that NRM intensity is very variable, even in rocks which appear lithologically and petrologically very similar. The clearest distinction is between the intensities observed in typical igneous rocks (10^{-1}–10^{-4} A m^2 kg^{-1}) and sediments (10^{-4}–10^{-7} A m^2 kg^{-1}). This mainly reflects the greater quantity and stronger magnetization of the magnetic minerals in the former, namely ~ 1–10% of titanomagnetites compared with ~ 0.5–2% of haematite in typical red sandstones.

Much effort has been expended in attempting to derive information about the absolute or relative strength of the geomagnetic field from NRM intensity. The type of rock that is investigated determines the information that can be gained, e.g. a sample of lava provides a spot reading of the field intensity whereas a sample of red siltstone possessing a DRM would be expected to provide a field strength averaged over $\sim 10^3$–10^5 years. Some of the techniques that have been employed are now considered.

14.2 Sediments

The most profitable investigations of sediments have been in connection with lake and sea-bottom sediments as recorders of the fine structure of the recent Earth's field.

Johnson *et al.* (1948) were among the early investigators who realized that information on field intensity might be recoverable from certain sediments, such as those laid down rather rapidly and possessing an unaltered DRM, the intensity of which would depend on the ambient field strength. In their investigations of varved clays from New England (annually deposited layers from glacial melt water) the authors redeposited the crushed material in water in a range of ambient field intensities and found that the original NRM intensity was reproduced in a field of $\sim$0.2 mT, not too different from the present field strength. This was encouraging, but later work on the re-deposition of Swedish varves by Griffiths *et al.* (1960) revealed some of the difficulties and uncertainties involved in this type of investigation. Among the effects involved are bedding and inclination errors and their dependence on particle size and stream velocity, which may significantly modify the DRM. Other, largely unknown influences are drying out and consolidation.

Following the pioneering work of MacKereth (1971) on the NRM of English lake sediments and the development of techniques for obtaining deep-sea sediment cores, several investigators have examined methods of extracting data on the intensity of the recent Earth's field from these sources. The specific magnetization of these materials is usually in the range 10^{-7}–10^{-5} A m^2 kg^{-1}.

The variation in NRM intensity down a lithologically homogeneous core is often due in part to variations in content of the remanence-carrying mineral. This may obscure changes in intensity due to ambient field variations during the sediment deposition, which influences the NRM intensity through greater or lesser alignment of particles in the DRM process. Since determination of absolute magnetic mineral content is difficult, attempts are made to establish the relative mineral content in different samples by comparing the magnitude of another magnetic property in each and 'normalizing' the NRM intensity (after any necessary cleaning) accordingly.

The assumption in this technique is, of course, that the magnitude of the property measured is proportional to the amount of remanence-carrying mineral present. This assumption is clearly open to question. Of the properties that have been employed, initial susceptibility is probably the least promising, depending as it does not only on 'ferromagnetic' but also on paramagnetic and diamagnetic mineral content. It has the advantage of simplicity of measurement, and if a strongly magnetic constituent, e.g. magnetite dominates, as indicated perhaps by strong NRM and susceptibility, its use might be profitable. Harrison (1966) and Creer, Anderson and Lewis (1976) found no evidence for magnetic field variations when values of NRM intensity were plotted against depth in a Pacific sediment core and a lake sediment respectively. However, variation in χ itself is often a useful indicator of mineralogical changes down a core, which may affect the derivation of field strength data. This is demonstrated by measurements of susceptibility in Lake Erie sediments (Creer *et al.*, 1976).

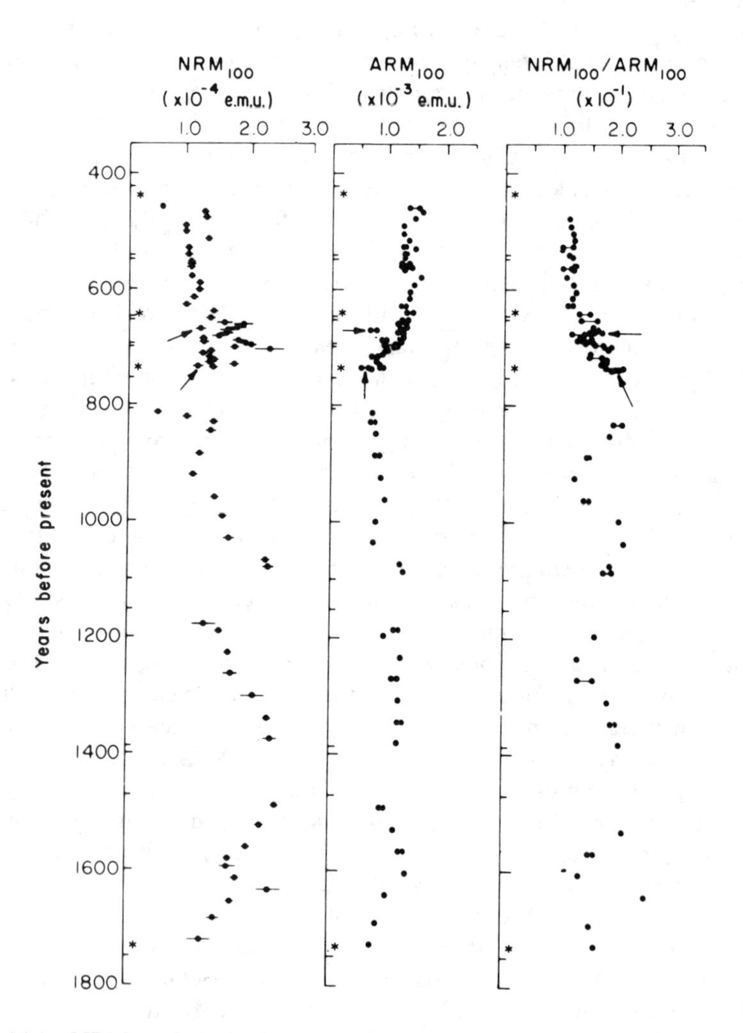

Fig. 14.1 NRM and ARM in Lake St Croix Minnesota sediment core, after demagnetization in 10 mT (100 G). (From Levi and Bannerjee, 1976)

A normalizing parameter which depends only on remanent properties is clearly desirable and isothermal remanent magnetization (IRM), saturated IRM and anhysteretic remanence (ARM) have all been used. The optimum property to use can be assessed by comparing the coercivity spectrum of the NRM with that of the imparted property. This has been done by Levi and Banerjee (1976) for lake sediments from Minnesota, USA in which ARM and NRM a.f. demagnetization curves were closely similar in one core (Lake St Croix) but IRM was a better fit in a core from another lake (Kylan). After cleaning the NRM in 10 mT to remove secondary magnetization a plot of NRM/ARM against depth in the Lake St Croix core gave a curve (Fig. 14.1) which bears some similarity to geomagnetic field intensity changes derived for the same period (0–1800 years B.P.) from archaeomagnetism. Other work of a similar nature is that of Opdyke *et al.* (1973) and Hillhouse and Cox (1976) who showed that a decrease in NRM intensity during a reversal of magnetization in a deep-sea core and ancient lake sediment respectively was most likely to be due to a weakening of the ambient field: no correlation was observed between NRM intensity and initial susceptibility or saturated IRM. The very marked variations in intensity found in a Pacific core by Johnson, Kinoshita and Merrill (1975) remained essentially unaltered after normalizing with ARM, indicating again that quantity of remanence-carrying mineral was not an important factor in causing the variations.

If lake and sea-bottom sediments possess depositional remanent magnetization carried by a single mineral and there are no major variations in lithology or grain size distribution down a core, then variations in the normalized NRM intensity may be expected to broadly reflect variations in the ambient field during deposition. However, in many cases it is clear that the magnetization history of these sediments is rather complex with, for example, evidence of post-depositional and chemical remanence and both magnetite and haematite-carrying NRM (Creer *et al.*, 1972). DRM itself is dependent on grain size and shape, bottom conditions, intensity of grain magnetization, etc. and re-deposition experiments have pointed to some of the associated problems (Lovlie 1974, 1976). The validity of another basic assumption should also not be assumed, that DRM intensity is proportional to ambient field strength: this is only true when particle alignment is well below the saturation value. Theory suggests, however, that small magnetite particles ($< \sim 1\,\mu$m) with $J_r \approx 1.0\,\mathrm{A\,m^2\,kg^{-1}}$ are expected to be completely aligned by a $50\,\mu$T field in falling through very modest depths of water (~ 10 cm or less), and therefore the DRM intensity would bear little relation to field strength (Stacey, 1963; Collinson, 1965a; King and Rees, 1966). However, more recent work shows that some randomization of particle magnetization directions takes place on contact with the bottom deposit and probably also during drying out and consolidation.

It is clear from the foregoing that a greater understanding of the magnetization processes in recent sediments is required before reliable data on geomagnetic field intensity over the past million years or so can be derived from them. It may also be said that the rewards of successful research in this area would be considerable and would contribute substantially to our understanding of the geomagnetic dynamo.

Little progress has been made in deriving field intensity information from other types of sediments. In spite of the large amounts of directional data obtained from red sandstones and siltstones, the acquisition of their NRM is still imperfectly understood and there is usually little information on the time interval represented in a given thickness of a formation. Regardless of the process of magnetization secular variation would be expected to be measured out in a sample thickness of the order of 1 cm, and variations in the mean value of the field intensity over typically a few million years would be obtained from different collection sites.

Nesbitt (1966) normalized the intensity of a wide range of red sandstones by using the ratio of NRM intensity to initial susceptibility, and used the ratio to assess changes in geomagnetic field intensity over geological time. Although apparently significant changes were observed, the use of susceptibility as a normalizing parameter for red sandstones is not promising for reasons already given and there is also the likelihood of the NRM residing in only part of the haematite content. Collinson (1974) has shown that in some representative redbeds with stable NRM from the western USA the NRM is carried by black haematite particles (specularite) and is almost certainly depositional in origin. However, normalizing the intensity by using J_r or ARM would not be profitable because of the acquisition of it by the pigment form of haematite as well as the specularite. In the Bonito Canyon quartzite (USA), which contains virtually no pigment, Collinson (1972) was able to show that the NRM was a DRM carried by the specularite and that intensity variations were mostly due to changes in the geomagnetic field during deposition. Comparatively precise normalization of NRM intensity, using IRM to derive specularite quantities, was possible, using measurements of IRM of extracted material with a small-scale magnetometer (Collinson, 1972 and Section 9.6): the IRM in 0.5 T of each sample, some extracted specularite grains and specularite-free matrix (pigment, quartz, etc.) was measured, from which the amount of specularite in each sample could be determined.

14.3 Igneous rocks

Because of the availability of methods for the absolute determination of palaeointensities from igneous rocks, described in the next section, little effort has been expended on more qualitative determinations.

If it is supposed that the NRM of rocks from several rock units is a primary TRM carried by all the magnetizable material in the rock, then a simple comparison of the mean NRM intensity of each unit may reflect changes in the ancient geomagnetic field intensity. A large number of samples is necessary in order to average out as far as possible variations in quantity and uniformity of magnetic minerals and other variables. Such a simple procedure may have more success with igneous rocks than with sediments because of the fewer variables involved in the igneous magnetization process. As with sediments, a more satisfactory method is to use a normalizing factor for magnetic mineral content as described in the last section.

The forerunner of the Thellier method of palaeointensity determination was a method originally applied by Königsberger (1938), who compared the ratio Q_n/Q_T in the rocks of different ages. Q_n is the ratio of the primary NRM intensity (J_r) of a sample to the intensity induced by the present field at the site, and Q_T is the ratio of the total TRM (J_T) acquired in the present field to that induced at room temperature in that field. If mineralogical changes during heating have been minimal and if the primary NRM is a TRM little affected by decay during geological time, then $Q_n/Q_T = J_r/J_T = B_a/B_p$ where B_a and B_p are the strength of the ancient field and present field respectively. Thus Q_n/Q_T is a measure of B_a. Königsberger (1938), with the limited data then available, showed there was a tendency for Q_n/Q_T and B_a to increase with decreasing age of the rocks tested, by roughly an order of magnitude since the Carboniferous. This result is not confirmed by the larger body of data now available.

14.4 Palaeointensity techniques

14.4.1 Introduction

The foregoing techniques provide information on relative changes in the ancient geomagnetic field intensity (palaeointensity) during the time interval in which remanent magnetism was acquired by the rock unit. However, the determination of absolute field intensities has always been an interesting and potentially informative undertaking and in principle at least is possible to achieve with rocks which possess suitable mineralogical and magnetic properties.

It is not possible to directly derive ancient field intensities from a theoretical (or even practical) knowledge of TRM or CRM intensity acquired by magnetic minerals in a given field. The magnetic (NRM-carrying) mineral content of the rock is rarely known with sufficient accuracy and the observed NRM intensity depends on grain-size distribution and SD and MD content, none of which are normally determinable with any precision. Nor is it possible where DRM is the relevant magnetization process, in which there is the additional complicating factor of the degree of alignment of the particle magnetic along the ambient field. In DRM the effect of grain size and water turbulence are among additional factors which make the theoretical approach to palaeointensity determinations impracticable.

The principle of absolute palaeointensity techniques, therefore, is the comparison of the observed NRM intensity J_r of a rock with the intensity of magnetization J_l acquired when the rock is given a remanence in the laboratory in a known field B_l by the process assumed to be responsible for the original NRM. Then on various assumptions, one of which is the proportionality of ambient field strength and resulting intensity of magnetization, the ancient field strength B_a is given by

$$B_a = B_l \left(\frac{J_r}{J_l} \right) \tag{14.1}$$

Because of the characteristics of the three processes of primary magnetization, Equation 14.1 can only be applied satisfactorily to rocks possessing thermoremanence. It is usually difficult to establish that the remanence carried by a rock is of chemical origin, and it is clear that the low temperature and often irreversible processes of long time scale involved in CRM in rocks cannot be reproduced in the laboratory. For rocks possessing a DRM an approach to the original magnetizing process is possible in the laboratory, by disaggregation and redeposition in water in an applied field, but it is not usually possible to reproduce the original, largely unknown conditions of deposition with sufficient accuracy to give reliable results. Imperfect disaggregation, the presence of diagenetically-produced minerals, a different and probably much smaller depth of water in which laboratory deposition takes place and different drying-out and compaction conditions affecting post-depositional RM are among factors which are expected to produce different DRM intensities from laboratory and natural deposition in the same ambient field. Verosub (1977b) has reviewed various aspects of the depositional magnetization of sediments.

Redeposition experiments carried out by Johnson *et al.* (1948) on recent glacial clays from north-east America and Collinson (1974) on haematite particles separated from red sandstones, gave DRM intensity approximately proportional to applied field up to $\sim$2 Oe, but Rainbow (1968) found somewhat inconsistent and not always repeatable results in the same field range using synthetic mixtures of magnetite and aluminium oxide. Johnson *et al.* (1948) estimated the ancient field intensity to be 100–200 μT at the deposition sites (based on the ratio of natural and laboratory DRM intensities) but King (1955), working with varved clays of a similar age from Sweden, obtained a palaeointensity of 20 μT. Clegg, Almond and Stubbs (1954) obtained a similar value from some Triassic sandstones from England and Khramov (1967) also reports some results from Palaeozoic red clays. These apparently discordant results for varved clays illustrate the suspect nature of the technique, although it should be noted that the above investigators' work was not primarily directed towards determining palaeointensities. With appropriate samples, careful redeposition experiments carried out specifically for field intensity determinations might well be profitable: this is an area of palaeomagnetism which to date has not been studied in a systematic way.

An interesting palaeointensity technique involving the use of sun-dried mud bricks has recently been described by Games (1977). The author demonstrates that the wet mud of which the bricks are made acquires a remanent magnetism when it is thrown into the mould and that in low fields the intensity of magnetization is proportional to ambient field strength. The technique therefore consists basically of measuring the NRM intensity of a sample taken from a brick and comparing it with the remanent intensity acquired on reforming the brick material in a known magnetic field from the wet, disaggregated mud sample, using the original moulding process.

The mechanism of remanence acquisition is not clear but appears to be associated with shear forces as the mud distorts to fill the mould. Laboratory and field investigations on Peruvian and Egyptian bricks show strikingly consistent results both internally and in comparison with other archaeological field intensity data derived from different techniques.

As might be anticipated, Equation 14.1 can best be applied to rocks magnetized by thermoremanence. Apart from the cooling time, and assuming no thermally-induced alteration of remanence-carrying minerals, a TRM very similar to the original process is easily imparted to a rock sample in the laboratory, and many investigations of palaeointensity have now been carried out on igneous rocks. However, since mineral alteration is often the factor which limits the usefulness of the technique, efforts have recently been made to utilize the similarities of anhysteretic remanent magnetization (ARM) and TRM to develop methods involving less or even no heating of the rock samples. It should perhaps be noted here that the original TRM acquisition in a rock may be a unique event which is not reproducible in the laboratory, even with the most stringent precautions against mineral alteration. This is because irreversible alteration of magnetic minerals may have occurred during the original cooling through the blocking temperature range, and also subsequently by exsolution in titanomagnetites during geological time. This possible 'thermochemical' remanent magnetism and its effect on palaeointensity determinations has been investigated by Skovorodkin (1969).

14.4.2 The Thellier–Thellier method

After demonstration of the laboratory acquisition of TRM in rocks by early researchers, systematic investigations by Königsberger (1938) on igneous rocks and Thellier (1937, 1941) on baked clays laid the foundations for the application of rock magnetic studies to the determination of ancient geomagnetic field intensities. The classical paper of Thellier and Thellier (1959) on the use of baked clays and geological samples for the investigation of palaeointensities led to the adoption of these eminent authors' names to designate the technique they used, which has since formed the basis of many successful palaeointensity determinations.

The principle of the Thelliers' method is the comparison of NRM intensity removed from a rock when it is heated in zero magnetic field through a succession of temperature intervals up to the Curie point with the PTRM

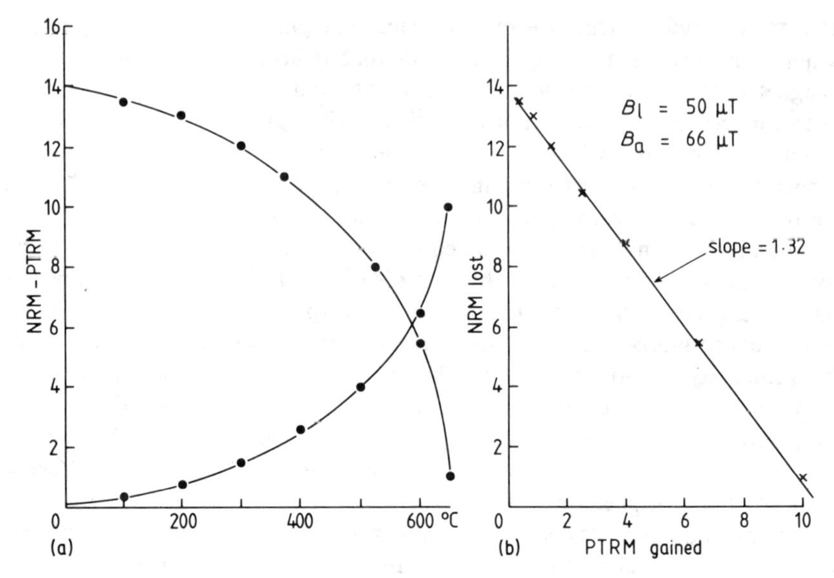

Fig. 14.2 Ideal behaviour of rock sample in the Thellier palaeointensity method: (a) Plot of NRM lost and PTRM gained after heating to and cooling from the indicated temperatures; (b) cumulative plot of NRM lost against PTRM gained in $B_l = 50\ \mu T$. Magnetizations in arbitrary units.

gained by cooling the rock through the same temperature intervals in a known field. Equation 14.1 is then applied by plotting the cumulative NRM lost against PRTM gained as the temperature increases: on certain assumptions the plot is expected to be a straight line, the slope of which is proportional to the ancient field intensity (Fig. 14.2). The basic assumptions inherent in the method are

(a) In the range covered by the geomagnetic and laboratory fields PRTM intensity is proportional to the applied field
(b) A PTRM acquired by cooling through a certain temperature interval is lost when the rock is heated through the same interval
(c) PTRM is an additive phenomenon, i.e. the sum of the intensities aquired on cooling from T_3 to T_2 and T_2 to T_1 is equal to the intensity acquired on cooling from T_3 to T_1
(d) TRM intensity is independent of cooling time

A further assumption is that the heating promotes no physical or chemical alteration of the remanence-carrying mineral which causes a change in its intensity of magnetization, nor results in the production of new magnetic minerals.

Although there is not yet an entirely satisfactory general theory of TRM, assumption (a) appears to be theoretically valid for non-interacting single domains (Néel, 1955; Stacey, 1963) and true multidomain grains (Everitt, 1962a; Dickson *et al.*, 1966) for fields up to 100 μT, and has been experimentally demonstrated in some terrestrial materials (Nagata, 1961; Wilson, 1961a; Everitt, 1962b; Coe, 1967; Dunlop, 1968; McElhinny and Evans, 1968; Dunlop and Waddington, 1975) and by Dunn and Fuller (1972) in lunar samples. The proportionality of TRM intensity and applied field can, of course, be tested experimentally. If non-linearity is observed and there is reliable evidence that no magnetic mineral alteration has occurred during heating, an appropriate correction can be applied to Equation 14.1. The concept of blocking temperature is strong evidence for (b) and also, in an assembly of non-interacting grains, for (c): these aspects of TRM have been demonstrated by Thellier (1951) and Nagata (1961). The problem of mineral alteration is often encountered in igneous rocks and also in lunar and meteoritic samples and procedures for its detection are commonly an integral part of the technique used. These are noted below in connection with the Thellier method and its variations. In lunar samples physical and chemical alteration of the finely divided iron grains by heating is a severe problem, and has caused difficulty even in testing the proportionality of TRM intensity and applied field (Dunn and Fuller 1972; Brecher, Vaughan, Burns and Morash, 1973) and is the prime reason for the search for non-thermal techniques to use on these rocks.

The original Thellier–Thellier technique was applied to archaeological samples (baked clays and bricks) in which haematite carries the NRM, and the full procedure is as follows.

The sample is first tested for acquistion of VRM in the present Earth's field B_p. The procedure is: storage in the laboratory for two weeks in the same orientation as when the sample was in place, to restore soft viscous components which may have altered since collection, measurement of NRM, rotation of 180° about a horizontal axis normal to the magnetic meridian, a further two weeks in this position and then remeasurement of NRM. The vector difference between the two NRMs is the VRM acquired in B_p in two weeks, and the VRM acquired since the sample cooled from its original heating is estimated on the assumption that VRM intensity is proportional to the logarithm of time. If this estimate is more than a few per cent of the NRM, the sample is rejected as probably being too unstable magnetically to give reliable results.

The sample is then heated to a temperature T_1 (60°C was first used but later 100°C), allowed to cool in B_p with a known orientation and the resulting NRM, $J(T_1, 0)$, measured. It is then rotated as above, reheated to T_1 and cooled in B_p again and the NRM, $J(T_1, 180)$, measured. The vector sum of $J(T_1, 0)$ and $J(T_1, 180)$, is twice the PTRM acquired by the sample when it cooled from above the Curie point to T_1 (100°C) in the ancient field, $J(T_c - 100)$, and the vector difference of the two NRMs is twice the PTRM acquired in B_p between T_1 and room temperature (J_T, Fig. 14.3). The sample is then heated to above the Curie point T_c, cooled in B_p and the resulting TRM

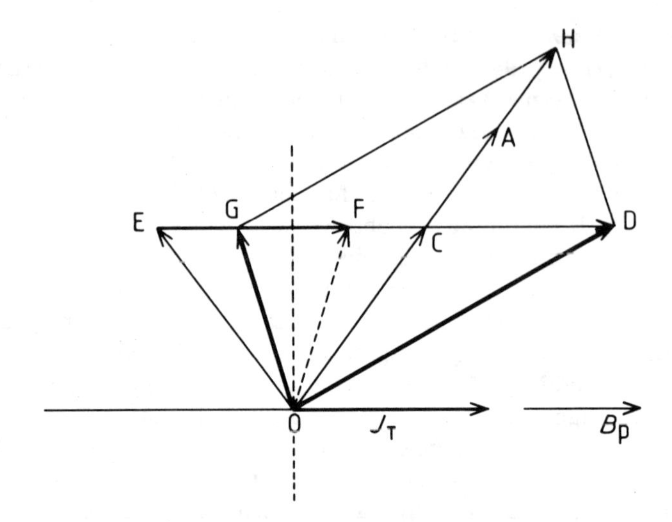

Fig. 14.3 Vector diagram to illustrate the original Thellier method. OA is the NRM of the sample. CA is removed on heating to 100°C while J_T = CD is added on cooling, leaving the sample with an NRM = OD. After rotation through 180° and the second heating and cooling, OF is the resulting NRM vector in space, but its direction in the rock relative to OD is OG. Then GD = 2CD = $2J_T$ and OH = 2OC = $2J(T_c - 100)$.

$J(T_c, 0)$ measured: after rotation through 180° it is heated to T_1, cooled in B_p and the NRM $J(T_c, 180)$ measured. The vector sum of $J(T_c, 0)$ and $J(T_c, 180)$ is twice the PTRM acquired in the present field on cooling from T_c to T_1 and the vector difference is the PTRM acquired in B_p between T_1 and room temperature. Equation 14.1 is then applied

$$B_a = B_p \left(\frac{J(T_1, 0) + J(T_1, 180)}{J(T_c, 0) + J(T_c, 180)} \right) \tag{14.2}$$

The temperature T_1 was chosen as being sufficient to eliminate any original VRM in the sample.

This is the simplest form of the original Thellier method, using a single temperature interval $T_c \rightarrow T_1$ and with all heating and cooling carried out in the Earth's field. The NRM vector differences will be of the same magnitude if no magnetically-important mineralogical changes have occurred on heating to T_c, and thus serve as a pointer to the reliability of the result.

Refinements of the method were later introduced by the Thelliers. These included dividing the temperature range into two intervals, 100–300°C and 300–670°C and later into 100°C or 50°C intervals, repeat heatings to 300°C and 670°C and cooling in B_p at the conclusion of the test, and when smaller intervals were used to 300°C after each pair of heatings, and prolonged heating of the samples (~10 hours). It is clear that these procedures improve the technique by providing several values of the ancient field from one sample, the consistency or otherwise of which are a check on the validity of the method in that sample, and by providing further tests for mineralogical change.

The complete Thellier–Thellier method, as described, is somewhat laborious but thorough and systematic in its approach and elegant in the way the results are obtained without benefit of zero magnetic field. However, although some of the modifications of the technique that have been developed involve fewer heatings and the use of more rapid a.f. demagnetization, palaeointensity determinations remain a time-consuming endeavour if reliable results are to be obtained. This is particularly true with geological samples, when there is a greater probability of mineralogical change in the iron oxides of lower oxidation state which carry NRM in igneous rocks than is the case with the haematite in most archaeological material and baked sediments.

The two features common to most Thellier-derived methods are the comparison of NRM and artificial PTRM in several blocking temperature or coercivity intervals and testing for changes in the minerals in which the remanence resides. Although both these procedures will warn of unsuitable mineralogical behaviour in the sample, the former may also indicate the presence of low coercivity secondary components of NRM through inconsistency ratios of natural and artificial PTRM intensities and therefore also indicate the range of PTRMs from which a valid palaeointensity can be derived.

Some investigators have found it convenient to follow Thellier and make all measurements in the Earth's or other applied field, using basically the same technique for separating natural and laboratory PTRM (Schwarz and Christie, 1967; Schwarz and Symons, 1969).

A common modification of the Thellier method is to heat the sample twice to the required temperature in zero magnetic field, the first cooling taking place also in zero field and the second in the applied field. The additive law of PTRM then allows the calculation of NRM lost and PTRM gained in the chosen temperature intervals (Nagata, Arai and Momose, 1963; Briden, 1966; Coe, 1967; Collinson *et al.*, 1973). Alternatively, the PTRM acquired in each temperature interval can be directly determined by setting the ambient field to zero during cooling except during the temperature interval in which the PTRM is to be acquired. There may be practical difficulties with this method, because of the thermal inertia of sample and furnace, in ensuring that the PTRM is only acquired between the desired temperatures. Some examples of results obtained by the Thellier method are shown in Fig. 14.4.

In a method described by Walton (1977) the magnitude of applied field is adjusted so that there is no change in the NRM of the sample on heating to a given temperature and cooling in the field, i.e. the NRM lost is equal to the PTRM gained. The average value of this field over several temperature intervals is equal to the ancient field, provided chemical or other changes do not affect the PTRM acquired in a given temperature interval. The technique was developed for archaeological samples and the use of a sensitive cryogenic

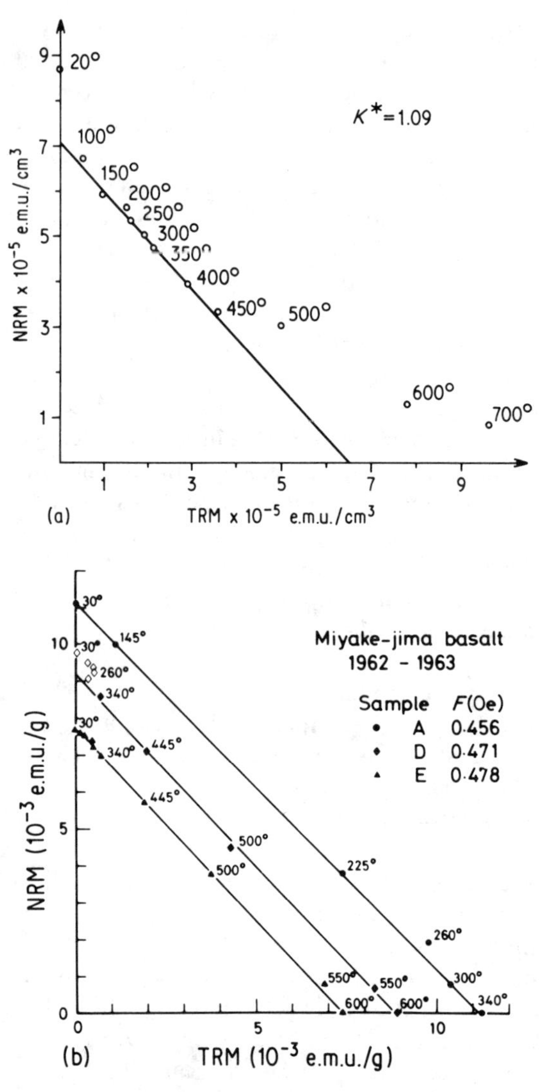

Fig. 14.4 Examples of Thellier–Thellier palaeointensity determinations: (a) Archaeological sample from Bulgaria, in which there is evidence of mineralogical alteration above 450°C. K^* is the slope of the line (from Kovacheva, 1977); (b) samples of an historic lava flow (1962–63) from Japan, where F is the derived palaeointensity; the observed field at the site in 1962 was $(46 \pm 1)\,\mu$T (0.46 ± 0.1 Oe) (from Kono, 1978).

magnetometer (Section 9.4) enables very small samples ($\sim$0.3 cm) to be measured using short heating times.

There are reasons for believing that the presence or absence of the applied field during heating as well as cooling in the PTRM step of the double heating method may influence the results. Theory suggests that high temperature

VRM may be significant in some rocks and also that an external field can modify particle relaxation times (Coe, 1967; Néel, 1955).

Levi (1975) investigated such an effect in laboratory magnetite samples possessing a TRM acquired in a known field by comparing plots of NRM lost/PTRM gained obtained in two ways, with the applied field present (a) and absent (b) during the heating part of the PTRM acquisition steps. Improved results were obtained in single domain (and pseudo-single domain) samples with the field applied during heating. This appears to be due to the second heating in zero field in method (b) causing further decay of NRM, whereas in method (a) this decay is offset by acquisition of high temperature VRM in the applied field. The author shows curves that are consistent with this hypothesis, in that NRM lost in a given temperature interval is proportionately greater than the PTRM gained. The author also points out in this context that a difference in the ancient and laboratory field intensities may influence palaeointensity results through the effect of an external field on relaxation times (Néel, 1955). With samples containing a significant fraction of multi-domain magnetite neither method gave a linear plot of NRM lost against PTRM gained.

An exhaustive discussion of the possible influence on the Thellier method of sample demagnetizing field, secondary magnetization, VRM, multidomain grains and cooling rates is given by Coe (1967), and the same author has examined theoretically the effect of grain interaction, showing that it is unlikely to be important. Doell and Smith (1969) discuss the effect of a.f. cleaning prior to using the Thellier method and Aitken *et al.* (1981) investigate the effect of susceptibility anisotropy on palaeointensity determinations in ceramic archaeological material. Other aspects of palaeointensity investigations based on the Thellier method are described in a series of papers (*Phys. Earth Planet. Inter.* (1977), 13, No. 4). Some criteria that may be used to extract the most reliable data from 'NRM lost/PTRM gained' plots and the associated statistics relevant to the precision of derived palaeointensities are discussed by Coe, Grommé and Mankinen (1978).

The question of whether the magnitude of PTRM acquired in a given temperature interval depends on the rate of cooling has been examined by Halgedahl, Day and Fuller (1980) in single domain magnetite, Fox and Aitken (1980) in baked clays and Dodson and McClelland-Brown (1980). Theory and experiment suggest that in both haematite- and titanomagnetite-bearing materials the more rapid the rate of cooling the lower the PTRM magnitude. Since in both archaeological and geological material laboratory cooling is more rapid than the natural cooling, palaeointensities tend to be over-estimated. The amount of possible enhancement appears to be of the order of 10–20 %, but since the appropriate tests are rarely carried out it is not known whether such enhancement always occurs.

Since laboratory PTRMs are usually acquired in fields of the order of 50 μT a residual field of ± 100 nT or less is usually quite sufficient for 'zero' field heatings. To null the field it is convenient to use a two- or three-component

Helmholtz coil system, with which the applied field can also be generated. Computation may be simpler if the PTRM is acquired in a direction perpendicular to the NRM at each stage, but since most magnetometers measure and display orthogonal components of magnetization, vector subtraction is not difficult.

A feature of the chemical and physical changes in rocks is the difficulty of predicting the type of thermal treatment that will most promote them. In some samples the repeated heatings at moderate temperatures may produce mineralogical changes which a single heating to the Curie point will not. Rapid heating and cooling of samples (15–20 minutes) was employed by Tanguy (1975) in an attempt to reduce thermal changes, and the author also plotted TRM intensity acquired as a function of total heating time. These curves were then used to extrapolate back to estimate the TRM intensity that would be acquired in 'zero' heating time, i.e. without mineralogical change. The method can be criticized in that extrapolation to short heating times may not always be valid but the author reports improved results from some Mt Etna lavas using this technique. Some methods involving a single heating only are now described.

14.4.3 Single-heating methods

The most direct of these techniques is that of Wilson (1961a) and Smith (1967). Using an astatic magnetometer which could measure NRM at elevated temperatures (Section 11.2), they recorded the decay of NRM with increasing temperature up to the Curie point, and the acquisition of PTRM as the sample cooled in an applied field. A recent application of this technique using a cryogenic magnetometer is described by Day *et al.* (1977) and some anomalous results are discussed. Other methods rely on comparison of NRM and laboratory TRM, either after each has been cleaned in the same alternating field to allow for secondary components in the original NRM, or after cleaning in several increasing values of alternating field. The basic procedure is as follows. The sample NRM is a.f. demagnetized up to the chosen maximum field, heated to the Curie point and cooled in the applied field, followed by a.f. demagnetization of the laboratory TRM up to the same peak field. The same type of diagram as used for the Thellier method can then be constructed to compare NRM and TRM ratios after treatment at different demagnetizing fields. Applications of the simplified technique, in which NRM and laboratory TRM are compared at one demagnetizing field, include that of van Zijl, Graham and Hales, (1962) on the Stormberg lavas from South Africa (21.9 mT) and Strangway *et al.* (1968c) on recent basalts from Arizona (50 mT). Comparisons of NRM and TRM at different levels of coercivity were carried out by Carmichael (1967), Smith (1967) and Lawley (1970) on basalts and McElhinny and Evans (1968) on the Modipe gabbro from Botswana (Fig. 14.5).

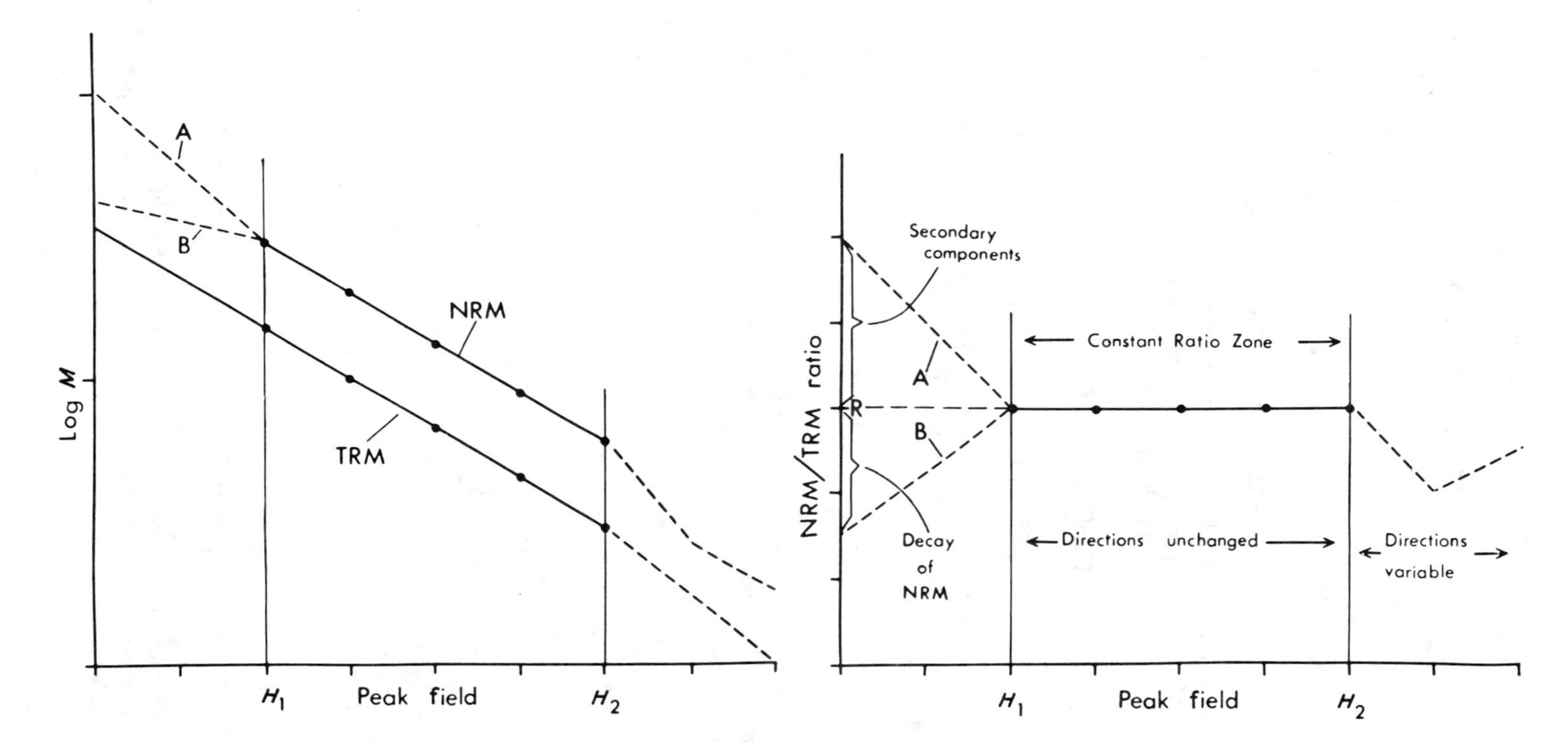

Fig. 14.5 Idealized example illustrating the comparison of a.f. demagnetization curves of NRM and TRM. H_1 and H_2 are the fields between which palaeointensity data can be deduced. Curves A and B represent two possible effects on the NRM which are removed in fields below H. (From McElhinny and Evans, 1968)

In a method due to Domen (1977b) two samples with their NRMs aligned parallel and antiparallel to the applied field are heated simultaneously. On heating to a temperature T and cooling, the NRMs observed on remeasurement are

$$\text{NRM} - \text{PTRM}(T_1, B_a) + \text{PTRM}(T_1, B_l) \tag{14.3}$$

and

$$\text{NRM} - \text{PTRM}(T_1, B_a) - \text{PTRM}(T_1, B_l) \tag{14.4}$$

for the parallel and antiparallel samples respectively. On the usual assumption that the NRM is a TRM, the second and third terms are the PTRMs acquired in the ancient field B_a and the laboratory field B_l.
Then

$$\text{PTRM}(T_1, B_a) = \text{NRM} - I_+/2 \tag{14.5}$$

and

$$\text{PTRM}(T_1, B_l) = I_-/2 \tag{14.6}$$

where I_+ and I_- are respectively the sum and difference of expressions (14.3) and (14.4). Assuming proportionality of PTRM and applied field, dividing Equation 14.5 by Equation 14.6 gives

$$\text{NRM} - (B_a/B_l)I_-/2 = I_+/2 \tag{14.7}$$

from which a mean value of B_a/B can be obtained over different temperature intervals, most conveniently from the slope of the line obtained by plotting $I_-/2$ against $I_+/2$. A feature of this method is that neither the absolute value of NRM nor the exact temperature need to be known.

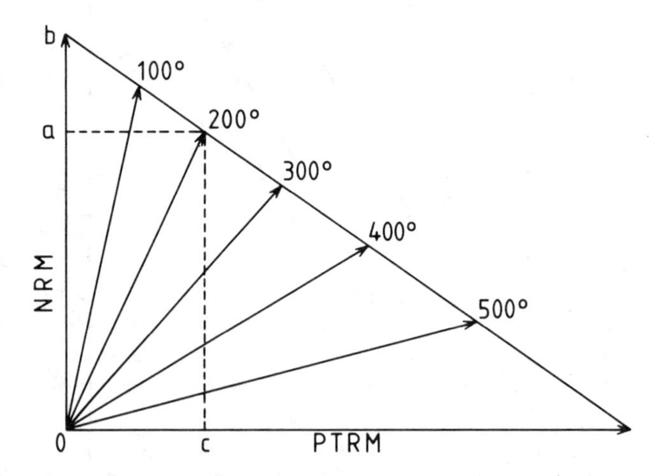

Fig. 14.6 Diagram illustrating the palaeointensity method of Kono and Ueno (1977). 0b is the primary NRM direction, ab is the NRM lost at 200°C and 0c the PTRM gained when cooled from 200°C in the laboratory field.

For rocks which possess only a very small secondary NRM, Kono and Ueno (1977) describe a method in which the known field is applied perpendicular to the NRM during cooling from a series of temperatures up to T_c. If the ratio NRM lost/PTRM gained is constant in each temperature interval and the original NRM vector does not change significantly in direction, then after each cooling the NRM vectors will lie in a plane and the vector end-points (relative to a common origin) will lie on a straight line, the slope of which is proportional to B_a (Fig. 14.6). Note that, although a straight line can be obtained after any secondary NRM (of different direction to the primary) has been removed, the slope will not be correct because the applied field is no longer perpendicular to the primary NRM.

The full Thellier method requires as wide a range of blocking temperatures as possible for its successful application, but in some rocks the TRM is acquired in a relatively restricted temperature interval just below the Curie point. However, as McElhinny and Evans (1968) observe, such rocks, of which the Modiple gabbro is an example, are expected to possess a highly stable remanence, a desirable property for palaeointensity work. Single-heating and coercivity comparison of NRM and TRM should be particularly suited to these rocks.

A disadvantage of the single heating method is that if the heating irreversibly alters the original magnetic mineral or produces a new, remanence-carrying mineral, no useful data can be recovered from the experiment. The Thellier method often allows the recovery of valid intensity data even if mineralogical changes occur, from readings obtained below the temperature of onset of the changes.

Schwarz (1969) emphasizes that neither natural nor artificial TRM in rocks can normally be completely removed by a.f. demagnetization, and therefore NRM and laboratory PTRM cannot be compared over all coercivity intervals unless the demagnetization curves can be reliably extrapolated to zero intensity.

14.4.4 Mineralogical alteration

Considerable effort has been expended in devising tests for the detection of chemical and physical changes promoted in rocks by heating, in characterizing rocks for likelihood of alteration, and in attempts to inhibit changes.

The most common change in igneous rocks is oxidation of titanomagnetites, which is often associated with a rise in Curie temperature: thus, an initially high Curie point may be a pointer to thermal stability (Ade-Hall, Wilson and Smith, 1965) while a rise after heating is evidence of alteration. Examples of stable rocks are the class 5, highly oxidized lavas of Wilson and Watkins (1967). Some of the most difficult igneous material for palaeointensity investigations are submarine basalts. These rocks are usually in a low oxidation state and undergo irreversible changes on heating to temperatures as low as 200°C, and the Thellier method can only be applied in a restricted temperature range (Dunlop and Hale, 1975). Archaeological material usually has a small alteration potential, since in most cases the remanence is carried by haematite which is at the end of the iron oxidation chain and is thermally stable. At the other extreme the very fine iron grains in lunar samples make them particularly prone to oxidation.

Oxidation and other possible changes, such as phase homogenization (which may occur in rocks which originally underwent rapid cooling), ionic ordering or disordering and the formation of new magnetic minerals, are generally reflected in changes in bulk magnetic properties, e.g. initial susceptibility and saturation-induced magnetization (J_{is}) and their variation with temperature. A common test in rocks for their suitability for palaeointensity determinations is repeatability of J_{is}–T curves obtained during both heating to the Curie point and cooling to room temperature. This is not a definitive test since the remanence-carrying mineral may form only a small proportion of the total magnetic mineral content and changes in J_{is} do not necessarily indicate alteration of the particles in which the NRM resides. Coe (1967) describes tests in which J_{is}–T curves and initial susceptibility before and after heating are compared with (NRM lost/PTRM gained) data, and which illustrate the fact that anomalous behaviour can be apparent in the former but not in the latter, and vice versa. However, rocks in which non-coincidence of J_{is}–T heating and cooling curves is observed are usually considered to be suspect for palaeointensity purposes.

A sound principle in rock-magnetic studies is that if some aspect of NRM is being investigated then the tests that are carried out should interact only with remanence-carrying minerals. This principle applies particularly in the present context because heat treatment can promote changes in magnetic minerals which can affect their remanent but not their bulk magnetic properties. An example of such a change is when heating affects barriers determining domain wall positions of multidomain grains, and it may be possible for single-domain particles to be affected by local, thermally-induced stresses.

McElhinny and Evans (1968) compared the coercivity spectra of the saturated isothermal remanence J_{rs} before and after heating, as well as those of the NRM and artificial TRM, as already described. In some samples they found that although the coercivity spectra of J_{rs} before and after heating were similar, J_{rs} had increased as a result of heating – an example of a change of magnetic 'texture'. Carmichael (1967) used the ratio of J_{rs} before and after heating to correct for a supposed change in the amount of NRM-carrying mineral: however, in view of the other possible causes of a change in J_{rs} and the uncertainty as to where the NRM resides this may be a somewhat suspect procedure.

X-ray analysis and microscopic examination of rock sections before and after heating was used by van Zijl *et al.* (1962) to provide contributory evidence for alteration, and Schwarz and Symons (1968) used the microscope to assess alteration potential in their diabase and gabbro samples prior to selection for palaeointensity work. As an example they point out that the pyrrhotite (FeS_{1+x}) seen in most of their samples, which thermomagnetic analysis indicated was not in the ferrimagnetic form, could become ferrimagnetic after heat treatment and acquire TRM. Thus significant quantities of this mineral in a rock would render it unsuitable for a field intensity determination.

The prevention of the most important chemical change, oxidation, has been attempted by heating in vacuum or an atmosphere other than air. Havard (1965) showed that oxidation and reduction can take place even when igneous rocks are heated in vacuum and Coe (1967) found little change in behaviour in a variety of rock types when heating was carried out in air, nitrogen or a vacuum of $\sim 10^{-5}$ torr. Khodair and Coe (1975) obtained improved data from low and intermediate Curie point Hawaiian basalts when heated in vacuum rather than air. Levi (1977) used a slightly reducing atmosphere of nitrogen at $\sim 10^{-1}$ torr and carbon, to prevent alteration of magnetite. The TRM intensity acquired by lavas after heating in an atmosphere containing different partial pressures of oxygen has been investigated by Kono and Tanaka (1977). They also tested the effect of heating samples in air, carbon dioxide, hydrogen and air, and hydrogen and oxygen on the results of palacointensity determinations. Among their results, from Hawaiian and Japanese basalts, were the observations that TRM acquired in a given field increased as the samples were heated in more oxidizing atmospheres and that the stability of the TRM also depended on the atmosphere.

Meteoritic and lunar samples pose a particularly difficult problem in palaeointensity work because of the readiness with which the iron or nickel–iron grains oxidize. In meteorites there may also be problems associated with changes in the kamacite–taenite composition produced by heating (Stacey and Banerjee, 1974; Nagata and Sugiura, 1977). Most lunar sample experiments have been carried out in vacuum to inhibit oxidation, Strangway, Larson and Pearce (1970) showing that oxidation was apparently prevented in some lunar fines when heated in a vacuum of 3×10^{-5} torr but most investigators have found some oxidation and other changes occurring above 400–500°C. Among changes observed by Larson (1978) in lunar basalts were reddening of iron-bearing silicate grains, decomposition of troilite (producing iron) and partial oxidation of ulvöspinel and iron grains. The alloying of iron with other elements and the consequent decrease in the α–γ transition temperature is another potential source of discrepant behaviour in lunar material (Schwarz, 1970; Strangway *et al.*, 1970). A novel approach to the problem has been investigated by Cisowski, Hale and Fuller (1977) who employed microwave induction heating ($\sim 2.5 \times 10^9$ Hz), using a commercial heating unit and waveguide. The potential advantage of this technique is that the heat is localized in the iron grains of high electrical conductivity, the matrix minerals remaining essentially at room temperature because of their much lower conductivity (Section 11.2.2).

The application of Thellier's method to lunar samples is described by Gose, Strangway and Pearce (1973), Collinson *et al.* (1973) and Sugiura and Strangway (1981).

14.4.5 Anhysteretic remanent magnetization methods

Anhysteretic remanent magnetization (ARM) is used in two ways in palaeointensity investigations, as a check on alteration by heating of the mineral in which remanence resides and as an analogue of TRM in palaeointensity methods which potentially involve no heating of the sample.

ARM has been used as a test for mineralogical alteration by Shaw (1974), whose basic technique is the comparison of NRM and TRM (single heating) at

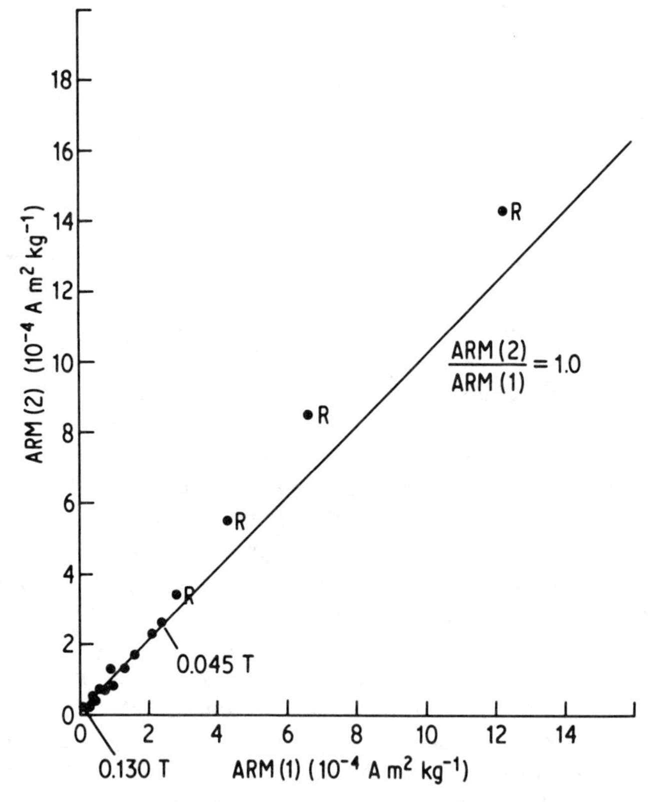

Fig. 14.7 Graph of ARM (2), acquired after heating, against ARM (1), acquired before heating, in sample of Mt Etna lava. The line has gradient = 1.0. The points are ARM intensities after demagnetization in peak alternating fields up to 130 mT, and those marked R indicate alteration of the ARM coercivity spectrum. In this sample the palaeofield is determined by comparison of NRM and TRM intensities in the range 45–130 mT. (From Shaw, 1974)

different levels of coercivity. The coercive force region in which the comparison is considered valid is determined by comparing a.f. demagnetization curves of saturation ARM acquired by the sample before and after heating and selecting the coercivity range in which they are identical (Fig. 14.7). Because of the similarity between ARM and TRM this criterion of reliability seems rather soundly based and the results quoted by Shaw are encouraging. The technique has also been used by Banerjee and Mellema (1976) on a lunar breccia, and by Games (1977) in his palaeointensity determinations using mud bricks (Section 14.4.1).

In some cases there is a linear relationship between ARM at different coercivity levels before and after heating, but the slope of the corresponding graph is not unity. This implies an unchanged coercivity spectrum but a change in magnitude. Kono (1978) shows that a valid palaeointensity can be derived from such samples if the deduced palaeointensity is corrected by a factor equal to the slope of the ARM graph. Rigotti (1978) also uses this correction factor if necessary and an additional ARM test is also done. Instead of only demagnetizing a saturation ARM before and after the single heating, the acquisition of ARM in a constant direct field and increasing alternating field is monitored as well. This procedure has certain advantages and in principle enables the effects of viscous components to be distinguished from mineralogical changes.

The difficulties arising from mineralogical and magnetic alteration due to heating in the Thellier method and its variations provided the incentive for developing an alternative technique for palaeointensity determinations. In lunar palaeointensity work the readiness with which the iron in the samples oxidizes emphasizes the alteration problem and there are the additional factors of a strictly limited amount of sample and the need for sample conservation. Ideally the technique would involve no heating at all and the properties of ARM suggest it as a basis of a viable method.

The properties of ARM and certain similarities between ARM and TRM have been described in Section 6.3.2. The relevant characteristics in the present context are

(a) The additive nature of partial ARMs
(b) The proportionality of ARM intensity and applied low field (0–200 μT)
(c) Proportionality of ARM and TRM intensity acquired in the same field
(d) Similar coercivity spectra of ARM and TRM

(a), (b) and (d) have been generally established in different rocks and minerals but there is currently no satisfactory general theory relating ARM and TRM intensity acquired in the same field and some relevant experimental work on magnetite provided ambiguous results (Dunlop, Bailey and Wescott-Lewis, 1975).

A knowledge of the ratio $J_{\mathrm{TRM}}/J_{\mathrm{ARM}} = R$, for the same applied field, is crucial to any palaeointensity technique based on ARM and its uncertainty has been the main obstacle in the development of such methods. The problem has been investigated by Dunlop and West (1969), Gillingham and Stacey (1971), Jaep (1971) Banerjee and Mellema (1974a) and Stephenson and Collinson (1974). If R can be reliably established for a particular rock, then Equation 14.1 is modified to

$$B_a = B_p \left(\frac{J_{NRM}}{J_{ARM}} \right) \frac{1}{R} \qquad (14.8)$$

The value of R in single and multidomain grains is closely associated with the ratio of the spontaneous magnetizations J_{S0}, J_{SB} at room temperature T_0 and the blocking temperature T_B and is therefore greater than unity. In fact, Gillingham and Stacey (1971) showed theoretically that $R = J_{S0}/J_{SB}$ for multidomain magnetite grains and Dunlop and West (1969) gave the same relation for single-domain magnetite: however, as noted above, experimental determinations of R on sub-micron size magnetite particles by Dunlop *et al.* (1975) and also Bailey and Dunlop (1977) showed that R is poorly predicted by the theory, and also by an expression derived by Banerjee and Mellema (1974a) based on earlier work by Jaep (1971). Their expression, an approximation of a more complex one, is

$$R = (J_{S0}/J_{SB})(T_B/T_0)^{1/2} \qquad (14.9)$$

Whereas the theoretical expressions predict values of R for the experimental magnetite samples to be approximately in the range 3–10, experimental values observed by Bailey and Dunlop were 4–20, with poor agreement between theory and experiment for samples of a given T_B: there was also evidence for variation of R with applied field and grain size. Bailey and Dunlop (1977) also point out a difficulty in applying Equations 14.8 and 14.9, namely the existence of a range of blocking temperatures in a rock and the uncertainty of the average value of T_B (seldom measurable to better than 20°C) and therefore of J_{SB} because of its rapid variation in the range of T_B.

Banerjee and Mellema (1974a) checked Equation 14.9 experimentally, using low-concentration mixtures of single-domain chromium oxide in (diamagnetic) thorium oxide, and found reasonable agreement. The authors (1974b) also applied the technique to palaeointensity determinations (by applying Equation 14.8) in three lunar samples, on which Stephenson and Collinson (1975) commented.

A drawback of the above technique is that it lacks the test of internal consistency inherent in methods where NRM and TRM are compared over a range of coercivities or blocking temperatures. Stephenson and Collinson (1974) approach the problem by comparing coercivity spectra of NRM (TRM) and ARM, obtained from a.f. demagnetization of the former and the intensity acquired in increasing alternating field of the latter. 'NRM lost' is then plotted against 'ARM gained' for the same coercivity intervals and the slope of the straight line (ideally obtained) corrected by the appropriate R-factor

(Equation 14.8) to derive the palaeointensity. The value of R was determined experimentally (for lunar samples) by imparting ARM and TRM to a 1% mixture of iron grains in a non-magnetic matrix, and also to a lunar crystalline rock sample, 10050,33. The coercivity spectra of the TRM and ARM were compared as above and good correspondence was observed. Saturation ARM, i.e. the ARM intensity acquired in infinite alternating field (the analogue of total TRM) is determined by extrapolation from the observed ARM intensity–alternating field curve. The values of R obtained, from the slope of the straight line, were 1.28 and 1.40 respectively, a mean of 1.34.

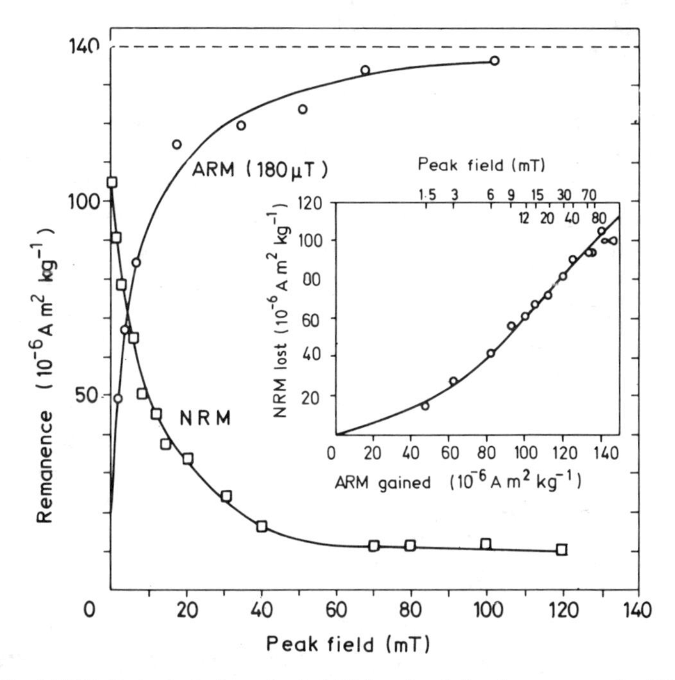

Fig. 14.8 NRM lost plotted against ARM gained for lunar sample 62235. The straight line part of the curve gives an ancient field of 140 μT.

An alternative determination of R is derived from

$$R = [p(T_B)f(T_B)]_{\text{average}} \tag{14.10}$$

where

$$f(T_B) = J_{S0}/J_{SB} \tag{14.11}$$

and $p(T_B)$ is the proportion of grains with blocking temperature T_B. A value of $R \approx 1.3$ was obtained for a lunar crystalline sample (62235,53) by this method (Stephenson, Collinson and Runcorn, 1975), $p(T_B)$ being obtained from thermal demagnetization of the NRM and $f(T_B)$ from the $J_s - T$ curve for iron (Bozorth, 1951). Figure 14.8 shows the result of a palaeointensity determination based on ARM for lunar sample 62235,53. A further value of R for lunar samples has been calculated by Stephenson *et al.* (1976) from data given in Banerjee and Mellema (1976) for sample 72215, a recrystallized breccia. Comparison of a.f. demagnetization curves of TRM and ARM gives $R = 1.6$ for this sample. Hoffman, Baker and Banerjee (1979) describe another method of deriving R and obtain a value of 0.94 from lunar sample 10017,35. The agreement among values of R so far determined for lunar samples is encouraging, and suggests that the use of an average value of R for samples in which it is not directly determinable may be a valid procedure at the present stage of lunar palaeointensity research.

Other aspects and results of the ARM method applied to lunar material are discussed by Stephenson *et al.* (1975) and Collinson and Stephenson (1977).

If results of maximum reliability are to be achieved, the above techniques still require at least one heating to the Curie point and several workers have proposed palaeointensity methods in which all measurements are made at room temperature. In the method proposed by Markert and Heller (1972) and applied to basalts from Mt Etna, the sample is first demagnetized in an alternating field and then the growth of ARM is plotted against the applied direct field using a maximum alternating field of 80 mT. The ARM acquired in each direct field is demagnetized at the same field values as the NRM and a family of curves is thus obtained each of which is a (partially demagnetized) ARM versus applied direct field curve. The intercepts on each of these curves of horizontal lines corresponding to the NRM intensity observed at the same time as the demagnetizing step define a line which, the authors state, intercepts the applied field axis at a value corresponding to the ancient field. Although the authors show an example of the method which gives a derived field intensity close to the known value, the technique appears to be insecurely based in that the assumption $R = 1$ is inherent in it. Theoretical and experimental evidence indicates much higher (and uncertain) values in sub-micron titanomagnetites (Bailey and Dunlop, 1977), although the same authors obtained indirect experimental values closer to unity for multidomain magnetite (2–15 μm).

A similar technique is described by Bagina and Petrova (1977) who compare the NRM and ARM intensities removed in the range of demagnetizing fields in which the coercivity spectra of the two magnetizations coincide. The ancient field intensity is calculated from the ratio of intensity changes of NRM and ARM in the chosen demagnetizing field range and the field in which the ARM was acquired, again carrying the implication that $R = 1$. The method appears to have achieved some success when results of the ARM and Thellier methods applied to the same rocks are compared. Another method reported by Soviet workers is based on the deviation from linearity of the ARM intensity–applied field relationship in a sample already possessing a magnetization (Shashkanov and Metallova, 1977).

To avoid some of the problems associated with lunar palaeointensity investigations and to achieve a greater sample coverage Cisowski *et al.* (1975, 1976, 1977) applied a simple normalization technique to a variety of lunar rocks. NRM intensities after 10 mT a.f. cleaning are normalized against saturated IRM: the ancient field intensity is then calculated from the results of prior heating experiments in which two samples were given a TRM in known fields and their TRM (after 10 mT)/IRM_s ratios measured as a function of the inducing fields. The authors do not claim accuracy for the method but their results are not inconsistent with the range of values found by other workers using different techniques. It is possible to argue that 10 mT cleaning is an arbitrary choice of field and that, if used, it should be applied to the IRM_s as well.

Another approximate palaeointensity method has been developed for lunar samples by Stephenson (Stephenson, Runcorn and Collinson, 1977). Although values of saturated IRM (IRM_s) and saturated ARM in 180 μT (ARM_s) in eight lunar samples vary by up to two orders of magnitude, the mean value of the ratio IRM_s/ARM_s lies in the range 14–53 with a mean of 33. Assuming that ARM_s intensity is linear with applied field B μT, then

$$\frac{IRM_s}{ARM_s} = \frac{33 \times 180}{B} = \frac{5940}{B} \qquad (14.12)$$

If TRM/ARM_s = R

$$\frac{IRM_s}{TRM} = \frac{5940}{BR} \qquad (14.13)$$

Thus for rocks in which the observed NRM is an original unchanged TRM, Equation 14.13 can be used to estimate B, which is now the ancient field intensity. Using $R = 1.34$ and rearranging the terms of Equation 14.13 an estimate of B_a(in μT) is given by

$$B_a = \frac{4400 \times NRM}{IRM_s} \qquad (14.14)$$

This expression is similar to one which can be derived from an expression given by Everitt (1962a) for the ratio TRM/IRM_s based on experiments on multidomain magnetite, and also to a value of IRM_s/TRM derived from two examples of Cisowski *et al.*'s method of palaeointensity estimation, described previously. Stephenson's method applied to *Appollo 11* sample 10020 gives a result consistent with the apparent trend of lunar palaeointensity with time (Stephenson *et al.*, 1977).

14.4.6 Presentation of data

The accumulation of palaeointensity data from archaeological and geological material from different sites and of different ages made it desirable to express the results in a form suitable for direct comparison. On the assumption of an ancient dipolar geomagnetic field, with its intensity dependent on latitude, comparison of palaeointensities derived from sample sites differing in latitude is clearly unsatisfactory.

Thellier and Thellier (1959) referred their ancient field values to a common latitude (or isocline) using the palaeomagnetically determined inclination and the variation in field intensity B with latitude for a dipole field, given by

$$B = \frac{P}{r^3}(4 - 3\cos^2\lambda)^{1/2} = \frac{2P}{r^3}(1 + 3\cos^2\phi)^{-1/2} \qquad (14.15)$$

where P is the geomagnetic dipole moment, r is the geocentric distance of the site at latitude λ where the magnetic inclination is ϕ. Doell and Cox (1961) proposed a similar method, in which all intensities are reduced to the same geographical position using the relation

$$B'_a = B_a(1 + 3\cos^2 p_0)^{1/2}(1 + 3\cos^2 p)^{-1/2} \qquad (14.16)$$

where B_a, B'_a are the measured and reduced field and p and p_0 are the angular distances from the pole of the sampling site and reference site respectively. A simpler procedure has been used by Burlatskaya (1962) and Nagata *et al.* (1963) in which B_a/B_p is given, where B is the present field intensity at the site, and by van Zijl *et al.* (1962) and Momose (1963) who quote values of the ratio J_N/J_T, where J_N is the natural moment and J_T the artificial TRM acquired in the present field.

Probably the most satisfactory method of representing palaeointensities is that proposed by Smith (1967), in which the geocentric dipole moment P_a is calculated (Equation 14.15) which would produce the measured ancient field intensity at the magnetic palaeolatitude of the sample site. By analogy with the calculation of virtual geomagnetic poles from directional palaeomagnetic data the author termed P_a the 'virtual dipole moment' (VDM). The present value of the geomagnetic dipole moment is 8.0×10^{22} A m^2.

Scatter in palaeointensity values and associated VDMs derived from samples within a site is expected for essentially the same reasons that scatter occurs in directional data. In archaeological investigations the aim is usually to delineate the ancient secular variation of intensity of the field with the aid of archaeologically-determined sample ages, whereas with geological samples it is usually a value of the average dipole field intensity that is required. Since a spot reading of the ancient field will be obtained from a single sample, it is necessary to average the results from several samples suitably distributed if secular variation is to be meaned out. The non-dipole field components will contribute to the scatter of results among the samples, but dipole 'wobble' will not be seen if the results are expressed as VDMs. It should be emphasized that for the calculation of a VDM a knowledge of the magnetic palaeolatitude is required, derived from the mean inclination observed in the group of samples.

In his compilation of archaeological and historic geological palaeointensity results, Smith (1967) noted that inclination data were rarely given (or known). He therefore made the assumption that the best fitting geomagnetic dipole has not moved in this period (10 000 years) and calculated dipole moments based on the present magnetic latitude of the sites: the author termed these moments 'reduced dipole moments' (RDMs) to distinguish them from those from palaeomagnetic latitudes.

A question which may reasonably be asked of palaeointensity determinations is 'which is the best technique?' Unfortunately, there are many examples in the literature which show that there is no general 'best' technique but for particular rocks there may be one technique which appears to give the best results. There have been few reports of comparison of methods, the most thorough being that of Coe and Grommé (1973), who determined palaeointensities by three different methods on adjacent samples taken from each of five basaltic lavas: the methods used were basically those of Thellier, Wilson and van Zijl and the field intensities in which the lavas acquired their TRM were independently known to $\pm 6\%$. For these rocks the Thellier method gave the best results and this technique is often found the most satisfactory for archaeological samples and high-Curie point geological samples, although careful interpretation of the results is sometimes required (see, for example, Barbetti, McElhinny, Edwards and Schmidt (1977) on the effects of weathering in baked clays and sediments).

Additional discussion of theoretical and experimental aspects of the different palaeointensity techniques are given by various authors (*Phys. Earth Planet. Inters.* (1977), 13, No. 4). References to lunar palaeointensity research have already been given and results up to 1977 have been summarized by Cisowski *et al.* (1977).

Appendix 1 Conversion factors between SI and c.g.s. units

The recommendations adopted at the 1973 meeting of the International Association of Geomagnetism and Aeronomy regarding the use of SI units in geomagnetic studies are based on the following quantities (IAGA Bulletin 35):

1. Magnetic field, to be expressed as the magnetic induction, or flux density B, the unit being the tesla (T) equal to 1 weber per square metre.
2. Volume intensity of magnetization, to be expressed as the magnetization M derived from $B = \mu_0(H + M)$. Thus, the magnetizing field H, and M have the same units, amperes per metre.

Quantity	Symbol used in this book	SI	c.g.s.	Magnitude, f (c.g.s. $= f \times$ SI)
Magnetic field	B*	$\mathrm{Wb\,m^{-2}}$ (tesla)	Gauss (G)	10^4
	H	$\mathrm{A\,m^{-1}}$	Oersted (Oe)	$4\pi \times 10^{-3}$
Magnetic moment	P	$\mathrm{A\,m^2}$	$\mathrm{G\,cm^3}$	10^3
Magnetic moment per unit volume	M	$\mathrm{A\,m^{-1}}$	G	10^{-3}
Magnetic moment per unit mass	J	$\mathrm{A\,m^2\,kg^{-1}}$	$\mathrm{G\,cm^3\,g^{-1}}$	1
Total susceptibility	–	$\mathrm{m^3}$	$\mathrm{G\,cm^3\,Oe^{-1}}$	$10^6/4\pi$
Susceptibility per unit volume	k	–	$\mathrm{G\,Oe^{-1}}$	$1/4\pi$
Susceptibility per unit mass	χ	$\mathrm{m^3\,kg^{-1}}$	$\mathrm{G\,cm^3\,g^{-1}\,Oe^{-1}}$	$10^3/4\pi$

$\mu_0 = 4\pi \times 10^{-7}$ (henries per metre)

* Note that $10^{-9}\,\mathrm{T} = 1\mathrm{n\,T} = 10^{-5}$ Gauss $= 1\,\gamma$, the commonly used c.g.s. unit for weak magnetic fields. Thus the 'gamma' is an acceptable unit in SI. For a comprehensive account of SI units and conversion factors the reader is referred to a paper by Payne (1981).

3. Volume susceptibility, expressed as the ratio between M and H and therefore dimensionless.

In rock and palaeomagnetism, magnitudes of magnetic properties per unit mass are somewhat more informative than per unit volume, and mass units are preferred in this book. The appropriate conversion factors between SI and c.g.s. units are given in the table.

Unless otherwise stated, where B is used in this book it is an external field and $B = \mu_0 H$.

Appendix 2 Demagnetizing factor of ellipsoids

When a magnetizing field H_e is applied to a particle the field inside the particle H_i is less than H_e because of the back field arising from the induced magnetization M. The demagnetizing factor N of the particle is defined by the relation

$$H_i = H_e - NM$$

where N is appropriate to the direction in the particle along which H_e is applied, and in general will vary with direction in non-spherical particles.

Calculations of N for most particle geometries is very complex because the internal field is generally not uniform. However, H_e is uniform inside bodies of ellipsoidal shape and in many instances in magnetic studies the shape of particles can be approximated to ellipsoids of revolution, for which demagnetizing factors can be comparatively easily determined. Stoner (1945) describes the analytical method for ellipsoids of revolution and Osborn (1945) analyses the more general case of the general ellipsoid, i.e. with three unequal axes.

The two cases of interest are the values of N along the polar and equatorial axes of the ellipsoid, i.e. the long (short) and short (long) axes in prolate (oblate) ellipsoids. A useful property of N is that $N_1 + N_2 + N_3 = 1$, where the subscripts 1, 2 and 3 denote the demagnetizing factors along three mutually perpendicular axes in a body. Therefore, in an ellipsoid of revolution with $N_1 = N_2 = N_e$ in the equatorial plane

$$N_e = \tfrac{1}{2}(1 - N_p)$$

where N_p is the polar demagnetizing factor. If the dimension ratio of the ellipsoid is β (= polar axis/equatorial axis) then (Stoner, 1945)

$$N_p = \frac{1}{\beta^2 - 1}\left[\frac{\beta}{(\beta^2 - 1)^{1/2}} \ln\left(\beta + (\beta^2 - 1)^{1/2}\right) - 1\right] \quad \text{for } \beta > 1$$

$$N_p = \frac{1}{1 - \beta^2}\left[1 - \frac{\beta}{(1 - \beta^2)^{1/2}} \cos^{-1}\beta\right] \quad \text{for } \beta < 1$$

i.e. for prolate and oblate ellipsoids respectively. Figure A.2.1 shows values of N_p and N_e plotted for $0.1 < \beta < 100$.

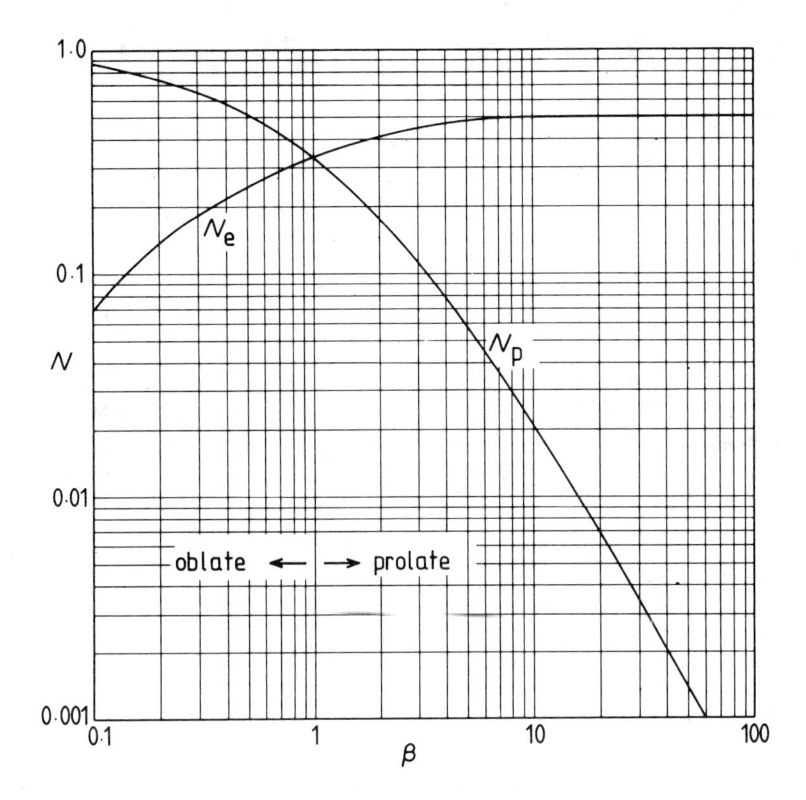

A.2.1

Appendix 3 Copper wire data

The data refer to wires of the SWG numbers shown. The AWG numbers are for wires of the closest equivalent size to that of the corresponding SWG numbers. The data are based on values given in wire tables supplied by the Kent Electric Wire Company, Bracknell, England.

SWG	AWG	Nominal diameter (mm)*	Resistance per 1000 m (Ω)	Weight per 1000 m (kg)	Turns per cm†
10	8	3.251	2.07	72.6	3.0
12	10	2.642	3.15	47.9	3.6
14	12	2.032	5.32	28.4	4.7
16	14	1.626	8.31	18.2	5.9
18	17	1.219	14.8	10.2	7.7
20	19	0.914	26.3	5.74	10.2
22	21	0.711	43.4	3.48	13.0
24	23	0.559	70.3	2.15	16.0
26	25	0.457	105	1.44	19.8
28	27	0.376	155	0.97	24.0
30	28	0.315	221	0.68	28.5
32	30	0.274	292	0.52	32.5
34	31	0.234	402	0.38	37.9
36	32	0.193	589	0.26	45.8
38	34	0.152	946	0.16	57.0
40	36	0.122	1477	0.10	70.3
42	38	0.102	2127	71 g	83.8
44	40	0.081	3324	45	103.6
46	42	0.061	5910	26	135.7
48	46	0.041	13296	11	196.8
50	50	0.025	34039	5	281.2

* Without insulation.
† With normal enamel insulation.

Appendix 4 Preparation of colloidal magnetite

In the method of Elmore (1938a), 5.4 g of hydrated ferric chloride ($FeCl_3 \cdot 6H_2O$) and 2 g of hydrated ferrous chloride ($FeCl_2 \cdot 4H_2O$) are dissolved in 300 cm^3 of water at 70°C. A solution of 5 g of sodium hydroxide in 50 cm^3 of water is then added with constant stirring. The precipitate of magnetite particles which is formed is recovered by filtering and washed thoroughly in distilled water and then washed once with 0.01 N hydrochloric acid. The magnetite is then stirred into one litre of a 0.5% solution of sodium oleate and the suspension is boiled for a short time to disperse the particles.

Baker *et al.* (Bozorth, 1951) prepare the magnetite precipitate in the same way. The wet precipitate is added to 10 cm^3 of N hydrochloric acid to which n-dodecylamine has been added until a pH value of 7 is obtained. The mixture is then made up to 150 cm^3 with distilled water and stirred vigorously to disperse the particles. Blackman and Gustard (1962) used an acetone-based colloid for investigating domain structure in haematite below the Monin transition (−20°C), but give no details of its preparation.

The general procedure for obtaining Bitter patterns is to place a drop of the colloid on the rock surface and form it into a thin layer by placing a microscope cover slip on it. It is important not to let the colloid dry on the surface, and if it remains on the surface too long it may stain it and necessitate repreparation of the surface. The liquid colloid can be removed with soap and water, followed by alcohol. It may be satisfactory to first cover the rock surface with a thin protective layer of collodion or transparent varnish.

Appendix 5 Correction of NRM directions to field reference systems

Before describing the graphical and mathematical procedures for making field corrections of NRM data, it is necessary to describe the reference systems used and the method of specifying the reference planes.

In the absence of evidence to the contrary it is usual to assume that rock formations were laid down horizontally and have not been tilted since formation, or since the acquisition of their primary NRM. In these cases the NRM declination is referred to true North at the site and inclination to the horizontal plane. If there is regional dip of the rock strata in the collection area a second correction procedure may be applied to refer the NRM to the plane of the tilted beds, assumed to have been horizontal when the NRM was acquired. This is the normal procedure when dealing with regionally dipping sediments. The correction of the NRM direction to the site horizontal and geographic North is referred to here as the 'field' correction and the correction for the regional bedding plane as the 'bedding' correction.

The following conventions are used for the correction procedures. The NRM declination, D, as determined in a specimen is measured from 0° to 360° clockwise from the fiducial direction marked on the top surface of the cylindrical (or cubical) specimen, and the inclination of the NRM, I, is positive (negative) when directed below (above) the plane of the top surface of the specimen. If the specimen comes from a hand-sample it is assumed that the sample was cored with the core axis perpendicular to the marked surface of the sample or to the surface of the orientation device used to orient the sample. The strike of this reference surface is parallel to the fiducial direction on the specimen and is measured clockwise (0°–360°) from geographic North. The strike direction, A, is chosen such that the dip B of an upward-facing reference surface is downward when measured in a direction 90° clockwise from A. B is measured from 0° to 180°, a value of B between 90° and 180° implying that the reference surface is on the underside of an exposure, i.e. facing downward. The regional bedding is specified in the same way, with strike direction C and dip E.

For cores drilled in the field the following orientation procedure is assumed. The point of intersection of the line scratched on the upper curved surface of the core section with the top of the core (Section 8) defines the fiducial direction for the specimens cut from the core. The recorded dip is the angle between the horizontal plane and the top surface of the core (i.e. the plane perpendicular to the core axis) measured in a direction 180° from the fiducial direction. There are other ways in which the field data can be recorded, but in this system by adding 90° to the fiducial direction and subtracting 90° from D the values of $(A+90°)$, $(D-90°)$ and B can be used in the correction procedures in the same way as A, D, and B for hand samples. The procedure for correcting hand-sample data is now described.

Correction of NRM directions using the graphical method

From this method an equatorial stereographic (Fig. 12.2) or equal area net is required, depending on which projection is favoured, with great and small circles marked at 2° intervals. A suitable diameter is 20 cm, a commonly used size for the plotting of NRM directions, and a circle of this diameter printed on tracing paper with 10° declination intervals and a reference (North) point marked on its circumference, is also required.

Field correction

Place the traced circle concentrically on the net with the North point at the top over one of the poles of the net and mark the strike position A on the circumference, clockwise from North. Then mark the declination D of the NRM clockwise from the strike direction. Rotate the circle or net until the declination is at the end of a straight line on the net (i.e. at N, E, S or W point) and mark the inclination I on the line, 0° on the circumference to 90° at the centre. Use solid dot (open circle) for positive (negative) inclination. Then put the strike to the pole of the net and move the marked point by the amount of dip B along the small circle passing through the point. If the inclination is positive move the point westward (to the left), or eastward (to the right) if I is negative. If the point goes past the circumference return along the same small wide circle and change the sign of I. Return the N-point on the projection to the pole of the net. The corrected values of D can then be read clockwise from North and I along one of the radii, as before.

Bedding correction

If they are not already marked, mark field corrected D and I on projection relative to North. Mark strike direction C of regional bedding on circumference clockwise from North. Put strike to pole of net and move marked point by the amount of dip E along the appropriate small circle in the *opposite* direction to that given for field correction, i.e. to the right (left) for positive (negative) inclination. Read off corrected D and I as for field correction.

Data from field-drilled cores oriented according to the method given are corrected in the same way as above, after the values of D and A have been modified as described.

Correction of NRM directions using trigonometrical analysis

This is essentially the problem of determining the change in co-ordinates of a point when its orthogonal reference axes are rotated through some angle about the origin. If x, y, z and x', y', z' are the co-ordinates of the point before and after rotation, then

$$\begin{aligned} x &= l_1 x' + m_1 y' + n_1 z' \\ y &= l_2 x' + m_2 y' + n_2 z' \\ z &= l_3 x' + m_3 y' + n_3 z' \end{aligned} \tag{1}$$

where (l_1, m_1, n_1), (l_2, m_2, n_2), (l_3, m_3, n_3) are the direction cosines of the new axes with respect to the old, i.e. l_1, m_1, n_1 are the direction cosines of the new x-axis with respect to the old x, y and z axes, and similarly for the new y-axis and z-axis. The new set of axes is that to which the uncorrected D and I are referred, and if we consider the direction of NRM as a unit vector, then x', y' and z' are the direction cosines of the unit vector, given by

$$\begin{aligned} x' &= \cos D \cos \mathrm{I} \\ y' &= \sin D \cos I \\ z' &= \sin I \end{aligned} \tag{2}$$

For the field correction, the direction cosines (l_i, m_i, n_i) can be obtained from the local strike direction A and dip B

$$\begin{array}{lll} l_1 = \cos A & l_2 = -\sin A \cos B & l_3 = \sin A \sin B \\ m_1 = \sin A & m_2 = \cos A \cos B & m_3 = -\sin B \cos A \\ n_1 = 0 & n_2 = \sin B & n_3 = \cos B \end{array} \tag{3}$$

Thus, substituting the appropriate terms from Equations 2 and 3 in Equation 1, we get

$$\begin{aligned} x &= \cos A \cos D \cos I - \sin A \cos B \sin D \cos I + \sin A \sin B \sin I \\ y &= \sin A \cos D \cos I + \cos A \cos B \sin D \cos I - \sin B \cos A \sin I \\ z &= \sin B \sin D \cos I + \cos B \sin I \end{aligned} \tag{4}$$

The field-corrected values of declination and inclination of NRM, i.e. referred to the local horizontal plane and true North, are given by

$$\tan D_f = y/x$$

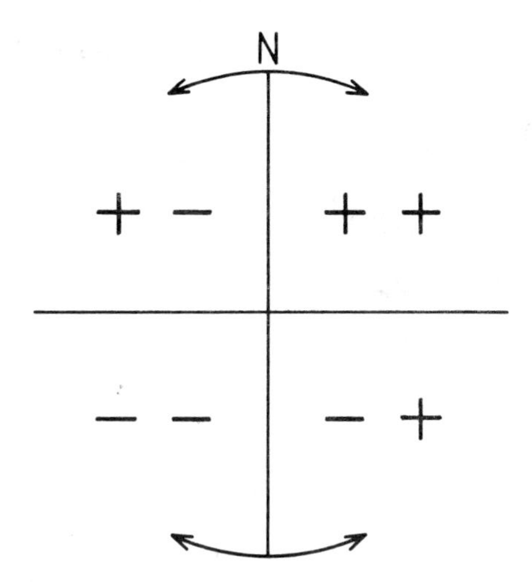

A.5.1

and

$$\sin I_f = z$$

The sign of the inclination is the sign of z and the quadrant in which D_f occurs can be obtained from the signs of x and y, as shown in Fig. A.5.1. The first of each pair of signs is that of x and the second of y, and the arrows indicate the axis from which D is measured.

The expressions in Equation 4 can also be used to apply the bedding correction to D_f and I_f. If the strike and dip of the regional bedding are C and E respectively, then equation 4 is modified by substituting $D_f - C$ for D and $-E$ for B.

Then

$$x = \cos C \cos (D_f - C) \cos I_f - \sin C \cos E \sin (D_f - C) \cos I_f - \sin C \sin E \sin I_f$$

$$y = \sin C \cos (D_f - C) \cos I_f + \cos C \cos E \sin (D_f - C) \cos I_f + \sin E \cos C \sin I_f$$

$$z = -\sin E \sin (D_f - C) \cos I_f + \cos E \sin I_f$$

and

$$\tan D_b = y/x$$

$$\sin I_b = z$$

where D_b and I_b are the declination and inclination of the NRM referred to regional bedding, i.e. to the ancient horizontal plane.

Appendix 6 Measurement of the magnetic field in a coil

If a magnet of regular geometry and known magnetic moment is available which is small compared with the volume enclosed by the coil, the period of oscillation of the magnet may be timed when a suitable direct current is passed through the coil. Since the magnet is suspended at the centre of the coil, with its axis horizontal, on a vertical torsionless suspension, this method is more suitable for a 'Helmholtz'-type coil with central access. The period of oscillation T is given by

$$T = 2\pi\sqrt{\left(\frac{I}{PB}\right)}$$

where I is the moment of inertia of the magnet about the vertical axis, P is its magnetic moment and B the coil field. I is calculated from the magnet mass and dimensions as follows

(a) Square cross-section magnet of mass m, length l and side d, about an axis perpendicular to its length and parallel to a side

$$I = m\left[\frac{l^2 + d^2}{12}\right]$$

(b) Circular cross-section magnet of mass m, length l and diameter d, about an axis perpendicular to its length

$$I = m\left[\frac{l^2}{12} + \frac{d^2}{16}\right]$$

If the Earth's horizontal field is not insignificant compared with the coil field, the measurement can be carried out in field-free space or with the horizontal field cancelled. Alternatively, the method for determining the magnetic moment of a magnet, described in Section 9.2.4(a), can be used.

The field may be determined by passing an alternating current through the coil and measuring the voltage induced in a small coil of known area-turns placed at the coil centre with its plane perpendicular to the field. The r.m.s. voltage V is given by

$$V = B(AN)\omega$$

where B is the r.m.s. value of the alternating field, (AN) is the area-turns of the coil and ω is the circular frequency of the current. V can be measured with an oscilloscope or digital voltmeter.

Appendix 7 The off-centre measurement method with the astatic magnetometer–correction terms for cylindrical samples and the upper magnet

Papapetrou's functions, quoted by Blackett (1952), are correcting factors which can be used in the off-centre measuring method to allow approximately for the shape (and non-dipole effects) in cylindrical samples. The functions F_H and F_V modify Equations 9.28 and 9.29 as follows

$$B_H = \frac{\mu_0 P_H F_H}{4\pi z^3} \sin\theta$$

and

$$B_V = \frac{3\mu_0 P_V x F_V}{4\pi z^4}$$

For a cylinder of height $2l$ and radius r, Papapetrou gives expressions for F_H and F_V in terms of the ratios $u = l/z$ and $v = r/z$: where z is the distance from the centre of the cylinder to the centre of the lower magnet.

$$F_H = \frac{1}{2uv^2}\left\{\frac{1-u}{[(1-u)^2+v^2]^{1/2}} - \frac{1+u}{[(1+u)^2+v^2]^{1/2}}\right\}$$

and

$$F_V = \frac{1}{6u}\left\{\frac{1}{[(1-u)^2+v^2]^{3/2}} - \frac{1}{[(1+u)^2+v^2]^{3/2}}\right\}$$

The calculations are based on the assumption that the off-centre traverse distance $x \ll (r^2 + (z-l)^2)^{1/2}$, i.e. x is very much less than the distance from the magnet to the top edge of the cylinder.

Because the magnets of an astatic magnet system are of opposite polarity the torque exerted on the system by the sample field acting on the upper magnet opposes the torque on the lower magnet. Thus there is an apparent decrease in the magnetometer sensitivity relative to that calculated if the lower magnet only is considered. From Equation 9.28, omitting the sin θ term

$$(B_{\mathrm{H}})_{\mathrm{l}} = \frac{\mu_0 P_{\mathrm{H}} (F_{\mathrm{H}})_{\mathrm{l}}}{4\pi z_{\mathrm{l}}^3}$$

and

$$(B_{\mathrm{H}})_{\mathrm{u}} = \frac{\mu_0 P_{\mathrm{H}} (F_{\mathrm{H}})_{\mathrm{u}}}{4\pi z_u^3}$$

The effective deflecting field is then

$$(B_{\mathrm{H}})_{\mathrm{l}} - (B_{\mathrm{H}})_{\mathrm{u}} = \frac{\mu_0 P_{\mathrm{H}}}{4\pi} \left[\frac{(F_{\mathrm{H}})_{\mathrm{l}}}{z_{\mathrm{l}}^3} - \frac{(F_{\mathrm{H}})_{\mathrm{u}}}{z_{\mathrm{u}}^3} \right]$$

where the subscript l and u refer to the lower and upper magnets. A similar expression involving F_{V} is obtained from Equation 9.29, and Equation 9.30 can then be obtained in the corrected form. The effect of the upper magnet is small if $z_{\mathrm{l}} < z_{\mathrm{u}} \quad z_{\mathrm{l}}$, the separation of the magnets.

Appendix 8 Detection of mains asymmetry (Hailwood and Molyneux, 1974)

The instrument consists essentially of two rectifiers connected back to back, and the current entering the instrument is divided so that exactly half passes through each rectifier. The difference between the mean peak values of the positive and negative half-cycles is then measured and displayed on a centre-zero output meter as a fraction of the input voltage. The principle of the circuit is illustrated in Fig. A.8.1. If it is supposed that A_1 and A_2 are perfect rectifiers (i.e. the forward voltage drop is zero and the reverse impedance infinite), then the voltage across the capacitor C_1 will charge to the peak value of the positive half-cycle and that across C_2 to the peak value of the negative half-cycle. If the resistances R_1 and R_2 have identical values, which combined with the capacitances C_1 and C_2 give a time constant that is long compared with the

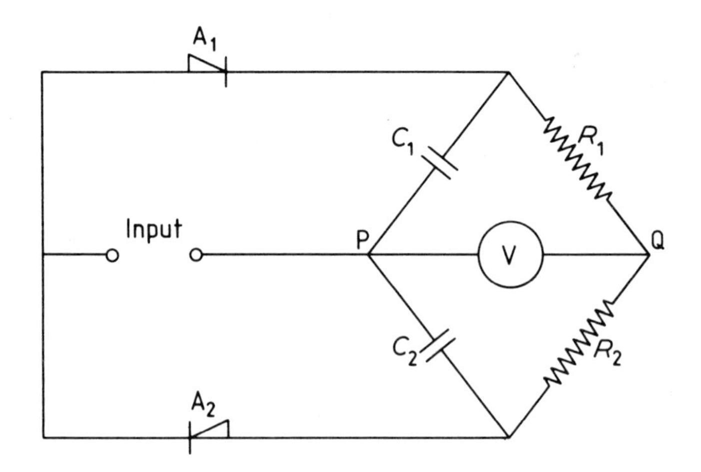

A.8.1

period of the input waveform, then if the peak value of the negative half-cycle is equal to that of the positive half-cycle the voltmeter (assumed to have infinite impedance) will register zero volts. If the peak values of the positive and negative half-cycles are not equal then a potential difference will result across PQ, and this can be shown to be proportional to the percentage asymmetry. The instrument can be calibrated for a given mean input voltage so that the percentage asymmetry is given directly from the meter reading.

In practice perfect rectifiers do not exist, but these are simulated in this instrument by the use of operational amplifiers. High sensitivity is achieved by using an output amplifier to increase the strength of the signal applied to the output meter, and a 'set zero' control is provided to compensate for the offset voltage and current of the output amplifier. Other refinements include a built-in calibration circuit, a balance control to allow for the effects of variations in temperature and condition of battery supply, and an input reversing switch. In its present form this instrument is capable of measuring asymmetries in the range 0–2.5 % with an accuracy of 0.02 %.

References

Abe, K. and Chikazumi, S. (1976), Computer controlled torque magnetometer for automatic determination of crystal orientation. *Jpn. J. Appl. Phy.*, **15**, 619–25.

Adcock, F. and Bristow, C. A. (1935), Iron of high purity. *Proc. R. Soc. London*, **A153**, 172–200.

Ade-Hall, J. M. (1964), A correlation between remanent magnetism and petrology and chemical properties of tertiary basalt lavas from Mull, Scotland. *Geophys. J. R. Astron. Soc.*, **8**, 403–23.

Ade-Hall, J. M (1969), Opaque petrology and the stability of natural remanent magnetism in basaltic rocks. *Geophys. J. R. Astron. Soc.*, **18**, 93–107.

Ade-Hall, J. M., Wilson, R. L. and Smith, P. J. (1965), The petrology, Curie points and natural magnetizations of basic lavas. *Geophys. J. R. Astron. Soc.*, **9**, 323–36.

Ade-Hall, J. M. and Watkins, N. D. (1970), Absence of correlations between opaque petrology and natural remanence polarity in Canary Island lavas. *Geophys. J. R. Astron. Soc.*, **19**, 351–60.

Ade-Hall, J. M., Aumento, F., Ryall, P. J. C., Gerstein, R. E., Brooke, J. and McKeown, D. L. (1973), The mid-Atlantic ridge near 45°N, XXI: Magnetic results from basalt drill cores from the median valley. *Can. J. Earth Sci.*, **10**, 679–96.

Aitken, M. J. (1974), *Physics and Archaeology*, Clarendon Press, Oxford.

Aitken, M. J., Alcock, P. A., Bussell, G. D. and Shaw, C. J. (1981), Archaeomagnetic determination of the past geomagnetic intensity using ancient ceramics: allowance for anisotropy. *Archaeometry*, **23**, 53–64.

Aitken, M. J., Harold, M. R., Weaver, G. H. and Young, S. A. (1967), A 'big-sample' spinner magnetometer and demagnetizing oven, in *Methods in Palaeomagnetism* (eds D. W. Collinson, K. M. Creer and S. K. Runcorn), Elsevier, Amsterdam.

Akimoto, S. (1955), Magnetic properties of ferromagnetic minerals contained in igneous rocks. *Jpn. J. Geophys.*, **1**, 1–31.

Akimoto, S., Horai, K. and Boku, T. (1958), Magnetic susceptibility of orthopyroxenes. *J. Geomagn. Geoelectr.*, **10**, 7–11.

Akimoto, S. and Kushiro, I. (1960), Natural occurrence of titanomagnetite and its relevance to the unstable magnetization of rocks. *J. Geomagn. Geoelectr.*, **11**, 94–110.

Aldenkamp, A. A., Marks, C. P. and Zijlstra, H. (1960), Frictionless recording torque magnetometer. *Rev. Sci. Instrum.*, **31**, 544–6.

Alldred, J. C. and Scollar, I. (1967), Square cross-section coils for the production of uniform fields. *J. Sci. Instrum.*, **44**, 755–60.

Allman, M. and Lawrence, D. F. (1972), *Geological Laboratory Techniques*, Blandford Press, London.

Andel, J. (1970), Negative analysis of the stability index. *Stud. Geophys. Geod.*, **14**, 310–24.

Anderson, J. C., Birss, R. R. and Scott, R. A. M. (1964), Linear magnetostriction in haematite. Proceedings of the International Conference on Magnetism, Nottingham, 597–9.

As, J. A. (1967a), The measurement of the anisotropy of susceptibility with an astatic magnetometer, in *Methods in Palaeomagnetism* (eds D. W. Collinson, K. M. Creer and S. K. Runcorn), Elsevier, Amsterdam.

As, J. A. (1967b), The a.c. demagnetization technique, in *Methods in Palaeomagnetism* (eds D. W. Collinson, K. M. Creer and S. K. Runcorn), Elsevier, Amsterdam.

Bagin, V. I. (1966), Certain characteristics of maghemite. *Phys. Solid Earth*, (*USSR*) 805–8.

Bagina, O. L. and Petrova, G. N. (1977), Determination of palaeomagnetic field intensity using anhysteretic magnetization. *Phys. Earth Planet. Inter.*, **13**, 363–7.

Bailey, M. E. and Dunlop, D. J. (1977), On the use of anhysteretic remanent magnetization in palaeointensity determinations. *Phys. Earth Planet. Inter.*, **13**, 360–62.

Balsley, J. R. and Buddington, A. F. (1958), Iron–titanium oxide minerals, rocks and aeromagnetic anomalies in the Adirondack area, New York. *Econ. Geol.*, **53**, 777–805.

Banerjee, S. K. (1963), An attempt to observe the basal plane anisotropy of haematite. *Philos. Mag.*, **8**, 2119–20.

Banerjee, S. K. (1965), On the transition of magnetite to haematite and its implications to rock magnetism. *J. Geomagn. Geoelectr.*, **17**, 357–61.

Banerjee, S. K. (1966), Exchange anisotropy in intergrown maghemite and haematite. *Geophys. J. R. Astron. Soc.*, **10**, 449–50.

Banerjee, S. K. (1971), New grain size limits for palaeomagnetic stability in haematite. *Nature* (*London*) *Phys. Sci.*, **232**, 15–16.

Banerjee, S. K. and Bartholin, H. (1970), Critical point behaviour of γ-Fe_2O_3. *IEEE Trans. Magn.*, **6**, 299.

Banerjee, S. K., Hoffman, K. and Swits, G. (1974), Remanent magnetization directions in a layered boulder from the South Massif. Proceedings of the 5th Lunar Science Conference, **3**, 2873–81.

Banerjee, S. K., King, J. and Marvin, J. (1981), A rapid method for magnetic granulometry with application to environmental studies. *Geophys. Res. Lett.*, **8**, 333–36.

Banerjee, S. K. and Mellema, J. P. (1974a), A new method for the determination of palaeointensity from the ARM properties of rocks. *Earth Planet. Sci. Lett.*, **23**, 177–84.

Banerjee, S. K. and Mellema, J. P. (1974b), Lunar palaeointensity from three Apollo 15 crystalline rocks using an ARM method. *Earth Planet. Sci Lett.*, **23**, 185–8.

Banerjee, S. K. and Mellema, J. P. (1976), A solar origin for the large lunar magnetic field 4.0×10^9 yr ago? *Proceedings of the 7th Lunar Science Conference*, **3**, 3259–70.

Banerjee, S. K. and Stacey, F. D. (1967), The high field torque meter method of measuring magnetic anisotropy of rocks, in *Methods in Palaeomagnetism* (eds D. W. Collinson, K. M. Creer and S. K. Runcorn), Elsevier, Amsterdam.

Barbetti, M. (1972), A furnace for archeomagnetic and paleomagnetic experiments. Research School of Earth Sciences, Australian National University, Report 997.

Barbetti, M. F., McElhinny, M. W., Edwards, D. J. and Schmidt, P. W. (1977), Weathering processes in baked sediments and their effects on archaeomagnetic field intensity measurements. *Phys. Earth Planet. Inter.*, **13**, 346–54.

Barton, C. E. (1978), Magnetic studies of some Australian lake sediments. PhD Thesis, Australian National University, Canberra.

Barton, C. E., McElhinny, M. W. and Edwards, D. J. (1980), Laboratory studies of depositional DRM. *Geophys. J. R. Astron. Soc.*, **61**, 355–77.

Basta, E. Z. (1960), Natural and synthetic titanomagnetites (the system Fe_3O_4–Fe_2TiO_4–$FeTiO_3$). *Neues Jahrb. Mineral.*, **94**, 1017–48.

Beams, J. W. and Pickels, E. G. (1935), High speed centrifuging. *Rev. Sci. Instrum.*, **6**, 299–308.

Bénard, J. (1939), Etude de la decomposition du protoxyde de fer et de solutions solide. *Anna. Chim.* (*Paris*), **12**, 5–92.

Berg, G. A. (1936), Notes on the dielectric separation of mineral grains. *J. Sediment. Petrol.*, **6**, 23–7.

Berlincourt, T. G. (1963), High magnetic fields by means of superconductors. *Br. J. Appli. Phys.*, **14**, 9–18.

Bezuglaya, L. S. and Skovorodkin, Yu. P. (1972), Thermoremanent magnetization of rocks in uniaxial compression at temperatures of 20°–200°. *Phys. Solid Earth* (*USSR*), 385–7.

Bhathal, R. S. (1971), Magnetic anisotropy in rocks. *Earth Sci. Rev.*, **7**, 227–53.

Bhathal, R. S. and Stacey, F. D. (1969), Frequency independence of low-field susceptibility of rocks. *J. Geophys. Res.* **74**, 2025–7.

Bidgood, D. E. T. and Harland, W. B. (1959), Rock compass: a new aid for collecting oriented specimens. *Bull. Geol. Soc. Am.*, **70**, 641–4.

Bingham, C. (1974), An antipodally symmetric distribution on the sphere. *Ann. Statist.*, **2**, 1201–25.

Biquand, D. and Prevot, M. (1971), A.F. demagnetization of viscous remanent magnetization in rocks. *Z. Geophys.*, **37**, 471–85.

Bitter, F. (1931), On inhomogeneities in the magnetization of ferromagnetic materials. *Phys. Rev.*, **38**, 1903–5.

Bitter, F. (1932), Experiments on the nature of ferromagnetism. *Phys. Rev.*, **41**, 507–15.

Bitter, F. (1936), The design of powerful electromagnets. *Rev. Sci. Instrum.*, **7**, 479–88.

Blackett, P. M. S. (1952), A negative experiment relating to magnetism and the Earth's rotation. *Philos. Trans. R. Soc. London*, **A245**, 309–70.

Blackett, P. M. S. (1956), *Lectures in Rock Magnetism*, Weizmann Science Press, Jerusalem.

Blackett, P. M. S. and Sutton (1956), in *Lectures on Rock Magnetism* (ed. P.M.S. Blackett), Weizmann Science Press, Jerusalem.

Blackman, M. and Gustard, B. (1962), Domains in haematite. *Nature* (*London*), **193**, 360–1.

Blow, R. A. and Hamilton, N. (1978), Effect of compaction on the acquisition of a detrital remanent magnetization in fine grained sediments. *Geophys. J. R. Astron. Soc.*, **52**, 13–23.

Boetzkes, P. C. (1973), A spinner magnetometer for susceptibility anisotropy in rocks. PhD Thesis, University of Alberta, Edmonton, Canada.

Boetzkes, P. C. and Gough, D. I. (1975), A spinner magnetometer for susceptibility anisotropy in rocks. *Can. J. Earth Sci.*, **12**, 1448–64.

Bol'shakov, V. A. (1975), On the mechanism of viscous magnetization of rocks. *Phys. Solid Earth (USSR)*, 251–5.

Bol'shakov, V. A. and Faustov, S. S. (1976), On temperature-dependent magnetic cleaning. *Phys. Solid Earth (USSR)*, 224–7.

Bond, W. L. (1951), Making small spheres. *Rev. Sci. Instrum.*, **22**, 344–55.

Bowden, G. J. (1972), Detection coil systems for vibrating sample magnetometers. *J. Phys. E.*, **5**, 1115–9.

Bozorth, R. M. (1951), *Ferromagnetism*, Van Nostrand, New York.

Braddick, H. J. J. (1963), *The Physics of Experimental Method*, Chapman and Hall, London.

Bragg, E. E. and Seehra, M. S. (1976), Analysis of induced e.m.f. in vibrating sample magnetometers. *J. Phys. E*, **8**, 216–22.

Brecher, A. (1976), Textural remanence: a new model of lunar rock magnetism. *Earth Planet. Sci. Lett.*, **29**, 131–45.

Brecher, A. and Arrhenius, G. (1974), The palaeomagnetic record in carbonaceous chondrites: natural remanence and magnetic properties. *J. Geophys. Res.*, **79**, 2081–106.

Brecher, A., Vaughan, D. J., Burns, R. G. and Morash, K. R. (1973), Magnetic and Mössbauer studies of Apollo 16 rock chips 60315, 51 and 62295, 27. *Proceedings of the 4th Lunar Science Conference*, **3**, 2991–3001.

Briden, J. C. (1965), Ancient secondary magnetization in rocks. *J. Geophys. Res.*, **70**, 5205–21.

Briden, J. C. (1966), Estimates of direction and intensity of the palaeomagnetic field from the Mugga Mugga Porphyry, Australia. *Geophys. J. R. Astron. Soc.*, **11**, 267–8.

Briden, J. C. (1972), A stability index of remanent magnetism. *J. Geophys. Res.*, **77**, 1401–5.

Brock, A. and Iles, W. (1974), Some observations of rotational remanent magnetization. *Geophys. J. R. Astron. Soc.*, **38**, 431–3.

Brooks, P. J. and O'Reilly, W. (1970), Magnetic rotational hysteresis characteristics of red sandstones. *Earth Planet Sci. Lett.*, **9**, 71–6.

Brown, H. C. and Khan, M. A. (1963), Portable apparatus for collecting small oriented cores in the field. *Geol. Mag.*, **100**, 451–5.

Brown, W. F. and Johnson, C. E. (1962), Temperature variation of saturation magnetization of gamma ferric oxide. *J. Appl. Phys.*, **33**, 2752–4.

Bruce, C. R. (1967), A spinner magnetometer with a magnetically shielded coil. *Geophysics*, **32**, 893–8.

Bruckshaw, J. McG. and Rao, B. S. (1950), Magnetic hysteresis of igneous rocks. *Proc. Phys. Soc. London*, **63**, 931–8.

Bruckshaw, J. M. and Robertson, E. I. (1948), The measurement of the magnetic properties of rocks. *J. Sci. Instrum.*, **25**, 444–6.

Brunhes, B. (1901), See Chevallier (1925) for this and other references to Brunhes' work.

Brynjolfsson, A. (1957), Studies of remanent magnetization and viscous magnetism in the basalts of Iceland. *Adv. Phys.*, **6**, 247–54.

Buchan, K. (1979), Palaeomagnetic studies of bulk mineral separates from the Bark Lake diorite, Ontario. *Can. J. Earth Sci.*, **16**, 1558–65.

Buchan, K. and Dunlop, D. J. (1976), Palaeomagnetism of the Haliburton intrusions: superimposed magnetization, metamorphism and tectonics in the later Precambrian. *J. Geophys. Res.*, **81**, 2951–67.

Burakov, K. S. (1977), A thermomagnetometer. *Phys. Solid Earth, USSR*, **13**, 365–8.

Burek, P. J. (1969), Device for chemical demagnetization of redbeds. *J. Geophys. Res.*, **74**, 6710–12.

Burek, P. J. (1971), An advanced device for chemical demagnetization of redbeds. *Z. Geophys.*, **37**, 493–8.

Burlatskaya, S. P. (1962), The ancient magnetic field of the Earth. *Bull. Acad. Sci. (USSR), Geophys. Ser.*, 524–8.

Burmester, R. F. (1977), Removal of drilling-induced magnetization by acid treatment. *Trans. Am. Geophys. Union* (abstract) **51**, 277.

Butler, R. F. and Banerjee, S. K. (1975a), Single-domain grain size limits for metallic iron. *J. Geophys. Res.*, **80**, 252–9.

Butler, R. F. and Banerjee, S. K. (1975b), Theoretical single-domain grain size range in magnetite and titanomagnetite. *J. Geophys. Res.*, **80**, 4049–58.

Carey, R. and Isaac, E. D. (1966), *Magnetic Domains and Techniques for their Observation*, English Universities Press, London.

Carmichael, C. M. (1961), The magnetic properties of ilmenite–haematite crystals. *Proc. R. Soc. London*, **A263**, 508–30.

Carmichael, C. M. (1967), An outline of the intensity of the palaeomagnetic field of the Earth. *Earth Planet. Sci. Lett.*, **3**, 351–354.

Chamalaun, F. H. (1964), Origin of the secondary magnetization of the Old Red Sandstones of the Anglo-Welsh cuvette. *J. Geophys. Res.*, **69**, 4237–8.

Chamalaun, F. H. and Creer, K. M. (1964), Thermal demagnetization studies of the Old Red Sandstone from the Anglo–Welsh Cuvette. *J. Geophys. Res.*, **69**, 1607–1616.

Chamalaun, F. H. and Porath, H. (1968), A continuous thermal demagnetizer for rock magnetism. *Pure Appl. Geophys.*, **70**, 105–9.

Channell, J. E. T. (1977), Palaeomagnetism of limestones from the Gargano Peninsula, (Italy), and the implications of these data. *Geophys. J. R. Astron. Soc.*, **51**, 605–16.

Chevallier, R. (1925), L'aimantation des laves de l'Etna et l'orientation du champ terrestre en Sicile du XIIe siecle. *Ann. Phys. (Paris)*, **4**, 5–162.

Chiron, G. and Delapierre, G. (1979), Description of a thin film directional magnetometer. *IEEE Trans. Magn.*, **15**, 1815–17.

Chiron, G., Laj, C. and Pochachard, J. (1981), A high sensitivity portable spinner magnetometer. *J. Phys. E.*, **14**, 977–80.

Christie, K. W. and Symons, D. T. A. (1969), Apparatus for measuring magnetic susceptibility and its anisotropy. *Geol. Surv. Can. Pap.* 69–41, 1–10.

Cialdea, R. (1966), Un nuovo tipo di magnetometro per le misure magnetiche sulle rocco, *Ann. Geofis.* **19**, 399–414.

Cisowski, S. M., Dunn, J. R., Fuller, M. D., Rose, M. F. and Wasilewski, P. J. (1974), Impact processes and lunar magnetism. Proceedings of the 5th Lunar Science Conference, **3**, 2841–58.

Cisowski, S. M., Dunn, J. R., Fuller, M. D., Wu, Y. M., Rose, M. F. and Wasilewski, P. J. (1976), Magnetic effects of shock and their implications for lunar magnetism. *Proceedings of the 7th Lunar Science Conference*, **3**, 3299–320.

Cisowski, S. M. and Fuller, M. D. (1978), The effect of shock on the magnetism of terrestrial rocks. *J. Geophys. Res.*, **83**, 3441–58.

Cisowski, S. M., Fuller, M. D., Wu, Y. M., Rose, M. F. and Wasilewski, P. J. (1975), Magnetic effects of shock and their implications for magnetism of lunar samples. Proceedings of the 6th Lunar Science Conference, **3**, 3123–41.

Cisowski, S. M., Hale, C. and Fuller, M. (1977), On the intensity of ancient lunar fields. Proceedings of the 8th Lunar Science Conference, **1**, 725–50.

Clark, J. W. (1938), New method for obtaining a uniform magnetic field. *Rev. Sci. Instrum.*, **9**, 320–2.

Clark, H. C. (1967), A fused-quartz Curie point balance, in *Methods in Palaeomagnetism* (eds D. W. Collinson, K. M. Creer and S. K. Runcorn), Elsevier, Amsterdam.

Clark, R. M. and Thompson, R. (1978), An objective method for smoothing palaeomagnetic data. *Geophys. J. R. Astron. Soc.*, **52**, 205–14.

Clark, R. M. and Thompson, R. (1979), A new statistical approach to the alignment of time series. *Geophys. J. R. Astron. Soc.*, **58**, 593–608.

Clegg, J. A., Almond, M. and Stubbs, P. H. S. (1954), The remanent magnetism of some sedimentary rocks in Britain. *Philos. Mag.* **45**, 583–98.

Clerk-Maxwell, J. (1892), *A Treatise on Electricity and Magnetism*, Clarendon Press, Oxford.

Cockroft, J. D. (1928), The design of coils for the production of strong magnetic fields. *Phil. Trans. R. Soc. London*, **A227**, 317–43.

Coe, R. S. (1967), The determination of palaeointensities of the Earth's magnetic field with emphasis on mechanisms which could cause non-ideal behaviour in Thellier's method. *J. Geomagn. Geoelectr.*, **19**, 157–79.

Coe, R. S. and Grommé, C. S. (1973), A comparison of three methods of determining geomagnetic palaeointensities. *J. Geomagn. Geoelectr.*, **25**, 415–35.

Coe, R. S., Grommé, C. S. and Mankinen, E. A. (1978), Geomagnetic palaeointensities from radio-carbon dated lava flows on Hawaii and the question of the Pacific non-dipole low. *J. Geophys. Res.*, **83**, 1740–56.

Cohen, D. (1967a), A shielded facility for low-level magnetic measurements. *J. Appl. Phys.*, **38**, 1295–6.

Cohen, D. (1967b), Enhancement of ferromagnetic shielding against low frequency magnetic fields. *Appl. Phys. Lett.*, **10**, 67–

Cohen, D. (1970), Large-volume conventional magnetic shields. *Rev. Phys. Appl.*, **5**, 53–8.

Collinson, D. W. (1965a), Depositional remanent magnetization in sediments. *J. Geophys. Res.*, **70**, 4663–8.

Collinson, D. W. (1965b), Origin of remanent magnetization and initial susceptibility of certain red sandstones. *Geophys. J. R. Astron. Soc.*, **9**, 203–217.

Collinson, D. W. (1968), An estimate of the haematite content of sediments by magnetic analysis. *Earth Planet. Sci. Lett.*, **4**, 417–20.

Collinson, D. W. (1970), An astatic magnetometer with rotating sample. *Geophys. J. R. Astron. Soc.*, **19**, 547–9.

Collinson, D. W. (1972), Micromagnetometer investigation of the Bonito Canyon quartzite. *Earth Planet. Sci. Lett.*, **15**, 430–5.

Collinson, D. W. (1974), The role of pigment and specularite in the remanent magnetism of red sandstones. *Geophys. J. R. Astron. Soc.*, **38**, 253–64.

Collinson, D. W. (1977), Experiments relating to the measurement of inhomogeneous remanent magnetism in rock samples. *Geophys. J. R. Astron. Soc.*, **48**, 271–5.

Collinson, D. W. (1979), On the possibility of using lunar fines to determine the intensity of the ancient lunar magnetic field. *Phys. Earth Planet. Inter.*, **20**, 312–16.

Collinson, D. W. (1980), An investigation of the scattered remanent magnetization of the Dunnet Head sandstone. *Geophys. J. R. Astron. Soc.*, **62**, 393–402.

Collinson, D. W. and Creer, K. M. (1960), Measurements in palaeomagnetism, in *Methods and Techniques in Geophysics*, (ed. S. K. Runcorn), Interscience, New York.

Collinson, D. W., Creer, K. M., Irving, E. and Runcorn, S. K. (1957), Palaeomagnetic investigations in Great Britain. I. The measurement of the permanent magnetization of rocks. *Philos. Trans. R. Soc. London*, **A250**, 73–82.

Collinson, D. W. and Girdler, R. W. (1967), in *Dictionary of Geophysics* (ed. S. K. Runcorn), Pergamon Press, Oxford.

Collinson, D. W., Molyneux, L. and Stone, D. B. (1963), A total and anisotropic magnetic susceptibility meter. *J. Sci. Instrum.*, **40**, 310–12.

Collinson, D. W., Runcorn, S. K., Stephenson, A. and Manson, A. H. (1972), Magnetic properties of Apollo 14 rocks and fines. Proceedings of the 3rd Lunar Science Conference, **3**, 2343–61.

Collinson, D. W. and de Sa, A. (1971), A new magnetometer for small scale magnetic studies. *J. Phys. E*, **4**, 337–41.

Collinson, D. W. and Stephenson, A. (1977), Palaeointensity determinations of lunar samples. *Phys. Earth Planet. Inter.*, **13**, 380–5.

Collinson, D. W., Stephenson, A. and Runcorn, S. K. (1973), Magnetic properties of Apollo 14 and 15 rocks. *Proceedings of the 4th Lunar Science Conference*, **3**, 2963–76.

Cooke, M. P. and de Sa, A. (1981), A radio frequency method for the measurement of initial magnetic susceptibility. *J. Phys. E*, **14**, 1192–6.

Cowan, B. K. and O'Reilly, W. (1972), The effect of heat treatment on magnetic minerals in red sandstones, studied using the technique of rotational hysteresis. *Geophys. J. R. Astron. Soc.*, **29**, 263–74.

Cox, A. (1961), Anomalous remanent magnetization of basalt. *US Geol. Surv. Bull.* 1083-E, 131–60.

Cox, A. and Doell, R. R. (1962), Magnetic properties of the basalt in hole EM7, Mohole project. *J. Geophys. Res.*, **67**, 3997–4004.

Cox, A. and Doell, R. R. (1967), Recording magnetic balance, in *Methods in Palaeomagnetism*, (eds D. W. Collinson, K. M. Creer and S. K. Runcorn), Elsevier, Amsterdam.

Creer, K. M. (1957), Palaeomagnetic investigations in Great Britain. V. The remanent magnetization of unstable Keuper Marls. *Philos. Trans. R. Soc. London*, **A250**, 130–43.

Creer, K. M. (1959), A.C. demagnetization of unstable Triassic Keuper Marls from S.W. England. *Geophys. J. R. Astron. Soc.*, **2**, 261–75.

Creer, K. M. (1961), Superparamagnetism in red sandstones. *Geophys. J. R. Astron. Soc.*, **5**, 16–28.

Creer, K. M. (1962a), On the origin of the magnetization of redbeds. *J. Geomagn. Geoelectr.*, **13**, 86–100.

Creer, K. M. (1962b) A statistical enquiry into the partial remagnetization of folded Old Red Sandstone rock. *J. Geophys. Res.*, **67**, 1899–906.

Creer, K. M. (1967), The production of high magnetic fields for experiments in rock magnetism, in *Methods in Palaeomagnetism* (eds D. W. Collinson, K. M. Creer and S. K. Runcorn, Elsevier), Amsterdam.

Creer, K. M., Anderson, T. W. and Lewis, C. F. M. (1976). Quaternary geomagnetic stratigraphy recorded in Lake Erie sediments. *Earth Planet. Sci. Lett.*, **31**, 37–47.

Creer, K. M., Hedley, I. G. and O'Reilly, W. (1967). Apparatus for magnetic measurements at low temperatures, in *Methods in Palaeomagnetism* (eds D. W. Collinson, K. M. Creer and S. K. Runcorn), Elsevier, Amsterdam.

Creer, K. M. and Like, C. B. (1967), A low temperature investigation of the natural remanent magnetization of several igneous rocks. *Geophys. J. R. Astron. Soc.*, **12**, 301–13.

Creer, K. M., Irving, E. and Nairn, A. E. M. (1959), Palaeomagnetism of the Great Whin Sill. *Geophys. J. R. Astron. Soc.*, **2**, 306–23.

Creer, K. M. and de Sa, A. (1970), An automatic translation balance for recording variation of magnetization. *J. Phys. E*, **3**, 74–5.

Creer, K. M., de Sa, A. and O'Reilly, W. (1967), New vibrating sample magnetometers for use between 4 and 1000 K. *J. Sci. Instrum.*, **44**, 133–5.

Creer, K. M. and Sanver, M. (1967), The use of the sun compass, in *Methods in Palaeomagnetism*, (eds D. W. Collinson, K. M. Creer and S. K. Runcorn), Elsevier, Amsterdam.

Creer, K. M., Thompson, R., Molyneux, L. and MacKereth, F. J. H. (1972), Geomagnetic secular variation recorded in the stable magnetic remanence of recent sediments. *Earth Planet. Sci. Lett.*, **14**, 115–27.

Crook, T. (1910). Electrostatic separation of minerals. *Mineral. Mag.*, **15**, 260–4.

Crownfield, F. R. (1964), Optimum spacing of coil pairs. *Rev. Sci. Instrum.*, **35**, 240.

Dagley, P. and Ade-Hall, J. M. (1970), Cretaceous Tertiary and Quaternary palaeomagnetic results from Hungary. *Geophys. J. R. Astron. Soc.*, **20**, 65–88.

Daly, L. F. (1967), Anisotropy measurements with a translation inductometer, in *Methods in Palaeomagnetism* (eds D. W. Collinson, K. M. Creer, and S. K. Runcorn), Elsevier, Amsterdam.

Daniels, J. M. (1950), A 100 kW water-cooled solenoid. *Proc. Phys. Soc. London*, **63**, 1028–34.

Day, R. (1977), TRM and its variation with grain size: a review. *J. Geomagn. Geoelectr.*, **29**, 233–65.

Day, R., Dunn, J. R. and Fuller, M. D. (1977), Intensity determinations by continuous thermal cycling. *Phys. Earth Planet. Inter.*, **13**, 301–4.

Day, R., O'Reilly, W. and Banerjee, S. K. (1970), Rotational hysteresis study of oxidized basalts. *J. Geophys. Res.*, **75**, 375–86.

Deaver, B. S. and Goree, W. S. (1967), Some techniques for sensitive magnetic measurements using superconducting circuits and magnetic shields. *Rev. Sci. Instrum.*, **38**, 311–18.

De Boer, F. and Selwood, P. W. (1954), The activation energy for the solid state reaction γ-$Fe_2O_3 \rightarrow \alpha$-$Fe_2O_3$. *J. Am. Chem. Soc.*, **76**, 3665–70.

Deer, W. A., Howie, R. A. and Zussman, J. (1962), *Rock Forming Minerals*, Vol. 5; Longmans, London.

Deutsch, E. R. (1956a), The measurement of magnetic hysteresis in rocks and minerals at high temperatures. *J. Geomagn. Geoelectr.*, **8**, 108–17.

Deutsch, E. R. (1956b), The magnetic hysteresis of rocks and minerals at high temperatures. *J. Geomagn. Geoelectr.*, **8**, 118–28.

Deutsch, E. R., Kristjansson, L. G. and May, B. T. (1971), Remanent magnetism of Lower Tertiary lavas on Baffin Island. *Can. J. Earth Sci.*, **8**, 1542–52.

Deutsch, E. R., Rao, K. V., Laurent, R., and Seguin, M. K. (1972), New evidence and possible origin of native iron in ophiolites of eastern Canada. *Nature* (*London*), **269**, 684–5.

Deutsch, E. R., Roy, J. L. and Murthy, G. S. (1967), An improved astatic magnetometer for palaeomagnetism. *Can. J. Earth Sci.*, **5**, 1270–3.

Dianov-Klokov, V. I. (1960), An apparatus for measuring small remanent magnetization of rocks. *Bull. Izv. Acad. Sci. USSR*, **1**, 91–5.

Dianov-Klokov, V. I. and Anasov, M. N. (1967), A sensitive meter for remanent rock magnetism. *Phys. Solid Earth* (*USSR*), **3**, 189–90.

Dickson, G. O., Everitt, C. W. F., Parry, L. G. and Stacey, F. D. (1966), Origin of thermoremanent magnetization. *Earth Planet. Sci. Letts.*, **1**, 222–4.

Dodson, R. E., Fuller, M. D. and Kean, W. F. (1977), Palaeomagnetic records of secular variation from Lake Michigan sediment cores. *Earth Plan. Sci. Letts.*, **34**, 387–95.

Dodson, R. E., Fuller, M. and Pilant, W. (1974), On the measurement of the remanent magnetism of long cores. *Geophys. Res. Lett.*, **1**, 185–8.

Dodson, M. H. and McClelland-Brown, M. E. (1980), Magnetic blocking temperature of single domain grains during slow cooling. *J. Geophys. Res.*, **85**, 2625–37.

Doell, R. R. and Cox, A. (1961), Palaeomagnetism. *Adv. Geophys.*, **8**, 221–313.

Doell, R. R. and Cox, A. (1962), Determination of the magnetic polarity of rock samples in the field. *US Geol. Surv. Prof. Pap.*, 450-D, 105–8.

Doell, R. R. and Cox, A. (1965), Measurement of the remanent magnetization of igneous rocks. *US Geol. Surv. Bull.*, 1203-A, 1–32.

Doell, R. R. and Cox, A. (1967a), Analysis of spinner magnetometer operation, in *Methods in Palaeomagnetism* (eds D. W. Collinson, K. M. Creer and S. K. Runcorn), Elsevier, Amsterdam.

Doell, R. R. and Cox, A. (1967b), Analysis of alternating field demagnetization equipment, in *Methods in Palaeomagnetism* (eds D. W. Collinson, K. M. Creer and S. K. Runcorn), Elsevier, Amsterdam.

Doell, R. R. and Cox, A. (1967c), Measurement of natural remanent magnetization at the outcrop, in *Methods in Palaeomagnetism* (eds D. W. Collinson, K. M. Creer and S. K. Runcorn), Elsevier, Amsterdam.

Doell, R. R. and Smith, P. J. (1969), On the use of magnetic cleaning in palaeointensity studies. *J. Geomagn. Geoelectr.*, **21**, 579.

Domen, H. (1968), A simply designed home-made microbalance. *Bull. Fac. Educ. Yamaguchi Univ.*, **17**, 19–21.

Domen, H. (1967), Chemical demagnetization of the natural remanent magnetization of rocks. *Bull. Fac. Educ. Yamaguchi Univ.*, **16**, 25–8.

Domen, H. (1977a), On a hand-made semi-automatic thermomagnetic balance. *Bull. Fac. Educ. Yamaguchi Univ.*, **27**, 17–23.

Domen, H. (1977b), A single heating method of palaeomagnetic field intensity determination applied to old roof tiles and rocks. *Phys. Earth Planet. Inter.*, **13**, 315–18.

DuBois, R. L. (1965), Some investigations of the remanent magnetism and domain structures of iron meteorites. *J. Geomagn. Geoelectr.*, **17**, 381–90.

Dunlop, D. J. (1965), Grain distributions in rocks containing single domain grains. *J. Geomagn. Geoelectr.*, **17**, 459–71.

Dunlop, D. J. (1968), Monodomain theory: experimental verification. *Science*, **162**, 256–8.

Dunlop, D. J. (1969), Comments on a paper by R. L. Wilson and P. J. Smith: The nature of secondary natural magnetization in igneous and baked rocks. *J. Geomagn. Geoelectr.*, **21**, 797–9.

Dunlop, D. J. (1970), Hematite: instrinsic and defect ferromagnetism. *Science*, **169**, 858–60.

Dunlop, D. J. (1971), Magnetic properties of fine-particle haematite. *Ann. Geophys.*, **27**, 269–93.

Dunlop, D. J. (1972), Magnetic mineralogy of unheated and heated red sediments by coercivity spectrum. *Geophys. J. R. Astron. Soc.*, **27**, 37–55.

Dunlop, D. J. (1973a), Superparamagnetic and single-domain threshold sizes in magnetite. *J. Geophys. Res.*, **78**, 1780–93.

Dunlop, D. J. (1973b), Thermoremanent magnetization in submicroscopic magnetite. *J. Geophys. Res.*, **78**, 7602–13.

Dunlop, D. J. (1973c), Theory of magnetic viscosity in lunar and terrestrial rocks. *Rev. Geophys. Space Phys.*, **11**, 855–901.

Dunlop, D. J. (1974), Thermal enhancement of magnetic susceptibility. *J. Geophys.*, **40**, 439–51.

Dunlop, D. J. (1976), Thermal fluctuation analysis: a new technique in rock magnetism. *J. Geophys. Res.*, **81**, 3511–17.

Dunlop, D. J. (1979), On the use of Zijderveld vector diagrams in multicomponent paleomagnetic studies. *Phys. Earth Planet. Inter.*, **20**, 12–24.

Dunlop, D. J., Bailey, M. E. and Wescott-Lewis, M. F. (1975), Lunar paleointensity determination using anhysteretic remanence (ARM): a critique. *Proceedings of the 6th Lunar Science Conference*, **3**, 3063–9.

Dunlop, D. J., Gose, W. A., Pearce, G. W. and Strangway, D. W. (1973), Magnetic properties and granulometry of metallic iron in lunar breccia 14313. *Proceedings of the 4th Lunar Science Conference*, **3**, 2977–90.

Dunlop, D. J. and Hale, C. J. (1975), A determination of paleomagnetic field intensity using submarine basalts drilled near the mid-Atlantic ridge. *J. Geophys. Res.*, **81**, 4166–72.

Dunlop, D. J. and Stirling, J. M. (1977), 'Hard' viscous remanent magnetization (VRM) in fine-grained hematite. *Geophys. Res. Lett.*, **4**, 163–6.

Dunlop, D. J. and Waddington, E. D. (1975), The field dependence of thermoremanent magnetization of igneous rocks. *Earth Planet. Sci. Lett.*, **25**, 11–25.

Dunlop, D. J. and West, G. F. (1969), An experimental evaluation of single-domain theories. *Rev. Geophys. Space Phys.*, **7**, 709–57.

Dunn, J. R. and Fuller, M. D. (1972), On the remanent magnetism of lunar samples with special reference to 10048,55 and 14053,48. *Proceedings of the 3rd Lunar Science Conference*, **3**, 2363–86.

Dürschner, H. (1954), Ein Magnetometer zur Bestimmung der magnetischen Eigenschaften von Gesteinen. *Ann. Geophys.*, **10**, 152–4.

Dzyaloshinsky, I. (1958), A thermodynamic theory of 'weak' ferromagnetism of antiferromagnetics. *J. Phys. Chem. Solids*, **4**, 241–55.

Eaton, J. A. and Morrish, A. H. (1971), Magnetic domain structure in haematite. *Can. J. Phys.*, **49**, 2768–72.

Edwards, J. (1965), Reversals of natural magnetization within the iron grit of Sussex. *Geophys. J. R. Astron. Soc.*, **9**, 389–97.

Edwards, J. (1980), Comparisons between the generation and properties of rotational remanent magnetization and anhysteretic remanent magnetization. *Geophys. J. R. Astron. Soc.*, **62**, 379–92.

Ellwood, B. B. (1978), Measurement of anisotropy of magnetic susceptibility: a comparison of the torque and spinner magnetometer systems for basaltic specimens. *J. Phys. E*, **11**, 71–5.

Elmore, W. C. (1938a), Ferromagnetic colloid for studying magnetic structures. *Phys. Rev.*, **54**, 309–10.

Elmore, W. C. (1938b), The magnetization of ferromagnetic colloids. *Phys. Rev.*, **54**, 1092–5.

Elston, D. P. and Purucker, M. E. (1979), Detrital magnetization in red beds of the Moenkapi formation (Triassic), Gray Mountain, Arizona. *J. Geophys. Res.*, **84**, 1653–66.

Embleton, B. J. J. and Edwards, D. J. (1973), A field instrument for orienting rock samples. Research School of Earth Sciences, Australian National University. Report No. 998.

Emery, G. R. and Broussard, D. E. (1954), A modified Kullenberg piston corer. *J. Sed. Pet.*, **24**, 207–11.

Emery, K. O. and Dietz, R. S. (1941), Gravity coring instrument and mechanics of sediment coring. *Bull. Geol. Soc. Am.*, **52**, 1685–714.

Engebretson, D. C. and Beck, M. E. (1978), On the shape of directional data sets. *J. Geophys. Res.*, **83**, 5979–82.

Evans, M. E. and McElhinny, M. W. (1969), An investigation of the origin of stable remanence in magnetite-bearing igneous rocks *J. Geomagn. Geoelectr.*, **21**, 757–73.

Everett, J. E. and Osmeikhian, J. E. (1966), Spherical coils for uniform magnetic fields. *J. Sci. Instrum.*, **43**, 470–4.

Everitt, C. W. F. (1962a), Self-reversal of magnetization in a shale containing pyrrhotite. *Philos. Mag.*, **7**, 831–42.

Everitt, C. W. F. (1962b), Thermoremanent magnetization. Theory of multidomain grains. *Philos. Mag.*, **7**, 599–616.

Everitt, C. W. F. (1962c), Thermoremanent magnetization. Experiments on multidomain grains. *Philos. Mag.*, **7**, 583–97.

Everitt, C. W. F. and Clegg, J. A. (1962), A field test for palaeomagnetic stability. *Geophys. J. R. Astron. Soc.*, **6**, 312–19.

Fales, C. L., Breckenbridge, R. A. and Debnam, W. J. (1974), A magnetometer instrumentation technique using a miniature transducer. *Rev. Sci. Instrum.*, **45**, 1009–11.

Fanselau, G. (1929), Die Engeugung weitgehend homogener Magnetfelder durch Kreisströme. *Z. Phys.*, **54**, 260–9.

Farrell, W. E. (1967), The resonance magnetometer, in *Methods in Palaeomagnetism* (eds D. W. Collinson, K. M. Creer and S. K. Runcorn), Elsevier, Amsterdam.

Ference, M., Shaw, A. E. and Stephenson, R. J. (1940), The production of annular magnetic fields of great uniformity. *Rev. Sci. Instrum.*, **11**, 57–62.

Firester, A. H. (1966), Design of square Helmholtz coil systems. *Rev. Sci. Instrum.*, **37**, 1264.

Fisher, R. A. (1953), Dispersion on a sphere. *Proc. R. Soc. (London) A*, **217**, 295–305.

Fletcher, E. J. and O'Reilly, W. (1974), Contribution of Fe^{2+} ions to the magnetocrystalline anisotropy constant K_1 of $Fe_{3-x}TiO_4$ $(0<x<0.1)$. *J. Phys. C*, **7**, 171–8.

Fletcher, E. J., de Sa, A., O'Reilly, W. and Banerjee, S. K. (1968), A digital vacuum torque magnetometer for the temperature range 300–1000 K. *J. Phys. E*, **2**, 311–14.

Flintner, B. H. (1959), A magnetic separation of some alluvial minerals in Malaya. *Am. Mineral.*, **44**, 738–51.

Folgheraiter, G. (1894), *see* Chevallier (1925).

Foner, S. (1959), Versatile and sensitive vibration magnetometer. *Rev. Sci. Instrum.*, **30**, 548–57.

Fonner, S. and McNiff, E. J. (1968), Very low frequency integrating vibrating sample magnetometer (VLFVSM) with high differential sensitivity in high d.c. fields. *Rev. Sci. Instrum.*, **39**, 171–9.

Foss, C. A. (1981), Graphical methods for rapid vector analysis of demagnetization data. *Geophys. J. R. Astron. Soc.*, **65**, 217–222.

Foster, J. H. (1966), A palaeomagnetic spinner magnetometer using a fluxgate gradiometer. *Earth Planet. Sci. Lett.*, **1**, 463–7.

Fox, J. M. W. and Aitken, M. J. (1980), Cooling-rate dependence of thermoremanent magnetization. *Nature (London)*, **283**, 462–3.

Franzen, W. (1962), Generation of uniform magnetic fields by means of air-core core coils. *Rev. Sci. Instrum.*, **33**, 933–8.

Fritz, J. J. and Johnson, H. L. (1950), Design and operation liquid nitrogen-cooled solenoid magnets. *Rev. Sci. Instrum.*, **21**, 416–20.

Fromm, K. (1967), Measurement of NRM with fluxgate unit, in *Methods in Palaeomagnetism* (eds D. W. Collinson, K. M. Creer and S. K. Runcorn), Elsevier, Amsterdam.

Fuller, M. D. (1961), Magnetic anisotropy of rocks. Ph.D. thesis, University of Cambridge.

Fuller, M. D. (1963), Magnetic anisotropy and palaeomagnetism. *J. Geophys. Res.*, **68**, 293–309.

Fuller, M. D. (1967), The a.c. bridge method. in *Methods in Palaeomagnetism* (eds D. W. Collinson, K. M. Creer and S. K. Runcorn), Elsevier, Amsterdam.

Fuller, M. D. (1969), Magnetic orientation of borehole cores. *Geophysics*, **34**, 772–4.
Fuller, M. D. (1970), Geophysical aspects of palaeomagnetism, CRC. *Crit. Rev. Solid State Sci.*, **1**, 137–219.
Fuller, M. D. (1974), Lunar magnetism. *Rev. Geophys. Space Phys.*, **12**, 23–70.
Fuller, M. D. and Kobayashi, K. (1967), Identification of magnetic phases in certain rocks by low temperature analysis, in *Methods in Palaeomagnetism* (eds D. W. Collinson, K. M. Creer and S. K. Runcorn), Elsevier, Amsterdam.
Fuller, M. D., Meshkov, E., Cisowski, S. M. and Hale, C. J. (1979), On the natural remanent magnetism of certain mare basalts. *Proceedings of the 10th Lunar and Planetary Science Conference*, **3**, 2211–34.
Gallop, J. C. and Petley, B. W. (1976), SQUIDS and their application. *J. Phys. E*, **9**, 417–29.
Games, K. (1977), The magnitude of the palaeomagnetic field: a new non-thermal, non-detrital method using sun-dried bricks. *Geophys. J. R. Astron. Soc.*, **48**, 315–29.
Garber, M., Henry, W. G. and Hoeve, H. G. (1960), A magnetic susceptibility balance and the temperature dependence of the magnetic susceptibility of Ca, Ag, Au. *Can. J. Phys.*, **38**, 1595–613.
Geissman, J. W. (1980), Palaeomagnetism of ash-flow tuffs: microanalytical recognition of TRM components. *J. Geophys. Res.*, **85**, 1487–99.
Giddings, J. W. (1976), Precambrian palaeomagnetism in Australia, I: basic dykes and volcanics from the Yilgarn block. *Tectonophysics*, **30**, 91–108.
Giddings, J. W. and Embleton, B. J. J. (1976), Precambrian palaeomagnetism in Australia, II: basic dykes from the Bawler block. *Tectonophysics*, **30**, 109–18.
Giddings, J. W. and McElhinny, M. W. (1976), A new index of palaeomagnetic stability for magnetite bearing rocks. *Geophys. J. R. Astron. Soc.*, **44**, 239–52.
Gidskehaug, A. (1976). Statistic on a sphere. *Geophys. J. R. Astron. Soc.*, **45**, 657–76.
Gillingham, D. E. W. and Stacey, F. D. (1971), Anhysteretic remanent magnetization (ARM) in magnetite grains. *Pure Appl. Geophys.*, **8**, 160–5.
Girdler, R. W. (1961), The measurement and computation of anisotropy of magnetic susceptibility of rocks. *Geophys. J. R. Astron. Soc.*, **5**, 34–44.
Girdler, R. W. (1963), Sur l'application de pressions hydrostatiques a des aimentations thermoremanentes. *Ann. Geophys.*, **19**, 118–22.
Goree, W. S. and Fuller, M. D. (1976), Magnetometers using R.F.-driven squids and their application in rock magnetism and palaeomagnetism. *Rev. Geophys. Space Phys.*, **14**, 591–608.
Gose, W. A. and Carnes, J. G. (1973), The time-dependent magnetization of fine-grained iron in lunar breccias. *Earth Planet. Sci. Lett.*, **20**, 100–6.
Gose, W. A., Strangway, D. W. and Pearce, G. W. (1973), A determination of the intensity of the ancient lunar magnetic field. *Moon*, **7**, 196–201.
Gough, D. I. (1956), A study of the palaeomagnetism of the Pilansberg dykes. Monthly Notices of the Royal Astronomical Society, *Geophysical Supplement*, **7**, 196–213.
Gough, D. I. (1964), A spinner magnetometer. *J. Geophys. Res.*, **69**, 2455–63.
Gough, D. I. and van Niekerk, C. B. (1959), On the palaeomagnetism of the Bushveld Gabbro. *Philos. Mag.*, **4**, 126–36.
Grabovsky, A. and Brodskaya, S. Ya (1958), Normal magnetization and thermal magnetization of anisotropic rocks. *Bull. Acad. Sci. USSR, Geophys. Ser.*, 560–6.

Graham, J. W. (1949), The stability and significance of magnetism in sedimentary rocks. *J. Geophys. Res.*, **58**, 243–60.

Graham, J. W. (1954), Magnetic susceptibility anisotropy, an unexploited petrofabric element. *Bull. Geol. Soc. Am.*, **65**, 1257–8.

Graham, J. W. (1955), Evidence of Polar shift since Triassic times. *J. Geophys. Res.*, **60**, 329–47.

Graham, J. W. (1967a), Preliminary account of a refined technique for magnetic susceptibility anisotropy measurement of rocks, in *Methods in Palaeomagnetism* (eds D. W. Collinson, K. M. Creer and S. K. Runcorn), Elsevier, Amsterdam.

Graham, J. W. (1967b), Preliminary evaluation of a new resonance magnetometer, in *Methods in Palaeomagnetism* (eds D. W. Collinson, K. M. Creer and S. K. Runcorn), Elsevier, Amsterdam.

Graham, J. W. (1968), Elimination of static charges for spinner magnetometers. *J. Geophys. Res.*, **73**, 4788.

Graham, J. W., Buddington, A. F. and Balsley, J. R. (1957), Stress-induced magnetizations of some rocks with analysed magnetic minerals. *J. Geophys. Res.*, **62**, 465–74.

Graham, K. W. T. and Keiller, J. A. (1960), A portable drill rig for producing short oriented cores. *Trans. Geol. Soc. S. Afr.*, **63**, 71–3.

Granar, L. (1958), Magnetic measurements on Swedish varved sediments. *Ark. Geofys.*, **3**, 1–40.

Gray, A. (1921), *Absolute Measurements in Electricity and Magnetism*, Dover Publications, New York (publ. 1967).

Green, R. (1960), Remanent magnetization and the interpretation of magnetic anomalies. *Geophys. Prospect.*, **8**, 98–110.

Grenet, G. (1933), Un appareil pur déterminer les propriétés magnétiques des roches. *C. R. Acad. Sci. Paris*, **196**, 874.

Grenet, G. (1966), A coil of improved sensitivity for rock magnetism. *Geophysics*, **31**, 149–52.

Griffiths, D. H. (1955), The remanent magnetism of varved clays from Sweden. *Mon. Not. R. Astron. Soc., Geophys. Suppl.*, **7**, 103–14.

Griffiths, D. H., King, R. F., Rees, A. I. and Wright, A. E. (1960), The remanent magnetization of some recent varved sediments. *Proc. R. Soc. London*, **A256**, 359–83.

Grover, F. W. (1947), *Inductance Calculations*, Van Nostrand, New York.

Guertin, R. P. and Foner, S. (1974), Application of vibrating sample magnetometer to magnetic measurements under hydrostatic pressure. *Rev. Sci. Instrum.*, **45**, 863–4.

Gumbel, E. J., Greenwood, J. A. and Durand, D. (1953), The circular normal distribution: theory and tables. *J. Am. Stat. Assoc.*, **48**, 131–6.

Gustafson, W. G. (1938), Magnetic shielding of transformers at audio frequencies. *Bell Syst. Tech. J.*, **17**, 416–37.

Guy, C. N. (1976), A simple approach to coil design for vibrating sample magnetometers. *J. Phys. E*, **9**, 790–1.

Hague, B. (1962), *Alternating Current Bridge Methods*, Pitman, London.

Haigh, G. (1957), Observations of the magnetic transition in haematite at -15° C. *Philos. Mag.*, **2**, 877–90.

Haigh, G. (1958), The process of magnetization by chemical change. *Philos. Mag.*, **3**, 267–86.

Hailwood, E. A. (1972), Palaeomagnetic studies on rock formations in the High Atlas and Anti-Atlas regions of Morocco. Ph.D. Thesis, University of Newcastle upon Tyne, England.

Hailwood, E. A. and Molyneux, L. (1974), Anhysteretic remanent magnetization due to asymmetrical alternating fields. *Geophys. J. R. Astron. Soc.*, **39**, 421–34.

Hale, C. J., Fuller, M. and Bailey, R. C. (1978), On the application of microwave heating to lunar paleointensity determination. *Proceedings of the 9th Lunar and Planetary Science Conference*, **3**, 3165–79.

Halgedahl, S. L., Day, R. and Fuller, M. D. (1980), The effect of cooling rate on the intensity of weak-field TRM in single domain magnetite. *J. Geophys. Res.*, **85**, 3690–8.

Halgedahl, S. L. and Fuller, M. D. (1981), The dependence of magnetic domain structure upon magnetization state in polycrystalline pyrrhotite. *Phys. Earth Planet. Inter.*, **26**, 93–7.

Hall, H. T. (1960), Ultra high pressure, high temperature apparatus: the 'belt'. *Rev. Sci. Instrum.*, **31**, 125–31.

Hall, D. H. (1963), The maximization of sensitivity of an a.c. method of measuring small magnetic moments. *Geophysics*, **28**, 767–77.

Halls, H. C. (1976), A least-squares method to find a remanence direction from converging remagnetization circles. *Geophys. J. R. Astron. Soc.*, **45**, 297–304.

Halls, H. C. (1978), The use of converging remagnetization circles in palaeomagnetism. *Phys. Earth Planet. Inter.*, **16**, 1–11.

Halls, H. C. (1979), The Slate Islands meteorite impact site: A study of shock remanent magnetization. *Geophys. J. R. Astron. Soc.*, **59**, 553–91.

Halls, H. C. (1978), Separation of multicomponent NRM: combined use of difference and resultant magnetization vectors. *Earth Planet. Sci. Lett.*, **43**, 303–8.

Halvorsen, E. (1973), The magnetic fabric of some dolerite intrusions, Northeast Spitzbergen: implications for their mode of emplacement. *Earth Planet. Sci. Lett.*, **21**, 127–33.

Hamilton, N. and Rees, A. I. (1970), The use of magnetic fabric in palaeocurrent estimation, in *Palaeogeophysics* (ed. S. K. Runcorn), Academic Press, London.

Hamilton, N. and Rees, A. I. (1971), The anisotropy of magnetic susceptibility of the Franciscan Rocks of the Diablo range, Central California. *Geol. Rundsch.*, **60**, 1103–24.

Hanneken, J. W., Carnes, J. G. and Vant-Hull, L. L. (1976), The frequency dependence of the viscous component of the magnetic susceptibility of lunar rocks and soil samples. *Earth Planet. Sci. Lett.*, **32**, 45–50.

Hanss, R. E. (1964), Thermochemical etching reveals domain structure in magnetite. *Science*, **146**, 398–9.

Hargraves, R. B. and Perkins, W. E. (1969), Investigations of the effect of shock on natural remanent magnetism. *J. Geophys. Res.*, **74**, 2576–89.

Hargraves, R. B., Collinson, D. W., Arvidson, R. E. and Spitzer, C. R. (1977), The Viking magnetic properties experiment; primary mission results. *J. Geophys. Res.*, **82**, 4547–58.

Hargraves, R. B. and Young, W. M. (1969), Source of stable remanence magnetism in Lambertville diabase. *Am. J. Sci.*, **267**, 1161–77.

Harrison, F. W. (1956), Some aspects in the design of torque magnetometer. *J. Sci. Instrum.*, **33**, 5–6.

Harrison, C. G. A. (1966), The palaeomagnetism of deep-sea sediments. *J. Geophys. Res.*, **71**, 3033–43.

Harrison, C. G. A. (1980), Analysis of the magnetic vector in a single rock specimen. *Geophys. J. R. Astron. Soc.*, **60**, 489–92.

Harrison, C. G. A. and Somayajula, L. K. (1966), Behaviour of the Earth's magnetic field during a reversal. *Nature (London)*, **212**, 1193–5.

Hart, P. J. (1967), *Universal Tables for Magnetic Fields of Filamentary and Distributed Circular Currents*, Elsevier, New York.

Havard, A. D. (1965), A thermal-magnetic investigation of the magnetite–ulvöspinel series. Ph.D. Thesis, University of London.

Havard, A. D. and Lewis, M. (1965), Reversed partial thermomagnetic remanence on natural and synthetic titanomagnetites. *Geophys. J. R. Astron. Soc.*, **10**, 59–68.

Hatfield, H. S. and Holman, B. W. (1924), Dielectric separation: a new method for the treatment of ores. *Bull. Inst. Min. Met.*, 234–9.

Hauptman, Z. (1974), High temperature oxidation range of non-stoichiometry and Curie point variation of cation deficient titanomagnetite, $Fe_{2\cdot4}Ti_{0\cdot6}O_{4+\gamma}$. *Geophys. J. R. Astron. Soc.*, **38**, 29–47.

Hauptman, Z. and Stephenson, A. (1968), A technique for growing ulvöspinel single crystals from the melt. *J. Phys. E*, **1**, 1236–7.

Hedley, I. G. (1968), Chemical remanent magnetization of the FeOOH, Fe_2O_3 system. *Phys. Earth Planet. Inter.*, **1**, 103–21.

Heiniger, C. and Heller, F. (1976), A high temperature vector magnetometer. *Geophys. J. R. Astron. Soc.*, **44**, 281–7.

Helbig, K. (1965), Optimum configuration for the measurement of magnetic moment of samples of cubical shape with a fluxgate magnetometer. *J. Geomagn. Geoelectr.*, **17**, 373–80.

Hellbardt, G. (1958), Das astatische Magnetometer und seine Verwendung zur Messung sehr schwacher Gesteinmagnetisierungen. *Geol. Jahrb.*, **75**, 319–46.

Heller, F. (1973), Magnetic anisotropy of granitic rocks of the Bergell massif (Switzerland). *Earth Planet. Sci. Lett.*, **20**, 180–8.

Heller, F. (1977), Palaeomagnetism of the Upper Jurassic limestones from Southern Germany. *J. Geophys.*, **42**, 475–88.

Heller, F. and Channell, J. E. T. (1979), Palaeomagnetism of Upper Cretaceous limestones from the Münster Basin, Germany. *J. Geophys.*, **46**, 413–27.

Heller, F. and Markert, H. (1973), The age of viscous remanent magnetization of Hadrian's Wall (Northern England). *Geophys. J. R. Astron. Soc.*, **31**, 395–406.

Heller, F., Scriba, H. and Weber, M. (1971), A furnace for magnetic investigations of rocks. *Geophys. J. R. Astron. Soc.*, **21**, 531–4.

Helsley, C. E. (1967a), Advantages of field-drilling samples for palaeomagnetic studies, in *Methods in Palaeomagnetism* (eds D. W. Collinson, K. M. Creer, S. K. Runcorn), Elsevier, Amsterdam.

Helsley, C. E. (1967b), Design of a portable transistorized air-turbine spinner magnetometer, in *Methods in Palaeomagnetism* (eds D. W. Collinson, K. M. Creer and S. K. Runcorn), Elsevier, Amsterdam.

Henry, S. G. (1979), Chemical demagnetization: methods, procedures and applications through vector analysis, *Can. J. Earth Sci.*, **16**, 1832–41.

Henshaw, P. C. and Merrill, R. T. (1979), Characteristics of drying remanent magnetization in sediments. *Earth Planet. Sci. Lett.*, **43**, 315–20.

Henyey, T. L., Pike, S. J. and Palmer, D. F. (1978), On the measurement of stress sensitivity of NRM using a cryogenic magnetometer. *J. Geomagn. Geoelectr.*, **30**, 607–18.

Herndon, J. M., Rowe, M. W., Larson, E. E. and Watson, D. E. (1972), Magnetism of meteorites: a review of Russian studies. *Meteoritics*, **7**, 263–84.

Heyding, R. D., Taylor, D. J. and Hair, M. L. (1961), Four-inch shaped pole caps for susceptibility measurements by the Curie method. *Rev. Sci. Instrum.*, **32**, 161–3.

Heye, D. and Meyer, H. (1972), A method for palaeomagnetic measurement of deep sea cores aboard a ship. *Z. Geophys.*, **38**, 837–947.

Hillhouse, J. and Cox, A. (1976), Brunhes–Matuyama polarity transition. *Earth Planet. Sci. Lett.*, **29**, 51–64.

Hillhouse, J. W. (1977), A method for the removal of rotational remanent magnetization acquired during alternating field demagnetization. *Geophys. J. R. Astron. Soc.*, **50**, 28–34.

Hodych, J. P. (1967), A two-component magnetometer used to measure magnetization changes of rock under uniaxial compression, in *Methods in Palaeomagnetism* (eds D. W. Collinson, K. M. Creer and S. K. Runcorn), Elsevier, Amsterdam.

Hoffman, K. A. (1975), Cation diffusion processes and self-reversal of thermoremanent magnetization in the ilmenite–haematite solid solution series. *Geophys. J. R. Astron. Soc.*, **41**, 65–80.

Hoffman, K. A., Baker, J. R. and Banerjee, S. K. (1979), Combining paleointensity methods: a dual-valued determination on lunar sample 10017,135. *Phys. Earth Planet. Inter.*, **20**, 317–323.

Hoffman, K. A. and Banerjee, S. K. (1975), Magnetic 'zig-zag' behaviour in lunar rocks. *Earth Planet. Sci. Lett.*, **25**, 331–7.

Hoffman, K. A. and Day, R. (1978), Separation of multi-component NRM: a general method. *Earth Planet. Sci. Lett.*, **40**, 433–8.

Homonko, P. (1978), A palaeomagnetic study of cave and lake deposits in Britain. MSc Thesis, University of Newcastle upon Tyne, England.

Hopkinson, J. (1889), Magnetic and other physical properties of iron at high temperatures. *Philos. Trans. R. Soc. London*, **A180**, 443.

Hoselitz, K. and Sucksmith, W. (1943), A magnetic study of the two-phase iron-nickel alloys. *Proc. R. Soc. London*, **A181**, 303–13.

Housley, R. M., Grant, R. W. and Abdel-Garwad, M. (1972), Study of excess Fe metal in the lunar fines by magnetic separation, Mössbauer spectroscopy and microscopic examination. *Proceedings of the 3rd Lunar Science Conference*, **1**, 1065–76.

Hoye, G. S. (1972), A study of the magnetic properties of synthetic ferromagnesian olivines and their oxidation products. PhD thesis, University of Newcastle upon Tyne, England.

Hrouda, F. and Janak, F. (1971), A study of the haematite fabric of some red sediments on the basis of their magnetic susceptibility anisotropy. *Sediment. Geol.*, **6**, 187–99.

Humboldt, A. von (1797), Uben die merkwürdige magnetische Polarität einer Gebirgskuppe von Serpentinstein. *Greus Neues J. Phys.*, **4**, 136–40.

Hummervoll, R. (1974), A method to reduce the measuring time for spinner magnetometers. *Z. Geophys.*, **40**, 265–8.

Hummervoll, R. (1976), Further utilization of the fluxgate magnetometer in the palaeomagnetic laboratory. *J. Geophys.*, **42**, 85–87.

Hummervoll, R. and Totland, O. (1979), Automatic a.f. demagnetizer for igneous rock specimens. *J. Phys. E.*, **12**, 212–15.

Humphrey, F. B. and Johnston, A. R. (1963), Sensitive automatic torque balance for thin magnetic films. *Rev. Sci. Instrum.*, **34**, 348–58.

Hutchings, A. M. J. (1966), A note on alternating field demagnetization apparatus. *Geophys. J. R. Astron. Soc.*, **11**, 555–6.

Hutchings, A. M. J. (1967), Computations of the behaviour of two- and three-axis rotation systems, in *Methods in Palaeomagnetism* (eds D. W. Collinson, K. W. Creer and S. K. Runcorn), Elsevier, Amsterdam.

Hvorsley, M. J. and Stetson, H. C. (1946), Free-fall coring tube: a new type of gravity bottom sampler. *Bull. Geol. Soc. Am.*, **57**, 935–50.

Irving, E. (1957), The origin of the palaeomagnetism of the Torridonian sandstone series of North-West Scotland. *Philos. Trans. R. Soc. London*, **A250**, 100–10.

Irving, E. (1964), *Palaeomagnetism and its application to Geological and Geophysical Problems*, John Wiley, New York.

Irving, E. and Major, A. (1964), Post-depositional detrital remanent magnetization in a synthetic sediment. *Sedimentology*, **3**, 135–43.

Irving, E., Molyneux, L. and Runcorn, S. K. (1966), The analysis of remanent magnetization intensities and susceptibilities of rocks. *Geophys. J. R. Astron. Soc.*, **10**, 451–64.

Irving, E. and Opdyke, N. D. (1965), The palaeomagnetism of the Bloomsburg redbeds and its possible application to the tectonic history of the Appalachians. *Geophys. J. R. Astron. Soc.*, **9**, 153–67.

Irving, E. and Park, J. K. (1973), Palaeomagnetism of metamorphic rocks: errors owing to intrinsic anisotropy. *Geophys. J. R. Astron. Soc.*, **34**, 489–93.

Irving, E. and Runcorn, S. K. (1957), Analysis of the palaeomagnetism of the Torridonian sandstone series of North-West Scotland. *Philos. Trans. R. Soc. London*, **A250**, 83–99.

Irving, E., Stott, P. M. and Ward, M. A. (1961), Demagnetization of igneous rocks by alternating magnetic fields. *Philos. Mag.*, **6**, 225–41.

Ishikawa, Y. (1962), Magnetic properties of the ilmenite-haematite system at low temperatures. *J. Phys. Soc. Japan*, **17**, 1835–44.

Ishikawa, Y. and Akimoto, S. (1957), Magnetic properties of the $FeTiO_3$–Fe_2O_3 solid solution series. *J. Phys. Soc. Jpn*, **12**, 1083–98.

Ishikawa, Y. and Akimoto, S. (1958), Magnetic property and crystal chemistry of ilmenite ($MeTiO_3$) and haematite (α-Fe_2O_3) system, II. Magnetic property. *J. Phys. Soc. Jpn*, **13**, 1298–310.

Ishikawa, Y. and Syono, Y. (1963), Order–disorder transformation and reverse thermoremanent magnetism in the $FeTiO_3 . Fe_2O_3$ system. *J. Phys. Chem. Solids*, **24**, 517–28.

Ising, G. (1942), On the magnetic properties of varved clay. *Ark. Mat. Astron. Fys.*, **29**, 1–37.

Jacquet, P. A. (1956), Electrolytic and chemical polishing. *Metall. Rev.*, **1**, 157–211.

Jaep, W. F. (1971), Role of interactions in magnetic tapes. *J. Appl. Phys.*, **42**, 2790–4.

Jahn, R. A. (1973), The effect of thermal cycling on the remanent magnetization of iron grains. MSc Thesis, University of Newcastle upon Tyne, England.

Jeffrey, P. G. (1978), *Chemical Methods of Rock Analysis*, Pergamon Press, Oxford.

Jeffreys, H. and Jeffreys, B. S. (1956), *Methods of Mathematical Physics*, Cambridge University Press.

Jelinek, V. (1973), Precision a.c. bridge set for measuring magnetic susceptibility of rocks and its anisotropy. *Stud. Geophys. Geod.*, **17**, 36–45.

Jenkin, B. M. and Mortimer, C. H. (1938), Sampling lake deposits. *Nature* (*London*), **142**, 834.

Jenkin, B. M., Mortimer, C. H. and Pennington, W. (1941), The study of lake deposits. *Nature* (*London*), **147**, 496–500.

Jensen, S. D. and Shive, P. N. (1973), Cation distribution in sintered titanomagnetites. *J. Geophys. Res.*, **78**, 8474–80.

Johnson, H. P., Kinoshita, H. and Merrill, R. T. (1975), Rock magnetism and palaeomagnetism of some North Pacific deep-sea sediments. *Bull. Geol. Soc. Am.*, **86**, 412–20.

Johnson, H. P., Lowrie, W. and Kent, D. V. (1975), Stability of anhysteretic remanent magnetization in fine and coarse magnetite and maghemite particles. *Geophys. J. R. Astron. Soc.*, **41**, 1–10.

Johnson, E. A. and McNish, A. G. (1938), An alternating current apparatus for measuring small magnetic moments. *Terr. Magn. Atmos. Electr.*, **43**, 393–9.

Johnson, E. A., Murphy, T. and Torreson, O. W. (1948), Pre-history of the Earth's magnetic field. *Terr. Magn. Atmos. Electr.*, **53**, 349–72.

Johnson, E. A., Murphy, T. and Michelson, P. F. (1949), A new high sensitivity remanent magnetometer. *Rev. Sic Inst.*, **20**, 429–34.

Johnston, M. J. S. and Stacey, F. D. (1969), Volcano-magnetic effects observed on Mt. Ruapehu, New Zealand. *J. Geophys. Res.*, **74**, 6541–4.

Jones, D. L., Robertson, I. D. M. and McFadden, P. L. (1975), A palaeomagnetic study of Precambrian dyke swarms associated with the Great Dyke of Rhodesia. *Trans. Geol. Soc. S. Afri.*, **78**, 57–65.

Kaczer, J. and Gemperle, R. (1956), Vibrating Penmalloy probe for mapping magnetic fields. *Czech. J. Phys.*, **6**, 173–83.

Kawai, N. (1963), Chemical demagnetization of natural remanent magnetization of rocks. Annual Progress Report, Rock Magnetism Research Group, Tokyo, pp. 8–10.

Kawai, N., Nakajima, T., Yasukawa, K., Hirooka, K. and Kobayashi, K. (1973), The oscillation of field in the Matuyama geomagnetic epoch. *Proc. Jpn. Acad.*, **49**, 619–22.

Keefer, C. M. and Shive, P. N. (1981), Curie temperature and lattice constant reference contours for synthetic titanomaghemites. *J. Geophys. Res.*, **86**, 987–98.

Kent, D. V. (1973), Post-depositional remanent magnetization in deep sea sediments. *Nature* (*London*), **246**, 32–4.

Kent, D. V. and Lowrie, W. (1975), On the magnetic susceptibility and anisotropy of deep-sea sediments. *Earth Planet. Sci. Lett.*, **28**, 1–12.

Kern, J. W. (1961), Effects of moderate stresses on directions of thermoremanent magnetization. *J. Geophys. Res.*, **66**, 3801–5.

Khan, M. A. (1962), The anisotropy of magnetic susceptibility of some igneous and metamorphic rocks. *J. Geophys. Res.*, **67**, 2873–85.

Khodair, A. A. and Coe, R. S. (1975), Determination of geomagnetic palaeointensities in vacuum. *Geophys. J. R. Astron. Soc.*, **42**, 107–15.

Khramov, A. N. (1958), Palaeomagnetism and Stratigraphic Correlation. Gostoptechizdat, Leningrad. English translation by A. J. Lojkine, published by Australian National University (1960).

Khramov, A. N. (1967), The Earth's magnetic field in the late Palaeozoic. *Bull. Acad. Sci. (USSR), Earth Phys. Ser.*, 86–108.

King L. V. (1933), Electromagnetic shielding at radio frequencies. *Philos. Mag.*, **15**, 201–23.

King, R. F. (1955), The remanent magnetism of artificially deposited sediments. *Mon. Not. R. Astron. Soc. Geophys. Suppl.*, **7**, 115–34.

King, R. F. (1967), Errors in anisotropy measurements with the torsion balance, in *Methods in Palaeomagnetism* (eds D. W. Collinson, K. M. reer and S. K. Runcorn), Elsevier, Amsterdam.

King, R. F. and Rees, A. I. (1962), The measurement of the anisotropy of magnetic susceptibility of rocks, by the torque method. *J. Geophys. Res.*, **67**, 1565–72.

King, R. F. and Rees, A. I. (1966), Detrital magnetism in sediments: an examination of some theoretical models. *J. Geophy. Res.*, **71**, 561–71.

Kinoshita, H. and Nagata, T. (1967), Dependence of magnetostriction and magnetocrystalline anisotropy on hydrostatic pressure. *J. Geomagn. Geolectr.*, **19**, 77–9.

Kirk, P. L. and Craig, R. (1948), Reproducible construction of quartz fiber devices. *Rev. Sci. Instrum.*, **19**, 777–84.

Kirschvink, J. L. (1980), The least-squares line and plane and the analysis of palaeomagnetic data. *Geophys. J. R. Astron. Soc.*, **62**, 699–718.

Klootwijk, C. T. (1971), Palaeomagnetism of the Upper Gondwana-Rajmahal traps, North East India. *Tectonophysics*, **12**, 449–67.

Klootwijk, C. T. and van der Berg, J. (1975), The rotation of Italy: preliminary palaeomagnetic data from the Umbrian sequence, Northern Apennines. *Earth Planet. Sci. Lett.*, **25**, 263–73.

Kobayashi, K. (1959), Chemical remanent magnetization of ferromagnetic minerals and its application to rock magnetism. *J. Geomagn. Geoelectr.*, **10**, 99–117.

Kobayashi, K. and Fuller, M. D. (1967), Vibration magnetometer, in *Methods in Palaeomagnetism* (eds D. W. Collinson, K. M. Creer and S. K. Runcorn), Elsevier, Amsterdam.

Kolm, H., Lax, B., Bitter, F. and Mills, R. (eds) (1962), *High Magnetic Fields*, MIT Press and John Wiley, New York.

Königsberger, J. G. (1938), Natural residual magnetism of eruptive rocks, I. *Terrestrial Magnetism and Atmospheric Electricity*, **43**, 119–27.

Kono, M. (1978), Reliability of palaeointensity methods using alternating field demagnetization and anhysteretic remanence. *Geophys. J. R. Astron. Soc.*, **54**, 241–61.

Kono, M., Hamano, Y., Nishitani, T. and Tosha, T. (1981), A new spinner magnetometer: principles and techniques. *Geophys. J. R. Astron. Soc.*, **67**, 217–28.

Kono, M. (1982), Design of coils for alternating field demagnetization. *J. Geophys. Res.*, **87**, 1142–8.

Kono, M. and Tanaka, H. (1977), Influence of partial pressure of oxygen on thermoremanent magnetization of basalts. *Phys. Earth Planet. Inter.*, **13**, 276–88.

Kono, M. and Ueno, N. (1977), Palaeointensity determination by a modified Thellier method. *Phys. Earth Planet. Inter.*, **13**, 305–14.

Kovacheva, M. (1977), Archaeomagnetic investigations in Bulgaria: field intensity variations. *Phys. Earth Plan. Interiors*, **13**, 355–9.

Krause, B. R. (1963), Device for measuring the change in remanent magnetic moments of rocks with temperature. *Can. J. Phys.*, **41**, 750–7.

Kroon, D. J. (1968), *Laboratory Magnets*, Cleaver-Hume Press, London.

Krumbein, W. C. and Pettijohn, F. T. (1938), *Manual of Sedimentary Petrology*, Appleton-Century-Crofts, New York.

Kumagai, N. and Kawai, N. (1953), A resonance-type magnetometer. *Mem. Coll. Sci., Kyoto Imp. Univ.*, **20**, 306–9.

Kume, S. (1962), Changements d'aimantation remanente de corps ferrimagnetiques soumis a des pressions hydrostatiques. *Ann. Géophys.*, **18**, 18–22.

Kuster, G. (1969), Effect of drilling on rock magnetization. *Trans. Am. Geophys. Union* (abstract), **50**, 134.

Lam, Y. W. (1965), An improved grinding and polishing apparatus for ferrimagnetic spheres. *J. Sci. Instrum.*, **42**, 761–2.

Larochelle, A. and Black, R. F. (1965), The design and testing of an alternating field demagnetizing apparatus. *Can. J. Earth Sci.*, **2**, 684–96.

Larochelle, A. and Christie, K. W. (1967), An automatic 3-magnet or biastatic magnetometer. *Geol. Surv. Can. Pap.* 67-28, 1–28.

Larochelle, A. and Pearce, G. W. (1969), A possible source of error in determining the remanent magnetization of cylindrical rock specimens with a biastatic magnetometer. *Geol. Surv. Can. Pap.* 68-62, 1–13.

Larson, E. E. (1978), Degradation of lunar basalts during thermal heating in vacuum and its relation to paleointensity measurements. *Lunar Planet. Sci.* IX (abstracts), **1**, 633.

Larson, E. E. (1981), Selective destructive demagnetization–another microanalytical technique in rock magnetism. *Geology*, **9**, 350–5.

Larson, E. E., Hoblitt, R. P. and Watson, D. E. (1975), Gas-mixing techniques in thermomagnetic analysis. *Geophys. J. R. Astron. Soc.*, **43**, 607–20.

Larson, E. D. and Walker, T. R. (1975), Development of chemical remanent magnetization during early stages of red-bed formation in Late Cenozoic sediments, Baja, California. *Geol. Soc. Am. Bull.*, **86**, 639–50.

Lawley, E. (1970), The intensity of the geomagnetic field in Iceland during Neogene polarity transitions and systematic deviations. *Earth Planet. Sci. Lett.*, **10**, 145–9.

Leith, C. J. (1950), Removal of iron oxide coating from mineral grains. *J. Sediment. Petrol.*, **20**, 174–6.

Levi, S. (1975), Comparison of two methods of performing the Thellier experiment. *J. Geomagn. Geoelectr.*, **27**, 245–55.

Levi, S. (1977), The effect of magnetite particle size on palaeointensity determinations of the geomagnetic field. *Phys. Earth Planet. Inter.*, **13**, 245–59.

Levi, S. and Banerjee, S. K. (1976), On the possibility of obtaining relative palaeointensities from lake sediments. *Earth Planet. Sci. Lett.*, **29**, 219–26.

Lewis, M. (1968), Some experiments on synthetic titanomagnetites. *Geophys. J. R. Astron. Soc.*, **16**, 295–310.

Likhite, S. D. and Radhakrishnamurty, C. (1965), An apparatus for the determination of susceptibility of rocks in low fields at different frequencies. *Bull. Nat. Geophys. Res. Inst. (India)*, **3**, 1–5.

Likhite, S. D. and Radhakrishnamurty, C. (1966), Initial susceptibility and constricted Rayleight loops of some basalts. *Curr. Sci.*, **35**, 534–6.

Likhite, S. D., Rahdhakrishnamurty, C. and Sahasrabudhe, P. W. (1965), Alternating current electromagnet type hysteresis loop tracer for minerals and rocks. *Rev. Sci. Instrum.*, **36**, 1558–64.

Livingstone, D. A. (1955), A lightweight piston sampler for lake deposits. *Ecology*, **36**, 137–9.

Lovlie, R. (1974), Post-depositional remanent magnetization in a redeposited deep sea sediment. *Earth Planet. Sci. Lett.*, **21**, 315–20.

Lovlie, R. (1976), The intensity pattern of post-depositional remanence acquired by some marine sediments deposited during a reversal of the external magnetic field. *Earth Planet. Sci. Lett.*, **30**, 209–14.

Lovlie, R. (1978), Remagnetization in Tertiary rocks from the Faeroe Islands. *Phys. Earth Planet. Inter.*, **16**, 59–64.

Lovlie, R., Lowrie, W. and Jacobs, M. (1972), Magnetic properties and mineralogy of four deep-sea cores. *Earth Planet. Sci. Lett.*, **15**, 157–68.

Lowrie, W. and Alvarez, W. (1977), Late Cretaceous geomagnetic polarity sequence: detailed work and palaeomagnetic studies of the Scaglïa Rossa limestone at Gubbio, Italy. *Geophys. J. Roy. Astr. Soc.*, **51**, 561–81.

Lowrie, W., Channell, J. E. T. and Heller, F. (1980), On the credibility of remanent magnetization measurements. *Geophys. J. R. Astron. Soc.*, **60**, 493–6.

Lowrie, W. and Fuller, M. D. (1971), On the alternating field demagnetization characteristics of multidomain thermoremanence in magnetite. *J. Geophys. Res.*, **76**, 6339–49.

Lyddane, R. H. and Ruark, A. E. (1939), Coils for the production of a uniform magnetic field. *Rev. Sci. Instrum.*, **10**, 253–7.

MacKereth, F. J. H. (1958), A portable core sampler for lake sediments. *Limnol. Oceanogr.*, **3**, 181–9.

MacKereth, F. J. H. (1969), A short core sampler for subaqueous deposits. *Limnol. Oceanogr.*, **14**, 145–51.

MacKereth, F. J. H. (1971), On the variation in direction of the horizontal component of remanent magnetization in lake sediments. *Earth Planet. Sci. Lett.*, **12**, 332–8.

Mager, A. J. (1968), Magnetic shielding efficiencies of cylindrical shells with axis parallel to the field. *J. Appl. Phys.*, **39**, 1914.

Mager, A. J. (1970), Magnetic shields. *IEEE Trans. Magn.*, **16**, 67–75.

Mallinson, J. (1966), Magnetometer coils and reciprocity. *J. Appl. Phys.*, **37**, 2514–15.

Manson, A. J. (1971), Rotational hysteresis measurements on oxidized synthetic and natural titanomagnetites. *Z. Geophys.*, **37**, 431–42.

Manson, A. J., O'Donovan, J. B. and O'Reilly, W. (1979), Magnetic rotational hystersis loss in titanomagnetites and titanomaghemites–application to non-destructive mineral identification in basalts. *J. Geophys.*, **46**, 185–99.

Markert, H. and Heller, F. (1972), Determination of palaeointensities of the geomagnetic field from anhysteretic magnetization methods. *Phys. Status Solidi*, **14**, 47–50.

Markert, H., Trissl, K. H. and Zimmerman, G. J. (1974), On a high frequency method for the measurement of susceptibilities and hysteresis losses in rocks and minerals between nitrogen temperature and 700° C. *Z. Geophys.*, **40**, 303–28.

Martelli, G. and Newton, G. (1977), Hypervelocity cratering and impact magnetization of basalts. *Nature (London)*, **270**, 478–80.

Martin, R. J., Habermann, R. E. and Wyss, M. (1978), The effect of stress cycling and inelastic volumetric strain on remanent magnetization. *J. Geophys. Res.*, **83**, 3485–96.

Matsuda, J., Hyodo, M., Inokuchi, H., Isezaki, N. and Yaskawa, K. (1981), A new alternating field demagnetizer for paleomagnetic studies without rotation of specimen. *J. Geomagn. Geoelectr.*, **33**, 161–70.

Mauritsch, H. J. and Turner, P. (1975), The identification of magnetite in limestones using the low temperature transition. *Earth Planet. Sci. Lett.*, **24**, 414–18.

Maxwell, J. A. (1968), *Rock and Mineral Analysis*, Interscience, New York.

McElhinny, M. W. (1964), Statistical significance of the fold test in palaeomagnetism. *Geophys. J. R. Astron. Soc.*, **8**, 338–40.

McElhinny, M. W. (1966), An improved method for demagnetizing rocks in alternating fields. *Geophys. J. R. Astron. Soc.*, **10**, 369–74.

McElhinny, M. W. (1973), *Palaeomagnetism and Plate Tectonics*, Cambridge University Press.

McElhinny, M. W. and Evans, M. E. (1968), An investigation of the strength of the geomagnetic field on the early Precambrian. *Phys. Earth Planet. Interiors*, **1**, 485–97.

McElhinny, M. W., Luck, G. R. and Edwards, D. (1971), A large volume magnetic field-free space for thermal demagnetization and other experiments in palaeomagnetism. *Pure Appl. Geophys.*, **90**, 126–30.

McElhinny, M. W. and Néel, T. J. (1967), Portable field sampling equipment, in *Methods in Palaeomagnetism* (eds D. W. Collinson, K. M. Creer and S. K. Runcorn), Elsevier, Amsterdam.

McFadden, P. L. (1980), The best estimate of Fisher's precision parameter, κ. *Geophys. J. R. Astron. Soc.*, **60**, 397–407.

McFadden, P. L. (1981), A theoretical investigation of the effect of individual grain anisotropy in alternating field demagnetizations. *Geophys. J. R. Astron. Soc.*, **67**, 35–52.

McFadden, P. L. and Jones, D. L. (1981), The fold test in palaeomagnetism. *Geophys. J. R. Astron. Soc.*, **67**, 53–8.

McFadden, P. L. and Lowes, F. J. (1981), The discrimination of mean directions drawn from Fisher distributions. *Geophys. J. R. Astron. Soc.*, **67**, 19–33.

McKeehan, L. W. (1934), Pendulum magnetometer for crystal ferromagnetism. *Rev. Sci. Instrum.*, **5**, 265–9.

McKeehan, L. W. (1936), Combinations of circular currents for producing uniform magnetic field gradients. *Rev. Sci. Instrum.*, **7**, 178–9.

McKeehan, L. W. and Elmore, W. C. (1934), Surface magnetization in ferromagnetic crystals. *Phys. Rev.*, **46**, 226–32.

von Meitzner, W. (1965), Zur Anwerrdung der Förster-sonde bei gesteinsmognetischen Arbeiten. *Z. Geophys.*, **31**, 332–44.

Melloni, M. (1853), *see* Chevallier (1925).

Mercanton, P. L. (1926), Inversion de l'inclination magnétique terrestre aux âges geologique. *Terr. Magn. Atmos. Electr.*, **31**, 187–90.

Mercanton, P. L. (1932), Inversion inclinaison magnétique aux âges geologique. *C. R. Acad. Sci. Paris*, **194**, 1371.

Merrill, R. T. (1970), Low temperature treatment of magnetite and magnetite-bearing rocks. *J. Geophys. Res.*, **75**, 3343–9.

Merrill, R. T. and Kawai, N. (1969), A method for detecting self-reversals using etching. *J. Geomagn. Geoelectr.*, **21**, 507–12.

Michel, A. and Chaudren, G. (1935), Etude du sesqui-oxide de fer cubique stabilisé. *C. R. Acad. Sci. Paris*, **201**, 1191–3.

Molyneux, L. (1971), A complete result magnetometer for measuring the remanent magnetization of rocks. *Geophys. J. R. Astron. Soc.*, **24**, 429–33.

Molyneux, L. and Thompson, R. (1973), Rapid measurement of the magnetic susceptibility of long cores of sediments. *Geophys. J. R. Astron. Soc.*, **32**, 479–81.

Molyneux, L., Thompson, R., Oldfield, F. and McCallan, M. E. (1972), Rapid measurement of the remanent magnetization of long cores of sediment. *Nature*, **237**, 42–3.

Momose, K. (1963), Studies on the variation of the Earth's magnetic field during Pliocene time. *Bull. Earthquake Res. Inst. Tokyo*, **41**, 487–534.

Montgomery, D. B. (1963), The generation of high magnetic fields. *Rep. Prog. Phys.*, **26**, 69–104.

Morin, F. J. (1950), Magnetic susceptibility of α-Fe_2O_3 and α-Fe_2O_3 with added titanium. *Phys. Rev.*, **78**, 819–20.

Morris, P. (1970), An air-suspended tumbler for the a.c. demagnetization of rock samples. *J. Phys. E*, **3**, 819–21.

Morris, P. (1971), Magnetic measurements at low temperatures using a fluxgate magnetometer. *J. Phys. E*, **4**, 920–1.

Morris, W. A. and Carmichael, C. M. (1978), Paleomagnetism of some late Precambrian and lower Paleozoic sediments from L'Adran de Mauritanie, West Africa. *Can. J. Earth. Sci.*, **15**, 253–62.

Moskowitz, B. M. (1981), Methods for estimating Curie temperatures of titanomagnemites from experimental J_s–T data. *Earth Planet. Sci. Lett.*, **53**, 84–8.

Muirhead, F. R. (1962), A torsion balance for individual magnetic particles. *J. Sci. Instrum.*, **39**, 633–5.

Murthy, G. S. (1971), Paleomagnetism of diabase dikes from the Grenville province. *Can. J. Earth Sci.*, **8**, 802–12.

Nagata, T. (1979), Meteorite magnetism and the early solar system magnetic field. *Phys. Earth Planet. Inter.*, **20**, 324–41.

Nagata, T. (1961), *Rock Magnetism*, Maruzen, Tokyo.

Nagata, T. (1967a), Identification of magnetic minerals in rock using methods based on their magnetic properties, in *Methods in Palaeomagnetism* (eds D. W. Collinson, K. M. Creer and S. K. Runcorn), Elsevier, Amsterdam.

Nagata, T. (1967b), Principles of the ballistic magnetometer for the measurement of remanence. In *Methods in Palaeomagnetism* (ed D. W. Collinson, K. M. Creer and S. K. Runcorn), Elsevier, Amsterdam.

Nagata, T., Arai, Y. and Momose, K. (1963), Secular variation of the total geomagnetic force during the last 5000 years. *J. Geophys. Res.*, **68**, 5277–81.

Nagata, T., Ishikawa, Y., Kinoshita, H., Kono, N., Syono, Y. and Fisher, R. M. (1970), Magnetic properties and natural remanent magnetization of lunar materials. *Proceedings of the 1st Lunar Science Conference*, **3**, 2325–40.

Nagata, T. and Kinoshita, H. (1965), Magnetization of titaniferous magnetite under uniaxial compression. *J. Geomagn. Geoelectr.*, **17**, 135.

Nagata, T. and Sugiura, N. (1977), Palaeomagnetic field intensity derived from meteorite magnetization. *Phys. Earth Planet. Inter.*, **13**, 323–9.

Nagata, T., Uyeda, S. and Akimoto, S. (1952), Self-reversal of thermoremanent magnetism of igneous rocks. *J. Geomagn. Geoelectr.*, **4**, 22–38.

Néel, L. (1949), Théorie du trainage magnétique des ferromagnétiques en grain fins avec application aux terres caites. *Ann. Geophys.*, **5**, 99–136.

Néel, L. (1955), Some theoretical aspects of rock magnetism. *Adv. Phys.*, **4**, 191–243.

Nesbitt, J. D. (1966), Variation of the ratio of intensity to susceptibility in red sandstones. *Nature* (*London*), **210**, 618.

Nicholls, G. D. (1955), The mineralogy of rock magnetism. *Adv. Phys.*, **4**, 113–90.

Nishida, J. and Sasajima, S. (1974), Examination of self-reversal due to N-type magnetization in basalt. *Geophys. J. R. Astron. Soc.*, **37**, 453–60.

Noël, M. and Molyneux, L. (1975), Rapid demagnetization of palaeomagnetic samples using a rotating magnetic field. *Geophys. J. R. Astron. Soc.*, **43**, 1017–21.

Noltimier, H. C. (1964), Calibration of a spinner magnetomer with a wire loop. *J. Sci. Instrum.*, **41**, 55.

Noltimier, H. C. (1971), Determining magnetic anisotropy of rocks with a spinner magnetometer giving in-phase and quadrature data output. *J. Geophys. Res.*, **76**, 4849–54.

Noltimier, H. C. (1972), Numerical evolution of the scatter of principal magnetic susceptibilities from the uncertainties in experimental measurements. *Tellus*, **24**, 65–71.

Nye, J. F. (1957), *Physical Properties of Crystals*, Oxford University Press, London.

O'Donovan, J. B. (1975), Studies of synthetic analogues of some carriers of the palaeomagnetic record. PhD Thesis, University of Newcastle upon Tyne.

O'Donovan, J. B. and O'Reilly, W. (1977a), The preparation, characterization and magnetic properties of synthetic analogues of some carriers of the palaeomagnetic record. *J. Geomagn. Geoelectr.*, **29**, 331–44.

O'Donovan, J. B. and O'Reilly, W. (1977b), Monodomain behaviour in multiphase oxidized titanomagnetite. *Earth Planet. Sci. Lett.*, **34**, 396–402.

O'Donovan, J. B. and O'Reilly, W. (1978), Cation distribution in synthetic titanomaghemites. *Phys. Earth Planet. Inter.*, **16**, 200–8.

Ohnaka, M. and Kinoshita, H. (1968), Effect of uniaxial compression on remanent magnetization. *J. Geomagn. Geoelectr.*, **20**, 93–9.

Olsen, J. L. (1964), Strong magnetic fields. *Contemp. Phys.*, **5**, 161–79.

Onstott, T. C. (1980), Application of the Bingham distribution function on paleomagnetic studies. *J. Geophys. Res.*, **85**, 1500–10.

van Oorschot, B. P. J. (1976), The design of pick-up coil detectors for spinner magnetometers. *Geophys. J. R. Astron. Soc.*, **45**, 557–68.

van Oorschot, B. P. J. and Ridler, P. F. (1976), A sensitive spinner magnetometer using a coil detector. *Geophys. J. R. Astron. Soc.*, **45**, 569–82.

Opdyke, N. D. (1961), The palaeomagnetism of the New Jersey Triassic: a field of study of the inclination error in red sediments. *J. Geophys. Res.*, **66**, 1941–9.

Opdyke, N. D. (1967), A large sampling drill. In *Methods in Palaeomagnetism*, (eds D. W. Collinson, K. M. Creer and S. K. Runcorn), Elsevier, Amsterdam.

Opdyke, N. D. (1972), Palaeomagnetism of deep-sea cores. *Rev. Geophys. Space Phys.*, **10**, 213–50.

Opdyke, N. D., Kent, D. V. and Lowrie, W. (1973), Details of magnetic polarity transitions recorded in a high deposition rate deep-sea core. *Earth Planet. Sci. Lett.*, **20**, 315–24.

O'Reilly, W. (1976), Magnetic minerals in the crust of the Earth. *Rep. Prog. Phys.*, **39**, 857–908.

Osborn, J. A. (1945), Demagnetizing factors of the general ellipsoid. *Phys. Rev.*, **67**, 351–7.

Oshima, S., Watanabe T. and Fukui, T. (1971), High sensitivity plated-wire sensor using second harmonic oscillation and its applications. *IEEE Trans. Magn.*, **7**, 436–7.

Otofuji, Y. and Sasajima, S. (1981), A magnetization process in sediments: laboratory experiments on post-depositional remanent magnetization. *Geophys. J. R. Astron. Soc.*, **66**, 241–60.

Özdemir, Ö. and O'Reilly, W. (1981), Laboratory synthesis of aluminium-substituted titanomaghemites and their characteristic properties. *J. Geophys.*, **49**, 93–100.

Ozima, M. and Kinoshita, H. (1964), Magnetic anisotropy of andesites in a fault zone. *J. Geomagn. Geoelectr.*, **16**, 194–200.

Ozima, M., Ozima, M. and Nagata, T. (1963), Low temperature characteristics of remanent magnetization of magnetite: inverse type of thermoremanent magnetization. *Geofis. Pura Appl.*, **55**, 77–90.

Ozima, M., Ozima, M. and Nagata, T. (1964), Low temperature treatment as an effective means of 'magnetic cleaning' of natural magnetization. *J. Geomagn. Geoelectr.*, **16**, 37–40.

Ozima, M. and Sakamoto, N. (1971), Magnetic properties of synthesized titanomaghemite. *J. Geophys. Res.*, **76**, 7035–46.

Park, J. K. (1970), Acid leaching of redbeds and its application to the relative stability of the red and black components. *Can. J. Earth Sci.*, **7**, 1086–92.

Parker, R. L. and Denham, C. R. (1979), Interpolation of unit vectors. *Geophys. J. R. Astron. Soc.*, **58**, 685–8.

Parkhomenko, E. I. (1967), *Electrical Properties of Rocks*, Plenum Press, New York.

Parkinson, D. H. and Mulhall, B. E. (1967), *The Generation of High Magnetic Fields*, Heywood, London.

Parry, J. H. (1957), The magnetic properties of the ore minerals. PhD Thesis, University of Newcastle upon Tyne.

Parry, J. H. (1967), Helmholtz coils and coil design, in *Methods in Palaeomagnetism* (eds D. W. Collinson, K. M. Creer and S. K. Runcorn), Elsevier, Amsterdam.

Parry, L. G. (1965), Magnetic properties of dispersed magnetite powders. *Philos. Mag.*, **11**, 303–12.

Patton, B. J. (1967), Magnetic shielding in *Methods in Palaeomagnetism* (eds D. W. Collinson, K. M. Creer and S. K. Runcorn), Elsevier, Amsterdam.

Patton, B. J. and Fitch, J. L. (1962), Design of a room-size magnetic shield. *J. Geophys. Res.*, **67**, 1117–21.

Payne, M. A. (1981), SI and Gaussian units, conversions and equations for use in geomagnetism. *Phys. Earth Planet. Inter.*, **26**, 10–16.

Pearce, G. W., Gose, W. A. and Strangway, D. W. (1973), Magnetic studies on Apollo 15 and 16 lunar samples. *Proceedings of the 4th Lunar Science Conference*, **3**, 45–76.

Pearce, G. W., Hoye, G. S., Strangway, D. W., Walker, B. M. and Taylor, L. A. (1976), Some complexities in the determination of lunar palaeointensities. *Proceedings of the 7th Lunar Science Conference*, **3**, 3271–97.

Penoyer, R. F. (1959), Automatic torque balance for magnetic anisotropy measurements. *Rev. Sci. Instrum.*, **30**, 711–14.

Petersen, N. (1967), A high frequency method for the measurement of Curie temperatures of ferrimagnetic minerals, in *Methods in Palaeomagnetism* (eds D. W. Collinson, K. M. Creer and S. K. Runcorn), Elsevier, Amsterdam.

Petherbridge, J., de Sa, A. and Creer, K. M. (1972), An astatic magnetometer with electronic feedback for the measurement of remanence of heated rock samples. *J. Phys. E*, **5**, 579–81.

Petley, B. W. (1980), The ubiquitous SQUID. *Contemp. Phys.*, **21**, 607–30.

Petrova, G. N. (1961), Various laboratory methods of determining the geomagnetic stability of rocks. *Bull. Acad. Sci. (USSR), Geophys. Ser.*, 703–9.

Philips, J. D. and Kuckes, A. F. (1967), A spinner magnetometer *J. Geophys. Res.*, **72**, 2209–12.

Pickles, A. T. and Sucksmith, W. (1940), A magnetic study of the two-phase iron-nickel alloys. *Proc. R. Soc. London*, **A175**, 331–44.

Porath, H. (1968), Magnetic studies on specimens of intergrown maghemite and haematite. *J. Geophys. Res.*, **73**, 5959–65.

Porath, H., Stacey, F. D. and Cheam, A. S. (1966), The choice of specimen shape for magnetic anisotropy measurements on rocks. *Earth Planet. Sci. Lett.*, **1**, 92.

Pozzi, J. P. (1967), Recent improvements carried out on astatic magnetometers, in *Methods in Palaeomagnetism* (eds D. W. Collinson, K. M. Creer and S. K. Runcorn), Elsevier, Amsterdam.

Pozzi, J. P. (1977), Effects of stresses on magnetic properties of volcanic rocks. *Phys. Earth Planet. Inter.*, **14**, 77–85.

Pozzi, J. P., Godefroy, A. and Legoff, M. (1970), Sur un despositif de mesure d'aimantations rémanentes et inductes de corps très faiblement magnétiques sous pression. *C. R. Acad. Sci. Paris*, **271**, 188–9.

Pozzi, J. P. and Thellier, E. (1963), Sur des perfectionments récents apportés aux magnétomètre a très haute sensibilité utilisés en minéralogie magnétique et an paléomagnétisme. *C. R. Acad. Sci. Paris*, **257**, 1037–40.

Prevot, M. Lecaille, A. and Mankinen, E. A. (1981), Magnetic effects of maghemitization of oceanic crust. *J. Geophys. Res.*, **86**, 4009–20.

Pullaiah, G., Irving, E., Buchan, K. L. and Dunlop, D. J. (1975), Magnetization change caused by burial and uplift. *Earth Planet. Sci. Lett.*, **28**, 133–43.

Purucker, M. E. (1974), Magnetic record of lightning strikes in sandstone. *Trans. Am. Geophys. Union* (abstract), **55**, 1112.

Purucker, M. E., Elston, D. P. and Shoemaker, E. M. (1980), Early acquisition of characteristic magnetization in red beds of the Moenkopi Formation (Triassic). *J. Geophys. Res.*, **85**, 997–1001.

Puschner, H. (1966), *Heating with Microwaves*, Springer, New York.

Radhakrishnamurty, C. and Sahasrabudhe, P. W. (1965), Instruments and techniques for the study of magnetic stability in rocks. *J. Ind. Geophys. Union*, **2**, 5–15.

Radhakrishnamurty, C. and Deutsch, E. R. (1974), Magnetic techniques for ascertaining the nature of iron oxide grains in basalts. *J. Geophys.*, **40**, 453–65.

Radhakrishnamurty, C. and Likhite, S. D. (1970), Hopkinson effect blocking temperature and Curie point in basalts. *Earth Planet. Sci. Lett.*, **7**, 389–96.

Radhakrishnamurty, C. and Sahasrabudhe, P. W. (1967) A new type of magnetic memory phenomenon in rocks. *Curr. Sci.*, **36**, 251–5.

Rainbow, R. (1968), Depositional remanent magnetization of synthetic sediments. M.Sc. Thesis, University of Newcastle upon Tyne.

Ramdohr, P. (1969), *The Ore Minerals and their Intergrowths*, Pergamon Press, Oxford.

Rathore, J. S. (1975), Studies of magnetic susceptibility anisotropy in rocks. Ph.D. Thesis, University of Newcastle upon Tyne.

Rathore, J. S. (1979), Magnetic susceptibility anisotropy in the Cambrian slate belt of North Wales and correlation with strain. *Tectonophysics*, **53**, 83–97.

Rathore, J. S. (1980), A study of secondary fabrics in rocks from the Lizard peninsula and adjacent areas in South West Cornwall, England. *Tectonophysics*, **68**, 147–60.

Readman, P. W. and O'Reilly, W. (1972), Magnetic properties of oxidized (cation deficient) titanomagnetities $(FeTi)_3O_4$. *J. Geomagn. Geoelectr.*, **24**, 69–90.

Rees, A. I. (1961), The effects of water currents in the magnetic remanence and anisotropy of susceptibility of some sediments. *Geophys. J. R. Astron. Soc.*, **6**, 235–51.

Rees, A. I. (1965), The use of anisotropy of susceptibility in the estimation of sedimentary fabric. *Sedimentology*, **4**, 257–71.

Rees, A. I. (1966), The effect of depositional slopes on the anisotropy of magnetic susceptibility of laboratory deposited sands. *J. Geol.*, **74**, 856–67.

Reeve, S. C. (1975), Observations on the chemical demagnetization of red beds. *Geology*, **3**, 90.

Reval, J., Day, R. and Fuller, M. D. (1977), Magnetic behaviour of magnetite and rocks stressed to failure in relation to earthquake prediction. *Earth Planet. Sci. Lett.*, **37**, 296–306.

Richards, J. G. W., O'Donovan, J. B., Hauptman, Z., O'Reilly, W. and Creer, K. M. (1973), A magnetic study of titanomagnetites substituted by magnesium and aluminium. *Phys. Earth Planet. Inter.*, **7**, 437–40.

Rigotti, P. A. (1978), The A.R.M. correction method of palaeointensity determination. *Earth Planet. Sci. Lett.*, **39**, 417–26.

Rimbert, F. (1956), Sur l'action de champs alternatifs sur des roches portant une aimantation remanente isotherme de viscosité. *C. R. Acad. Sci. Paris*, **242**, 2536–8.

Rimbert, F. (1959), Contribution à l'étude de l'action de champs alternatifs sur les aimantations rémanentes des roches. *Rev. Inst. Fr. Pét.*, **14**, 17–54, 123–55.

Robertson, D. J. (1978), Absolute calibration of a spinning magnetometer. *J. Phys. E*, **11**, 393–4.

Roy, J. L. (1963). The measurement of the magnetic properties of rock specimens. *Publ. Dom. Obs., Ottawa*, **27**, 420–39.

Roy, J. L. and Lapointe, P. L. (1978). Multiphase magnetizations: problems and implications. *Phys. Earth Planet. Inter.*, **16**, 20–37.

Roy, J. L. and Park, J. K. (1972), Red beds: DRM or CRM? *Earth Planet. Sci. Letts.*, **17**, 211–6.

Roy, J. L. and Park, J. K. (1974), The magnetization process of certain redbeds: vector analysis of chemical and thermal results. *Can. J. Earth Sci.*, **11**, 437–71.

Roy, J. L., Reynolds, J. and Sanders, E. (1972), An astatic magnetometer with negative feedback. Earth Physics Branch, Dept of Energy, Mines and Resources, Ottawa, publication 42, 167–82.

Roy, J. L., Reynolds, J. and Sanders, E. (1973), An alternating field demagnetizer for rock magnetism studies. Earth Physics Branch, Dept of Energy, Mines and Resources, Ottawa, Publication 44, 37–45.

Roy, J. L., Robertson, W. A. and Keeping, C. (1969), Magnetic 'field free' spaces for palaeomagnetism, rock magnetism and other studies. *Can. J. Earth Sci.*, **6**, 1312–16.

Roy, J. L., Sanders, E. and Reynolds, J. (1972), Un four électrique pour l'étude des proprietés magnétiques des roches. Earth Physics Branch, Dept of Energy, Mines and Resources, Ottawa, Publication 42, 229–37.

Rubens, S. M. (1945), Cube-surface coil for producing a uniform magnetic field. *Rev. Sci. Instrum.*, **16**, 243–5.

Rücker, A. W. (1894), On the magnetic shielding of concentric spherical shells. *Philos. Mag.*, **37**, 95–130.

Rudd, M. E. and Craig, J. R. (1968), Optimum spacing of square and circular coil pairs. *Rev. Sci. Instrum.*, **39**, 1372.

Runcorn, S. K. (1960), Statistical methods in rock magnetism. *Philos. Mag.*, **5**, 523–4.

Runcorn, S. K. (1967a), The anisotropy of magnetization of rocks, in *Methods in Palaeomagnetism* (eds D. W. Collinson, K. M. Creer and S. K. Runcorn), Elsevier, Amsterdam.

Runcorn, S. K. (1967b), Statistical discussion of magnetization of rock samples, in *Methods in Palaeomagnetism* (eds D. W. Collinson, K. M. Creer and S. K. Runcorn), Elsevier, Amsterdam.

Runcorn, S. K., Collinson, D. W., O'Reilly, W., Bettey, M. H., Stephenson, A., Jones, J. M., Manson, A. J. and Readman, P. W. (1970), Magnetic properties of Apollo 11 lunar samples. *Proceedings of the Apollo 11 Lunar Science Conference*, **3**, 2369–87.

Runcorn, S. K., Collinson, D. W., O'Reilly, W., Stephenson, A., Battey, M. H., Manson, A. J. and Readman, P. W. (1971), Magnetic properties of Apollo 12 lunar samples. *Proc. R. Soc. London*, **A325**, 157–74.

Runcorn, S. K., Collinson, D. W. and Stephenson, A. (1981), Palaeomagnetism and planetology. *Phys. Earth Planet. Inter.*, **24**, 205–17.

Ryall, P. J. C. and Ade-Hall, J. M. (1975), Radical variation of magnetic properties in submarine pillow basalts. *Can. J. Earth Sci.*, **12**, 1959–69.

Rybak, R. S. (1971), Diagnostics of maghemite by the thermomagnetic method. *Phys. Solid Earth* (*USSR*), 289–91.

de Sa, A. (1965), A digital method of measurement of amplitude and phase at low frequencies. *J. Sci. Instrum.*, **42**, 265–7.

de Sa, A. (1966), Digital integration and recording of low frequency signals. *J. Sci. Instrum.*, **43**, 614.

de Sa, A. (1968), A radio frequency method for determining Curie point temperatures. *J. Phys. E*, **1**, 1136–7.

de Sa, A. (1980), Amplitude control for electro-mechanical vibrators. *New Electron.*, **13**, 44.

de Sa, A. (1981), An automatic microbalance for recording variations in magnetization. *J. Phys. E*, **14**, 923–4.

de Sa, A. and Molyneux, L. (1963), A spinner magnetometer. *J. Sci. Instrum.*, **40**, 162–5.

de Sa, A. and Widdowson, J. W. (1974), An astatic magnetometer with negative feedback. *J. Phys. E*, **7**, 266–8.

de Sa, A. and Widdowson, J. W. (1975), A digitally controlled a.f. demagnetizer for peak fields up to 0.1 tesla. *J. Phys. E*, **8**, 302–4.

de Sa, A., Widdowson, J. W. and Collinson, D. W. (1974), The signal to noise ratio of astatic magnetometers with negative feedback. *J. Phys. E*, **7**, 1015–19.

Sakamoto, N., Ince, P. and O'Reilly, W. (1968), Effect of wet grinding on the oxidation of titanomagnetites. *Geophys. J. R. Astron. Soc.*, **15**, 509–16.

Sallomy, J. T. and Briden, J. C. (1975), Palaeomagnetic studies of Lower Jurassic rocks in England and Wales. *Earth Planet. Sci. Lett.*, **24**, 369–76.

Sampson, W. B. (1968), Superconducting magnets. *IEEE Trans. Magn.*, **4**, 99–107.

Schmidbauer, E. and Petersen, N. (1968), Some magnetic properties of two basalts under uniaxial compression measured at different temperatures. *J. Geomagn. Geoelectr.*, **20**, 169–80.

Schmidlin, H. (1937), Über entmagnetisierende Winkung der Änderungen des magnetischen Endfeldes. *Beit. Angew. Geophys.*, **7**, 94–111.

Schmidt, V. A. (1974), On the use of orthogonal transformations in the reduction of palaeomagnetic data. *J. Geomagn. Geoelectr.*, **26**, 475–86.

Schult, A. (1968a), Self-reversal of magnetization and chemical composition of titanomagnetites in basalts. *Earth Planet. Sci. Lett.*, **4**, 57–63.

Schult, A. (1968b), The effect of pressure on the Curie temperatures of magnetite and some other ferrites. *Z. Geophys.*, **34**, 505–11.

Schult, A. (1976), Self-reversal above room temperature due to N-type magnetization in basalt. *J. Geophys.*, **42**, 81–4.

Schwarz, E. J. (1968a), Magnetic phases in natural pyrrhotite, $Fe_{0\cdot 89}S$ and $Fe_{0\cdot 91}S$. *J. Geomagn. Geoelectr.*, **20**, 67–74.

Schwarz, E. J. (1968b), A recording thermomagnetic balance. *Geol. Surv. Can.* Pap. 68–37.

Schwarz, E. J. (1969), A discussion of thermal and alternating field demagnetization methods in the estimation of palaeomagnetic field intensities. *J. Geomagn. Geoelectr.*, **21**, 669–77.

Schwarz, E. J. (1970), Thermomagnetics of lunar dust sample 10084,88. *Proceedings of the Apollo 11 Lunar Science Conference*, **3**, 2389–97.

Schwarz, E. J. (1975), Magnetic properties of pyrrhotite and their use in applied geology and geophysics. *Geol. Surv. Can. Pap.* 74–59.

Schwarz, E. J. and Christie, K. W. (1967), Original remanent magnetization of Ontario potsherds. *J. Geophys. Res.*, **72**, 3263–9.

Schwarz, E. J. and Symons, D. T. A. (1968), On the intensity of the palaeomagnetic field between 100 million and 2500 million years ago. *Phys. Earth Planet. Inter.*, **1**, 122–8.

Schwarz, E. J. and Symons, D. T. A. (1969), Geomagnetic intensity between 100 million and 2500 million years ago. *Phys. Earth Planet. Interiors*, **2**, 11–18.

Schwarz, E. J. and Vaughan, O. J. (1972), Magnetic phase relations of pyrrhotite. *J. Geomagn. Geoelectr.*, **24**, 441–58.

Schwarz, E. J. and Whillans, T. (1973), A ballistic magnetometer for the measurement of rock magnetic properties. Paper 73-19. *Geological Survey of Canada*, 29–34.

Schweizer, F. (1962), Magnetic shielding factors of a system of concentric shells. *J. Appl. Phys.*, **33**, 1001–3.

Scott, G. G. (1957), Compensation of the Earth's magnetic field. *Rev. Sci. Instrum.*, **28**, 270–3.

Scriba, H. and Heller, F. (1978), Measurements of anisotropy of magnetic susceptibility using inductive magnetometers. *J. Geophys.*, **44**, 341–52.

Shapiro, V. A. and Ivanov, N. A. (1967), The stability parameters of dynamic magnetization compared with other types of remanent magnetization. *Phys. Solid Earth (USSR)*, 681–5.

Sharma, P. V. (1965), On the point dipole representation of a uniformly magnetized cylinder. *Helv. Phys. Acta*, **38**, 234–40.

Sharma, P. V. (1966), Rapid computation of magnetic anomalies and demagnetization effects caused by bodies of arbitrary shape. *Pure Appl. Geophys.*, **64**, 89–109.

Sharma, P. V. (1968), Choice of configuration for measurement of magnetic moment of rock specimens with a fluxgate unit. *Geoexploration*, **6**, 101–8.

Shashnakov, V. A. (1970), Method of obtaining demagnetization curves by means of an alternating field. *Phys. Solid Earth (USSR)*, 675–7.

Shashnakov, V. A. and Metallova, V. V. (1977), Determination of palaeointensity from sedimentary and igneous rocks by the method of alternating field remagnetization. *Phys. Earth Planet. Inter.*, **13**, 368–72.

Shaw, J. (1974), A new method of determining the magnitude of the palaeomagnetic field; application to five historic lavas and five archaeological samples. *Geophys. J. R. Astron. Soc.*, **39**, 133–41.

Sherwood, J. and Watt, J. H. (1968), The use of a portable fluxgate magnetometer for the estimation in the field of the direction and stability of remanent magnetization of rock specimens. *Pure Appl. Geophys.*, **70**, 88–93.

Shimada, M., Kume, S. and Koizumi, M. (1968), Demagnetization of unstable remanent magnetization by application of pressure. *Geophys. J. R. Astron. Soc.*, **16**, 369–74.

Shive, P. N. and Butler, R. F. (1969), Stresses and magnetostrictive effects of lamellae in the titanomagnetite and ilmenohaematite series. *J. Geomagn. Geoelectr.*, **21**, 781–96.

Sholpo, L. Y., Afremov, L. M. and Pershin, V. L. (1972), Interpretation of the angular dependence of secondary magnetization and demagnetization of rocks. *Phys. Solid Earth (USSR)*, 415–17.

Sholpo, L. Y. and Shcekin, M. N. (1977), The study of composition and structure of ferromagnetic components from the coercive spectra of rock specimens. *Phys. Solid Earth (USSR)*, 651–8.

Sholpo, L. Y. and Yanoviskii, B. M. (1967), The problem of the palaeomagnetic stability of rocks. *Phys. Solid Earth (USSR)*, 834–5.

Skovorodkin, Yu. P. (1969), The study of lavas as a means for determining the strength of the palaeomagnetic field. *Phys. Solid Earth (USSR)*, 127–30.

Smith, P. J. (1967), The intensity of the ancient geomagnetic field: a review and analysis. *Geophys. J. R. Astron. Soc.*, **12**, 213–62.

Smith, R. L. (1979), A high pressure cell for chemical demagnetization of sediments. *Geophys. J. R. Astron. Soc.*, **59**, 605–8.

Smyth, W. R. (1950), *Static and Dynamic Electricity*, 2nd Edn, McGraw-Hill, New York.

Snape, C. (1967), A 500 c/s alternating field demagnetization apparatus, in *Methods in Palaeomagnetism* (eds D. W. Collinson, K. M. Creer and S. K. Runcorn), Elsevier, Amsterdam.

Snape, C. (1971), An example of anhysteretic moments being induced by alternating field demagnetization apparatus. *Geophys. J. R. Astron. Soc.*, **23**, 361–4.

Soffel, H. (1963), Untersuchungen an einigen ferrimagnetischen Oxyd- und Sulfidmineralien mit der Methode der Bitterschen streifen. *Z. Geophys.*, **29**, 21–34.

Soffel, H. (1965), Magnetic domains of polycrystalline natural magnetite. *Z. Geophys.*, **6**, 345–61.

Soffel, H. C. (1966), Stress dependence of the domain structure of natural magnetite. *Z. Geophys.*, **32**, 63–77.

Soffel, H. (1968), Die Beobachtung von Weiss'schen Bezirken auf einem Titanomagnetitkorn mit einem Durchmesser von 10 Mickron in einem Basalt. *Z. Geophys.*, **34**, 175–81.

Soffel, H. (1971), The single-domain – multidomain transition in natural intermediate titanomagnetites. *Z. Geophys.*, **37**, 451–70.

Soffel, H. (1977), Domain structure of titanomagnetites and its variation with temperature. *Adv. Earth Planet. Sci.*, **1**, 45–52.

Soffel, H. (1978), The reliability of palaeomagnetic data from basalts with low Curie temperatures ($T_c < 90°$ C). *Phys. Earth Planet. Inter.*, **16**, 38–44.

Soffel, H. (1981), Domain structure of natural fine-grained pyrrhotite in a rock matrix (diabase). *Phys. Earth Planet. Inter.*, **26**, 98–106.

Soffel, H. and Petersen, N. (1971), Ionic etching of titanomagnetite grains in basalts. *Earth Planet. Sci. Lett.*, **11**, 312–16.

Springford, M., Stockton, J. R. and Wampler, W. R. (1971), A vibrating sample magnetometer for use with a superconducting magnet. *J. Phys. E*, **4**, 1036–40.

Srnka, L. (1977), Spontaneous magnetic field generation in hypervelocity impacts. *Proceedings of the 8th Lunar Science Conference*, **1**, 785–92.

Stacey, F. D. (1959), Spinner magnetometer for thermal demagnetization experiments on rocks. *J. Sci. Inst.*, **36**, 355–9.

Stacey, F. D. (1960a), Magnetic anisotropy of dispersed powders. *Aust. J. Phys.*, **13**, 196–201.

Stacey, F. D. (1960b), Magnetic anisotropy of igneous rocks. *J. Geophys. Res.*, **65**, 2429–42.

Stacey, F. D. (1961), Theory of the magnetic properties of igneous rocks in alternating fields. *Philos. Mag.*, **6**, 1241–60.

Stacey, F. D. (1962), A generalized theory of thermo-remanence, covering the transition from single domain to multidomain grains. *Philos. Mag.*, **7**, 1887–900.

Stacey, F. D. (1963), The physical theory of rock magnetism. *Adv. Phys.*, **12**, 45–133.

Stacey, F. D. (1964), The seismomagnetic effect. *Pure Appl. Geophys.*, **58**, 5–23.

Stacey, F. D. (1967), The Königsberger ratio and the nature of thermoremanence in igneous rocks. *Earth Planet. Sci. Lett.*, **2**, 67–8.

Stacey, F. D. and Banerjee, S. K. (1974), *The Physical Principles of Rock Magnetism*, Elsevier, Amsterdam.

Stacey, F. D., Barr, K. G. and Robson, G. R. (1965), The volcanomagnetic effect. *Pure Appl. Geophys.*, **62**, 96–104.

Stacey, F. D., Lovering, J. E. and Parry, L. G. (1961), Thermomagnetic properties, natural magnetic moments and magnetic anisotropies of some chondritic meteorites. *J. Geophys. Res.*, **66**, 1523–34.

Stacey, F. D. and Parry, L. G. (1957), A modification of the method of E. R. Deutsch for the magnetic hysteresis of rocks. *J. Geomagn. Geoelectr.*, **9**, 157–61.

Starkey, J. and Palmer, H. C. (1971), The sensitivity of the conglomerate test in paleomagnetism. *Geophys. R. Astron. Soc.*, **22**, 235–40.

Stephenson, A. (1969), The temperature-dependent cation distribution in titanomagnetites. *Geophys. J. R. Astron. Soc.*, **18**, 199–210.

Stephenson, A. (1971a), Single domain grain distributions. I. A method for the determination of single domain grain distributions. *Phys. Earth Planet. Inter.*, **4**, 353–60.

Stephenson, A. (1971b), Single domain grain distributions. II. The distribution of single domain iron grains in Apollo 11 lunar dust. *Phys. Earth Planet. Inter.*, **4**, 361–9.

Stephenson, A. (1972), Spontaneous magnetization curves and Curie points of cation deficient titanomagnetites. *Geophys. J. R. Astron. Soc.*, **29**, 91–107.

Stephenson, R. W. (1976), A study of rotational remanent magnetization. *Geophys. J. R. Astron. Soc.*, **47**, 363–73.

Stephenson, A. (1980a), Rotational remanent magnetization and the torque exerted on a rotating rock in an alternating magnetic field. *Geophys. J. R. Astron. Soc.*, **62**, 113–32.

Stephenson, A. (1980b), Gyromagnetism and the remanence acquired by a rotating rock in an alternating field. *Nature (London)*, **284**, 48–9.

Stephenson, A. (1980c), The measurement of the magnetic torque acting on a rotating sample using an air turbine. *J. Phys. E*, **13**, 311–14.

Stephenson, A. and Collinson, D. W. (1974), Lunar magnetic field palaeointensities determined by an anhysteretic remanent magnetization method. *Earth Planet. Sci. Lett.*, **23**, 220–8.

Stephenson, A. and Collinson, D. W. (1975), A note on the magnetic properties of lunar samples 15535,28 and the implications regarding palaeointensity determinations. *Earth Planet. Sci. Lett.*, **27**, 360–1.

Stephenson, A., Collinson, D. W. and Runcorn, S. K. (1975), On changes in the intensity of the ancient lunar magnetic field. Proceedings of the 6th Lunar Science Conference, **3**, 3049–62.

Stephenson, A., Collinson, D. W. and Runcorn, S. K. (1976), On the intensity of the ancient lunar magnetic field. *Proceedings of the 7th Lunar Science Conference*, **3**, 3373–82.

Stephenson, A., Runcorn, S. K. and Collinson, D. W. (1977), Palaeointensity estimates from lunar samples 10017 and 10020. *Proceedings of the 8th Lunar Science Conference*, **1**, 679–87.

Stephenson, A. and de Sa, A. (1970), A simple method for the measurement of the temperature variation of initial magnetic susceptibility between 77 and 1000 K. *J. Phys. E.*, **3**, 59–61.

Stewart, A. D. and Irving, E. (1974), Palaeomagnetism of Pre-cambrian sedimentary rocks from N.W. Scotland and the apparent polar wandering path of Laurentia. *Geophys. J. R. Astron. Soc.*, **37**, 51–72.

Stober, J. C. and Thompson, R. (1979), Magnetic remanence acquisition in lake sediments. *Geophys. J. R. Astron. Soc.*, **57**, 727–39.

Stone, D. B. (1963), Anisotropic magnetic susceptibility measurements on a phonolite and on a folded metamorphic rock. *Geophys. J. R. Astron. Soc.*, **7**, 375–90.

Stone, D. B. (1967a), An anisotropy meter, in *Methods in Palaeomagnetism* (eds D. W. Collinson, K. M. Creer and S. K. Runcorn), Elsevier, Amsterdam.

Stone, D. B. (1967b), Torsion balance method of measuring anisotropic susceptibility, in *Methods in Palaeomagnetism* (eds D. W. Collinson, K. M. Creer and S. K. Runcorn), Elsevier, Amsterdam.

Stone, D. B. (1967c), A sun compass for the direct determination of geographic North. *J. Sci. Instrum.*, **44**, 661–2.

Stoner, E. C. (1945), The demagnetizing factors for ellipsoids. *Phil. Mag.*, **36**, 803–21.

Stott, P. M. and Stacey, F. D. (1960), Magnetostriction and palaeomagnetism of igneous rocks. *J. Geophys. Res.*, **85**, 2419–24.

Strangway, D. W. (1965), Magnetic anomalies over some Precambrian dikes. *Geophysics*, **30**, 783–96.

Strangway, D. W., Honea, R. M., McMahon, B. E. and Larson, E. D. (1968), The magnetic properties of naturally occurring goethite. *Geophys. J. R. Astron. Soc.*, **15**, 35–59.

Strangway, D. W., Larson, E. E. and Goldstein, M. (1968), A possible cause of high magnetic stability in volcanic rocks. *J. Geophys. Res.*, **73**, 3787–96.

Strangway, D. W. and McMahon, B. M. (1973), Palaeomagnetism of annually banded Eocene Green river sediments. *J. Geophys. Res.*, **78**, 5237–45.

Strangway, D. W., McMahon, B. E., Honea, L. M. and Larson, E. E. (1967), Superparamagnetism in haematite. *Earth Planet. Sci. Lett.*, **2**, 37–71.

Strangway, D. W., McMahon, B. E. and Larson, E. E. (1968), Magnetic paleointensity studies on a recent basalt from Flagstaff, Arizona. *J. Geophys. Res.*, **73**, 7031–8.

Strangway, D. W., Larson, E. E. and Pearce, G. W. (1970), Magnetic studies of lunar samples – breccia and fines. *Proceedings of the Apollo 11 Lunar Science Conference*, **3**, 2435–51.

Street, R. and Woolley, J. C. (1949), A study of magnetic viscosity. *Proc. Phys. Soc. London*, **A62**, 562–72.

Stupavsky, M. and Symons, D. T. A. (1978), Separation of magnetic components from a.f. step demagnetization data by least squares computer methods. *J. Geophys. Res.*, **83**, 492–32.

Sucksmith, W. (1929), An apparatus for the measurement of magnetic susceptibility. *Philos. Mag.*, **8**, 158–65.

Sugiura, N. and Strangway, D. W. (1981), Comparison of magnetic palaeointensity methods using a lunar sample. *Proceedings of the 11th Lunar and Planetary Science Conference*, **3**, 1801–13.

Suguira, N., Wu, Y. M., Strangway, D. W., Pearce, G. W. and Taylor, L. A. (1979), A new magnetic palaeointensity value for a "young lunar glass". *Proceedings of the 10th Lunar and Planetary Science Conference*, **3**, 2189–97.

Symons, D. T. A. and Stupavsky, M. (1974), A rational paleomagnetic stability index. *J. Geophys. Res.*, **79**, 1718–20.

Syono, Y. (1960), Magnetic susceptibility of some rock forming silicate minerals. *J. Geomagn. Geoelectr.*, **11**, 85–93.

Syono, Y. and Ishikawa, Y. (1963), Magnetocrystalline anisotropy of $x Fe_2TiO_4 \cdot (1-x) Fe_3O_4$. *J. Phys. Soc. Jpn*, **18**, 1230–31.

Tanguy, J. C. (1975), Intensity of the geomagnetic field from recent Italian lavas using a new palaeointensity method. *Earth Planet. Sci. Lett.*, **27**, 314–20.

Tarling, D. H. (1971), *Principles and Applications of Palaeomagnetism*. Chapman and Hall, London.

Tarling, D. H. (1975), Archaeomagnetism: the dating of archaeological materials by their magnetic properties. *World Archaeol.*, **7**, 185–97.

Tarling, D. H. (1983), *Palaeomagnetism*, Chapman and Hall, London.

Tarling, D. H. and Symons, D. T. A. (1967), A stability index of remanence in palaeomagnetism. *Geophys. J. R. Astron. Soc.*, **12**, 443–8.

Taylor, L. A. (1979), Palaeointensity determinations at elevated temperatures: sample preparation technique. *Proceedings of the 10th Lunar and Planetary Science Conference*, **3**, 2183–7.

Terman, F. E. (1943), *Radio Engineers' Handbook*, McGraw-Hill, New York.

Thellier, E. (1933), Magnétometre insensible aux champs magnétiques troubles des grandes villes. *C. R. Acad. Sci. Paris*, **197**, 224–34.

Thellier, E. (1936), Determination de la direction de l'aimantation permanente des roches. *C. R. Acad. Sci. Paris*, **203**, 743–4.

Thellier, E. (1937), Aimantation des terres cuites: application à la recherche de l'intensité du champ magnétique terrestre dans le passé. *C. R. Acad. Sci. Paris*, **204**, 184–6.

Thellier, E. (1938), Sur l'aimantation des terres cuites et ses applications geophysiques. *Ann. Inst. Physique du Globe, Univ. Paris*, **16**, 157–302.

Thellier, E. (1941), Sur le vérification d'une method permettant de determiner l'intensité du champ magnétique terrestre dans le passé. *C. R. Acad. Sci. Paris*, **212**, 281–3.

Thellier, E. (1951), Propriétés magnétiques des terre cuites et des roches. *J. Phys. Radium*, **12**, 205–18.

Thellier, E. (1966), Methods of alternating current and thermal demagnetization, in *Methods and Techniques in Geophysics*, Vol. 2 (ed. S. K. Runcorn), Interscience, London.

Thellier, E. (1967), A 'big sample' spinner magnetometer, in *Methods in Palaeomagnetism* (eds D. W. Collinson, K. M. Creer and S. K. Runcorn), Elsevier, Amsterdam.

Thellier, E. and Thellier, O. (1959), Sur l'intensite du champ magnetique terrestre dans le passe historique et geologique. *Ann. Geophys.*, **15**, 285–376.

Thomas, A. K. (1968), Magnetic shielded enclosure design in the d.c. and VLF region. *IEEE Trans. Electromagn. Compat.*, **10**, 142–52.

Tickell, F. G. (1965), *The Techniques of Sedimentary Mineralogy*, Elsevier, Amsterdam.

Trukhin, V. I. (1966), An experimental investigation of magnetic viscosity. *Phys. Solid Earth* (*USSR*), 342–6.

Trukhin, V. I. (1967), The possibility of determining the absolute age of rocks from the amounts of viscous magnetization. *Phys. Solid Earth* (*USSR*), 138–41.

Tucker, P. (1978), Thermoremanence in multidomain single crystal titanomagnetites. Effects of deuteric oxidation. Ph.D. Thesis, University of Newcastle upon Tyne.

Tucker, P. (1980), A grain mobility model of post depositional realignment. *Geophys. J. R. Astron. Soc.*, **63**, 149–63.

Tucker, P. and O'Reilly, W. (1980), The acquisition of thermoremanent magnetization by multidomain single crystal titanomagnetite. *Geophys. J. R. Astron. Soc.*, **60**, 21–36.

Twenhofel, W. H. and Tyler, S. A. (1941), *Methods of Study of Sediments*, McGraw-Hill, New York.

Urquhart, H. M. A. and Goldman, J. E. (1956), Magnetostrictive effects in an antiferromagnetic haematite crystal. *Phys. Rev.*, **101**, 1443–50.

Urrutia-Fucugauchi, J. (1981), Some observations on short-term magnetic viscosity behaviour at room temperature. *Phys. Earth Planet. Inter.*, **26**, No. 3, 1–5.

Uyeda, S. (1957), Thermoremanent magnetism and coercive force of the ilmenite-haematite series. *J. Geomagn. Geoelectr.*, **9**, 61–78.

Uyeda, S., Fuller, M. D., Belshe, J. C. and Girdler, R. W. (1963), Anisotropy of magnetic susceptibility in rocks and minerals. *J. Geophys. Res.*, **68**, 279–91.

Uytenbogaardt, W. and Burke, E. A. J. (1971), *Tables for Microscopic Identification of Ore Minerals*, 2nd Edn, Elsevier, Amsterdam.

Valeyev, K. A. (1975), Remanent magnetization of rocks under conditions of high uniaxial pressure. *Phys. Solid Earth* (*USSR*), **11**, 373–7.

Vallentyne, J. R. (1955), A modification of the Livingstone piston sampler for lake deposits. *Ecology*, **36**, 139–40.

Vant-Hull, L. L. and Mercereau, J. E. (1963), Magnetic shielding by a superconducting cylinder. *Rev. Sci. Instrum.*, **34**, 1238–42.

Verhoogen, J. (1959), The origin of thermoremanent magnetization. *J. Geophys. Res.*, **64**, 2441–9.

Verosub, K. L. (1977a), A poor person's sun compass. *Geology*, **5**, 319.

Verosub, K. L. (1977b), Depositional and post-depositional processes in the magnetization of sediments. *Rev. Geophys. Space Phys.*, **15**, 129–44.

Verwey, E. J. W. and Haayman, P. W. (1941), Electronic conductivity and transition point in magnetite. *Physica*, **8**, 979–82.

Vincenz, S. A. (1965), Frequency dependence of magnetic susceptibility of rocks in weak alternating fields. *J. Geophys. Res.*, **70**, 1371–7.

Vlasov, A. Ya and Bogdanov, A. A. (1964), Domain structure in a single crystal of magnetite. *Bull. Acad. Sci. USSR, Geophys. Ser.*, 231–4.

Vlasov, A. Ya and Kovalenko, G. V. (1964), Magnetic anisotropy of artificial sediments. *Bull. Acad. Sci. USSR, Geophys. Ser.*, 732–7.

Vlasov, A. Ya, Kovalenko, G. V. and Chikhacher, V. A. (1967), The superparamagnetism of α-FeOOH. *Phys. Solid Earth* (*USSR*), 460–4.

Vlasov, A. Ya, Kovalenko, G. V. and Fedoseeva, N. V. (1967), The rotational magnetic hysteresis of single hematite crystals in artificial sediments containing magnetite and hematite particles. *Phys. Solid Earth* (*USSR*), 129–33.

Vlasov, A. Ya., Kovalenko, G. V. and Tropin, Yu.D. (1961), Effect of compression of artificially deposited sediments upon remanent magnetization. *Bull. Acad. Sci. USSR, Geophys. Ser.*, 775–7.

Vlasov, A. Ya., Zvegintsev, A. G. and Pavlov, V. F. (1964), Self-reversal of the magnetization of synthetic sediments. *Bull. Acad. Sci. USSR, Geophys. Ser.*, 332–5.

Vollstädt, H. (1968), On the determination of rock-magnetic parameters by differential thermal analysis. *Geophys. J. R. Astron. Soc.*, **16**, 71–8.

van der Voo, R. (1969), Palaeomagnetic evidence for the rotation of the Iberian peninsula. *Tectonophysics*, **7**, 5–56.

van der Voo, R. and Klootwijk, C. T. (1972), Paleomagnetic reconnaissance study of the Flammaville granite, with special reference to the anisotropy of its susceptibility. *Geol. Mijnbouw*, **51**, 609–17.

Wadey, W. G. (1956), Magnetic shielding with multiple cylindrical shells. *Rev. Sci. Instrum.*, **27**, 910–16.

Walton, D. (1977), Archaeomagnetic intensity measurements using a SQUID magnetometer. *Archaeometry*, **19**, 192–200.

Ward, J. C. (1970), The structure and properties of some iron sulphides. *Rev. Pure Appl. Chem.*, **20**, 175–206.

Wasilewski, P. J. (1981), Possible experimental verification of the Butler–Banerjee prediction: no stable single domain size range for iron spheres. *Abstracts, 12th Lunar and Planetary Science Conference*, **3**, 1157–8.

Wasson, J. T. (1974), *Meteorites*, Springer, New York.

Watkins, N. D. and Haggerty, S. E. (1968), Oxidation and magnetic polarity in single Icelandic lavas and dykes. *Geophys. J. R. Astron. Soc.*, **15**, 305–15.

Watson, D. (1981), Development of a small-scale magnetometer for measurements in rock magnetism. M.Sc. Thesis, University of Newcastle upon Tyne.

Watson, D. E., Larson, E. E., Herndon, J. M. and Rowe, M. M. (1975), Thermomagnetic analysis of meteorites; 2. C2 chondrites. *Earth Planet. Sci. Lett.*, **27**, 101–7.

Watson, G. S. (1956), A test for randomness of directions. *Mon. Not. R. Astron. Soc. Geophys. Suppl.*, **7**, 160–1.

Watson, G. S. (1960), More significant tests on the sphere, Biometrika, **47**, 87–91.

Watson, G. S. and Irving, E. (1957), Statistical methods in rock magnetism. *Mon. Not. R. Astron. Soc., Geophys. Suppl.*, **7**, 289–300.

West, G. F. and Dunlop, D. J. (1971), An improved ballistic magnetometer for rock magnetic experiments. *J. Phys. E*, **4**, 37–40.

Westcott-Lewis, M. F. and Parry, L. G. (1971), Magnetism in rhombohedral iron–titanium oxides. *Aust. J. Phys.*, **24**, 719–34.

Wheeler, H. A. (1928), Simple inductance formulas for radio coils. *Proc. Inst. Radio Engrs*, **16**, 1398–405.

Widdowson, J. W. (1974), Some physical aspects of palaeomagnetic instrumentation. Ph.D. Thesis, University of Newcastle upon Tyne.

Williamson, P. and Robertson, W. A. (1976), Iterative method of isolating primary and secondary components of remanent magnetization illustrated by using the Upper Devonian Catombal Group of Australia. *J. Geophys. Res.*, **81**, 2531–8.

Wills, A. P. (1899), On the magnetic shielding effect of trilamellar spherical and cylindrical conducting shells. *Phys. Rev.*, **9**, 193–213.

Wilson, D., Heron, K., de Sa, A. and O'Reilly, W. (1977), An automatic rotating-head torque magnetometer. *J. Phys. E*, **10**, 1214–16.

Wilson, I. T. (1941), A new device for sampling lake sediments. *J. Sediment. Petrol.*, **11**, 73–9.

Wilson, R. L. (1959), Remanent magnetism of late Secondary and early Tertiary British rocks. *Philos. Mag.*, **4**, 750–5.

Wilson, R. L. (1961a), Palaeomagnetism in Northern Ireland, I: The Thermal demagnetization of natural magnetic moments in rocks. *Geophys. J. R. Astron. Soc.*, **5**, 45–58.

Wilson, R. L. (1961b), Palaeomagnetism in Northern Ireland, II. On the reality of a reversal in the earth's magnetic field. *Geophys. J. R. Astron. Soc.*, **5**, 59–69.

Wilson, R. L. (1962), The palaeomagnetism of baked contact rocks and reversals of the Earth's magnetic field. *Geophys. J. R. Astron. Soc.*, **7**, 194–202.

Wilson, R. L., Haggerty, S. E. and Watkins, N. D. (1968), Variation of palaeomagnetic stability and other parameters in a vertical traverse of a single Icelandic lava. *Geophys. J. R. Astron. Soc.*, **16**, 79–96.

Wilson, R. L. and Lomax, R. (1972), Magnetic remanence related to slow rotation of ferromagnetic material in alternating magnetic fields. *Geophys. J. R. Astron. Soc.*, **30**, 295–304.

Wilson, R. L. and Smith, P. J. (1968), The nature of secondary natural magnetization in some igneous and baked rocks. *J. Geomagn. Geoelectr.*, **20**, 367–80.

Wilson, R. L. and Watkins, N. D. (1967), Correlation of petrology and natural magnetic polarity in Columbia Plateau basalts. *Geophys. J. R. Astron. Soc.*, **12**, 405–24.

Wohlleben, D. and Maple, M. B. (1971), Application of the Faraday method to magnetic measurements under pressure. *Rev. Sci. Instrum.*, **42**, 1573–8.

Wu, Y., Fuller, M. D. and Schmidt, V. A. (1974), Micro-analysis of NRM in a granodiorite intrusion. *Earth Planet. Sci. Lett.*, **23**, 275.

Yamamoto, N. (1968), The shift of the spin flip temperature of α-Fe_2O_3 fine particles. *J. Phys. Soc. Jpn*, **24**, 23–8.

Yu, S. P. and Morrish, A. H. (1956), Torsion balance for a single microscopic magnetic particle. *Rev. Sci. Instrum.*, **27**, 9–11.

Zahn, C. T. (1963), A new absolute null method for the measurement of magnetic susceptibilities in a weak low frequency field. *Rev. Sci. Instrum.*, **34**, 285–91.

Zijderveld, J. D. A. (1967), A.C. demagnetization of rocks: analysis of results, in *Methods in Palaeomagnetism* (eds D. W. Collinson, K. M. Creer and S. K. Runcorn), Elsevier, Amsterdam.

van Zijl, J. S. V., Graham, K. W. T. and Hales, A. L. (1962), The palaeomagnetism of the Stormberg lavas, II. The behaviour of the magnetic field during a reversal. *Geophys. J. R. Astron. Soc.*, **7**, 169–82.

Zijlstra, H. (1970), A vibrating reed magnetometer for microscopic particles. *Rev. Sci. Instrum.*, **41**, 1241–3.

Zotkevich, I. A. (1972), Reduction of the natural remanent magnetization of a plunging fold to the ancient coordinate system in palaeomagnetic studies. *Phys. Solid Earth* (*USSR*), 125–7.

Zussman, J. (ed.) (1967), *Physical Methods in Determinative Mineralogy*, Academic Press, New York.

Index